Too Smart for Our Own Good

The Ecological Predicament of Humankind

We are destroying our natural environment at a constantly increasing pace, and in so doing undermining the preconditions of our own existence. Why is this so? This book reveals that our ecologically disruptive behaviour is in fact rooted in our very nature as a species.

Drawing on evolution theory, biology, anthropology, archaeology, economics, environmental science and history, this book explains our ecological predicament by placing it in the context of the first scientific theory of humankind's development, taking over where Darwin left off.

The theory presented is applied in detail to the whole of our seven-million-year history. Due to its comprehensiveness, and in part thanks to its extensive glossary and index, this book can function as a compact encyclopædia covering the whole development of *Homo sapiens*. It would also suit many courses in the life and social sciences. Most importantly, *Too Smart for Our Own Good* makes evident the very core of the paradigm to which our species must shift if it is to survive.

Anyone concerned about the future of humankind should read this groundbreaking work.

CRAIG DILWORTH, Canadian by birth, received his PhD in Sweden in 1981, and is presently Reader in Theoretical Philosophy at Uppsala University. A true generalist, his work includes creating and running various environmental projects, as well as purely academic studies in metaphysics, philosophy of science, human ecology, theoretical physics, theoretical biology and the social sciences. He is the author of two majors works in the philosophy of science, *Scientific Progress* (1981; 4th edn. 2008) and *The Metaphysics of Science* (1996; 2nd edn. 2007), and an earlier book in human ecology, *Sustainable Development and Decision Making* (1997).

Too Smart for Our Own Good

The Ecological Predicament of Humankind

CRAIG DILWORTH

Department of Philosophy
Uppsala University, Sweden

CAMBRIDGE
UNIVERSITY PRESS

CAMBRIDGE UNIVERSITY PRESS
Cambridge, New York, Melbourne, Madrid, Cape Town, Singapore,
São Paulo, Delhi, Dubai, Tokyo

Cambridge University Press
The Edinburgh Building, Cambridge CB2 8RU, UK

Published in the United States of America by Cambridge University Press, New York

www.cambridge.org
Information on this title: www.cambridge.org/9780521764360

First published 2010

Printed in the United Kingdom at the University Press, Cambridge

A catalogue record for this publication is available from the British Library

Library of Congress Cataloguing in Publication data
Dilworth, Craig.
  Too smart for our own good : the ecological predicament of humankind /
  Craig Dilworth.
     p.   cm.
  Includes bibliographical references and index.
  ISBN 978-0-521-76436-0 (hardback)
  1. Human evolution.   2. Human ecology.   I. Title.
  GN281.D56 2009
  599.93'8–dc22   2009028743

ISBN 978-0-521-76436-0 Hardback
ISBN 978-0-521-75769-0 Paperback

It is highly probable that with mankind the intellectual faculties have been mainly and gradually perfected through natural selection.

Charles Darwin

Although the brain of *Homo sapiens* is no larger than that of Neanderthal man, the indirect evidence strongly suggests that the first *Homo sapiens* was a much more intelligent creature.

Sherwood Washburn

If there is one thing of which we can be certain it is of the high adaptive value of intelligence as a factor in both the mental and physical evolution of man. Instinct does not permit the emergence of novelty, of innovation, or of originality. Intelligence does.

Ashley Montagu

We can see, that in the rudest state of society, the individuals who were the most sagacious, who invented and used the best weapons or traps, and who were best able to defend themselves, would rear the greatest number of offspring.

Charles Darwin

There are no criteria except adaptation. Intelligence was never an end in itself: it developed because of its adaptive advantages.

Richard Wilkinson

The structure of modern man must be the result of the change in the terms of natural selection that came with the tool-using way of life. It was the success of the simplest tools that started the whole trend of human evolution and led to the civilizations of today.

Sherwood Washburn

The world has again and again approached the condition of being
saturated with human inhabitants, only to have the limit raised by
human ingenuity.

William Catton, Jr.

It appears that we must regard the growth of intellect as having
enabled man to avoid the serious consequences which a fecundity
in excess of that necessary to ensure our species' survival would
otherwise have brought about.

A. M. Carr-Saunders

An increase in efficiency by natural selection may endanger the
whole population if it reaches the point where the source of
food is wiped out. A curb upon the presumed evolutionary trend
towards greater hunting skill would therefore be of advantage. For
human beings we thus reach the paradoxical conclusion that in
times of the pressure of population on food resources any process
which tended to *lower* the mental capacity, physical dexterity or
perceptual acuity of a certain number of individuals might mean
the saving of the race.

D. H. Stott

We have failed to take into account the long-run consequences
of just doing what we have always done – but better and better.
The further our cleverness departs from nature's well worked out
patterns, the greater the likelihood that the clever action will have
unintended consequences – ones likely to injure humans and the
environment.

Lester W. Milbrath

The very aspect of human nature that enabled *Homo sapiens* to become
the dominant species in all of nature was also what made human
dominance precarious at best, and perhaps inexorably self-defeating.

William Catton, Jr.

One is tempted to believe that every gift bestowed on man by his
power of conceptual thought has to be paid for with a dangerous evil
as the direct consequence of it.

Konrad Lorenz

Man is far too clever to be able to survive without wisdom.

E. F. Schumacher

"Paul here had some questions," said Kroner.
"Questions? Questions, my boy?"
He wanted to know if we weren't doing something bad in the name
of progress.

Kurt Vonnegut, Jr.

It is because the reality of Progress can never be determined that the
nineteenth and twentieth centuries have had to treat it as an article
of religious faith.

Aldous Huxley

Or is there anybody who would seriously deny that during the past
hundred thousand years *Homo sapiens* has made progress and has
improved himself?

Max Planck

The growing threat to the planet and to humanity caused by the over-
success of technology has generated severe doubts as to the entire
notion of progress so popular in the Western world.

Sol Tax

Once again it appears that a formidable group of innovations should
not be regarded as the fruits of a society's search for progress, but
as the outcome of a valiant struggle of a society with its back to the
ecological wall.

Richard Wilkinson

Society must cease to look upon 'progress' as something desirable.
'Eternal Progress' is a nonsensical myth.

Aleksandr Solzhenitsyn

To deride the hope of progress is the ultimate fatuity, the last word in
poverty of spirit and meanness of mind.

Sir Peter Medawar

It does happen that what evil people achieve and pass on to evil
people following them brings about progress. If this were not so,
the world could never have attained its high level of technological
development.

Buddhadasa Bhikkhu

Rather than progressing, we have developed our technology as a means of approximating as closely as possible the old status quo in the face of our ever-increasing numbers.

Mark Nathan Cohen

One should not be ashamed of a belief in progress. It is painfully slow and intermittent, interspersed with catastrophes and reversals, but there is a strong case for believing that in the long run it is built into the system, provided there is not an ultimate and irretrievable catastrophe.

Kenneth Boulding

The vaunted 'progress' of modern civilization is only a thin cloak for global catastrophe.

Barry Commoner

Man is by nature a jeopardized creature.

Arnold Gehlen

Any group, or clade, that slowly becomes extinct must reach a stage with only one existing species. Humans have reached that stage, as has the aardvark.

Roger Lewin

What is wrong with the world is that many things are wrong with human nature.

R. M. Yerkes and A. W. Yerkes

It is as if the human species were determined to have a short but exciting life.

Nicholas Georgescu-Roegen

Contents

Figures and tables

Preface

This book is the development of an idea I got back in 1992, after I had been thinking about the nature of humans' ecological problems for some three years or so. I have called the idea *the vicious circle principle*, and my task in this book is to present and apply it as well as I can.

Richard Wilkinson's *Poverty and Progress* (1973) has been my main source of intellectual inspiration, though I have also benefited hugely from the works of Malthus and Darwin.

For comments on various drafts of this book, I would like to thank Dennis Meadows, Richard Douthwaite, David Pimentel, Richard Wilkinson, Herman Daly, Timothy Earle, Anthony McMichael and Allen Johnson.

Special thanks are due Matt Lloyd, Editorial Director of Earth and Life Sciences at Cambridge University Press, who has been enthusiastic and supportive of the book from when he first came in contact with it. Anna-Marie Lovett, Production Editor at the Press, has also been extremely kind and helpful. I also express my gratitude to both Sarah Price, the book's Project Manager, and Laila Grieg-Gran, the Copy-Editor, for their competence and professionalism in preparing the book for publication.

I would also like to thank the staff at the Gnesta Public Library (in Sweden) for their unfailing help in getting me books through interlibrary loan during the time I researched the book.

Udden

Introduction

There is no denying that the world is facing ecological changes that we ourselves have brought about, such as climate change, that are of great detriment to our own species as well as to others. And it must also be admitted that the longer we wait before wholeheartedly dealing with the situation, the worse it will be for us and our children. But where should we concentrate our efforts? What is the appropriate general strategy? To be able to answer these questions, late though they be, we should first consider more closely the nature of our ecologically disruptive behaviour. What exactly does it consist in; and how long has it been going on?

In this book answers are provided to these questions. As regards how long we have been behaving in an ecologically disruptive way, it will be found that we have been doing so as long as we have existed. And, as is suggested by this, what this behaviour consists in is intimately related to our nature as a species. To understand our negative impact on the environment we have to understand ourselves as a species.

For purely intellectual reasons we have since the time of Darwin been in need of an explanation of the development of *Homo sapiens* as distinct from other species. What has been lacking is a theory in which the *causal mechanism* behind our development is laid bare. For such an explanation to be acceptable, the theory must of course be in keeping with the results of science – an added bonus being that it also be in keeping with common sense. In this book I shall present a theory of *Homo sapiens'* development that attempts to meet these criteria, a theory based on what I call *the vicious circle principle*.

Darwin's theory of natural selection provides this sort of explanation of the development of *life*, the primary cause lying behind this development being "the mutability of species," i.e. the tendency for species to succeed one another

1

through the mutation of their chromosomal structures or karyotypes. In the explanation of humankind's development to be attempted here, I shall, like Darwin, also suggest that development is primarily the result of one key cause, which in Darwin's terms could be called "the mutability of technology," i.e. humans' tendency to *innovate*.

So where Darwin's theory of natural selection is based on the principle of evolution, the theory of *Homo sapiens'* development presented here, which presupposes Darwin's theory and involves similar reasoning, is based on the vicious circle principle. In fact, the present theory may be seen as an extension of Darwin's theory in such a way as to explain the development of humankind.

Where the principle of evolution came to constitute the core of biology, which is presupposed by all the life sciences, the vicious circle principle is intended to constitute the core of *human ecology*, which is presupposed by all the *social* sciences. So the vicious circle principle is here being advanced as the fundamental principle of the social sciences, against the background of which social change is to be understood.

From the above it may be seen that our species is special in being the only species to have constantly developed technology. And, as we shall see, it is just this technological innovativeness that is responsible for our present ecological predicament. In sum, we have simply been too smart for our own good.

1

Scientific ground rules

Principles of physics, chemistry and biology

Any attempt to explain a particular phenomenon – in the present case the development of humankind – must rest on certain *principles*. These are the basic presuppositions underlying the explanation; and they must be accepted as correct by those to whom the explanation is directed. The presuppositions on which the theory to be presented here are based are central principles of modern science,[1] each of which states something about the nature of reality as it is assumed to be in science and, thus, as it ought to be assumed to be generally. The relevant sciences include physics, chemistry, biology and ecology – as well as human ecology, the core of which is here suggested to be the vicious circle principle. In what follows I shall present the relevant principles explicitly, marking them with Roman numerals, it generally being the case that each principle presupposes others with a smaller number.

The most important principle of *physics* is:

I. *The principle of the conservation of energy*

This principle was first put forward by R. J. Mayer in 1842. It is also known as the first law of thermodynamics. It states that:

Quantity of energy is constant.

Thus energy can be neither created nor destroyed, but only *transformed*.

Another physical principle of consequence to the development of humankind is:

II. The principle of the equivalence of mass and energy

This principle has been expressed in the form of the equation:

$E = mc^2$

– *energy equals mass times the speed of light squared*. The aspect of this principle – due to Henri Poincaré and others during the first decade of the 1900s – that is of relevance for our investigation is that *matter is a form of energy*.

III. The principle of the conservation of matter

This principle underlies the science of *chemistry*, and was first advanced by Antoine Laurent Lavoisier in 1789. It states that:

Quantity of matter is constant.

In other words, *energy cannot be transformed from a material form to a non-material form or vice versa*. Matter can change from being in a state of high potential energy to being in one of low potential energy, or the other way round, but the total amount of matter in both states is the same. Though this principle does not apply to subatomic energy transformations, it is considered to apply to all other forms of energy change, thus generally making matter a form of energy which is itself like energy in that it can neither come into nor go out of existence.

The fundamental way in which energy is transformed is captured by:

IV. The entropy principle

This principle, first advanced by Sadi Carnot in 1824, is also known as the second law of thermodynamics. It states that:

Energy tends to dissipate.

*Over **time**, energy tends to spread in **space***. This principle may also be expressed in terms of systems, namely: *systems tend towards disorder*; or, *the amount of entropy in a system tends to increase*. It can also be expressed as: *the quality of the energy in a system tends to decline*; or, *the degree of 'organisation' of matter in a system will tend to decrease*; or, *the free or potential energy in a system will tend to become bound or kinetic energy*.

While the principle of the conservation of energy tells us that energy never really disappears, the entropy principle tells us that it constantly spreads out in space, which, from the point of view of its availability for human use, is equivalent to its disappearing. Energy from the sun counteracts the entropy principle in the case of life, such that the dissipation incumbent upon the principle is not evident for periods of time in the case of living organisms. But for this to be possible the entropy of the larger system – including the sun – in which the biological organism exists must itself increase.[2]

Here we might consider the relation between the entropy principle and human resource use. The entropy principle implies that nothing can be recycled indefinitely, so that a sustainable human society, should one ever come about, will have to do without non-renewable resources, including many stones. In this regard we must keep in mind the difference between applying the entropy principle solely to a *particular used entity*, applying it to the *larger system* in which the entity is used and which includes it, and applying it to the *total resources* of the same sort as the entity. At least in the second and third cases, and often or usually in the first, in the end there will be an increase in entropy. As regards the particular entity used, there may on occasion be a *decrease* in entropy vis-à-vis the material of which it is composed even when its life-cycle is complete (e.g. in the case of a discarded aluminium can), though such a decrease will have no practical implications, and only occurs in the case of certain sorts of entity under certain conditions. Before pursuing this topic further, however, we should distinguish some basic types of resource.

Types of resource

Some resources may be termed *potentially permanent*. These include air (or oxygen), which constitutes a potentially permanent 'stock,' and fresh water and solar radiation, which constitute potentially permanent *flows*. Air and water (matter) naturally move in *cycles*, while solar radiation (energy), after being used, is released as heat into space. The presence of all three (including water generally rather than just fresh water – salt water being a stock rather than a flow) of these resources is a precondition for the existence of virtually all life and thus all biotic material. What makes them only *potentially* permanent is their availability. Thus, air and fresh water may become less available by being polluted; or the availability of solar radiation could be lessened e.g. by an increase in cloud cover. In other words, these resources *are* permanent in a natural setting, but the more this setting is interfered with by humans, the less accessible they are both to us and to other species. Note that all non-nuclear energy used by humans stems from solar radiation, either directly, as in the case of solar panels, or indirectly, as in the case of virtually all other forms of energy provision.

Non-permanent resources include those which are *renewable*, i.e. semi-biotic materials such as soil, the presence of which is necessary for all non-marine life, and biotic materials such as forests and food; and those which are *not* renewable, such as metals, fossil fuels and species. While the *atoms* of metals and other chemical elements are permanent, the ability to separate the atoms from one another makes the elements themselves susceptible to the entropy principle and thus non-renewable, as will be seen below. And where biotic materials generally can renew themselves, once a *species* disappears it cannot be regenerated. In the case of soil, in order that it constantly be renewed, it cannot be eroded or leached of nutrients; and as regards biotic material, the conditions necessary for its reproduction cannot be undermined. The *sustainable* use of renewable resources implies that they are given time and space to regenerate.

The use of any non-permanent resource will mean its depletion, i.e. an increase in its entropy. If the resource is renewable, and is being used sustainably, then this depletion will only be temporary. If the resource is non-renewable, then this entropy increase will be permanent, and we can thus call the use of such resources *non-sustainable* (or unsustainable).

As used by humans, direct solar radiation is in fact *not* permanent, since its acquisition or use presupposes the acquisition or use of non-renewable resources such as metals or petroleum-based plastics. Similarly such energy sources as the tides, ocean thermal energy, and the undoubtedly unattainable nuclear fusion must all be considered non-permanent. Nor is hydroelectric power or geothermal energy for electric power generation a permanent resource. All dammed reservoirs eventually fill with silt, and all geothermal electric power facilities decline in their ability to supply energy.

Of the non-renewable resources, the potentially permanent resources of air and water are *reusable*, while solar radiation is not. Metals, paper and plastics, on the other hand, are *recyclable*, while e.g. fossil fuels and animal species are not. Any particular recyclable resource will increase in entropy with each recycling until it cannot be recycled any more or until recycling it isn't worth the effort.

Apart from plant and animal species (as well as genera, families, orders and so on), virtually all non-renewable resources are minerals, the term *mineral* correctly suggesting that such substances are obtained by *mining*. Minerals may be organic or inorganic, organic minerals *having been* biological – the primary instance being fossil fuels, while inorganic minerals include the chemical elements and thus metals, as well as stones.

In economic-thermodynamic terms, we tap *reserves* of various forms of low entropy available to us when we mine, while in using renewable resources we tap the *flow* of low entropy to be found in living beings (biota). Both fossil fuels and biotic resources constitute sources of low entropy in the form of concentrated solar energy.[3] The exploitation of renewable resources typically involves the *harvesting* of living beings, such as animals, vegetable foods, and trees, while the exploitation of potentially permanent (or quasi-permanent) resources involves the *harnessing* (for energy) of such non-living entities as wind and water. The use of a renewable resource in a way that does not allow it to renew itself essentially makes it a non-renewable resource, and eventually leads to its disappearance, the paradigm being the extinction of megafaunal species during the Palaeolithic.

Another potential form of resource – finite though renewable – pointed to by J. P. Holdren,[4] consists in the capacity of the environment to absorb the effluents and other impacts of modern technology.

Products' life-cycles

Looking at human resource use more closely, we note that the life-cycle of a product can be said to go through four basic stages (as well as certain intermediate ones, which can be assimilated to the basic ones): acquiring the resource, working it into a product, using the product, and then getting rid of the remains. The place from which the resource is acquired is its *source*; once acquired, the resource is turned into a *product*, which is then *used* and later disposed of, the place of its final disposal being its *sink*. In this context, the entropy principle says that *all* used resources will eventually end up in a sink. This implies, among other things, that quantity of waste cannot be reduced without reducing the quantity of materials used.

The principle of the conservation of matter tells us that the total quantity of matter that goes into a sink is the same as that which comes from the source, and that it is in fact the very same matter; and the principle of the conservation of energy tells us that the total amount of energy before and after using the material will also be the same. As regards renewable resources such as food and wood, we know that, as long as they are not over-exploited, their use as a whole can continue indefinitely; the constant influx of solar energy is sufficient to counter the operation of the entropy principle. As regards resources that are neither renewable nor recyclable, such as fossil fuels and animal species, their use or elimination is a one-time event: once e.g. oil has been burned its usable low entropy has been completely converted to useless high entropy. But the situation concerning *recyclable* resources is more complex.

In this regard we might consider the life-cycle of a particular recyclable product, such as a sheet of copper.[5] The *first* stage in the life-cycle of the sheet consists in the acquisition of copper ore from a mine – the *source* of the copper – and transporting it to where it can be made into a product, an operation that requires the transformation of free into bound energy through the use of machines. This means, at this stage, an increase in the *total* amount of entropy, i.e. an increase in the entropy in the larger system in which the relevant activities are taking place.

The *second* stage – the creation of a usable product in the form of a sheet of copper – may be seen as including the refining of the copper, the making of the product, and the transportation of the product to where it can be used. Here we note that the entropy of the resultant copper will have *decreased* as a result of these processes, particularly that of refining the copper: the pure copper has a greater degree of order than does the copper ore still in the ground. Following Nicholas Georgescu-Roegen, we see that this does not mean that human economic activity eludes the entropy principle however.[6] The processes of obtaining the ore, refining the copper, making the sheet, and transporting it to where it may be used, *all* require the transformation of free into bound energy, and this increase in entropy is much greater than the loss in entropy in the copper. In systems terms, the entropy principle tells us that it must be the case that more free energy or low entropy is used up in the larger system than the difference in the subsystem between the entropy of the original ore and that of the finished product.[7] As expressed by Herman Daly, the inevitable cost of arranging greater order in one part of the system (the human economy) is the creation of a more than offsetting amount of disorder elsewhere (the natural environment). And in increasing the entropy of the non-human part of the biosphere we interfere with its ability to function, since it also runs on low entropy.[8] This is a state of affairs that doesn't arise in the case of other species, in which there is no technological development.

As regards the *third* stage in the life-cycle of the copper sheet, its *use*, the entropy principle tells us that this use will increase the entropy of the copper the sheet is made of, as it wears away or is otherwise released into its surroundings. Any attempt at *recycling* will of course be part of a process in which the entropy of the copper constantly increases. Eventually the bits of copper become too spread out to be usable;[9] the free energy required to obtain them will be greater than that required to obtain copper from copper ore. Thus, as expressed by Georgescu-Roegen, "There is no free recycling just as there is no wasteless industry."

As regards the *fourth* stage, the disposal of the copper, what is left of the sheet may just end up in the dump, which would then constitute its *sink*. By this time, it would no longer be worthwhile to collect it once again for recycling, otherwise we would be mining dumps for copper (as I'm sure some poor people are doing). In any case, the free energy required to obtain copper is constantly increasing, and at some point it will no longer be worthwhile in energy and/or economic terms to continue mining it in any way.

Entropy and the economy

This state of affairs has important ramifications regarding the human economy. As made clear by Georgescu-Roegen, the economic process, like any systemic process, consists of a continuous transformation of low entropy into high entropy, the energy cost of any economic enterprise always being greater than the product, as intimated above. Economic activity neither creates nor consumes matter or energy, but only transforms low entropy into high. In the purely physical world the entropic process is automatic in the sense that it goes on by itself. The economic process, on the other hand, through its increasing order in the *social* system, *increases* the rate of entropy production in the larger biophysical system, i.e. in the *ecological* system (ecosystem), of which the social system is a part. In fact the economic system itself constitutes a biophysical system the existence of which leads to increased entropy production in the other biophysical systems with which it interacts,[10] through its producing waste at a rate greater than can be assimilated by the ecosystem. Such activity cannot continue in a sustainable society, in which energy use may not exceed the usable energy received from the sun on a year-to-year basis, which implies that waste must be neutralised at least once a year if the population is not to experience population pressure. The entropy principle also applies to the sun, which is itself a system that is constantly losing energy. But this loss is taking place so slowly that it is not taken into account with regard to the development of *Homo sapiens*. In all probability the earth will receive a flow of solar energy of sufficient intensity to counteract entropy and maintain life on this planet for another five billion years.

The existence of low entropy, or free energy, is necessary to life on earth. The nature of this life, as studied in biology and as presupposed by our investigation, is determined by:

V. The principle of evolution

This is the fundamental principle of *biology* – put forward by Charles Darwin in 1859. In its modern form it states that:

> *Life forms on earth have evolved from a common source, each surviving as a species as a result of its being karyotypically adapted to its biological and physical environment.*

One implication of the principle of evolution is that humans are animals, and constitute a part of nature just as other animals do. Accordingly, as regards the basis of our physical existence, we do not stand above other life forms, but among them. And, just as is the case with them, when the human species proves unable to adapt to its environment, it will become extinct.

Relative to other life forms, humans are an extremely recent arrival. The change in karyotype (see Glossary) resulting in the coming to be of our species occurred only about 200,000 years ago. This may be compared with other species, an extreme example being that of shellfish called *brachiopods* or *lamp shells*, some forms of which have remained unchanged for at least the last 500 million years, thus having existed on earth about 2500 times longer than we have.

Adapting to the environment means fitting into a system.

Physical and biological systems

A *system* in its simplest form is any group of entities amongst which there is an ongoing cause-and-effect relationship. The existence of the form of the relationship over time is what makes the system a system and not simply a one-time event. Thus a raindrop falling on a stone is not a system, but a constant dripping of water on one part of the stone is a system.

As implied above, what more is required of a state of affairs in order for it to be considered a system is that its constituents *interact*. Given the physical notion of cause, however, *any* continuing causal relation constitutes a system. This is thanks to Newton's third law – another principle, already expressed by Aristotle – which states that physical cause and effect are equal and opposite, or, to every physical *action* there corresponds and equal and opposite physical *reaction*.

This interaction may be more or less noticeable in different cases. A single stone constitutes a system, since its constituents interact, holding it together. (In fact, from one point of view, any situation that does not evince total entropy constitutes a system.) But their interaction is not obvious, in that it does not involve change. (In what follows the term *system* will be used only in referring to such systems as involve internal change.) In some systems the cause may be obvious while the effect is hidden, or vice versa. In the case of water dripping on a stone, the collision of the water drops with the stone causes the water molecules to disperse rapidly, while the stone molecules disperse much more slowly, eventually giving rise to a hollow in the stone. The effect of the stone on the water drops is more noticeable than the effect of the water drops on the stone, though they are physically equal.

The *solar system* constitutes a relatively simple system, where the gravitational force (cause) exerted by the sun on the planets is exactly equal and opposite to the force exerted by the planets on the sun. The effect of the sun on the planets appears to be larger however. One thinks of the planets' orbits as resulting from their attraction to the sun, when they are just as much a result of the sun's attraction to them.

The internal *spatio-temporal* relations of a system constitute its *structure*, and the interaction of its internal and external *causal* relations determine how it is *organised*, i.e. the nature of its *internal* causal relations (and, in turn, its structure).

As long as the causal relations of the system remain relatively unchanged, the system has the same organisation, even if the spatio-temporal relations within it, that is, its structure, change. Thus you can say that the structure of a system can change while it remains the same system, but with a change of organisation there is also a change of system. The distinction between structural change and organisational change is not a black-and-white affair, though it is vital to understanding the nature of systems.

Static vs. dynamic equilibrium

As long as the processes in a system are not leading to organisational change, the system is in a state of *dynamic equilibrium*. When there is no structural change the equilibrium is *static*, or the system is in *static equilibrium*: whatever processes as exist in the system repeat themselves. Static equilibrium can be lost and regained without losing dynamic equilibrium, in which case its being regained can end in a return to the original structure, or result in a new structure.

Loss of dynamic equilibrium, or a change in the arrangement among the causes, or *forces*, within the system, leads to the dissolution of the system unless it is counteracted by free energy originating from outside. In fact, due to the entropy principle, all instances of recovered equilibrium, whether they be of static or dynamic equilibrium, will require the changing of free energy originating from outside the system into bound energy within the system. A system's *resilience* is its ability to 'acquire' and 'use' such energy to regain dynamic equilibrium.

A system is *out of static equilibrium*, or has *lost its static equilibrium*, when its spatio-temporal structure has been altered and a new (or old) static equilibrium has yet to assert itself. A system is out of *dynamic* equilibrium, on the other hand, when there has been a change in its internal *causal* relations. If these relations do not come to be reasserted, we should say that there has been a change in the organisation of the system, and, consequently, that it is no longer the same system. In these terms, the entropy principle says that *all systems tend towards dynamic disequilibrium*. Every system, sooner or later, as long as no energy is coming into it, will lose its dynamic equilibrium and 'collapse' into a different system having a lower degree of organisation; and this process will continue until there is no system at all.

The concept of equilibrium can be further clarified with our example of the solar system. The solar system may be seen as being a system in static equilibrium: there is an interaction amongst its elements, while their relation to one another remains structurally unchanged; processes in the system are *repeated*. Imagine now a piece of matter the size of the earth entering the solar system and going into orbit around Jupiter. The static equilibrium of the system is lost. The orbits of the planets are affected by this new source of gravity, and become irregular. When they eventually stabilise, the system is once again in static equilibrium, and has a different structure: the orbits of all the planets have undergone a change. But we need not say that the *organisation* of the system has changed, so long as, say, the system has not experienced a notable increase in entropy, as it would were e.g. a planet larger than Jupiter's new moon to fall into the sun, or if the order of the planets had changed. If the organisation of the system *has* changed – depending on how we conceive of such a change – we might say that the old system has ceased to exist and a new system has come into being. The entrance of the new mass has had such an effect on the original system that it was unable to maintain dynamic equilibrium, and a new system resulted. In the event that the planets retain their relative positions with respect to the sun after settling into new orbits, we could say that dynamic equilibrium has been maintained despite the loss of static equilibrium: the system's organisation has remained the same despite structural change in the system.

So if the relative positions of the planets should change, we could say that there has been a change of system, since the causal relations between them will have altered. After the arrival of the piece of matter from outer space, the solar system lost not only its static equilibrium but also its dynamic equilibrium; this it did at one and the same time, namely, when the orbits of the planets started to change. The organisation of the system – and consequently the system itself – can be said to have changed when the planets adopted their new orbits. And the original system can be said to have been *out of dynamic equilibrium* between these two times.

Note that, as intimated above, what is taken to be static as compared to dynamic equilibrium is to some extent arbitrary. For example, you *could* say that the solar system would still be essentially the *same* system even if the relative positions of the planets were to change, depending on how you conceive of a change in causal relations. In the case of living beings however, the notion is much less arbitrary. In the case of *organisms* seen as systems, a loss of static equilibrium results in a change in the body's structure, and a loss of dynamic equilibrium results in illness and possibly death. In the case of *species*, a loss of static equilibrium results in a change in the size of the species' population, and the loss of dynamic equilibrium results in a constant growing or shrinking of the population (which implies a change in its biotic mass) – which may lead to the species' extinction. The final stage of species extinction consists of course in the shrinking not the growing of its total population, the minimum number of individuals required for dynamic disequilibrium of the population not to be irrevocable being its *critical number*.

In the case of an *ecosystem*, dynamic equilibrium may be said to have been lost when there is a constant diminution in the number of species in the system, or in its biomass. For example, if a tropical forest becomes a desert, the dynamic equilibrium of the original system has been lost – so there has been system change – even if some of the original species remain. It may however be difficult to specify at what point an ecosystem may be said to lose dynamic equilibrium – just as it may be difficult to specify when an organism is no longer healthy but ill. When one ecosystem undergoes structural and organisational change resulting in the formation of another system which is on a *lower* entropy level than the original, such as when a desert changes into a rainforest, it would seem that while structural equilibrium is lost, dynamic equilibrium is not, but rather comes to include a greater number of factors. In a similar way dynamic equilibrium can be said never to be lost in a healthy growing organism.

Three points are passed in the demise of any system. The first is that at which the structure of the system begins to change. The second is that at which the system's structural change becomes organisational (dynamic) change, after which the system is out of dynamic equilibrium. The third is the 'point of no return' after which dynamic equilibrium cannot be regained. Once the third point is reached, the system will sooner or later be replaced either by a different system – normally on a higher entropy level – or by no system at all. In purely physical systems the second and third points coincide; in biological systems they tend not to.

Here I should mention some other systems notions, namely those of open, closed, and isolated systems. As depicted in our example, the solar system is an *open* system, taking in matter and therefore energy from outside, as well as giving off solar energy to the outside. A *closed* system is one which *matter* neither enters nor leaves; and an *isolated* system is one which *energy* neither enters nor leaves.

The situation as regards *living* systems is different from that regarding non-living systems. Living systems tend to *maintain their entropy at a constant level*, i.e. to maintain dynamic equilibrium despite the entropy principle. As expressed by Georgescu-Roegen, to the extent that they succeed in doing this, it is by sucking low entropy from the environment to compensate for the increase in entropy to which they are otherwise continuously subject. All of this low entropy, apart from in the case of the vent faunae of the marine abyss, comes directly or indirectly from the sun.[11] More than this, a biological system's entropy can *decrease* through the behaviour of the system itself, which is not the case as regards physical systems unaffected by living beings.

As regards systems quite generally, if a new system has been created, it may be one having the *same* level of organisation as the old: in our non-biological solar system example the planets may continue to orbit the sun but follow radically different paths; or a new biotope may emerge which has the same biomass and variety of species as the old, though many of the species are different. Or it may have a *new* level of organisation: the planets might all fall into the sun; a rainforest may become a desert; a desert may become a rainforest.

The entropy principle says that systems-change generally will be towards *lower* levels of organisation. But in our solar-system example energy was *added* to the system in the form of new matter in motion, which means that the system's level of organisation could rise. Of course as regards e.g. the biosphere, the constant contribution of free energy by the sun means that such a shift as from desert conditions to rainforest conditions is quite possible. Nevertheless, as said earlier, all systems, including living systems, eventually cease to exist, thanks to the operation of the entropy principle. In the case of cells, organs and organisms, this occurs relatively quickly. In the case of *species*, it can in principle take as long as the sun provides energy.

The biosphere, for all intents and purposes, is a *closed* system (as is the solar system, normally considered). No appreciable quantities of matter enter or leave it. On the other hand, it is not an *isolated* system: there is a constant influx of solar energy (all of which is sooner or later reflected back into space). All biological entities on the other hand, including species, are *open* systems, each of which interacts with the other systems on or near the surface of the earth.

Due ultimately to the constant intake of energy from the sun, either directly through photosynthesis or in the form of food, the demise of a living system can often be avoided despite its losing dynamic equilibrium. Thus a living system can employ the free energy it acquires from without not only in growing, but also in supporting its equilibrium when it has stopped growing. So just as a system's *growth* demands an increase in the entropy of the biological systems with which it interacts (unless it is counteracted in those systems by an increase in their direct or indirect intake of free solar energy), so does its regaining lost equilibrium. And its regaining lost equilibrium, or simply attaining equilibrium, is something it must constantly be engaged in. You could define biological systems (which constitute *life*) in systemic terms as open systems in which entropy increase is constantly counteracted in a movement towards the maintenance or attaining of dynamic equilibrium.

The growth of living systems normally occurs via the replication of certain of their subsystems thanks to the functioning of DNA. This growth can only be manifest within certain system-limits however, beyond which the system either loses dynamic equilibrium or is unable to regain it once lost. *In the case of populations, loss of dynamic equilibrium is manifest in nothing other than runaway growth or shrinking.* (In this regard it is ironic that many people take the presently huge human population as a sign of our biological success.)[12] If equilibrium is not regained, the population will cease to exist; and if the population is the last of the species, the species will cease to exist.

Given the above analysis of the notion of a system, we are in the position to present the principle of ecology:

VI. *The principle of ecology*

If we were to suggest a principle for the discipline of *ecology*, it could well be that:

> *Populations of living beings constitute open systems, each of which tends to be in dynamic equilibrium with the other systems constituting its environment.*

The systems with which representatives of the human species interact are physical and (physico-)biological. DNA strands, cells and the organs of living beings are each biological systems. Individual living beings such as plants and animals are also biological systems. And populations of living beings, including cells and organisms, constitute systems. Populations of organisms are subsystems of even larger systems, namely biomes. Eventually we get to the *biosphere*, which has all other biological systems as subsystems.

Positive vs. negative feedback loops

Feedback loops pertain to both physical and biological systems, and are of two kinds, positive and negative. A *positive* feedback loop is a circle of cause-and-effect relations within the system in which an original increase or decrease in a particular activity leads to a further increase or decrease in that activity, and so on. If a positive feedback loop is not stopped by the application of some force (the employment of free energy), it leads to loss of dynamic equilibrium and, ultimately, to the self-destruction of the system. A *negative* feedback loop is a circle of internal cause-and-effect relations where the original increase or decrease in an activity leads to its opposite, thus checking the increase or decrease. (Such loops do not occur spontaneously in physical systems, but must be created by living systems.) Thus, in the case of populations for example, where positive feedback loops generate runaway growth (or shrinking), negative feedback loops regulate growth, thereby tending to maintain the equilibrium of the system. Another way of putting this latter point is to say that negative feedback loops function *homeostatically*, or that they tend to maintain the homeostasis of the system. Yet another way is to say that the system is *cybernetically* controlled.[13]

Examples of positive feedback loops involving *increase* are nuclear chain reactions, the effects of a leak in a dike, population explosion, industrial expansion, capital invested at compound interest, the wage–price inflation spiral, the proliferation of cancer cells, and arms escalation; examples involving *decrease* are bankruptcy, economic depression, the weakening of a person due to overwork, and letting a garden run to seed. Examples of *negative* feedback loops, which counteract increase or decrease, include that involving a thermostat, which controls temperature, the neural mechanism in animals that controls pupil dilation and contraction, the predator–prey relationship and, in a social context, the political phenomenon of balance of power. If a system is to maintain dynamic equilibrium, positive feedback loops, if and when they appear, must be countered by negative.

In the case of *biological* systems, whether they be cells, organs, organisms or populations, negative feedback loops are most notably manifest as *checks to growth*. Such checks may be either purely physical – birds of a particular species can lay only a certain number of eggs, which can function as a check to population growth – or they may, while still having a somatic basis, be manifest behaviourally, as in the case e.g. of territoriality.

Genetics and homeostasis

Species vs. populations; and karyotype vs. genotype

So populations of plant and animal species constitute systems. It is important in this regard to distinguish between, on the one hand, a particular *species*, and, on the other, *populations* of members of the species.[14] A species is not itself a physically existing thing, as are e.g. genes, chromosomes, organs, organisms and populations. A species is the *form*[15] a population takes, a form determined by its karyotype; or, since species are conceptually primary, we should say that the (total) population of a species is the *phenotypic manifestation* (Glossary) of the species' karyotype. Despite the formal nature of species, systems notions can be applied to them as well as to populations – but it is to be kept in mind that in so doing one is working on a higher level of abstraction.

It is only through a change in karyotype – karyotypic mutation – that species *succession* occurs, i.e. the replacement of one species by another that has evolved from it. There is a widespread misunderstanding on this point, associated with the views of Richard Dawkins, according to which species succession is solely a *genetic* phenomenon.[16] But species can and do change on the genetic level – undergo genetic mutation – and remain the same species. Such changes are *internal* to the species, involving a redistribution or mutation of gene *types* (allele types) (Glossary) in the population. Such change constitutes biological *development*, not evolution. Thus, for example, all domesticated dogs are members of the wolf species (*Canis lupus*), despite the fact that the individuals of many races of dog look quite different from wolves. This is because dogs have the same karyotype as wolves – no karyotypic mutation has occurred, and no succession could be said to have taken place if wolves were to disappear. What has happened, rather, is that at many places in the chromosomes of dogs, genes have been incorporated that are of a type different from the genes that occupy those places in wolves. The great divergence between some dogs and wolves and the great diversity of types of dog is facilitated by wolves' and dogs' having so many chromosomes – 76, as compared e.g. to humans' 46 and horses' 64.

One can say rather generally that all of the phenotypic characteristics that the members of a species share stem from their species' karyotype, while the characteristics that distinguish them stem from their genes. Over the last 500 million years few new species have arisen. Of all evolutionary lines that have ever existed, 99.999 per cent have by now become extinct.[17]

Part of what defines a species is that its populations are descended from the same ancestors, as in the case of dogs and wolves. However, one individual (or more) in such a population may be a chromosomal mutant, i.e. have a karyotype different from that of its parents, and go on to create a new species – in a sexually reproducing species this would virtually always be together with a member of the opposite sex of the original species.

The *total population* of a given species, then, is limited to living individuals sharing the same karyotype and descending directly from the same ancestors. (So the total population of the wolf species includes all domesticated dogs.) Though it is in principle possible that a different but similar species mutate and give rise to a new species with the same karyotype as the given species, it is highly improbable. And if it did occur, since the new species' members would not have the same direct ancestors as the given species, it would not be considered to be the same species, even if its members were to reproduce with members of that species. Note rather generally that organisms needn't have the same species karyotype in order to reproduce – only very similar ones – or the evolution of sexually reproducing organisms would not have been possible.

Also, possession of the same karyotype ought to be *sufficient* or close to sufficient to allow reproduction. Thus, for example, Chihuahuas can in principle breed with wolves. The failure of individuals from different groups of the same species to breed, though they could do so, is taken to indicate that they belong to different populations, as in the case of wolves and dogs.

The change in size of any population of organisms over an extended period of time constitutes a *demographic transition*. Normally employed, however, the notion of demographic transition is to apply to situations in which the size of a particular population is relatively constant, after which it changes, and then later becomes stable at a new level – a change through which it is to have maintained dynamic equilibrium. As the notion of demographic transition is intended, though the species' structure may have changed through the transition, its organisation has not, otherwise it would have either mutated into a new species and/or become extinct.

The food chain

The populations of all biological species are involved in a food chain, the number of links or *trophic* (nutritive) *levels* in the chain rarely exceeding five.[18] The food chain passes from plants, to herbivores, and on up to carnivores, including (omnivorous) humans, who are at the top. Green plants obtain all of their energy ('nutrition') via photosynthesis directly from the sun, while herbivores obtain solar energy indirectly from eating plants, and carnivores obtain it from eating herbivores. The food chain may also be seen as constituting a *system*, a subsystem of some particular ecosystem and of the biosphere.

From one point of view the food chain may be seen as making a loop. The bodies of the carnivores are consumed by micro-organisms, who turn them into an inorganic state, and these materials in turn constitute the building blocks employed in the creation and maintenance of new micro-organisms. Some of these new micro-organisms disappear through being consumed by higher life forms. But the material of which they consisted remains in the chain, in accordance with the principle of the conservation of matter. And this material, as well as the solar energy contained in energy-carrying molecules, is competed for on every trophic level.[19]

The total biomass produced on any one level in the food chain is necessarily less than that of the level below it, on which it depends, and greater than that of the level above it, which it supports. In order to maintain their size, populations on higher levels must cull from the populations on lower levels in such a way that the lower levels' biomass as a whole remains relatively constant. As expressed by G. A. Bartholomew and J. B. Birdsell, this means that each level must in the long run live on the interest, not the capital, of the level below it. In this way we not only have a food chain, system and circle, but also a food *pyramid*, with the less massive higher levels being dependent on the more massive lower ones. For each population, as for total populations and thereby species, this means being in dynamic equilibrium with both higher and lower trophic levels.[20] If a species is unable to maintain such equilibrium, it will of course become extinct.

Here the differences between energy *lost*, *obtained* and *used* should be noted. Due to the entropy principle, the free energy a system or systems *lose* will always be more than that *obtained* by some other system, which in turn will always be more than the energy actually *used* by the obtaining system. In this last regard, in the case of human beings for example, only about 18 per cent of the calories consumed as food are converted into mechanical energy; the rest is lost, mainly as heat.

As regards free energy *lost* and free energy *obtained*, and in keeping with the pyramidal aspect of the food chain, green plants capture about 0.1 per cent of the solar energy that strikes them, herbivores capture about ten per cent of the free energy in the plants they eat, while carnivores capture about ten per cent of the energy in the herbivores (or other carnivores) they eat,[21] these figures being known as the respective *percentage efficiencies of energy transfer*. This notion of efficiency is the same as that used with respect to engines, and is always less than 100 per cent. In terms of the *net energy ratio* (Glossary), an energy transfer of ten per cent constitutes a net energy ratio of 1:10. Each link upwards in the food chain involves a decrease in the net energy ratio, which means that greater amounts of solar energy are necessary to sustain organisms the higher their position.

Due to the fact that organisms higher on the food chain eat those lower on it, whatever matter that is ingested and accumulates in organisms gains in concentration the higher the organism is on the chain. Many artificial substances produced by humans, such as DDT, are of this kind, such that we find them in toxic concentrations in such animals as birds of prey and humans. It might also be mentioned here that humans' activities are everywhere leading to a reduction in the length of food chains.[22]

In the case of all animal species other than humans, virtually all energy taken in comes in the form of food, obtained by eating members of other species. In the case of humans, on the other hand, energy began to be consumed other than in the form of food beginning with the first fires 'fed' by wood. Next, also to produce fires of various sorts, we began first using fossilised coal, which is solid, then fossilised oil, which is liquid, then fossilised natural gas, which is gaseous. And finally we moved on to inorganic uranium. Thus, throughout human existence, technological development has meant a shifting away from the use of the *constant flow* of low entropy in the form of solar radiation and its more immediate products (food, wood), to the use of low entropy in the form of the earth's mineral resources,[23] both organic and inorganic. The obtaining of matter and energy from physical and fossilised biological systems involves *mining*, as suggested earlier, in the case of purely physical systems the mining mainly of metals; and in the case of fossilised biological systems the mining of coal, oil and natural gas. It may be further noted that this process involves moving from the near-in-time and near-at-hand biological, to the older and further removed organic, and on to the still-older physical.

Unlike other biological systems, the human population since its inception has been exponentially increasing the amount of free energy it takes in from other systems – at a rate much greater than population growth – and from physical and fossilised biological systems in particular.

Increasing biological complexity

The energy-accumulating aspect of life, in the form it takes for populations, lies behind the principle of population, to be discussed in Chapter 3. As applied to the biosphere as a whole, it means a tendency on the part of the physical aspect of life to increase not only in mass, but also in organisational complexity, in the form of the arising of an increasing number of biological subsystems on an increasing number of levels. This is essentially effected by species' tendency via mutation to inject potential new species into the ecosystem, coupled with the fact that successful mutation tends towards larger and more complex organisms (their increased complexity requiring their increased size). Part of this development consists in the coming to be of K-selected species (Glossary). In the same vein, as noted by Darwin,

> the more diversified in structure the descendants from any one species can be rendered, the more places they will be enabled to seize on .... In each genus, the species, which are already extremely different in character, will generally tend to produce the greatest number of modified descendants; for these will have the best chance of filling new and widely different places in the polity of nature.[24]

As Darwin also says: "the greatest amount of life can be supported by great diversification of structure." Applied to the evolutionary process vis-à-vis various ecosystems, this suggests that the biosphere and its ecosystems have the greatest possible biomass. Thus it is natural for the biosphere as a whole that the number of species should constantly tend to increase (until some physico-biological limit is reached) such that the one-celled organisms with which it began develop via mutation in such a way as to give rise to species of the complexity of humans,[25] and that the mass of the biosphere should reach a maximum given the physical preconditions.

A state of affairs in which the number of species and the biomass of an ecosystem has reached its practical limits is said to constitute a *climax* ecosystem, *all* ecosystems being such before being affected by humans, and *none* being so after.[26] The fact that a climax ecosystem contains a great number of species gives it greater stability, since the loss of any one species will tend to have less impact on the system as a whole,[27] and the complexity of the system will better allow it, e.g. through mutations, to fill the gap. Thus a climax ecosystem and an ecosystem experiencing the most stable form of equilibrium are essentially the same thing.

Species/population homeostasis

In the case of species, though their total populations have a tendency to grow through the procreation of individuals beyond replacement level (the principle of population – Chapter 3), only under exceptional circumstances will they in fact grow to the point of disequilibrating other systems on which their existence depends. Predatory animals do not chronically depress their stocks of prey, nor do herbivores impair the regeneration of their food plants.[28]

The loss of equilibrium of any one living system will tend to cause disequilibrium in other living systems with which it interacts, the effect being dependent on the nature and degree of the interaction. This applies to all biological systems, from ecosystems to individual cells. So, for example, the death of an animal will have a tendency to disequilibrate the population of which it is a part, and of course the animal's organs and cells will lose equilibrium and stop functioning. Likewise the disequilibrium of the population of e.g. a particular animal species, manifest in its dramatically increasing or decreasing in number, will affect the populations of other species, most directly those that eat it and those that it eats. In fact, in keeping with the principle of ecology, a population's being in equilibrium is essentially its being in equilibrium with other populations.

Vacillation about a mean

In the case of non-human species, as long as the number of individuals does not fall below the critical level, the lower the population density, the greater the quantity of food available to each individual. The spreading out of the population will also mean less opportunity for predators, and a decrease in parasites and disease. Conversely, when numbers increase, per capita food supplies decrease, and there is increased opportunity for the activity of the parasites and predators that live off the species in question. So, from each side, there is an automatic tendency towards the development of a balanced state of affairs for every species.[29]

The sizes of non-human populations tend to increase and decrease in cycles. As a population grows, its food consumption increases, which, unless some other check comes into play first, eventually leads to the population's experiencing scarcity. This in turn necessitates reduced consumption, which leads to a reduction in the size of the population – paradigmatically through an increase in infant mortality – allowing the resource to recover. Note that this process presupposes the sustainable use of resources. Once their unsustainable use is introduced, the chance of the system experiencing disequilibrium through overpopulation increases.

In Barry Commoner's terms, "stabilising cybernetic relations are built into an ecological cycle."[30] Part of what determines the nature of this cycle is *Liebig's law of the minimum* which, as expressed by Bartholomew and Birdsell, "states that a biological reaction at any level is controlled not by the factors which are present in excess, but by that essential factor which is present in minimal quantity." Thus, setting aside the question of parasites and predators, the maximum population size in this oscillation is limited not by the total resources available, but by those *vital* resources that are at a minimum. So, for example, the size of any population of K-selected species will be limited by the number of its breeding sites *or* the amount of food – whichever is in shortest supply; and an increase in the one without an increase in the other (unless there already exists an excess of the other) could not result in an increase in the size of the population. As expressed by Commoner, if the entire cyclical system is to remain in balance, the overall rate of turnover must be determined by the slowest step; and any external effect that causes part of the cycle to operate faster than the overall rate leads to disequilibrium.[31]

Checks to growth

Given the above, it is evident that the causes of both increase and decrease in size must be operative in order for a population to constitute a system in equilibrium. (The general effect of this tendency of populations towards equilibrium is the tendency of the biosphere as a whole towards equilibrium.) If the causes of growth increase without a concomitant increase in the manifestation of checks to growth, the system will grow until such checks come into play, as they must sooner or later since no population can expand forever. If, on the other hand, the external forces tending to reduce the size of the system are greater than the internal forces tending to expand it, the system will contract, in the worst of cases disappearing entirely. If this happens for all populations of a particular species, the species of course becomes extinct; and in order for this not to happen, counteracting causes of growth must be present.

Since the populations of all species have an innate tendency to grow, it is in the case of checks to growth that the notions of negative feedback and homeostasis are particularly relevant. This is not to deny, however, that there may exist particular checks to shrinking which become operative when the relevant population becomes too small.

Checks to growth are imposed either externally, by other biological or physical systems, or internally, by the population-system itself.

External checks

External checks to population growth can take the form of a scarcity of some kind – e.g. of food or breeding sites – or may take the form of disease or predators. Temporary periods of scarcity occur naturally for all living organisms given the normally occurring changes in their environments, such as increases or decreases in temperature or rainfall. To survive as a species the populations that constitute the species must be able to live through such changes, which they do, given the changes are not too great, thanks to their adaptability, which is built into their karyotypes.

Note that from a Darwinian point of view, scarcity leading to static or dynamic disequilibrium can never constitute the normal state of affairs for a population. The normal state for all species is a general sufficiency of resources.[32] However, major environmental changes resulting from changes in the climate, such as those at the beginning of or following an ice age, can bring about a situation of 'permanent scarcity' that leads to disequilibrium. Such a situation would involve higher mortality and/or lower fertility, and eventually, in the best of cases, result in the species' attaining a new equilibrium through a demographic shift to a permanently smaller population. Of course some environmental changes may be so great that the species cannot adapt, the result being that, seen as a system, it loses equilibrium to such an extent that it cannot be regained, so that it becomes extinct. Here of course it is possible that karyotypic mutation occur (or have occurred earlier), and that a successor species (a new system) come into existence that is better adapted to the changed conditions.

Internal checks

Internal checks to population growth are negative feedback loops within the population that counteract its basic tendency to accumulate as much solar energy as possible at any one time. Internal checks tend either to maintain a counter-entropic, i.e. living, system in a state of dynamic equilibrium,[33] or to return it to such a state after it has been lost. Such checks exist for a variety of biological systems, and not just for the populations of particular species. To be in equilibrium with other biological systems, one such system cannot grow indefinitely, but must be neither too large nor too small, some size within an even narrower range perhaps constituting its *optimum* size. If the system, whatever it may be, does not meet this requirement of neither growing nor shrinking too much, it will disappear; and, as suggested above, its disappearance will also affect the systems of which it forms a part, in certain cases also leading to the disappearance of some of them. In the case of *species* in particular, however, their total populations tend to be as large as possible within the allowed range, their size being that which allows them to incorporate the maximum amount of obtainable free energy without jeopardising the species' existence. (This is essentially the same as the fourth formulation of the principle of population to be given in Chapter 3.) Thus in the case of certain species we see internal homeostatic processes resulting in the maintenance of dynamic equilibrium in populations – ultimately for the 'good' of the maintenance of the system constituting the species.

Note however the difference between populations and organisms in this regard. Internal checks stop the growth of organisms – as well as cells and organs – when they are mature, and see to it that death ensues after a certain period of time ('inbuilt obsolescence'), if it has not been brought on earlier by purely external factors. With regard to equilibrium maintenance in the case of such entities, we may say that while growing they maintain (or tend to maintain) dynamic but not static equilibrium. (Both the quantity of free energy and the quantity of bound energy can be increasing in a growing system. If the system is in dynamic equilibrium while growing, the ratio of the two will tend to be constant.) When growing systems become mature they are in or tend towards a state of both static *and* dynamic equilibrium; and when 'malfunctioning' – or, in the case of organisms, when ill or dying – they have *lost* both static and dynamic equilibrium. Thus not only is the point after which a living system is doomed later than that at which it loses dynamic equilibrium, but it is at least as hard to determine.

When it comes to populations and ecosystems, on the other hand, there is no internal check equivalent to the death of an organism. You could say that where organisms mature, age, and die, ecosystems, and species as manifest in their populations, are potentially immortal – as long as they neither grow nor shrink to the point where the consequent loss of equilibrium cannot be rectified. (This of course does not prevent any particular ecosystem from being in or out of equilibrium.) Similarly as regards the biosphere as a whole, there exist no internal homeostatic mechanisms or negative feedback loops checking growth, all such checks being external.

In the case of species, as suggested by D. H. Stott, if a population were to experience a food shortage that was severe enough to impair individuals' functional efficiency, the population would be threatened with annihilation from predators. Thus, as argued by D. H. Chitty, a species which frequently exhausted its food supply might well be supplanted by one whose population density was controlled at a safer level.[34] Hence disequilibrium due at least in part to *internal* causes can be brought about by a period during which the checks to population growth are particularly *lax*. Overly lax checks lead to overpopulation and a consequent disarming of the species' environment, which could result in the eradication of its resources, and the extinction of the species for that reason.

Internal checks to growth exist in virtually all vertebrates[35] and even in sea-anemones.[36] And they may come into play before, after, or simultaneously with external checks. As Stott says regarding such checks, from the study of a number of widely dissimilar species of animal it is clear that when their population density reaches a certain point, even without a shortage of food, changes take place which have the effect of reducing the size of the population.[37] And these methods by which animal populations curb their own growth and promote the efficient long-term exploitation of food resources can and do differ for different species.[38]

In many less-complex life forms such as plants and r-selected species generally (Glossary), on the other hand, the internal homeostatic mechanism has not evolved at all or as far, and population size is determined essentially or wholly by external factors. In such cases there still exists a homeostatic mechanism, but it is *inter*- rather than *intra*specific.

> Somatic, constant, invariable, behavioural, instinctual, transitory, learned, variable and cultural checks

Here I shall clarify the nature of various kinds of internal check, noting first that *all* internal checks have their roots in the species' karyotype, and are to be found primarily if not exclusively in K-selected species.

> Somatic: Non-behavioural physical.
> Constant: Always acting; require no external stimulus.
> Invariable: Stem *directly* from the species karyotype (and are consequently the same for all populations of a species).
> Behavioural: Involving behaviour.
> Instinctual (or innate): Invariable behavioural.
> Transitory: Require an external stimulus.
> Learned: Behavioural, transitory, and stemming *in*directly from the individual's karyotype or genotype.
> Variable: Vary between populations of a social species. Stem indirectly from the karyotype.[39] Are mediated by *genes* on the micro level and, when behavioural, by *learning* on the macro. Variable checks stem from later-evolved parts of the brain.
> Cultural: Human, learned, and differ for various populations.[40]

Constant somatic internal population checks include senescence – which among other things sets limits on the time that mature organisms can devote to rearing either their own or others' offspring – and the number of offspring a female can physiologically produce during her lifetime (her fecundity), which is regulated by how often she ovulates, the number of eggs she produces, the onset of menopause, and so on.

Invariable checks are transitory, i.e. require a stimulus. What makes them invariable is that they are always manifest the same way given the same stimulus. Most transitory population checks in non-human species are invariable.

Behavioural checks include increases in infanticide and/or other non-succorant maternal behaviour, sexual abstention, coitus interruptus, the excluding of certain individuals from food or breeding, migration and, in the case of humans in particular, abortion, perhaps long lactation,[41] and the employment of various taboos.

The operation of invariable behavioural checks in nature is dramatically evinced in the periodic *lemming migrations* in Norway, which recur in cycles of three to four (or more) years when population peaks are reached, and result in the populations being drastically reduced. Lemmings also live in the Canadian Arctic; there, however, they do not migrate. Nevertheless it is of some relevance that in the Canadian case at peak levels lemming populations are about *50 times* as large as at low levels, and that, whenever populations peak, the lemmings strip the ground bare of vegetation.[42]

From one point of view these migrations can be seen as the restoring of the lemming population's equilibrium as the result of behavioural changes incumbent upon stress experienced by individuals.[43] Something 'clicks' in the head of at least one lemming which leads it to behave in a way that may function as a catalyst to more 'clicks' in the heads of other lemmings.

The mass migrations leading to population crashes amongst such animals almost invariably occur in the late summer and autumn, when food is plentiful, rather than in the winter, and in the case of the lemmings might be described as a behavioural aberration taking the form of always wandering downhill, which brings them to the sea, into which they plunge. Though none of the emigrants find their way back,[44] most of them would have died of starvation anyway, though some do find other places to live.[45]

The urge to participate in such migrations lies at a deeper micro level than that concerning the existence or non-existence of particular genes. That is to say, it is *invariable*, being determined *directly* by the species' karyotype; and in being behavioural, it is *instinctual*, comparable to e.g. the maternal instinct, or, for that matter, the survival instincts. Without such karyotypically-based instincts, species of this sort would not exist.[46]

Another example of an invariable behavioural check was manifest in a series of experiments by R. P. Silliman and J. S. Gutsell. Over a three-year period they showed that when stable populations of *guppies*, kept in tanks, were cropped by removal of a proportion of the fish at regular intervals, the remainder responded by producing more young that survived, with the result that the losses were compensated. In the controls, on the other hand, where the stocks were left untouched, the guppies went on breeding all the time, but by cannibalism they consistently removed at birth the whole of the surplus produced.[47]

Transitory checks come into operation in situations of crowding, and may be somatic or behavioural, as well as being invariable or variable. Somatic transitory checks include reduced ovulation, shortened time of sexual receptivity in the female, reduced conception, and the malformation of infants. Under appropriate circumstances in mice, for example, ovulation and reproduction can decline and even cease, as long as the ceiling density is maintained.[48] Such checks are psychosomatic, becoming effective through being triggered by stress.

The influence of crowding on transitory checks can be considered in analogy to a thermostat. As mentioned earlier, thermostats function as parts of negative feedback loops. They homeostatically regulate the temperature of a locale by connecting or disconnecting the power source feeding a heating or cooling apparatus when particular temperatures have been reached. In the case of populations, 'thermostats' become operative when the size of the population has reached a critical level beyond which it cannot be allowed to grow. Following V. C. Wynne-Edwards, we can call whatever phenomena as trigger these internal biological thermostats *epideictic*.[49] When epideictic phenomena appear, changes on the part of the populations of such species tend to rectify the situation.

As regards rate of recruitment, for example, which is a transitory behavioural check, groups can exclude particular individuals from food when there is a scarcity, as is the case for example with female lions. Thus each potential group member is governed by the population-system's homeostatic size-controlling mechanism, even when, as commonly happens in the case of recruitment, this means the exclusion of an individual from food even in the midst of apparent plenty, or its exclusion from reproduction when others are breeding. The annual rate of recruitment into the group must be determined by the contemporary relation between population density and resources. Thus under normal conditions only part of the fecundity of the group need be realised in a given year or generation.[50]

J. B. Calhoun showed in an experiment with wild *Norway rats* that both somatic and behavioural checks, and both instinctual and learned behavioural checks, played a role in limiting their numbers. Calhoun bred a colony from a few individuals in a pen of 1000 m^2, providing them with more than sufficient food at all times. If they had realised their breeding potential (fecundity) over the 28 months of the experiment, they would have numbered 50,000. If they had accepted the 0.2 m^2 per rat allowed for caged rats in laboratories there could have been 5000 of them. What actually happened was that the population stabilised itself at less than 200.[51] Note that this density was still much greater than that in the wild.

This reaction to the situation is an effect of crowding. Amongst its *somatic* effects were that "more and more individuals were stunted despite the availability of food. Such stunted rats seemed healthy – they simply failed to grow very large, and attained their mature weight very slowly."[52] Another was reduced conception. And among the behavioural checks was interference in proper maternal behaviour.[53]

The social behaviour of the colony limited population growth indirectly in three ways. First, the rats split themselves up into local sub-colonies, between which were maintained buffer zones without burrows. Second, the normal dominance hierarchy broke down, leading to unstable groups. This had the somatic effect of reducing the frequency of conception and decreasing the viability of the suckling young. Of the few which survived beyond weaning very few in turn had progeny of their own. Third, crowding caused increased attack upon the young, and those who received severe punishment were likely to succumb.[54] It may also be noted that where one male took command, allowing no others to mate, mating was successful and order reigned.

Variable checks presuppose a social species, and are much more evident in humans than in other life forms. They are such that different checks can be manifest given essentially the same stimulus, and thus potentially differ from population to population within the species. In such a case, when behavioural, in humans they are *cultural*. Thus overcrowding may lead to the use of different checks or the same checks to different degrees in different cultures.

The predator–prey relationship

As mentioned above, the influx of free energy from the sun has the effect of promoting an increase in the population sizes of all species. This expansion is directly affected by the expansion of the populations of other species – e.g. the expansion of 'prey' species will support the expansion of the species in question, and the expansion of 'predator' species will lead to its diminution. And in the absence of prey species, the species' own internal checks should (after a time) come into play. As taken up above, there is a point at which the size of the population of each non-human biological life form tends to stabilise or oscillate, depending on the nature of its environment, including the existence, number and type of predators and prey.

On the whole, healthy adult populations of potential prey species suffer little from predators, which suggests that the latter play but a small role in population control. *Adult muskrats*, for example, in their home ranges at normal densities, live in noticeable security. The larger *ungulates* (hoofed mammals) suffer from non-human predation chiefly when immature, aged, crippled, starved, sick or isolated from their fellows. According to P. L. Errington, it is basically intraspecific self-limiting mechanisms that determine the population-levels maintained by the prey.[55] In other words, it is built into the karyotypes of prey species to allow their own numbers to be culled to a certain extent by predator species, so that their food source doesn't become over-exploited. Thus the prey species 'allows' its members to be killed, but only to the extent that this supports the ecological equilibrium of its population, such that what appears to be a purely external check, predation, is at least partly an internal one. In fact, some species that would appear to be perpetually checked by predation have been shown to use homeostatic mechanisms to vary their fecundity according to whether predation is heavy or light;[56] thus we might expect that in such species, if for some reason predators were to take an extra-small toll, self-limiting mechanisms would come into operation to check the growth of numbers – if the change were not too sudden. On the other hand, if losses were great from external causes these mechanisms would not be operative – as in the case of the larger ungulates studied by Errington.[57]

From the other end, we also see why hunting skill in predators is not always favoured in a Darwinian survival-of-the-fittest situation. As noted above, predators do not chronically depress their stocks of prey, as is in keeping with the necessity for species to be in equilibrium with higher and lower levels in the food chain. Any predator group's overdevelopment of hunting prowess would eventually eliminate its source of food. As expressed by Stott, any major advantage gained by the 'consumer' species over its prey implies that it has been able to make significant inroads into the numbers of healthy adults. Even if only a minority of the predator-species are able to do this, their increased hunting capacity would spread by natural selection, and still further inroads, without limit, would be made into the numbers of the prey, until they, and the predators themselves, were exterminated – a positive feedback loop of the decreasing kind. Thus predators only take a marginal toll in the form of the weak and young animals, and the latter are only at risk for a critical few weeks of their lives[58] (as is in keeping with the nature of the food pyramid).

As expressed by David Lack:

> There is a tendency for self-regulation [homeostasis] in any system
> in which one species preys on another, though as shown by [G. F.]
> Gause's experiments, the predator may indeed exterminate its
> prey under very simple ecological conditions. Secondly, only those
> predatory species which have not exterminated their prey survive
> today, hence we observe in nature only those systems which
> have proved sufficiently stable to persist, and many others were
> presumably terminated in the past by extinction.

Such internal checks in both predator and prey species would have evolved karyotypically from the interaction between the species' ancestors and the environments in which they lived.

As pointed out by Jared Diamond, almost all known cases of species' exterminating other species involve two factors. First, predator species find themselves in new environments where they encounter prey populations naïve to their form of predation, and by the time the ecological dust settles and a new equilibrium is reached some of the new-found prey may have been exterminated. Second, the perpetrators of such exterminations are not dependent for their survival on the species they eradicate, but can move on to a new form of prey.[59] It may be noted that both of these factors are present in much of the human eradication of other species.

Population crises

When a population's situation changes sufficiently, the population may begin to lose its habitat, in which case a crisis ensues. This crisis is manifest in the population members' experiencing various degrees of *stress*; and it may or may not be accompanied by a decrease in their number. As expressed by Errington with regard to muskrats, during periods of population decline the animals are in a 'state of nerves.'[60] And as expressed by Chitty with regard to the fact that the offspring of the peak generation of a group of *voles* (mouse-like rodents) either died prematurely or were infertile: "We cannot at present be more precise about this supposed condition than to imagine some disturbance of the hormonal balance of the mother which in some way affected the foetus." That there is a *regular karyotypic provision* for the production of malformation, or poor viability, in the offspring in times of stress is consistent with the view that malformation must be considered one of the mechanisms for adapting population-numbers to the resources of the environment, and as such to have survival value for the species.[61]

There are two largely distinct methods of regulating reproductive output in K-selected species, both of which have been widely adopted in the animal kingdom. One is to limit the number of adults that are permitted to breed, as results from the adoption of a territorial or other system in which the number of permissible breeding sites is restricted. The other is to influence the number of young that each breeding pair produces.[62]

Territoriality

As mentioned, virtually all vertebrates, including primates, as well as a number of non-vertebrates, evince territoriality in some form or other. The fact that territoriality exists in so many species, some of which are not highly developed evolutionarily, suggests that its cause is deeply embedded in the karyotypes of those species that evince it, and that it must therefore be a powerful factor in their survival.[63]

Individual vs. group vs. species territoriality

Territoriality is expressed basically in one of three ways: individual males of a species vie with one another; social units conflict with other social units of the same species; or one species competes with another. From the micro point of view, all three expressions of territoriality support the continuing existence of the species' karyotype. Intraspecific rivalry between individual males does this through promoting fit *individual gene lines*, where each line has a characteristic combination of gene types. And intraspecific rivalry between groups supports fit *group* gene lines. Inter-specific rivalry supports the karyotype of the species by keeping other species (with other karyotypes) from occupying the species' niche (Glossary).

The survival, sexual and social instincts

We can get an overview of this situation and others related to territoriality by considering it in terms of the evolution of instincts. Here we can create a classification of instincts on the basis of phylogenetic succession, the earlier a type of instinct appears in a species' evolution, the more basic instincts of that type are. First then come the *survival* instincts, which exist in all animals, and include such instincts as to breathe and consume food, and to avoid being consumed. The 'fight or flight' response thus falls under the survival instincts.

Next are the *sexual* instincts found in all sexually reproducing animals, first among them being to impregnate or get impregnated; in more-developed species they include the maternal and other parental instincts. The influence of parental instincts will increase with the relative brain size of the members of the species, since infants will be progressively less mature when born and thus need more care. Both the survival and sexual instincts support the individual's gene line (its genetic fitness); and it is in the context of sexual instincts in particular that individual territoriality arises.

Virtually all animals have individual territories during part of the year, but social animals are characterised by having *group* territories. *Social* instincts arise in conjunction with group territory, and include those manifest in supporting one's group by e.g. defending it against attack from other groups of the same species, or killing one's own offspring for the benefit of the group. Similar instincts exist in non-social animals, as manifest e.g. in lemming migrations and guppy cannibalism. Social species include primates, amongst whom group territoriality is manifest in combative behaviour between groups, keeping them apart, while individual territoriality is manifest as combative behaviour among males within the group, which works towards the determination of the group's power structure and smooth organisation.[64]

The instinct in complex (K-selected) species to acquire and protect individual territory is more basic than that concerning group territory, since social species were preceded evolutionarily by non-social ones. All social instincts are evolved from lower-level instincts (particularly sexual instincts), just as all sexual instincts are evolved from survival instincts. For any social species, however, groups must exist if the species is to exist, and social instincts support the gene line of the group in particular.

The reason for the existence of all three sorts of instinct is to support the continuing existence of the species. Ultimately animals do not try to preserve their own lives 'for the sake of' their genetic fitness, but because through supporting their genetic fitness they are at the same time supporting the continuing existence of their species. The existence of the species is a precondition for the existence of its members. In a Darwinian perspective, karyotypes are more fundamental than genotypes, and species are more fundamental than individuals.

In most cases these three types of instinct work in concert. But in certain situations there can arise a 'conflict of interests' among them. The more-basic instincts will have a tendency to override the less-basic (e.g. hunger takes precedence over the sex drive), a phenomenon which, if it persists too far, could lead to disequilibrium and the demise of the species. In this regard, the social instincts differ from the survival and sexual instincts in that they may support the species' karyotype *at the expense* of the individual's gene line, at least in the short term. Thus there can arise situations in which an individual's supporting its own genetic fitness can *reduce* the karyotypic fitness of the species – such as when a pair rears more than a replacement number of offspring in an overpopulated group. It is here, in the context of the manifestation of *social* instincts, that altruism and morals come into existence.

As regards human social instincts we might cite Darwin:

> We have now seen that actions are regarded by savages, and were probably so regarded by primeval man, as good or bad, solely as they obviously affect the welfare of the tribe …. This conclusion agrees well with the belief that the so-called moral sense is aboriginally derived from the social instincts, for both relate at first exclusively to the community.[65]

And:

> [T]he social instincts which no doubt are acquired by man as by the lower animals for the good of the community, will from the

first have given him some wish to aid his fellows, some feeling of sympathy, and have compelled him to regard their approbation and disapprobation. Such impulses will have served him at a very early period as a rude rule of right and wrong.

Axel Hägerström[66] and Konrad Lorenz,[67] following Darwin, also maintain that it is from these instincts, which also exist in other animals, that morality is derived. As Wynne-Edwards says: "Social systems … entail codes of behaviour with which the individual members *instinctively* comply, even when compliance demands the resignation of rights to vital resources or to reproduction." Morality is a cultural manifestation of this, being an expression of the least basic type of instinct. Due to its potential 'competition' with other instincts, as Hägerström points out, "the social instinct does not infallibly act on its own account, but may be overcome by interests or passions which lead to antisocial action."[68] Thus we see that in order to operate properly, the social instincts require a stable species setting. Since they are the weakest – the most tenuous – anything that upsets them, or even changes the population's situation to a small degree, can lead to their being overridden by more basic instincts.

Being social is a prerequisite for the existence of the more sophisticated species in question – it stems from the species' karyotype as manifest in the species' instincts; and altruism, or morality in humans, is a necessary part of being social. Where individual territoriality (based on the sexual instincts) leads to smooth organisation through combative behaviour, group territoriality (based on the social instincts) gives *cohesion* to this organisation through altruistic behaviour. Such behaviour supports the group's survival as a group; and it includes infanticide and combative behaviour with the members of other groups.

With regard to territoriality then, we see that individual territoriality concerns one's sexual instincts while group territoriality concerns one's social instincts, and that sexual instincts are more basic than social. Here we can also see the germs of the development of economics and politics. The drive to attain economic power stems mainly from the survival instincts, but also from the sexual instincts, including those that support providing food and territory for one's family. The desire to acquire political power, on the other hand, is based primarily on the sexual instincts in the form of the male's maintaining or increasing the size of his individual territory. In this case, the sexual instincts supporting fighting with other males for a mate and/or individual territory have evolved into *social* instincts supporting the quest for leadership in the group. And the actual acquisition of such leadership at the same time presupposes that others in the group manifest social instincts supporting group cohesion. Economics concerns *owning consumables* or *individual* territory, which makes it the owner's *property*; and politics concerns *ruling over group* territory, which consists in, to a greater or lesser extent, *physically controlling the other members of the group.*

Individual territoriality and aggression

As has been argued by Lack, individual territoriality primarily concerns the availability of breeding sites, *not* food. This is particularly evident in the case of gregarious birds, whose territories include their breeding sites but not their source of food.[69] Ernst Mayr suggests further that, in the evolution of bird species, territoriality was originally developed in connection with mating, and only afterwards acquired a secondary significance related to the food-providing area.[70]

While granting these points, we might nevertheless note that the presence of territoriality is intimately related to food acquisition even when the territory does not include the area from which food is obtained. By checking population size, individual territoriality ensures that the population does not over-exploit its food base, no matter where the food base might be. (As regards lemmings, their individual territories overlap,[71] which can explain why they overpopulate and that their population check takes the form it does.) Here we should of course see the distinction between mating/breeding sites and food as the basis of territoriality as being that between sexual and survival instincts. The basis of territoriality lies in the male's sex drive, and is manifest in his acquisition and defence of familial property, including his acquisition of consumables (food) for his family.

In the case of non-human animals more generally, intraspecies sexual rivalry is normally the most intense form of competition, but it often becomes ritualised or formalised. As regards intraspecific individual fighting among *birds*, C. B. Moffat has suggested that it serves to parcel out territories, thereby limiting the number of pairs in an area to a fairly constant figure, condemning less powerful individuals to unreproductiveness. Fighting also ensures that the strong and not the weak reproduce – attack, encroachment and defence constituting important aspects of territoriality, as do challenge, vocalisation, song, and other display or signalling activities. The last, incidentally, are bivalent in the sense that they repel other males and at the same time attract females.

In an experiment with fish, it was found that the male *stickleback* would reproduce in its own territory, and at the same time inhibit other males' doing so within close visual distance. The territorial behaviour of fish closely resembles that of birds: territories are established, nests are built and guarded, and mating occurs within the territory.[72]

As regards the basis of territoriality lying in the sexual instincts, it is of interest to note the role played by the male sex hormone in this process. In work on the formation of (non-social) power rankings in *ring doves*, it was found that males injected with testosterone enlarged their previously held territories.[73] And Wynne-Edwards found more particularly that implanting a small pellet of male sex hormone beneath the skin of a *red grouse* cock resulted in its expanding its territory by 75 per cent at the expense of its weakest neighbours.[74]

When it comes to the nature of intraspecific conflict between individual males in the case of *mammals*, Darwin says:

> With mammals the male appears to win the female much more through the law of battle than through the display of his charms. The most timid animals, not provided with any special weapons for fighting, engage in desperate conflicts during the season of love. Two male hares have been seen to fight together until one was killed; male moles often fight, and sometimes with fatal results; male squirrels engage in frequent contests, 'and often wound each other severely.' ... The law of battle prevails with aquatic as with terrestrial mammals. ... All male animals which are furnished with special weapons for fighting [e.g. elephants] are well known to engage in fierce battles.

This leads one to suspect that intraspecific killing among mammals generally may well be greater than it is among today's humans.

> Territories containing food and breeding sites vs. territories
> containing only breeding sites
> Individual territoriality primarily functions so as to limit population

size by limiting the number of breeding sites and thereby indirectly access to food. In some species whose territories include their source of food it can also function so as to support the species' survival by directly denying less fit animals access to that food, which otherwise might also result in an over-exploitation of the resource base. As suggested by W. H. Burt with respect to individual territoriality amongst wild terrestrial vertebrates, territorial defence forces individuals out into marginal situations to which they must adapt or die, and thus the resources of the optimal habitat are not exhausted. Most of the displaced individuals do not survive, but some may find suitable unexploited areas and thus extend the range of the species (cf. the pioneering principle, below). The result is that a population tends to be maintained at or below the optimum density in the preferred habitat. Thus (individual) territoriality is one of the primary factors determining the density of the population. It organises it into a well-spaced array that allows adequate living conditions for all successful (fit) individuals. By limiting the breeding population that can exist in suitable habitats, it helps to prevent increase beyond the carrying capacity of the range. This dispersive effect of territoriality plays an important role both in migration and in the spread of genes through a population.[75] In the case of Wynne-Edwards' red grouse, the heavy winter mortality of outcasts follows the bumper months of heather growth and superabundant food, the shorter days forcing the established birds to become more selective in their diet. These individuals probably need the exclusive use of their feeding territories, and it could be this that finally precipitates the fixing of the boundaries.[76] (Cf. Liebig's law.)

In this way, given a relatively constant supply of food, territories constitute incompressible units that prevent crowding.[77] When more food is available, however, in the case of territories in which food is acquired, the size of the populations may be expected to increase and the size of the territories diminish. The decrease in the size of individual territories occurs, however, only to an extent that ensures sufficient food and breeding sites for mating pairs and their families, at which point they again become incompressible (cf. the penned Norway rats). What is key is that the population, through its *own* behaviour, not only avoid expanding so as to threaten its source of food, but actually allow the existence of a *surplus*.[78]

The negative feedback loop (internal homeostatic/cybernetic mechanism; internal population check) by which such territorial units come into existence is manifest in conflict between males – the denser the population, the greater the amount of conflict. The effects of such intraspecific fighting over mates and territory may well constitute the most important means of governing population size among mammals and birds,[79] if not among vertebrates more generally; and in social species such conflict occurs between groups as well as individuals.

With regard to the above reasoning, it may be pointed out that it is important to distinguish between males and females. In order to function as a population check, intraspecific territoriality must prevent at least some females from reproducing – the exclusion of males is not sufficient, since a small number of males could still impregnate all the females, and there need be no reduction in population. Thus it may be said that the way territoriality restrains or can restrain population growth is through the effect it has on the females of the species.

Sexual selection

Through evolution the competition between males over a mate, individual territory, and/or social rank has led to the karyotype of the male of a species differing from that of the female. The process by which this has occurred is called *sexual selection*, which, in Darwin's terms, is the (genetic) advantage acquired by certain individuals over their sexual rivals.[80] And, as pointed out by Darwin, there is a greater difference between the sexes in polygamous than in monogamous species: "What contrast is presented between the sexes of the polygamous peacock or pheasant, and the monogamous guinea-fowl or partridge!"

Thus in many species the male's aggressive dominance over other males, together with sexual attraction in the form of plumage or large antlers etc., allows him to mate with more than one female, while other males mate with none. In mammals, however, as noted above, display plays a smaller role, and the obtaining of a mate etc. depends more on fighting ability, sexual selection reinforcing this ability in males due to their being the ones that engage in fighting. Such selection thus results in males generally having larger bodies and canines or horns and less fat than females – the latter also being associated with the females having to bear the species' young. As Darwin says: "When the males [of any species] are provided with weapons which in the females are absent, there can be hardly a doubt that these serve for fighting with other males; and that they were acquired through sexual selection."[81] And, as in the case of evolution more generally: "Sexual selection will also be largely dominated by natural selection tending towards the general welfare of the species. Hence the manner in which the individuals of either or both sexes have been affected through sexual selection cannot fail to be complex to the highest degree."

As suggested by G. K. Nobel with regard to the domination of males over females in mammals, including humans, the male sex hormones are correlated with aggressive behaviour and contribute to body size, while the female sex hormones apparently inhibit such behaviour while making no contribution to body size. Consequently in most situations males tend to dominate over females. In the higher primates, as in many other social mammals, sexual dimorphism in size reinforces the greater aggressiveness of the male and ensures the superiority of his status to that of the female in situations where force is involved. In most social mammals, instinctive gregariousness (a social enstinct) overcomes the disruptive effect of this dominance–subordinance relation, however, and keeps the social unit intact. Among primates male aggressiveness can also function so as to support the maintenance of the social unit in that the dominant animal may serve its subordinates, who look to him for protection as well as leadership.[82]

Social vs. non-social species

In social species there is an interaction among the members of a population which actively supports the population's continuing survival. Among other things, this interaction allows the power within the group to be directed outwards, towards other groups, which is not the case with non-social species.

In this vein, male orang-utans defend their females against intruding males, and silverback male gorillas have been seen fighting savagely over control of a group.[83] The fit male members of social species are not only prepared to defend/ extend their personal territories, but also that of their group. It is only in social species, and in all of them, that altruism exists, i.e. action potentially supporting the gene line of the community while weakening that of the individual performing the action. In the case of K-selected species, increased sociability and increased intelligence go hand in hand, increased sociability requiring an increase in intelligence, and an increase in intelligence facilitating increased sociability. The more complex the group, however, the more likely it is that its territory be *ruled over* by some individual, in which case a ranking or hierarchy develops. In the case of humans this may be manifest in leaders constantly experiencing conflict between the drive to act for the best of the community and the drive to improve their personal position, the tendency towards the latter increasing with their rise in the hierarchy (power corrupts), and being largely responsible for that rise.

Intraspecific rankings and hierarchies

In a population such as that of the red grouse, while there is rivalry there is no ranking. Each grouse eats from and mates in its own territory. A *ranking* develops when the rivals *share* the food source or territory. Generally in social K-selected species power relations constitute either a ranking or a *hierarchy*, i.e. a *pyramidal* ranking. Both rankings and hierarchies exist only in social species; and determining one's position in the ranking increases in intensity with population density.

Rankings and hierarchies are power relations ('pecking orders') among individuals which, in the case of rankings, are based solely on the sexual instincts, while in the case of hierarchies are based *mainly* on the sexual instincts, but also involve the social instincts, including those manifest in altruism. Both rankings and hierarchies are also *social power structures*, structures part of the function of which is to defend the group's territory against other groups of the same species.

Social species could, in principle, live without hierarchies, as did modern hunter-gatherers. But it is endemic to every such species, and thus to every human group other than those of marginalised modern hunter-gatherers, that population growth be supported for defence, such that crowding result.

In many social species, as described above, the primary breeding male is also the leader of the group. In fact the paradigm of a social species could be said to be one in which a polygamous leader allows other males to mate in his territory. (In the case of chimpanzees, it is also the case that females mate with a number of males.)[84] As the population grows larger, a hierarchy develops out of a ranking, with the power structure of the group taking the aforementioned form of a pyramid, similar to that of the food chain, with the members on any one lower level essentially sharing the same pecking order, and with class societies eventually arising.

In *wild monkeys* – social animals – according to Claire and W. M. S. Russell, position in the power ranking is generally determined by influence and popularity, rather than by toughness and brute force. The dominant male, whether or not he be so in part due to his fighting prowess, is the leader of the group, has greatest access to females, and has the most *power* over the other monkeys and the highest *status* among them – both of these going into constituting his *social position*.

In wild monkeys, the dominant male's individual territory is coincident with his group's territory, as in the case of all other non-human social species. We might say that his economic and political interests coincide. In the case of 'sub-leaders,' while their geographical territory is the same as the leader's – as is that of all members of the group – they haven't the same degree of access to the contents of the territory as he does, due to his power over them. Nevertheless, the responsibility falls on all the leaders to guide and protect the monkey band in its relations with its surroundings. A monkey leader in the wild will always protect mothers and babies from others, and certainly not harm them himself; and when duty or danger calls, monkey leaders – essentially all fit males – are never wanting.[85]

The development of rankings and hierarchies in social groups is abetted not only by the degree of complexity of the species, but also by conditions of crowding. Thus, for example, *mice* and *moles*, faced with a shortage of space while food is abundant, can shift from an individual-territorial ('non-pecking order') to a ranking way of life.[86] The development of a ranking in social groups is mainly through individual conflict, at least in non-human animals, and is only temporary: once the ranking has been determined, and with it the fittest individual, life in the group goes on its peaceful way – in accordance with his wishes. This continues until the leader's position is challenged and the next occasion for conflict arises.

That the population of a species evince dynamic equilibrium is a precondition for the species' survival. This equilibrium can only be reached through the population's having some physical order, i.e. through its counteracting entropy. According to C. R. Carpenter, and as intimated above, individual territoriality (sexual instincts) combined with social organisation (social instincts) reduces stress, conflict, pugnacity and non-adaptive energy expenditure on the part of the members of the population. Thus any geographical or other physical changes which radically disturb the territorial order of a species' population may seriously affect the survival of the species itself.[87] Animals' various manifestations of territoriality intertwine to determine the structure of their societies, thereby maintaining social order. When for some reason territoriality is unable to manifest itself due to crowding, as in the above-mentioned case of mice and moles, dominance behaviour tends to play a greater role.[88] And in the case of voles, rankings and spacing arrangements function as territories in regulating population density.[89]

As Wynne-Edwards says with regard to species whose source of food lies within their territories, one could predict that if the food were openly exposed to an unruly scramble, there would be no safeguard against its over-exploitation. What actually happens is that, provided the size of each male's territory is adequate, the rate of exploitation of the food resources the habitat contains will automatically be prevented from exceeding the critical threshold for the group as a whole.[90] Note that, as in the red grouse case, this need not involve group territoriality or social instincts; all that is required are sexual instincts. When, in the case of *social* animals, the size of each 'citizen male's' territory is not adequate, intragroup conflict between sub-groups may result.

As regards wild monkeys and group territoriality, we can say that the territory of a monkey band is common property, consisting in those aspects or parts of its environment that its reproducing male members are prepared to defend. According to Russell and Russell, the monkey territorial system serves to *avert* conflicts between whole bands, each of which is familiar with one territory and fearless of and dominant over other bands which enter it. When it comes to both monkeys and apes in the wild, there is no inequality of health or well-being between the highest and the lowest in the band; and in their case individual territory is almost only personal space.[91]

As regards social stratification in the form of the domination of males 'owning' individual territory over those who do not, the power exerted by dominant males can not only exclude other animals from food or breeding, but also have the somatic effect of inhibiting their sexual development.[92] Other things being equal – which in the case of humans they are not – this should benefit the species, in that it promotes the survival of the fittest organisms.

Primate hunting, sharing and sophistication

Most non-human primates are omnivorous,[93] with *baboons* eating more meat than any other non-human primates.[94] Baboons cooperate in hunting,[95] capturing the young of several antelope species (Thomson's gazelle, impala, dikdik, steenbok), as well as cape hare, button quail, several smaller ground-nesting birds, and some tree-nesting birds.[96] Almost all of the captured prey is taken by males.[97] Baboons, like other non-human primates that prey on mammals, do not seem highly motivated to eat carrion,[98] and rarely scavenge.[99]

Despite adequate digestive systems, gorillas do not consume meat at all (though they eat insects), and orang-utans and bonobos (pygmy chimps) eat meat only rarely.[100] Of the four non-human species of great apes, *chimpanzees* are the only frequent predators. Though generally considered to be frugivores, chimps nevertheless prize meat, and hunt monkeys, young antelopes, and other mammals.[101] (For the sake of interest, it may be mentioned that Geza Teleki saw chimps eat a young baboon who was a regular play partner of several infant chimps.)[102] At the Gombe study area, hunting occurs throughout the year, but is much more frequent in the dry season.[103] Only male chimps hunt frequently, an activity in which they may cooperate. Similarly to the predominance of male hunting among baboons, over a 20-year period more than 90 per cent of all chimp kills at Gombe had been made by males.[104] Adult male chimps seldom scavenge,[105] other than freshly-killed animals;[106] female and juvenile chimps, on the other hand, scavenge even the long-dead.[107] Although meat is a small component of the overall diet in comparison with such modern hunter-gatherers as the !Kung,[108] its role in some chimpanzee societies is disproportionately great.[109]

In one of over 200 observations, a chimp threw a large stone at prey (an adult bush pig).[110] Chimps use and manufacture tools (though they don't manufacture *stone* tools), and engage in cooperative hunting, food sharing favouring kin (nepotism), Machiavellian social tactics, infanticide, sex for social bonding rather than reproduction, and inter-group fighting.[111]

Altruism

Altruism exists only in social animals, and consists in individuals behaving in such a way as jeopardises their own genetic fitness while benefiting the fitness of the group, and thereby the species. In other words, in cases of altruism a less-basic social instinct overrides a more-basic survival or sexual instinct. This happens in small ways in all social groups, and is manifest in such behaviour as fit males defending the group's territory.

According to Riccardo Baschetti, the selective advantages of altruism explain why *chimpanzees* display such altruistic acts towards non-kin as sharing resources with them, caring for and consoling them, and attempting to reconcile them. These and other altruistic forms of behaviour, such as succour, the expression of empathy, special treatment of the handicapped, protection of injured individuals, tolerance towards other members of the species, and community concern, can all make sense only in the context of an evolutionary strategy for promoting the survival of the social groups of chimpanzees.[112]

As regards the sharing of meat, C. B. Stanford points out that meat eating itself is a *learned* tradition among chimps as well as baboons.[113] Chimpanzees and capuchins (New World monkeys) are active and strategic meat sharers. In the case of chimps, where the males sometimes brutalise the females to mate, meat performs the same function. More generally, meat is a social currency among chimps.[114]

As Frans de Waal reasons, if attachment and bonding are at the root of succorant behaviour, parental care must be its ultimate evolutionary source. As explained by Irenäus Eibl-Eibesfeldt, with the evolution of parental care in birds and mammals came feeding, warming, cleaning, alleviation of distress, and grooming of the young, which in turn led to the development of infantile appeals to trigger these activities. Once tender exchanges between parent and offspring had evolved – with the one demanding and the other providing care – they could be extended to all sorts of other relationships, including those among unrelated adults.[115] Parental care and defence of the family, deriving from the sexual instincts, are the phylogenetic precursors of intragroup sharing and altruism more generally, deriving from the social instincts.

Crowding, stress and aggression

Monkey communities in the zoo, despite always being amply supplied with food, are full of tensions, and are capable of orgies of slaughter. As Russell and Russell emphasise, the monkeys experience stress due to their limited space, suggesting that in such cases we might speak in terms of 'affluent crowds.'[116]

But even in the wild the members of primate groups engage in lethal fighting.[117] In keeping with what was mentioned above regarding intraspecific killing among mammals, recent studies of *apes* suggest that a gorilla or common chimpanzee stands at least as good a chance of being killed by conspecifics as does the average human. Among gorillas, for instance, males fight each other for ownership of harems, and the victor may kill the loser's infants as well as the loser himself. Such intra-band fighting is a major cause of death for infant and adult male gorillas, some 40 per cent of infant gorilla deaths being due to infanticide.

In a study by Jane Goodall, chimpanzee bands were seen to eradicate other chimp bands. One such band, the Kasakela, which had eight mature males and occupied 15 km², exterminated another, the Kahama, which had six mature males and occupied 10 km². The first fatal incident occurred when six of the adult males in the Kasakela band, one adolescent male, and one adult female left behind the young chimps and moved silently towards the other band, increasing their speed when they heard chimp calls from that direction, until they surprised a male, Godi, from the Kahama band. One Kasakela male pulled the fleeing Godi to the ground, sat on his head, and held down his legs while the others spent ten minutes hitting and biting him. Finally, one attacker threw a large rock at him, after which they left. Although able to stand up, Godi was badly wounded, bleeding, and had puncture marks. He was never seen again and presumably died of his injuries.

The next month, three Kasakela males and one female again travelled south and attacked the Kahama male Dé, who was already weak from a previous attack or illness. The attackers pulled Dé out of a tree, stamped on him, bit and hit him, and tore off pieces of his skin. This attacking behaviour continued until after four years the Kahama band no longer existed, at which time an even larger band, the Kalande with at least nine adult males, began to encroach on Kasakela territory and may have accounted for several vanished or wounded Kasakela chimps. Similar inter-group assaults have been observed in the sole other long-term field study of common chimps, but not in long-term studies of bonobos. Note that in such cases however the conflicting groups taken as a unit might well have been experiencing crowding, due to humans' increasing removal of forests; and so population pressure could at least partly account for the killings. According to Diamond, of the two patterns of genocide commonest among humans, both have animal precedents: killing both men and women fits the common chimpanzee and wolf pattern, while killing men and sparing women fits the gorilla and lion pattern.

In keeping with Russell and Russell, it might be suggested that *increases* or *decreases* in intraspecific violence in any form are directly proportional to experienced population pressure. Thus, for example, a society might maintain ecological equilibrium through having a high level of infanticide or deaths in fights over territory, without this in itself being an expression of population pressure. It is when these levels *increase* that population pressure is manifest. Thus lemmings – which are r-selected – as a species may be said not to experience population pressure, while particular populations of lemmings do.

In the elaborate societies of higher primates, the effects of crowding may be quite complex, involving the replacement of friendly leaders with aggressive bullies, lethal mob attacks on persecuted individuals, and mortal combat between bands. In an overly dense population, cooperation and parental behaviour are replaced by competition, dominance and violence. But the end result is the same – the death of females and young, and a reduction in the size of the population.[118]

Eibl-Eibesfeldt maintains that aggressiveness is not learned by the individual but is innate to the species – i.e. stems *directly* from the species' karyotype – like the organs specially evolved for combat in many animals; and Lorenz stresses the instinctual basis and species-preserving function of animal aggression. ("*Instinct* is the system informing the organism of what to do in the interest of the survival of the species.")

There is no evidence from the behaviour of *mammals* that social violence is more prevalent or intense among carnivores than among herbivores; in both alike it increases with increasing stress, especially stress resultant upon crowding. Everything suggests that social violence evolved in relation to population control on the part of both.[119]

Cliques can develop in stressed populations.[120] For example Calhoun found that his Norway rats organised themselves into groups of about 12 individuals each, and that this was the upper limit of group membership with the maintenance of harmonious living. In this species as well, crowding implied social stress.[121]

The size of the local populations of such fast-breeding animals as voles and Norway rats was reduced due to violence in the face of crowding. There exists a varied set of social devices which act to reduce violence in spacious conditions and increase it in conditions of crowding. The whole complex of violence in response to stress produces effects over several generations, thereby reducing the population of such animals long enough to permit the complete recovery of natural resources.[122]

Territoriality as an internal population check

From the point of view of the survival of the species, both individual and group intraspecific territoriality, which are *internal* to the species, see to it that it is the more fit individuals and groups that breed, while individual territoriality helps ensure that the species' populations do not grow to such a size that they undermine their own resource base. Both functions strengthen the likelihood of the species' continued existence.

The function of intraspecific territoriality as a population check increases in importance the more complex and social the species is. According to Lack, individual territoriality in particular is primarily a mechanism for the regulation of population size by restraining the number of breeding pairs[123] – in the guppies example, by reducing the number of *potential* breeding pairs.

In the case of the red grouse studied by Wynne-Edwards, the exclusion of access to food is the result of a competition in which one individual confronts and dominates another. The winner of such a competition adopts a territory – which he can hold due to his superior fitness – the existence of which is requisite to the survival of an optimum number of reproducing offspring. When there is not enough room for all the candidates to win a territory, the losers are simply excluded. Over half of the August adults in fact die from starvation, predators or disease; and smallholder cocks simply fail to mate at the usual time, though not for want of available hens. The number that remains is such that the population density is balanced around the optimum level, at which the population achieves homeostasis and the highest sustainable use is made of food resources.[124] And since males take territories that are ample in terms of the resources they contain, an adequate standard of living is maintained by all biologically fit members of the group.

Note that this internal, behavioural population check is based on individual territory, not food; i.e. it is based on sexual rather than survival instincts – instincts needed for the *family* to exist. This makes individual territoriality central when it comes to internal checks, sexual instincts being more basic than social. This also applies to social species, whose social instincts support this population-curbing aspect of individual territoriality. The various types of instinct can and do get entwined in one another.

In the case of group territoriality the gene line of the more fit group is strengthened through the winning of inter-group conflicts, the loser group producing fewer offspring, if any, than the winner. And such inter-group conflicts serve to control the population size of both groups taken together.

Apart from functioning so as to limit population size, in mammals and birds individual territoriality performs the more positive function of providing a context in which the young can be fed and protected by their parents, allowing them to develop to maturity and themselves reproduce, thereby ensuring the continuing existence of the species.

Species territoriality

Species territoriality is manifest in the members of one species vying with those of another, and may have to do with the acquisition of either food or breeding sites. Such rivalry may be obvious or gradual, in the latter case the ability directly to acquire the resource in question being the determining factor. In the former case fighting may break out, as when gulls attempt to scare off other species of bird from their nesting sites, or chickadees and blue tits fight at a feeder. In such cases there may also be mortalities, particularly when the competing parties can function as food for one another, as in the case of gangs of rats[125] and armies of ants of different species – and perhaps Cro-Magnons and Neanderthals.

Migration

Criteria for the occurrence of experienced population pressure include the marginalisation and emigration of members of the population, and increased mortality; and, for population pressure quite generally in the case of humans, the extinction of other species in the area. In the case of r-selected species, population pressure is a recurrent phenomenon which the population and thereby the species overcomes through increased mortality, particularly among the young. It may also occur in K-selected species in the event e.g. of a 'positive' change in their surroundings, which gives rise to the pioneering phenomenon.

The pioneering phenomenon

In cases where both food *and* territory increase, or where predators are removed, the result will be an increase in population; and if the change is sudden and large, the population will undoubtedly grow beyond what the habitat can support, with external checks eventually coming into play and giving rise to a see-saw phenomenon until an equilibrium is reached, at which point the normally-functioning internal territorial checks once again become established. For example, in 1912 or 1913 a few moose crossed over the ice from the Michigan mainland to Isle Royale in Lake Superior. This island is 73 km long and some 14 km wide, and the moose had no serious enemies there, with the result that by 1930 their numbers had risen to between 1000 and 3000. Numbers then fell sharply to about 200 in 1935, then rose gradually to 800 in 1948, then fell again, so that in 1950 there were only 500.

Similar phenomena have been witnessed with reindeer and mule or black-tailed deer, both of whom over-exploited their habitats in response to a drastic weakening of external checks, and suffered marked die-back through starvation, the reindeer dying out completely, while the deer continued to destroy the vegetation, and continued to fall in numbers, for more than ten years after their initial heavy decrease.[126]

If or when a group does become overpopulated as a result of a change in its environment, its size will of course eventually be reduced by external checks if not by internal ones. Whether or not such checks are external or internal, they can be seen to be part of a homeostatic mechanism controlling the size of the population in question. However, as is evident from the above examples, and has been pointed out by Wynne-Edwards, some environments are so unstable or transitory that it is impossible for colonising animals to reach a ceiling density and invoke their regulatory machinery before the habitat becomes untenable or is destroyed. Populations in these conditions are always at the pioneering stage, freely increasing in size. Instability of this kind tends to appear around the fringes of the geographical range of all free-living animals, and especially in desert and polar regions. It is also very common in agricultural land due to the constant disturbance of ploughing, seeding, spraying, harvesting, and the rotating of crops.[127]

VII. The pioneering principle

The pioneering principle is implied by the principle of population to be presented in Chapter 3, when applied in a state of surplus food and breeding sites. It says that:

> *Any increase in food or space, given a surplus of the other, will tend to be consumed or occupied by a population, with the result that the population grows.*

The earlier examples of the manifestation of the pioneering principle are rather paradigmatic in that they capture the principle without the admixture of other causes. The primary other cause is *population pressure*, which often provides the push to population growth while the pioneering principle provides the pull. Thus even in the case of the moose, for example, population pressure due to human encroachment may well have lain behind their crossing over to the island in the first place. In the human case, the application of the pioneering principle virtually always involves the effects of population pressure. In other words, the pioneering principle may be, and de facto is, supported by *need*. In such cases as where the pioneering principle and population pressure interact, the question can become one of their relative strengths.

In keeping with the population principle, Bartholomew and Birdsell state that the reproductive potential of animals is such that under favourable conditions, such as access to a previously unexploited habitat, the size of a population can increase at an essentially logarithmic rate (cf. Malthus: "Population, when unchecked, increases in a geometrical ratio.").[128] This capacity for rapid increase makes possible the recovery of populations following drastic population reduction.[129]

VIII. The reaction principle

In keeping with the pioneering principle, the reaction principle suggests that:

> *The members of any species will tend to react to their immediate environment.*

This form of behaviour is an integral part of evolution.

From the pioneering and reaction principles, we arrive at:

IX. The overshoot principle

The overshoot principle is implied by the pioneering and reaction principles. It states that:

> *Given a pioneering situation, populations of slow-breeding animals will expand beyond the carrying capacity of their environment.*

Once a space is filled, due to the pioneering principle, overshoot becomes a fact, due to the reaction principle and the consequent delay involved in implementing internal population checks. And overshoot can be expected to be greater the more sudden and greater the increase in food and space.

Such behaviour as that in which the pioneering, reaction and overshoot principles are manifest is instinctual, stemming ultimately from the survival instincts but in certain cases being supported by the sexual and social instincts.

The equilibrium of the human species

Thanks to Darwin, we know that species evolve from other species; and thanks to research since the time of Darwin, we know that karyotype succession is an essential part of this process. The precursors of modern humans had karyotypes different from ours, and through those precursors' interaction with their environments, together with their karyotypic mutations, the human species with its characteristic karyotype has evolved. It is this species-specific karyotype that determines our physical structure and behavioural characteristics. In this way, from a biological point of view, the human karyotype constitutes the *essence* of what it is to be human[130] – it is what makes humans human, or what makes the human species the human species. And when we first came into existence, it was thanks to our particular karyotype that our species survived.

The modern-human karyotype is such that it supported the continued existence of our species *given the conditions when the species first came into existence*, i.e. given the state of the physical and biological systems with which we interacted at that time. (Cf. Commoner: "the structure of a present living thing or the organization of a current natural ecosystem is likely to be 'best' in the sense that it has been so heavily screened for disadvantageous components that any new one is very likely to be worse.")[131] Selection operates in the short term; as is implied by the reaction principle, species are selected for the immediate requirements of their environments. If environmental conditions change, attributes that were previously adaptive may no longer be so.[132]

As the subsequent development of our species has made clear, a central aspect of the human karyotype which allowed us to survive in the beginning was the plasticity of behaviour it implied. In other words, the human karyotype, to a much greater extent than the karyotypes of other organisms, favoured learning over instinct. This particular aspect, as it turned out, provided us with a great advantage over other species, both competitors and prey. For this advantage not to be too great and place our existence in jeopardy however, i.e. lead to overshoot and loss of dynamic equilibrium, it was thus also necessary that our karyotype be such that this possible development be successfully *checked*. Such checks could be learned, but their basis would still be instinctual, i.e. stem from our karyotype, as would our ability to learn. Given our intellectual endowment, that the total population of the human species would become too small was not a problem; but that it would become too large certainly was.

2

The new views in anthropology, archaeology and economics

The *traditional* or *Western* perspective on the development of humankind goes back at least to Thomas Hobbes, who believed that in humankind's natural condition there is "continual fear and danger of violent death; and the life of man [is] solitary, poor, nasty, brutish, and short. [It] is a condition of war of everyone against everyone."[133] We have an expression of this perspective in the case of a modern writer where we read reference being made to the horticultural revolution as:

> that revolution whereby man ceased to be purely parasitic and,
> with the adoption of agriculture and stock-raising, became a creator
> emancipated from the whims of his environment.[134]

And where, with reference to the rise of agriculture in Mesopotamia, the same writer states that:

> An all round advance is obvious. Emmer wheat and barley were now
> certainly cultivated, and two breeds of cattle as well as sheep, goats
> and pigs were kept.

A recent newspaper article expresses the same perspective:

> The emergence of settled agriculture was a boon to humanity. Life
> became less subject to the vagaries of weather and the availability
> of wild animals. Human beings began to control their environment
> rather than the other way around.[135]

On the traditional perspective it is taken as axiomatic that, as regards technological development and demographic and social change, the former has been the cause and the latter the result. Why do populations grow? Because new sources of food are constantly being discovered and made available by technological improvements. Why did village life replace mobile foraging at the time of the horticultural revolution? Because gardening is more secure and less arduous than constantly moving about. Why did iron tools replace stone tools? Simply because it was discovered that iron is more malleable, can hold a sharper edge, and can sustain more rough use than stone.[136]

The traditional perspective on *Homo sapiens'* development is that of the average educated person of today, and may even constitute part of what many people consider to be common sense. It is integral to the modern worldview in terms of which reality is conceived.

Beginning in the 1950s and 1960s however there emerged more or less independently in anthropology, archaeology and economics different views each of which had a strong ecological link, and each of which ran counter to the traditional perspective. In this regard, already in 1953 Bartholomew and Birdsell noted that:

> For several years it has been apparent that an ecological approach
> is imperative for all studies in population genetics, including those
> pertaining to man. It also offers a potentially useful point of view to
> the physical anthropologist, the ethnologist, and the archeologist,
> and it should provide an important integrative bridge between the
> various fields of anthropology.[137]

The beginnings of this point of view in fact go back to the early 1920s and the work of A. M. Carr-Saunders, who refuted the Hobbesian conception that in humankind's natural condition lives would be solitary, poor, nasty, brutish and short, and that there would be a condition of war of everyone against everyone. As in the case of other social animals, humans of course are not solitary but live in groups, and inside these groups there is not constant fighting but, for the most part, cooperation.

Taking hunter-gatherers' living in groups for granted, Carr-Saunders emphasised their practically universal employment of cultural population checks, which kept their numbers at a level optimal for the continuing survival of their groups, and meant that everyone was sufficiently well provided for to be healthy. Focusing on the ecological aspects of humankind's development, he noted further that all then-known hunters and gatherers evinced longevity.[138]

A number of factors have contributed to this rejection of the traditional perspective. Carr-Saunders' line has been corroborated and further developed in more recent anthropological studies of modern hunter-gatherers,[139] such as those of the Bushmen, Hadza and Pygmies of Africa, the Aborigines of Australia, and the Inuit/Eskimo tribes of North America, to which may be added studies of non-human primates.[140] Further, many important archaeological discoveries have been made over the last half century concerning the beginnings of agriculture, as well as life during the Upper Palaeolithic and Mesolithic eras, particularly as regards human predation.[141] And in economics the work of ecological economists (as distinct from *environmental* economists)[142] has been central. Finally, also having an important effect are the theoretical contributions synthesising some of these results, as will be dealt with in the next chapter.

Anthropology

20th-century hunter-gatherers

Food acquisition and population pressure

Anthropological studies of modern hunter-gatherers have shown that the time they spend – or used to spend, considering that there no longer exist hunter-gatherers uninfluenced by other peoples – acquiring food is small, and the food resources they exploit abundant. For example, despite living in a marginal area – as have all modern hunter-gatherers, being confined there by their agricultural neighbours – both men and women of the !Kung Bushmen of the Kalahari spent on the average less than three hours a day obtaining food. Even the hardest-working individual in a particular camp, a man who went out hunting on 16 of the 28 days of a study period, spent on each of those days an average of only five hours in the quest for food.[143] And since little other effort was required, the !Kung had much time for leisure. (As suggested by Richard Wilkinson, the long hours of leisure which were once possible in parts of the less-developed world must have been the source of the colonial European's belief that the indigenous populations of many countries were naturally lazy.)[144] Furthermore, modern hunter-gatherers as a whole have not undergone the periodic crises that have commonly been attributed to them in the past.[145]

Apart from plants, the Bushmen eat wart hog, kudu, duiker, steenbok, gemsbok, wildebeest, springhare, porcupine, ant bear, anteater, hare, guinea fowl, francolin (pheasants; two species), korhaan (like chicken), fish, tortoise, python and other snakes, bullfrogs and lizards.[146] Among insects they eat locusts, scorpions, beetles, young bees, termites, flying ants and ants' eggs.[147] Human foragers, like modern apes, particularly in dry tropical and subtropical regions[148] (including e.g. Australia),[149] eat insects and insect larvae whenever they can obtain them.[150]

When it comes to acquiring food, there is a clear division of roles between the sexes in hunter-gatherer communities. Men hunt and women gather, a male !Kung who does not hunt remaining a child who cannot marry.[151] As regards attaining manhood status, it requires a course of training and an initiation ceremony, such ceremonies having a profound behavioural effect. In Australian initiation ceremonies the novices are instructed in moral and religious teaching; they are admonished against selfishness, told to share all that they have with their friends, to live peaceably, to be strict in their relations towards women, to abstain from forbidden food, and to consider the advice of their elders and obey their voice.[152] By circumcision around the age of six to eight the Murngin Aboriginal male passes from the social status of a woman to that of a man. When at about the age of 18 he achieves parenthood and is shown his totems for the first time, he goes to another, higher status; and to a still higher when, at about 35, he sees the high totems.[153] The infringement of an Aboriginal moral code is attended by public reprobation and often punished with extreme severity, not infrequently with death.[154] Here we see in the relevant ceremonies, in their defining the basic moral code for the community, the beginnings of religion.

In the largest cross-cultural database that exists – a survey of 179 hunter-gatherer societies that examines how labour is divided – men alone hunt in 166, both men and women hunt in 13, and in none do women alone do the hunting.[155] Some of the few exceptions to exclusively male hunting are among the Agta in the Philippines, where women are/were expert bow and arrow hunters,[156] and among the Shoshoni of the north-western United States and the Copper Eskimo of northern Canada. But there is no hunter-gatherer society in which the individual hunting of larger animals is the regular occupation of women.[157] (This is in keeping with the fact that in baboons and chimpanzees hunting is largely a male activity.)[158] In the case of the !Kung, men's and women's work input is roughly equivalent in terms of hours spent, while the women provide two to three times as much food by weight as the men. Similarly in Australia: the Murngin women furnish the bulk of the daily food. The men's contribution of meat, although much more prized, is less certain.[159] And while the G/wi Bushmen, for example, do not make heroes of their hunters, hunting is nevertheless a prestigious activity. The presence of meat is itself a stimulus to excitement, as can be explained by its status as an article of exchange.[160]

People in the age-group 20 to 40 support a large proportion of non-productive younger and older people, with about 40 per cent of the population contributing little to the supply of food. This allocation of work to young and middle-aged adults allows for a relatively carefree childhood and adolescence and a relatively unstrenuous old age.[161] !Kung girls do not begin regular food gathering and water and wood collecting until they are about 14 years old, while boys are 16 or older before they begin serious hunting, there being little reason to train children to provide for themselves.[162] Hunting is usually at most a one-day operation, though there may sometimes be treks taking up to seven days, while the women's gathering expeditions are always completed within a day. Generally both men and women hunter-gatherers satisfy their hunger at the place where the food is obtained;[163] and stocks of food are not a prominent feature of their societies. Since everyone knows where the food is, and everyone knows the movements of everyone else, there is little concern that food resources will be appropriated by others.[164] In fact their lack of stores reflects conditions of plenty rather than scarcity.[165] As regards the hunter-gatherer Hadza of Tanzania:

> In spite of the fact that [they] make scarcely any attempt to conserve the food resources of their area, that they rapidly eat all the food which comes into camp without preserving it, that they do not cooperate very much or coordinate their food gathering activities with each other, that they make hunting difficult for themselves by using their arrows for gambling, that a high proportion of men are failures at hunting, they nonetheless obtain sufficient food without undue effort. Over the year as a whole probably an average of less than two hours a day is spent obtaining food.

In comparison with their agricultural neighbours the Hadza are well protected against the dangers of famine.

Hadza[166]

It should also be pointed out that the amount of energy expended on acquiring food, as compared with the energy in the food, is less in the case of hunting-gathering than in any other form of human food acquisition. It is truly energy-efficient in that the energy acquired is greater than the energy expended in acquiring it – as is the case with all other animals, and which is far from being the case in modern society, with its dependence on fossil fuels in agriculture. In comparison with other forms of society modern hunter-gatherers use ca. 5000 calories daily (2000 in the form of food, 3000 in the form of firewood), while horticulturists use 12,000, agrarians 77,000, and modern industrialised people 250,000. In terms of the net energy ratio, we see that the amount of energy utilised compared with energy expended (including both muscle-energy and fire) in the case of hunter-gatherers gives a figure of 0.4, in the case of horticulturists 0.16, agrarians 0.026, and industrialised people 0.008.

The question may be raised as to whether food acquisition is experienced as work by modern hunter-gatherers. It would seem that it certainly is so by women, and men may even experience hunting as hard work. But both hunting and gathering are rewarding for those involved: they are appreciated by the other members of the group as providing a service necessary for the group's existence; and people are not alienated from the products of their efforts. More than this, men at least get to experience the thrill of the kill, while the excitement had by women is more likely to be on the level of finding an unexpectedly large root.

More generally with regard to work, however, it should be kept in mind that some amount of physical activity is necessary for both our physical and psychological well-being. From a biological point of view the expenditure of *no* energy acquiring food is not beneficial to any organism nor thus to its species. There is an optimum level of energy expenditure, stress, etc. determined by the species' karyotype. This is evident in the case of human hunting by its being, in proper doses (unlike for the Eskimos), *fun*.[167] The notion of work, like that of being poor, can be related to population pressure. Humans in their least technological state are similar to wild animals; and just as we wouldn't say that wild animals work (except anthropomorphically), we wouldn't say so of the most primitive humans. But with increasing population pressure through human evolution, work and poverty have arisen and grown.

The amount of time and effort required to obtain food and wood is related to population pressure. For example, when a number of otherwise separated hunter-gatherer tribes live together for a period of time, the amount of per capita effort required to obtain food increases noticeably. There is some optimal band size, an important aspect of the determination of which being the number of men required to constitute a hunting party. (Another is the band's being large enough to defend itself.) As regards the Mbuti Pygmies of the Democratic Republic of the Congo, for example, there are two economic divisions: that of *net hunters* who live in large camps of seven to 30 families, based on communal or cooperative hunting, and that of *archers*, who live in much smaller groups and hunt individually with the bow and arrow[168] (though some Mbuti archers hunt in groups).[169] (Note that net hunting is in many respects more like gathering than hunting, including its involving the use of a receptacle, and being engaged in by women and children as well as men. The situation is of course similar when it comes to fishing with nets.) If there are too few men in the net-hunter group, it may not be possible to oversee a net; too many and it may be necessary to form two hunting parties, which would mean that the average distance travelled by both parties would have to be greater. But the question of optimal size of hunting party might also be related to population pressure. In the case of the Hadza, hunting is essentially an individual pursuit using the bow and arrow, and is done exclusively by men and boys.[170] But the species the group considers most advantageous to hunt, e.g. whether it be rabbits or buffalo, and thereby the size of the hunting party, can depend on the level of earlier exploitation resulting from population pressure.

Pygmy bowmen[171]

At the time of a 1960s study, some 466 Bushmen lived in the Dobe area, which covered approximately 6400 km²,[172] giving a population density of one person per 14 km². And, also in the 1960s, some 400 Hadza lived in the tsetse bush to the east of Lake Eyasi in Tanzania, an area of more than 2500 km²,[173] making the population density there one person per 6 km². In 1788, before European settlement, Australia had an estimated 300,000 inhabitants, divided into about 600 tribes of 500 members each.[174] Australia's area being 7.7 million km², this means only one person to every 25 km².[175]

Population control

As noted already in 1922 by Carr-Saunders, and as mentioned in Chapter 1, the evidence contained in the then-recent anthropological fieldwork on primitive societies showed that the use of internal population checks was so widespread as to have been practically universal.[176] These checks were *variable*, and took the form of abortion, infanticide, prolonged abstention from intercourse, and the postponement of marriage,[177] the result being an approach to the optimum number in each society. In the case of the Australian Aborigines for example, the first born were usually exposed to die; they were not supposed to be sufficiently 'mature,' and from the customs attendant on marriage there must always have been some doubt of their paternity. More generally, from one-third to one-half of the newly born were allowed to perish.[178] Abortion amongst the Murngin, in which the pregnant woman's sisters exert pressure on her abdomen with knees and hands, is not infrequent. When twins are born one is always killed. If they are a boy and girl, the girl is usually put to death. It was said that a boy made a people strong while a girl only caused trouble. A woman kills a twin because, she says, it makes her feel like a dog to have a litter instead of a single child.[179] Among the Auen and Heikum Bushmen one twin is invariably killed by being buried alive by the mother or one of her attendants immediately after the birth. Usually only every second or third child is weaned, the one or two born in the interval being killed.[180] Carr-Saunders sees the operation of these variable checks as dependent on the fact that humans are *social* animals, suggesting that in primitive cultures, where technological innovation is slow and social conditions more or less stationary, the optimum number of people may remain about the same over long periods of time.[181]

As compared with more advanced cultures – particularly our own – modern hunter-gatherers have had excellent control of the size of their populations, showing no trend towards an increase in numbers until recently. The Aborigines of Australia are special from the point of view of the limitation of numbers, since at the time of their discovery by Europeans they were the only extant instance of a human population at the hunter-gatherer stage covering a complete land mass, without anywhere to emigrate and with no immigrants arriving. To this I might add that neither did they undergo technological change relevant to their vital needs, not developing the bow and arrow, for example, living as they did in a warm climate where such needs are more easily met. Being subject to no predatory wild animals, if internal mechanisms for the limitation of human populations exist they should be found among them.[182] As expressed by L. R. Binford, such data suggest that while modern hunting-gathering populations may vary in density between different habitats in direct proportion to the relative size of the standing food crop, nevertheless within any given habitat the population is *homeostatically* regulated *below* the level of depletion of the local food supply.[183] In terms of population pressure, the fact that food is not scarce for hunter-gatherers suggests that the area in which they live is not overpopulated. As suggested by Carr-Saunders:

> It is thus clear that within any group in any primitive race, the members of which co-operate together to obtain their food from a definite area to which they are confined, the principle of the optimum number holds good. There is, that is to say, taking into account the abundance of game, the fertility of the land, the skilled methods [technology] in use, and all other factors, a density of population which, if attained, will enable the greatest possible average income per head to be earned; if the density is greater or if it is less than this desirable density, the average income will be less than it might have been. Obviously it must be a very great advantage for any group to approximate to this desirable density.

Carr-Saunders' reference to the greatest possible average income per head may also be seen as an expression of the species, through its populations, over the whole of its existence acquiring as much solar energy as possible.

Binford says:

> Most demographers agree that functional relationships between the normal birth rate and other requirements (for example, the mobility of the female) favor the *cultural* regulation of fertility through such practices as infanticide, abortion, lactation taboos, etc. These practices have the effect of homeostatically keeping population size below the point at which diminishing returns from the local habitat would come into play.[184]

Note the mention of taboos here.

The following quote of E. F. Moran includes, and perhaps pertains primarily to, horticulturists, though much of it is relevant to hunter-gatherers as well.

> Female infanticide, abortion, long periods of sexual abstinence after childbirth, warfare, and a strong male fear of too frequent

sexual contact with women are characteristic of many of the world's peoples, including rain forest dwellers. Among some populations, intercourse between husband and wife is forbidden from the onset of pregnancy until the child is weaned – often not until between ages three and five. Sexual continence is commonly required prior to ceremonies, raids, and hunting. The number of prohibitions that are practiced varies a great deal and may be related to other forms of population control. All these practices have had the net effect of controlling the size of aboriginal populations throughout the humid tropics.

As regards infanticide, often involving exposure,[185] a study of modern hunter-gatherers revealed that it was practised in 80 of the 86 societies examined; and it was estimated that between 15 and 50 per cent of all live births ended in infanticide in societies at this level of development. Another study found that abortion was practised in 13 of 15 such societies.[186]

!Kung women practise infanticide when in their opinion it is necessary. Sometimes a child is born that cannot be supported, in which case it is destroyed. If a woman bears a child that is crippled or badly deformed, she is expected to destroy it, and if the season is very hard and she already has an infant under a year old, depending on her milk, she is forced to kill her newborn child[187] (note the potential involvement of morals). Other instances of infanticide among the !Kung include in the case of the birth of twins, only one of which is allowed to survive, or when a woman feels she is too old to produce milk for another baby.[188] But with the Bushmen infanticide is purportedly rare. They have no mechanical form of contraception and do not know how to cause miscarriage or abortion, and, according to Elizabeth Marshall Thomas, prefer to abstain from intercourse for long periods rather than suffer such pain.[189]

It has also been claimed that amongst the G/wi Bushmen in particular community perception of overpopulation is sensitive, and symptoms are recognised *before* a crisis develops.[190] This being the case, as is implied by Carr-Saunders, suggests an ability on the part of humans not only to keep their population size at or under the carrying capacity of their environment, but to decrease that size through cultural checks when the carrying capacity is reduced. It may be, however, that the Bushmen constitute an exception among modern hunter-gatherers in that infanticide among them is uncommon.[191]

In the case of Australian Aboriginals, when drought or other natural catastrophe occurs, cooperation and reciprocity in the division of the few available resources on the tracts of land owned by the various groups is essential for the survival of *the community as a whole*. Here we have a manifestation of the social instincts. It is better for the community to destroy an infant or young child whose chances of survival are small anyway than to hinder the mother unnecessarily in her task of food collecting. Cooperation and reciprocity are a matter of life and death for aboriginal societies.[192]

As pointed out by Wilkinson and as intimated above, many systems of population limitation depend upon *taboos* or other superstitious beliefs. Such taboos include those on sexual intercourse during lactation, and others which have the effect of conserving resources, including taboos related to *totemism* (which, for example, existed among all Australian tribes),[193] the effect of which is to preserve species of particular animals or plants. As Wilkinson says, totemism clearly includes elements which serve to spread the load of a society's consumption over a wider range of animal and plant species than would otherwise have been the case.[194]

Territoriality

The Bushmen defend both nucleated territories, concentrated on a well, and areal ones. Even though there is no general political or social organisation and no clearly formulated laws or rights, every family knows the exact limits of its range and is careful not to cross foreign ground. Every trespasser runs the risk of being killed in an ambush.[195]

Here we might consider the difference between a *tribe* and a *band*, a tribe consisting of a number of bands and, among other things, constituting the source of the majority of the mates for the various bands' members (thereby avoiding in-breeding). Within the Bushman band, for example, the only division is into families, the family consisting of a man and his wife or wives, together with their dependent children, and the band itself being an extended family.[196] Most of the hunting peoples of the world live in bands, which consist of somewhere between 30 and a little over 100 individuals, who own or occupy a single territory and share food.[197] In the case of the Bushmen of the Kalahari, bands were most frequently reported to consist of 25 to 50 people. As regards tribes, Birdsell cites the earlier-noted figure of 500 as their typical size in Aboriginal Australia, such a population being large enough to ensure an adequate recruitment of mates and yet small enough for everyone to know everyone else.[198]

Paradigmatically, in primitive cultures *each tribe has its own language*. As expressed by Lewis Mumford: "Linguistically, each group is surrounded by an invisible wall of silence, in the form of a different language group." This fact reinforces the 'otherness' and thus 'enemy' conception of other tribes.

In central Australia in the mid-1800s the tribes had separate hunting grounds, which were further parcelled out among individual members, who could always point out the exact boundaries of their territory. In Australia,

Tasmania and New Zealand, land was held primarily by tribal right; but within this tribal right each warrior had his own rights over some portion. And as regards horticulturists on the Trobriand Islands east of New Guinea, it was believed that every man and woman by birth and descent was connected with a definite spot, and through this with a village community and a territory. In all cases the natives could point out the landmarks marking the edge of their land, and the stability of territorial boundaries was regarded almost as a natural fact. Territorial behaviour on the part of tribes, bands, and large and small families is universal among primitive societies, both hunter-gatherer and horticultural.[199]

Here we might note Nicolas Peterson's idea with respect to Australian Aborigines, but which is undoubtedly of relevance to humankind as a whole, namely that once a person or party has been through a rite of entry, they have the same access to the resources of the group's territory as do others who have been through the rite. Greeting ceremonies, which frequently incorporate combative displays and controlled aggression between males, are thus functionally analogous to boundary defence in that they prevent unregulated movement between territories and control access to food resources. The ceremony makes the visitor temporarily a member of the community, performing the same sort of function as initiation rites. In this way, rites of entry ensure that people are encapsulated in groups, and do not wander from one area of abundance to another: they help create and maintain local bands.[200] (As regards the Bushmen, it has been observed[201] that, as intimated above, several bands of the same tribe would come together on a seasonal basis – which would probably involve a choosing of mates – resulting in a division of the year into 'public' and 'private' periods.)

Though it is not emphasised by writers on the topic other than Peterson, we see here with rites of entry and the defence of boundaries a clear role for territoriality in limiting the populations of modern hunter-gatherers.

Sharing and social stratification

Also of interest is the degree of egalitarianism and altruism evinced by hunter-gatherer groups. There is much sharing and little social stratification. In the case of the Netsilik Eskimos for example, an unlucky hunter was always certain by right to receive a portion of the daily catch secured by any of his partners. Camp fellows share together and/or starve together. And as regards the Bushmen, all game is shared out among the members of the band.[202] As noted by Carr-Saunders: "When one feasted they all partook; and when one hungered they all equally suffered." This phenomenon may be related to the difficulty involved in acquiring food. As noted by Frederick Eggan, in Siberia food-sharing amongst hunter-gatherers increased with latitude,[203] and, we might add, with the proportion of meat in their diet.

Since people own only what they can carry, and everybody knows everybody else, there can be little theft; but when it does occur, it is severely punished.[204] The Bushman band's territory, on the other hand, is owned in common.[205] Among the G/wi, there are no competitive games and little competition generally; and no wealth is attached to individuals. Nor are there any exclusive social groupings; and goods move from the haves to the have-nots.[206]

Taking an example from another group, a dominant value in the culture of the hunter-gatherer Bateks of Malaysia is the moral obligation felt to share food with all other families in the camp.[207] Sharing in these groups probably derives from the need to share *meat* in the case of larger kills in such a way as best supports the continued existence of the group[208] – as is the case with all altruistic behaviour seen from an evolutionary point of view. In the case of the Hadza, for example, although vegetable foods form the bulk of their diet, they attach very little value to them. In addition to being the preferred food, meat is also intimately connected with *rituals* to which Hadza men attach great importance.[209]

The !Kung band really has no chief;[210] nor does the Hadza. The common affairs of the Bushman band, such as migrations and hunting parties, are regulated rather by the skilled hunters and the older, more experienced men generally. None of the hunter-gatherer bands that have been studied by anthropologists had dominating leaders, but where they had leaders at all they were more 'headmen' than chiefs, the slight authority they did have depending to a considerable extent on their personalities, as is reminiscent of the nature of the leaders of wild-monkey bands. If, for example, a headman has a strong character, or possesses a gun, he will have a correspondingly greater influence.[211] G/wi leadership is persuasive and authoritative, not authoritarian, and serves only to guide the band towards the consensus that is the real locus of decision.[212] Part of the reason for this may be everyone's having to be able to carry everything they own when the band migrates, which prevents wealth differences through the accumulation of individual property, and thereby supports a generally egalitarian system.[213] The Aboriginal Murngins' undeveloped sense of property seems to be associated with their general lack of interest in technological development, and both of these appear to stem from their desire to be free from the burdens of, and responsibilities towards, objects which would interfere with their itinerant existence.[214]

Perhaps the greatest point of contrast between hunter-gatherers and sedentary peoples such as the Tolowa of north-west California and south-west Oregon is that the former lack the motivation to achieve and maintain superior status through the production of a cache to last the winter.[215] The further north people live, the more they are driven to hoard due to the greater seasonal variation in crops and the better storing conditions given the colder climate, the ability to hoard itself being a precondition for the existence of horticultural and all more complex societies. As regards hunter-gatherers, food storage techniques are more common and more sophisticated in higher latitudes, with little or no storage occurring in equatorial regions where seasonality is less pronounced and stored food spoils easily.[216] Autumn hunting affords Eskimos with a large store of reindeer meat, which is dried and set aside as provision for the winter. If carefully protected from damp it will keep for several years.[217] Salmon fishing is also actively pursued by Eskimos, and large quantities of these fish are preserved for future use.[218]

As touched on earlier, social inequality is also to be found in the respective roles of men and women in the acquisition of food. For example, where in hunting men generally travel light, in gathering women can at the end of the day be carrying burdens – including firewood – almost as heavy as they are themselves.[219] But looking at the hunter-gatherer type of society as a whole, it may be said that social inequality is less there than it is in any other form of society. As pointed out by C. A. Reed, when the nomadic !Kung become sedentary, as is fast happening, members of both sexes become fatter and taller – and the men become more socially dominant over the women.[220] The living conditions and diet of all members of a hunter-gatherer community are essentially the same, while, apart from in the case of food acquisition, social differences manifest themselves more in terms of relative status and Darwinian individual fitness; in this latter regard, the better the hunter, the greater the likelihood of his having an attractive wife or wives, and a larger number of them than otherwise.

State of health, longevity

Modern hunter-gatherers have lived better lives than their farming neighbours, in comparison with whom they have had, for example, more well-rounded diets.[221] Although there may have been constant complaints of being hungry, e.g. among some Bushmen of the Kalahari, this did not show itself in the actual state of their health. In a study based on a large proportion of some 2500 !Kung in Botswana, there were no signs of protein deficiency; and dental inspection showed that while their teeth were worn down, they had no caries; nor did they have varicose veins, haemorrhoids or hernias. (Note the absence of carbohydrates in their diets, carbohydrates easily being broken down by α-amylase and bacteria in the mouth, producing acid which increases the risk of caries.) Furthermore there was no evidence of coronary disease, high blood pressure, hypertension, neurological disease or suicide; and with increasing age their hearing was well preserved.[222] Also, !Kung infants were advanced in motor development in general as compared with their American counterparts.[223]

Also in the case of the !Kung Bushmen, mortality beyond infancy is not high:

> Another indicator of the harshness of a way of life is the age at which people die. Ever since Hobbes characterized life in the state of nature as 'nasty, brutish and short,' the assumption has been that hunting and gathering is so rigorous that members of such societies are rapidly worn out and meet an early death. [But i]n a total population of 466 [!Kung Bushmen of the Dobe area], no fewer than 46 individuals (17 men and 29 women) were determined to be over 60 years of age, a proportion that compares favourably to the percentage of elderly in industrialized populations.

The results of a medical examination of the Hadza in 1960 indicate that the clinical nutritional status of all the children was good by tropical standards; and, in particular, protein-energy malnutrition (kwashiorkor and marasmus), rickets, infantile scurvy, and vitamin B deficiency syndromes were not seen. Furthermore, while agriculturalists were liable to suffer from recurrent famine in this area, hunter-gatherers were not.[224] And, among hunter-gatherers quite generally, diseases, other than perhaps malaria and sleeping sickness, do not contribute to mortality. This is so for a number of reasons, including that isolated populations are less likely to contract epidemic diseases, and that hunter-gatherers are well nourished.[225] Of the various hunter-gatherer groups studied, *all* functioned well as communities, communities whose members were for the most part both healthy and content. As Richard Lee says, "The Dobe-area Bushmen live well today on wild plants and meat, in spite of the fact that they are confined to the least productive portion of the range in which Bushman peoples were formerly found."[226]

As pointed out by Carr-Saunders:

> Longevity may be considered as evidence of good health, and of the presence of aged people in Australia many observers speak. 'From numerous instances it would appear that the former [sic] generations were fairly long aged. Almost every small community would have in it two or three men or women over seventy years of age, and here and there some centenarians would be met with.' So too Burchell records having noticed many old people among the Bushmen. Writing of the Eskimos in a medical journal Smith calls them 'uncommonly healthy.' This is the opinion that one gains from other accounts, some of which

Kalahari Bushmen.[227] The San have been living in southern Africa for at least 11,000 years, perhaps for as long as 40,000 years.[228] The Kalahari has been inhabited from very early times, and there is no reason to assume that it is so inhospitable as to be merely a last refuge.[229]

specially mention longevity as a characteristic. 'The North Americans are in general robust and of a healthful temperament, calculated to live to an advanced age.' ... Of the Shushwap we are told that they were formerly [sic] healthy and lived to a great age. Hill Tout sums up the situation with regard to the Salish as follows: 'the great age, to which both men and women formerly lived, shows the vigour of the race and the general wholesomeness of their lives and condition.' 'The Nootkas are generally a long-lived race.'

As regards health, we might pause to compare the situation of hunter-gatherers with that of other animals. If one were to perform a medical examination of any animals uninfluenced by humans one would expect to obtain similar findings. Thanks to natural selection not all would live beyond infancy, and those that did would be healthy and live as long as may be expected given the species' somatic checks. In the case of primitive humans one can see perhaps two main causes of non-maximum average life expectancy, but causes which would not interfere with health nor the proportion of old people in the community. The one is the check of infanticide, and the other that of inter-tribal killing. It is only in populations affected by technology that ill health is widely manifest.

Conflict

Conflict involving modern hunter-gatherers is certainly not unknown, and anthropological investigations indicate that it has in fact involved proportionally more killing than one finds in civilised societies, while at the same time major conflicts have been found to be more common among horticulturists than among hunter-gatherers. This latter finding parallels those of archaeology, according to which the per capita number of major conflicts increased substantially during the horticultural era.[230]

Amongst the Murngin, (proto)warfare (Glossary) is an important band activity, the band being the war-making group. Within this group no violent conflict ever takes place, no matter how much cause is given. Members may quarrel, but for bandsmen to fight one another would be considered an unnatural act, and it never occurs. When the band is at war the ceremonial leader almost always acts as the war leader. A band seldom goes to battle as a group against another band, but usually has an eternal feud with certain of its neighbours, which results in occasional ambushes in which a man or two is killed. Fear of death while in a fight is seldom seen; and a man who shows an unwillingness to fight is called a woman, and held in extreme contempt (note the relevance of morals). The two main causes leading to war amongst the Murngin are the killing of a bandsman by a member of another band, and inter-band rivalry over women. The latter is the usual cause of a first killing, while blood vengeance forces further killings.[231] In the case of the horticulturist West Africans, "Sometimes the most fiendish treachery is indulged in during a friendly feast, … and head-hunting was always regarded as a popular and manly sport."[232] Tribal blood-feuds could continue for years, for the children of a man killed even in battle would keep up the vendetta.[233] A Bushman boy whose father has been killed is obliged to undertake vengeance as soon as he is grown up. His mother impresses this obligation upon him during his childhood, and as soon as he has killed his first head of big game the obligation must be carried out.[234]

The Murngin have developed a number of types of warfare which form episodes in the several chains of feuds linking past generations to present. Murngin society being polygamic, the one important effect of warfare is the seasonal slaying of a small proportion of young men who have passed adolescence and are potential or eligible mates. Polygyny, which plays an important role in the social structure of the band, is possible only because of warfare, and, conversely, is a decided factor in stimulating open conflict because of the resultant scarcity of women.[235]

According to Lawrence Keeley, killing amongst the !Kung, often involving feuds between bands, was at a rate four times that of the United States from

1920 to 1955. This brings to mind the possible population-checking function of warfare, noted in the quote from Moran, above. The Bushmen also raided the pastoral Khoikhoi (Hottentots) at the time of first contact with Europeans, and both foragers and pastoralists showed a propensity for stealing crops as well as livestock from settled farmers.[236]

Amongst the Bushmen, homicide, trespass, theft, adultery and wife-stealing give rise to quarrels and conflicts which generally assume the form of blood feuds, and may even lead to protowar. The infringement of territorial rights is one of the main causes of dispute between neighbouring bands, and almost invariably leads to bloodshed, as all Bushmen readily attack those who encroach on their land. In fact blood vengeance is the principal, if not the only, recognised way of dealing with serious offences committed against a person, even by members of the same band. In the case of adultery, the injured husband attempts to kill the adulterer, if he possibly can, while the unfaithful wife is beaten and in some cases divorced. Theft also provokes blood vengeance when the thief refuses to return the stolen property.[237]

An example of intratribal killing amongst the Bushmen: An Auen man had a good deal of tobacco, most of which he hid in a tree. Returning later he found the tobacco gone, and human footprints round the tree. He followed the spoor to one of a cluster of huts, where he saw another Bushman sitting at the fire cutting the stolen tobacco. He had a spear with him, and rushed up and killed the thief with it. The man's wife, intervening, was also stabbed. As the assailant was escaping he was shot in the side with a poisoned arrow from the group round the next fire. He succeeded in reaching his home, but died of the poison a few days later.

Another example: One Bushman shot another in the leg with an arrow during a honey-beer drinking bout. The wounded man revenged himself on the spot by shooting his assailant in the heart. The dead man's brother pursued the killer but could not overtake him, and so returned to the camp, where he met the man's wife carrying a child. He tore it from her, threw it up in the air and caught it on the point of his spear. The mother saved herself by fleeing, but two other women related to the original killer were also killed by the dead man's brother.

Among the Auen there is no formal declaration of war, but when a man wears a cap of aardwolf skin, he is on the warpath. On one occasion several friendly Auen villages came together and decided to attack another camp. The war caps were put on, and after sunset the men marched in the direction of the enemy camp. In the neighbourhood of the latter a halt was made and, without lighting a fire despite its being the cold time of the year, they waited until the morning. After sunrise, when it was somewhat warmer, the camp was surrounded and attacked from all sides, the spear being the weapon chiefly used. In such attacks all the men are killed, as are all boys, for fear of a blood feud.

At the beginning of the 20th century in the central Kalahari there were, however, common hunting grounds in which during the rainy season members not only of different bands but also of several different tribes could be found hunting together peacefully, though at other times of the year each band carefully avoided going beyond its own area.

A member of a horticulturist tribe in New Guinea intimated to Diamond: "Of course the Fayus will kill any trespasser; you surely don't think they're so stupid that they'd admit strangers to their territory? Strangers would just hunt their game animals, molest their women, introduce diseases, and reconnoitre the terrain in order to stage a raid later." Each tribe consisted of a few thousand people occupying a particular valley and living within a radius of 15 km. And each tribe had its own genetic abnormalities and diseases, practised its own form of cannibalism and self-mutilation, and had its own language and culture. While most fighting in New Guinea consisted of skirmishes leaving no or few dead, groups sometimes did succeed in massacring their neighbours. Like other peoples, New Guineans tried to drive off or kill their neighbours on occasions when they found it advantageous, safe, or a matter of survival to do so.[238]

> [I]n New Guinea until recently, each tribe maintained a shifting pattern of warfare and alliance with each of its neighbors. A person might enter the next valley on a friendly visit (never quite without danger) or on a war raid, but the chances of being able to traverse a sequence of several valleys in friendship were negligible. The powerful rules about treatment of one's fellow 'us' did not apply to 'them,' those dimly understood, neighboring enemies. As I walked between New Guinea valleys, people who themselves practiced cannibalism and were only a decade out of the Stone Age routinely warned me about the unspeakably primitive, vile, and cannibalistic habits of the people whom I would encounter in the next valley.

As regards cannibalism etc. amongst modern primitive people, it may be mentioned that relatively recent skeletal remains found at Saunaktuk, an Eskimo village in the Canadian Arctic, suggest torture, violent death, dismemberment and probable cannibalism on the part of the nearby Dene Indians.[239] There is conclusive evidence that, amongst many tribes, cannibalism arose out of a belief that eating a man caused his good qualities to pass into the body of the host.[240] According to W. J. Sollas, human flesh is a great dainty, said by

Australian Aborigines to taste much better than beef;[241] the chief of the horticultural Miranhas of Brazil says: "I know of no game which tastes better than men;" and A. J. N. Tremearne claims that there is no doubt that one of the principal causes of cannibalism is a longing for the taste of human flesh.[242]

A survey of studies of 99 modern hunter-gatherer bands found that in the 68 bands in which fighting was current or had been engaged in during the past five years, the ratio of young males to young females was more than three to two, while the ratio of adult males to adult females was one to one. In the other 31 bands, where warfare had been stopped by Europeans for at least five years, the sex ratio was close to one to one for all age-groups. And similar results were found for modern horticulturist groups, with, in both cases, the sex ratio being closest to normal the longer the time since the relevant group had been involved in fighting. What we see in primitive societies engaged in fighting is a high rate of female infanticide, which accounts for there being more boys than girls, and a high rate of male death in fighting, which accounts for an evening out of the sex ratio amongst adults.[243]

As suggested by William T. Divale, warfare – and, we might add, the killing of males it involves – creates a need for male warriors-to-be, which leads to a high rate of female infanticide. And, from an ecological point of view, both the infanticide and the fighting have the function of restraining population growth;[244] or, as we should say, both are manifestations of internal population checks.

It has elsewhere been suggested that in at least some modern hunter-gatherer societies violent conflict is slight. In a 1950s study of the hunter-gatherer Mbuti Pygmies, members of a neighbouring tribe came into the area of the researcher's group and took honey from trees. A confrontation arose which consisted of arguing, and after a time the transgressors left, leaving the honey behind.[245] According to Lee and Irven DeVore:

> [W]hen arguments break out it is a simple matter to part company
> in order to avoid serious conflict. This is not to say that violence is
> unknown; both homicide and sorcery are found among a number
> of current hunter-gatherers. The resolution of conflict by fission,
> however, may help to explain how order can be maintained in a
> society without superordinate means of social control.

According to C. M. Turnbull, when disputes arise within the Mbuti band, the principals simply part company rather than allow the argument to cross the threshold of violence. Harmony is thereby maintained without recourse to fighting or formal modes of litigation. The essential condition seems to be the lack of exclusive rights to resources; thus it is a relatively simple matter for individuals and bands within the same tribe to separate when harmony is threatened.[246] Such a form of conflict resolution would not be possible in situations where social units are strictly defined and firmly attached to parcels of real estate.[247]

James Woodburn and Richard Lee respectively report a similar mode of conflict resolution among the Hadza and !Kung Bushmen; and Lee and DeVore suggest that, judging from their generally flexible group structure, resolution of conflict by fission may well be a common property of hunter-gatherer societies.[248]

The high rate of Bushman killing in the past however indicates that fission, if it is to resolve conflicts, has not been particularly effective. In this regard, Keeley, referring to R. C. Bailey and associates, suggests that the Mbuti Pygmies studied by Turnbull were in fact under a superordinate authority, being the 'servants' of the neighbouring Bantu horticulturists. It is this, including the Mbuti's dependence on the Bantu for 65 per cent of their caloric intake, that accounts for their apparent peacefulness. Reizo Harako, on the other hand, claims that only a limited number of Mbuti bands are truly dependent on villagers for vital commodities.[249] However, field studies by Bailey and N. R. Peacock have shown that the Efe Pygmies, who live south of the Mbuti, are dependent on the horticultural Lese for trading relations with villages outside the rainforest.[250] It would appear then that at least in the case of the Pygmies of the Ituri forest (the Bambuti) there may well be a superordinate means of social control (à la Hobbes) which would make it impossible for them to engage in violent conflict, at least with the Bantu (from whom they nevertheless stole crops).[251]

Ecological sustainability

As related to our treatment of the equilibrium of the human species in the previous chapter, there has been much discussion in recent years about the unsustainability of the present Western lifestyle. Thus it should be mentioned that the lifestyle of modern hunter-gatherers, to the extent that it does not involve the use of non-renewable resources and has existed for the last 10,000 years, may be considered as being ecologically sustainable. It is of course requisite to sustainability that the renewable resources that are used not be harvested beyond the point of their being able to regenerate. In this regard we cite what has been noted in the case of the G/wi Bushmen:

> Potentially, the predation efficiency of the G/wi hunter is very high and he could use that capacity to create a very unstable relationship between humans and prey populations in much the same way as did the hunting people of North America who engaged in the fur trade. But the G/wi apply their efficiency so as to keep the relationship between them and each prey species at a low level of intensity.[252]

In keeping with this, Wilkinson emphasises that some societies have managed to exist over long periods without experiencing population increase or food shortage, the latter implying that some of the checks to population growth have been internal to these societies, and the former that during such periods the societies have been in ecological equilibrium with their surroundings. "Only if a society is already out of equilibrium will it plough back increases in efficiency to achieve greater output and higher levels of environmental exploitation."

The original affluent society

Marshall Sahlins is responsible for a good portion of the thrust suggesting that the traditional, Western way of seeing our species' development is fundamentally mistaken:

> It will be extremely difficult to correct this traditional wisdom. Perhaps then we should phrase the necessary revisions in the most shocking terms possible: that this was, when you come to think of it, the original affluent society. By common understanding an affluent society is one in which all the people's wants are easily satisfied; and though we are pleased to consider this happy condition the unique achievement of industrial civilization, a better case can be made for hunters and gatherers .... Rather than anxiety, it would seem the hunters have a confidence born of affluence, of a condition in which all the people's wants (such as they are) are generally easily satisfied. This confidence does not desert them during hardship.[253]

Note that Sahlins is here speaking of hunter-gatherer societies generally, and what he says should not be taken to apply to hunter-gatherers living in extremely harsh conditions, as do the northernmost Eskimos such as the Netsilik and other Central Arctic peoples, who are perhaps unique in the near past in the almost total absence of vegetable foods in their diet. This factor, in combination with the great cyclical variation in the numbers and distribution of Arctic fauna, makes Eskimo life hard. In effect, the kinds of animals that are 'luxury goods' to many hunters and gatherers are to the Eskimos the absolute necessities of life. However, even this view should not be exaggerated, since most of the Eskimos in historic times (Neo-Eskimos) have lived south of the Arctic Circle, and many of the Eskimos at all latitudes have depended primarily on fishing, which is a much more reliable source of food than hunting land and sea mammals.[254]

Sahlins' way of expressing himself draws attention to the extent to which the traditional perspective is mistaken in its conception of the life of hunter-gatherers, and he did not intend to suggest that they were affluent in the modern sense which, among other things, implies being *wealthy*. In fact, just as the term *work* is not easily applied to humans at the nomadic hunter-gatherer stage, nor to other animals, neither are the terms *affluence* and *poverty*, unless we define them more broadly. (Note how the notion of affluence is most often related to standard of living rather than quality of life.) If we were to do so we might say, following Sahlins, that an affluent society is one in which everyone has an adequate diet and the opportunity to procreate – including housing in the case of sedentary societies – and that people do not have to work too much. But thinking in this way we would find that hunter-gatherers do not merely constitute the *original* affluent society, but, considering the presence of poverty in almost all civilised societies, as intimated by Sahlins, perhaps the *paradigm* of an affluent society. (As suggested by others, modern industrial society might better be termed the *effluent* society.) Whether there can exist e.g. *infanticide* in an affluent society then depends on what you count as being a member of society (e.g. as to whether it should require going through an initiation ceremony). On the other hand, however, a high level of violent deaths in a society – whether this need be the case amongst modern hunter-gatherers – should count against its being affluent (unless it is taken e.g. to perform a valuable existentialist function). What the issue concerns in broader terms is *quality of life*, and in this respect the lives of hunter-gatherers may well be superior even to those of people living in modern industrial societies in which there is no poverty or high rate of killing. This is certainly true as regards such aspects of life as stress and tedium, of which the hunter-gatherers experience very little. From a biological point of view, we see that what we are leading to here is the idea that true affluence is provided through living as close as possible to the sustainable way that wild animals do.

In a later work Sahlins adds what we partly noted earlier, namely that:

> A good case can be made that hunters and gatherers work less than we do; and, rather than a continuous travail, the food quest is intermittent, leisure abundant, and there is a greater amount of sleep in the daytime per capita per year than in any other condition of society.[255]

He further notes:

> It is interesting that the Hadza, tutored by life and not by anthropology, reject the neolithic [horticultural] revolution in order to *keep* their leisure. Although surrounded by cultivators, they have until recently refused to take up agriculture themselves, 'mainly on the grounds that this would involve too much hard work.'

Note also however that horticulture does not give the same return to effort spent in areas occupied by modern hunter-gatherers – otherwise the hunter-gatherers would not find themselves in just those areas.

And as Wilkinson suggests regarding leisure:

> A leisure preference is a clear indication of the relative sufficiency of a society's material means of subsistence and should be regarded as a feature of societies in ecological equilibrium. If a society is in equilibrium and has adequate subsistence then the leisure preference of its members will prevent increases in economic efficiency leading to what we would recognize as economic development.[256]

Sahlins again:

> The world's most primitive people have few possessions, *but they are not poor*. Poverty is not a certain small amount of goods, nor is it just a relation between means and ends; above all it is a relation between people. Poverty is a social status. As such it is the invention of civilization. It has grown with civilization, at once as an invidious distinction between classes and more importantly as a tributary relation – that can render agrarian peasants more susceptible to natural catastrophes than any winter camp of Alaskan Eskimo.

As suggested by Lee and DeVore:

> Sahlins' argument served to underline the point that anthropologists have tended to view the hunters from the vantage point of the economics of scarcity. Viewed on their own terms, the hunters appear to know the food resources of their habitats and are quite capable of taking the necessary steps to feed themselves. … Since a routine and reliable food base appears to be a common feature among modern hunter-gatherers, we suspect that the ancient hunters living in much better environments would have enjoyed an even more substantial food supply.[257]

And Lee:

> It is impossible to define 'abundance' of resources absolutely. However, one index of *relative* abundance is whether or not a population exhausts all the food available from a given area. By this criterion, the habitat of the Dobe-area Bushmen is abundant in naturally occurring foods. By far the most important food is the mongongo nut. Although tens of thousands of pounds of these nuts are harvested and eaten each year, thousands more rot on the ground for want of picking.

The mongongo nut, because of its abundance and reliability, alone accounts for 50 per cent of the vegetable diet by weight. In this respect it resembles a cultivated staple crop such as maize or rice. Nutritionally it is even more remarkable, for it contains five times the calories and ten times the proteins per cooked unit of the cereal crops. The average daily per-capita consumption of 300 nuts yields about 1,260 calories and 56 grams of protein. This modest portion, weighing only about 200 grams, contains the caloric equivalent of one kg of cooked rice and the protein equivalent of 400 grams of lean beef.

Furthermore the mongongo nut is drought resistant and it will still be abundant in the dry years when cultivated crops may fail. The extremely hard outer shell protects the inner kernel from rot and allows the nuts to be harvested for up to twelve months after they have fallen to the ground. A diet based on mongongo nuts is in fact more reliable than one based on cultivated foods, and it is not surprising, therefore, that when a Bushman was asked why he hadn't taken to agriculture he replied: 'Why should we plant, when there are so many mongongo nuts in the world?'[258]

This output of 2,140 calories and 93 grams of protein per person per day [in the !Kung diet] may be compared with the US Department of Agriculture's Recommended Daily Allowance (RDA) for persons of the small size and stature but vigorous activity regime of the !Kung Bushmen. The RDA for Bushmen has been estimated at 1,975 calories and 60 grams of protein per person per day. Thus the !Kung's food intake exceeds the RDA's requirements by 165 calories and 33 grams of protein. One can tentatively conclude that even a modest subsistence effort of two or three days' work per week is enough to provide an adequate diet for the !Kung Bushmen.[259]

The !Kung do not amass a surplus, because they conceive of the environment itself as their storehouse. The necessities of the hunter's life are in the bush no less surely than those of the agriculturalist are in the cultivated ground.

Note, however, that the Bushman's 'storehouse' is much less densely stocked than the agriculturists, necessitating, of course, a lower population density.

Similar comments have been made by Woodburn as regards the Hadza, generally regarded to be the most 'affluent' of modern hunter-gatherers: "For a Hadza to die of hunger, or even to fail to satisfy his hunger for more than a day or two, is almost inconceivable."[260]

According to Wilkinson, the issue at stake

> involves throwing out the whole popular idea of primitive societies as societies which exist in a perpetual state of hardship, scarcely able to scrape up a bare minimum of subsistence, with large families, children suffering from malnutrition, and a low life expectancy. In so far as this *is* the situation in parts of the underdeveloped world today, it is a comparatively recent phenomenon. Contact with Europeans, and with their alien values and practices, has led to the abandonment of customs such as abortion and infanticide which once helped to prevent overpopulation. As early as the third quarter of the 18th century a missionary in Paraguay claimed that Christianity had led to the abolition of abortion and infanticide among the Abipones. Ironically, the same missionary also remarked on how plentiful food had been, as if the fact was unrelated to the existence of these practices.

Archaeology

Prehistoric overkill

Archaeological research has also relatively recently taught us of another phenomenon that will show itself relevant to the theory to be advanced here, according to which our development has followed a route much different from that supposed on the traditional perspective. As was first emphasised by Paul S. Martin, particularly towards the end of the Pleistocene (ca. 10,000 BP), but covering the period from 100,000 to 1000 BP, throughout the world exceedingly many species and genera of large mammals became extinct. In fact, late Quaternary extinction swept away half of the 167 genera of large mammals (>44 kg) found on the continents.[261]

In *Africa*, over the past two million years, some 50 large mammalian genera disappeared, with the peak occurring around 50,000–40,000 BP. At least 26 of these genera became extinct during the last 100,000 years, with the result that the African game plains today contain only about 60 per cent of the genera that existed there at the beginning of this period.[262] In *South-east Asia* major extinctions started around 50,000 BP.[263] In *Australia*, some 46,000 years ago,[264] major extinction episodes resulted in the disappearance of all but one of the continent's 16 genera of megafauna weighing more than 100 kg. Some 28 genera and approximately 55 species of vertebrates are thought to have become extinct in Australia towards the end of the Pleistocene. These include approximately 16 per cent of the species and genera of the continent's

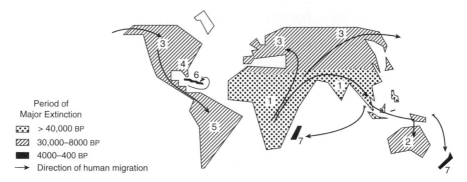

The global pattern of late-Pleistocene extinction in sequence: 1, Africa and
Southern Eurasia; 2, New Guinea and Australia; 3, Northern Eurasia and northern
North America; 4, Southeastern United States; 5, South America; 6, West Indies;
7, Madagascar and New Zealand. In each case, the major wave of late-Pleistocene
extinction does not occur until prehistoric hunters arrive.[265]

mammals, including all of the largest carnivores and herbivores. In *Eurasia,*
between 30,000 and 10,000 BP, many species of large herbivore and some species of carnivore became extinct as the ice sheets retreated and the tundra was
replaced by forest.[266] In *North America* 34 genera,[267] including 57 species[268] (each
ca. 70 per cent of the total), of the mammalian megafauna became extinct,
with a subsequent reduction of over 50 genera in *South America,*[269] all of this
occurring some 11,500 to 10,500 years ago. There were also major extinction
episodes in the *West Indies,* which started around 4000 BP, in *New Zealand* ca.
1000 BP, on *Madagascar* around 800 BP,[270] and on various other islands around
the world at various times.

In Africa between 1.8 million and 900,000 BP the *Paracolobus* – Africa's largest monkey, three species of sabretooth (which survived in Eurasia and
America), ten species of elephant-like Proboscidea, one or more 'forest hippo'
(*Hexaprotodon*) species, the large ground-sloth analogue *Ancylotherium* ('hooked
beast'), the lion-sized hyena *Pachycrocuta*, the cheetah-like 'running hyena'
Chasmaporthetes, a species of large dog, giant pigs, the 'super-aquatic' hippo *H.
gorgops*, the hipparion (a genus of horse), and the australopithecines all went
extinct. After 500,000 BP the giant *Hippotragus* antelope, Africa's giant baboon
species, and the *Sivatherium* ('Shiva's beast') giraffe/okapi became extinct; and
around 12,000 BP the giant wildebeest *Megalotragus*, the large zebra *Equus cap-
ensis*, the African musk-ox *Makapania*, and a small steenbok-like antelope also
went extinct.[271]

Among the species or genera that disappeared in *Australia* were giant marsupials. These included the marsupial lion, *Thylacoleo carnifex*, which may have
weighed up to 160 kg and have been comparable in ferocity to a sabre-toothed

Sivatherium (Africa; extinct ca. 8000 BP)

Megalotragus (Africa; extinct ca. 10,000 BP)

tiger, and its prey, the diprotodon or marsupial rhino, the size of an Indian elephant and the largest marsupial that has ever lived. A wombat (short-legged marsupial quadruped) the size of a cow is another of the extinct megafaunal marsupials, which also include the short-faced kangaroo. Other extinctions include gigantic reptiles (pythons, varanid lizards, crocodiles and horned turtles), the ostrich-like Thunderbird, weighing as much as 500 kg, and a number of other flightless or poorly flighted birds, including several weighing 1 kg or less.[272] Among the herbivore species that went extinct in *Eurasia* were the woolly mammoth, woolly rhino, hippopotamus, 'narrow-nosed' rhinoceros and the straight-tusked elephant, all of which weighed more than 1000 kg. Herbivore extinctions also included the giant Irish elk, musk-ox, long-horned steppe bison (*Bison priscus*) and the giant deer (*Megaloceros giganteus*); and carnivore extinctions included the cave bear (*Ursus spelaeus*) and the spotted hyena.[273] Many of these animals inhabited the Mammoth Steppe – a vast expanse of dry low grasslands just south of the glaciers, spanning northern Europe, Asia and Alaska.

Marsupial lion skeleton (Australia; extinct ca. 46,000 BP)

In *North America* the species and genera of large mammals that became extinct include the American mastodon, the Colombian, woolly, Jefferson's, imperial and pygmy mammoths, a species of tapir, the stag moose, two species of deer including the giant deer, the camel (*Camelops hesternus*), the llama, the native American horse, the long-horned and largest bison (*Bison latifrons*),[274] a species of pronghorn, the Asian antelope, the yak, four species of ox including the shrub-ox and the woodland musk-ox, the sabre-toothed and scimitar-toothed cats, the American cheetah, the spectacled and giant short-faced bears, the dire wolf (a giant wolf), the giant beaver (the size of a black bear), two species of peccary (related to swine), the anteater, the glyptodont (>1000 kg, one of the largest ancient armadillos), and various ground sloths, including Rusconi's and the Shasta (the smallest, but still the size of a black bear);[275] and non-mammalian extinctions included that of the giant tortoise.

Woolly mammoth (Eurasia, North America; extinct ca. 11,250 BP)[276]

Some animals, such as the ancient cats, had a recurring geographical range: Europe, Africa, Asia, and, over the Beringian land mass, North America. Large cats originally from the Old World evolved into now-extinct endemic North American species; while now-extinct camelids that came into existence in North America migrated to Asia, evolving into today's camels. Mammoths flourished in the open dry-tundra landscape characteristic of Beringia during the colder periods, while mastodons, which depended on shrubs for food, were uncommon.[277]

An even greater number of extinctions occurred in the tropics of the Americas. In *South America* they included the elephant-sized giant ground sloth (*Megatherium*), the glyptodont and the rhino-sized toxodont (all over 1000 kg), as well as the ground sloth (*Mylodon*), the camel-like litoptern (*Macrauchenia*), horses and the sabretooth. The major extinction episode in the *West Indies* included the

Bison latifrons (Eurasia, North America; extinct ca. 11,250 BP)[278]

Sabre-toothed cat (North and South America; extinct ca. 11,000 BP)[279]

disappearance of several species of monkey, ground sloths, a bear-sized rodent and several owls: the normal, colossal and titanic. In *New Zealand*, 11 species of ostrich-like moa together with 20 other types of bird became extinct. These other birds included the kiwi, the weka, a pelican, a swan, a giant raven, a flightless goose, a coot (waterbird) and Haast's eagle (the world's largest eagle). All of these extinctions occurred before the arrival of Europeans, Haast's eagle in particular becoming extinct about 500 years ago. In the past 2000 years at least 77 bird species have become extinct in New Zealand.[280]

Glyptodont (North and South America; extinct ca. 10,500 BP)[281]

In *Hawaii*, the Hawaiian flightless geese and dozens of smaller birds, totalling some 40 species of land bird, disappeared.[282] Apart from a species of giant tortoise, 16 genera of vertebrates became extinct on *Madagascar*,[283] including giant lemurs (the size of a gorilla), flightless struthios and elephant birds, two species of giant land tortoise, an aardvark, a large mongoose-related carnivore built like a short-legged puma, as well as two hippopotamus species.[284]

According to Martin's prehistoric overkill hypothesis, virtually all of these extinctions were directly or indirectly the result of human predation. This hypothesis receives strong support from the fact that we humans were present when all of them occurred, and most of them took place shortly after our arrival. It has been established that in the case of the birds in New Zealand, for example, the extinction of the smaller birds was due to rats arriving with the Polynesian settlers, and of the moas due to hunting pressure from settlers.[285]

The application of the hypothesis to America, where the greatest number of extinctions occurred over the shortest period, suggests that immigrants from

Giant ground sloth (South America; extinct ca. 10,500 BP)[286]

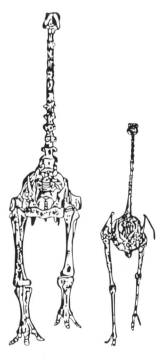

Elephant bird and ostrich skeletons. Elephant birds lived on Madagascar, and went extinct after 1200 AD. They were some 3 m tall, and weighed around 500 kg.

north-eastern Asia traversed the entire length of the Americas in less than a millennium, arriving at the tip of South America by 10,500 BP and causing the extinction of the various megafauna as they advanced.

As Martin reasons, we need only assume that a relatively innocent prey was suddenly exposed to a new and thoroughly superior predator, a hunter who preferred killing and persisted in killing as long as the prey were available. A rapid rate of killing would wipe out the more vulnerable species before there was time for the animals to learn defensive behaviour, and thus the hunters would not have needed to plan elaborate cliff drives or build clever traps.[287]

The weapons used by this wave of humans in North America were part of what is called the *Clovis* toolkit. Upon the demise of the megafauna that could be hunted using this toolkit, it was replaced by the *Folsom*, which had smaller points. Clovis hunters pursued large herbivores including the *mammoth* from 11,300 to 10,900 BP in western North America before being replaced by hunters who used Folsom points and killed *bison*.

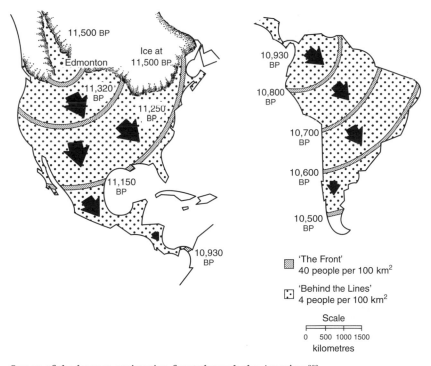

Sweep of the human emigration front through the Americas[288]

In the Americas such large-scale kills as those effected by means of stampedes occurred for the most part after the original wave of extinctions, the evidence of such kills in North America being more typically associated with later phases of the big-game hunting culture, and suggesting groups of humans that are both larger and better organised than their predecessors. But the mass slaughter of whole herds, presumably accompanied by attempts at meat preservation, would have required much more effort, which suggests that such resources were less reliable, humans having had to make the most of occasional herds.[289]

As dealt with in the preceding chapter, normal predator–prey relationships involve homeostatic mechanisms in which the size of predator populations is regulated by internal and/or external checks in the form of diminishing returns. In this way predators do not, in most cases, overextend their consumption or destroy their prey species. The overkill thesis implies that no such limiting mechanisms operated to restrain human hunters,[290] or that if they did, they were not sufficiently strong. As expressed by Juliet Clutton-Brock, it is highly unlikely that in the prehistoric period the world was peopled with noble savages who were consciously aware of the need to conserve their resources.[291] What we have, rather, is a classic manifestation of the pioneering (and overshoot) principle(s) in the case of the human species.

The climate-change hypothesis

Other reasons have been given for the massive extinctions of mega-fauna at the end of the Pleistocene, in particular the effect of climate change. There are a number of problems with this suggestion, however. One is that many of the extinctions took place at times other than at the end of the last ice age. In Africa, the peak in extinctions occurred there in the *middle* of the last ice age, as it did in South-east Asia.

Not only was the major occurrence of extinctions in both of these areas not related to climate change, but in Africa it took place at around the same time as the Mousterian hand-axe culture (Glossary) disappeared there roughly 50,000 years ago.[292] This suggests that these extinctions resulted from *sapiens'* employment of Mousterian technology in their hunting, which was *big* big-game hunting, and that this technology was no longer viable once such game disappeared.[293] And much of the *Australian* loss of large and medium-sized mammals took place some 36,000 years before the climate changed – in the case of the megafauna perhaps as early as 500 years after modern humans arrived on the continent.[294]

Here we must emphasise the aforementioned fact that these various extinctions all took place in the presence of humans. The immense wave of mammalian megafaunal extinctions in the *Americas*, though it did occur at the end of the Pleistocene, took place during a period shortly after what may well have been the arrival of the first and only wave of Palaeolithic modern humans around 11,500 BP. Thus if climate change were to be the cause of these extinctions, one would wonder why European and Ukrainian woolly mammoths became extinct 13,000 years ago, just prior to the emigration of people to North America, while in North America they became extinct some 2000 years later, just after the people arrived there.

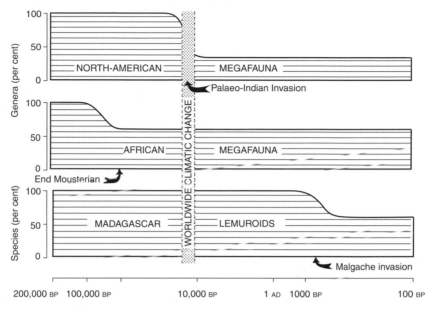

North American, African and Madagascar extinctions[295]

The major extinction episodes in the *West Indies* started with the arrival of humans; and the extinctions on *Madagascar* occurred directly after the arrival of immigrants from Indonesia and Africa (the Malagache invasion), and were not coincident with those in Africa 350 km away. On other islands we see the same phenomenon of widespread extinctions after the arrival of humans: within 600 years after the first human settlement of *New Zealand*, the various species of flightless bird had been hunted to extinction.[296] In *Hawaii* 39 species of land bird became extinct within 1000 years of settlement by Polynesians[297] some 1700 years ago. Similar extinctions occurred on the islands of *Crete* and *Cyprus* after the arrival of humans (between 10,500 and 6000 BP).[298] (As regards known mammalian extinctions that have occurred during the last 500 years, some 80 per cent have been among mammals endemic to island habitats.) None of these extinction waves occurred at the end of the Pleistocene. As expressed by Martin, in the case of the ice-age megafauna a sudden global extinction of all species for whatever reason is disproved, since the Late Pleistocene losses were diachronous between continents. Many millennia separated the extinctions of Australian diprotodons, American mammoths and Malagasy giant lemurs.[299]

Another problem faced by the climate-change hypothesis is that, as pointed out by John Alroy with regard to North America, there is almost no credible evidence for climate-driven extinction at any other point in the fossil record of the continent's mammals.[300] This can in fact be generalised: climatic changes have never produced such massive extinctions in the past.[301] On top of this, it may be pointed out that large mammalian fauna (K-selected species), under most circumstances, should be the species most tolerant of environmental changes. It is well established, for example, that the presence or absence of large mammals is among the least accurate of Palaeo-climatic indicators. Plants and microfauna (r-selected) are more sensitive to changes in temperature and humidity; yet relatively few of these life forms seem to have become extinct at the end of the Pleistocene.[302] The victims of the various extinction waves were not merely large, but *much* larger than one would have predicted, and in fact so large that they densely occupy a narrow range of the size spectrum. Earlier time intervals, on the other hand, consistently show a tendency for relatively severe extinction episodes to affect all parts of a broad spectrum of species/ genera.[303] In fact, as has been suggested by Martin and independently by P. J. Mehringer, the onset of postglacial conditions would have been generally *beneficial* to the megafauna, providing improved natural conditions. Mehringer argues that the recession of the ice would have enhanced rather than eliminated the megafaunal habitats, and that more such habitats would have been available in post-Pleistocene times.[304]

The Late Pleistocene extinctions are concentrated on the large, gregarious, herding, or slow moving (K-selected) animals – the ideal prey of human hunters.[305] It is only on oceanic islands that numerous small vertebrate genera were obliterated, again, after the arrival of modern humans.[306] Among the varieties that escaped in the *Americas* were smaller animals (which generally reproduce faster and are less visible), relatively solitary animals (less easily hunted), the occupants of areas not well suited to hunting and gathering (such as the boreal-forest moose, the North American Elk or wapiti, caribou, and, thanks to a refuge in the unglaciated eastern Canadian Arctic, the Arctic tundra caribou and musk-ox), the inhabitants of inaccessible terrain (e.g. mountain sheep), and the excessively formidable (such as the grizzly and black bears).[307] Although on the island of *Madagascar*, animals as small as dogs disappeared in the wave of extinctions, this may be attributed to the fact that until 1000 years ago the fauna there were protected not only from man but from advanced carnivores of virtually all kinds,[308] as well as to humans' having weapons intended for smaller prey.

Another finding of relevance to the question of human predation vs. climate change is that of Dan Fisher in his investigation of the tusks of Michigan mastodons. From an analysis of growth bands on the tusks of animals that died at the end of the Pleistocene he was able to determine that not only were they very well fed and not at all suffering from stress due to climate change, but they were giving birth every four to five years, thereby supporting the thesis that their demise was due to predators rather than a change in climate.[309]

The disease hypothesis

Another possible alternative explanation of megafaunal extinction is that the various species of animal were each killed off by newly introduced diseases. As regards America, for example, the opening of the Beringian land bridge would have re-established contact between previously isolated faunal groups, exposing them in both hemispheres to new disease organisms. As argued by Mark N. Cohen, however, the introduction of new diseases is likely to have been a regular phenomenon associated with the lowering of sea levels, and it seems improbable that this would have resulted in such a marked pattern of extinction only at the end of the Pleistocene.[310] For example, evidence in North America shows an interchange of mammalian species around 20 million years ago. As expressed by Alroy, the disease model invokes a class of killing agents that surely were introduced to the continent at various times throughout the Cenozoic – for mammalian diseases, there was no 'first contact,' but instead scores, or even hundreds of contacts, created by repeated mammalian immigration events over the Beringian land mass.[311]

If, on the other hand, it is thought that the diseases that should have killed off the various species of megafauna were uniquely carried by *humans*, one could then possibly account for their extinctions soon after our arrival – not only in the Americas, but everywhere. Here though one would wonder why in some cases only large mammals succumbed while in others smaller animals also became extinct – notably in those cases where we had suitable weapons with which to kill them. But the idea that disease should be the cause of their demise is in any case countered by Fisher's aforementioned finding that e.g. American mastodons were healthy just prior to their disappearance. Furthermore, as W. E. Edwards has pointed out, infectious diseases rarely lead to the eradication of a species, since their thinning-out effect on populations makes transmission between individuals prohibitively difficult before actual extinction occurs.[312]

The purported African counterexample

Another aspect of this phenomenon is that while many megafaunal extinctions occurred in the Americas and Australia, fewer occurred in Africa and Asia, despite the presence of modern humans there as well. Nevertheless, as pointed out by Martin, roughly 50 genera disappeared in Africa during the Pleistocene. Thus despite its extraordinary diversity, the presently living African fauna must be regarded as depauperate, albeit less so than that of America or Australia.[313]

Largely in keeping with the suggestion of Sherwood Washburn and C. S. Lancaster, we might say that prior to hunting with the lance, our ancestors would have moved closely among those herbivores who were too large or too quick to club to death; they would have fed beside them, and shared the same waterholes. (In game reserves animals of many different kinds soon learn not to fear humans, and they no longer flee.) But with the introduction of the lance by an archaic *Homo* species perhaps some 300,000 years ago, this peaceful relation to large herbivores was destroyed. In this way the whole human view of what is normal and natural in our relation to large animals is a product of hunting, and the world of flight and fear is the result of our efficiency as hunters.[314]

Using this line of thinking we can explain the overkill hypothesis by suggesting that it was particularly when we humans entered *new* areas, where our prey had not had the time to adapt to our constantly evolving predatory prowess, that they were driven to extinction.[315] Thus the chronology of extinction in Africa, in the Americas and in Madagascar – and the intensity of extinction – moderate in Africa, heavier in America, and extremely heavy in Madagascar where it affected much smaller species – seems clearly related to the spread of human beings, their technological/weapons development, and the vulnerability of the fauna they encountered.[316]

Essentially the same process of overhunting as occurred in Europe during the Upper Palaeolithic also occurred in the Americas; but where in Europe it took place over almost 30,000 years, in America it only took 1000. Not only did the overkill phenomenon occur over a shorter period in America, but it was more drastic in its effects, probably due to the fact that the animals there were completely unacquainted with human predation, whereas those in Europe would already have become acclimatised to Neanderthal hunting over some 200,000 years. Furthermore, since the Neanderthals used clubs and lances rather than javelins, this may also have meant a less severe impact on their prey species, while the people who migrated to America on the other hand had the use of high-quality points and perhaps spearthrowers, and hunted animals that had no previous experience of humans.

Homo erectus

Another point, argued for by Martin in a discussion, is that not only *modern* humans have precipitated extinction waves, but the same may be said of earlier humans. Evidence for this is that on such islands as Flores, Timor, and perhaps Sumba in the south-west Pacific, giant tortoises and pygmy elephants east of Wallace's line (the hypothetical boundary between the Oriental and Australasian zoogeographical regions, which runs through Indonesia) are thought to have gone extinct[317] coincident with the arrival of *Homo erectus* 750,000 years ago.[318]

Martin also notes generally that a surprisingly large number of species or genera appear to have become extinct without being replaced in their ecological niches, hence their extinction was not due to competitive displacement by immigrant or newly evolved species.[319]

The food crisis in prehistory

Archaeological evidence suggests that during the Upper Palaeolithic and Mesolithic eras in general (40,000–10,000 BP), though perhaps not in Europe for certain periods, the size of the human population increased dramatically. This period of population growth was followed by the horticultural revolution. As regards why the horticultural revolution began just at that time, or at all, archaeological evidence further suggests that food production developed in response to problems caused by the undue success of the hunter-gatherers, whose rate of food extraction came to exceed the carrying capacity of the environment.[320] As expressed by Cohen, "population growth and population pressure are essentially ubiquitous in the archaeological record and can readily be perceived as leading to economic and technological growth culminating in the origins of agriculture." It also suggests that, in keeping with Sahlins' remarks regarding the Hadza, some people resisted the introduction of horticulture for centuries after first making its acquaintance.[321] Here we have archaeological confirmation both of Martin's overkill hypothesis and of the view of technological change as being reactionary rather than opportunistic.

In keeping with the new view in anthropology, according to Cohen:

> The record appears to show that human populations on each continent first concentrated fairly heavily on the exploitation of large mammalian fauna – a prized but scarce resource – and then shifted gradually toward broader spectrum economies geared to more plentiful but less palatable resources. In each case, domestication techniques were then focused on plant species chosen not for their palatability but for their ability to provide large quantities of storable calories or storable protein in close proximity to human settlements.

Never before had the total number of humans been so great. And it was roughly at this time that the first wave of modern-human geographic expansion ceased. Hunters and gatherers had fully occupied those portions of the globe which would support their lifestyle with reasonable ease. They were forced to become increasingly eclectic in their food gathering, and to concentrate on foods lower on the food chain and of high areal density. In the period between about 9000 and 2000 BP populations throughout the world had to adjust to further increases in population not by increasing those resources which they preferred to eat, but those which responded well to human attention and could be made to produce the greatest number of edible calories per unit land.

In Mexico, "as in the Middle East, domestication appears to have involved, not choice resources, but low priority foods which were used extensively only after the range of the diet had been broadened to include a number of other items."[322] And in South America:

> Defying conventional wisdom, mummy studies revealed that when Andean civilization shifted to agriculture, health did not improve but declined. Hunter-gatherers were not stricken with many of the illnesses that plagued their descendants, and early cultures had a child-mortality rate roughly half that of later populations. As nomadic groups settled into sedentary societies, sanitation problems led to an increase in tuberculosis, pneumonia, and intestinal parasites.

That horticulture was not developed in an opportunistic or entrepreneurial spirit is further supported by Cohen's looking at the phenomenon from another angle: "It ... seems to me very unlikely," he says, "that a human population would have settled down near a field of wild wheat for the dubious pleasure of harvesting, threshing, and grinding the grain and eating the gruel year-round, thus forgoing some of the other portions of their diet."[323] And since the time of the horticultural revolution, according to Cohen, "rather than stabilizing at some optimum level or 'carrying capacity,' the human population as an aggregate has grown continuously, requiring more or less continuous redefinition of the ecology of the species as a whole." In the case of humans, thanks to our adaptability/intelligence, we have often been able to create new niches for ourselves, which other species, including to a large extent non-*sapiens* humans, can or could do only through speciation.[324] The need continuously to acquire a new – and/or broader – niche is a condition we have imposed on ourselves by our own activities. Thus we may say that on the new view in archaeology, the adoption of agriculture is basically the result of population pressure; and technological advance has essentially been the result not of our seeking a better life, but of people's simply trying to survive.

While we have suggested above that modern hunter-gatherers have been in ecological equilibrium with their environments, this cannot be said of Palaeolithic modern humans. As Cohen says, the process of ecological adjustment on the part of human populations has been a continuous one, and the archaeological data imply that we must break away from our assumption that human cultures are inherently stable systems until jostled by some 'outside event.'[325]

The lifestyle of humans living at the beginning of the Mesolithic period can be said to be similar to or perhaps even essentially the same as that of modern hunter-gatherers such as the Hadza. However, many of those societies began successively using more sophisticated technology. Modern hunter-gatherer societies, on the other hand, have not developed technologically nor grown in population since that time. (Note that virtually all such societies are in tropical areas, moreover such areas as where technological development in the form of horticulture would not provide an advantage.) Furthermore, the technology they do use does not involve the exploitation of non-renewable resources. (The same can apply to long-lived simple horticultural societies.) In this way the lifestyle during the Mesolithic might have stabilised without population growth, despite the prior permanent eradication of important sources of food. In this case such an elimination of non-renewable resources may have meant a smaller subsequent population however.

Economics

Parallel to the developments in anthropology and archaeology, an ecologically oriented view has also arisen in economics; but where the new views in the former disciplines have spread, the new view in economics has not.

The industrial revolution

Where those advocating the new views in anthropology and archaeology have concerned themselves mainly with hunter-gatherer and agricultural populations, Wilkinson applies the idea of ecological equilibrium to industrial populations as well. He points out the vicious circle (positive feedback loop) involved in the need constantly to take more from the environment: "[I]ndustrialisation requires a more extravagant lifestyle than the modes of production that preceded it. The problems it creates and the needs it sets up make increased consumption a necessity if people are to lead reasonably satisfactory lives."[326]

It is now generally accepted that at the time of the industrial revolution it was the inability to husband a potentially ecologically viable resource in the form of wood that led to the need to tap a non-ecologically viable resource in the form of coal. Thus the step taken to the large-scale mining of coal, needed to fuel the industrial revolution, and the development of such machines as the steam engine, needed to make that mining possible, resulted from scarcity.

As expressed by Wilkinson: "Once again it appears that a formidable group of innovations should not be regarded as the fruits of a society's search for progress, but as the outcome of a valiant struggle of a society with its back to the ecological wall." And against this background Wilkinson gives consideration to the nature of economic growth, favoured by the vast majority of modern economists, and suggests that it is needed to solve the problems created by industrialisation, and should perhaps be regarded more as a reflection of the rising real cost of living than as an indication of increasing welfare.[327] Here we see in the realm of economics a view very much in keeping with the new views in anthropology and archaeology, in which technological development is seen as a reaction to need rather than the result of an entrepreneurial spirit.

Economic growth

That economic growth should reflect the rising real cost of living rather than simply be the beneficial result of improved economic activity is antithetical to the views of virtually all major decision makers in the world today. Not only world leaders, but virtually everyone familiar with the concept considers gross national product to be a measure of social welfare. However, as expressed by Daly, the problem with using GNP as such a measure is that it adds together three very dissimilar categories: throughput, additions to stock, and services rendered by the stock. As pointed out by Irving Fisher and Kenneth Boulding, wants are satisfied by the third category, services.[328]

Thus, for example, the cost of the treatment of waste, e.g. in the form of sewage and garbage treatment plants, which increases with increasing production, is taken to be a social benefit. So are such things as the cost of maintaining military and police forces to defend property, the cost of fuel to transport food due to reliance on imports, and the manufacture of armaments. Each of these is a defensive move to assuage the effects of economic growth and concomitant growth in the population.

It is not always easy in this context to distinguish defensive expenditures or costs from benefits. If we compare the modern situation with that of hunter-gatherers, we may say that they had a GNP equivalent to zero, and that *every* development involving increasing energy use from that state to ours represents a cost. On the other hand, however, without such a development we would not have what we consider the positive benefits of e.g. modern science and classical music.

'The Wild West'

Boulding depicts the economy as seen in the traditional or Western perspective as the 'cowboy' economy, in which natural frontiers are non-existent, resources are inexhaustible, and wastes are innocuous. Such a view fosters the "reckless, exploitative, romantic, and violent behavior" to be expected of a cowboy.[329] E. F. Schumacher, for his part, speaks of 'the people of the forward stampede'; and he contrasts them with

> people in search of a new life-style, who seek to return to certain basic truths about man and his world; I call them home-comers. ... The slogans of the people of the forward stampede burst into the newspaper headlines every day with the message, 'a breakthrough a day keeps the crisis at bay.'

And Daly, in the same vein, has supplied us with the notion of *growthmania*,[330] which is to refer to the adoption of a religious-like unquestioning attitude to economic growth.

As in the case of overkill, we here see in the Wild West economy an instance of the pioneering phenomenon. Where during the Pleistocene humans consumed all that they could, thereby supporting population growth and at the same time eradicating non-renewable resources at the fastest possible rate, we are doing the same today in our pursuit of economic growth.

Entropy and natural capital

Both Boulding and Georgescu-Roegen see the notion of entropy as of central importance to economics. Boulding says:

> In regard to the energy system there is, unfortunately, no escape from the grim Second Law of Thermodynamics; and if there were no energy inputs into the earth, an evolutionary or developmental process would be impossible. The large energy inputs which we have obtained from fossil fuels are strictly temporary.[331]

Note that Boulding here shows that he realises that entropy is constantly being counteracted by energy inputs of solar radiation.

According to Georgescu-Roegen:

> First, there is the general practice of representing the material side of the economic process ... by a mathematical model in which the continuous inflow of low entropy from the environment is completely ignored. ... Even if only the physical facet of the economic process is taken into consideration, this process is not circular, but *unidirectional*. As far as this facet alone is concerned, the economic process consists of a continuous transformation of low entropy into high entropy, that is, into *irrevocable waste* .... The conclusion is that, from the purely

physical viewpoint, the economic process is entropic: it neither creates nor consumes matter or energy, but only transforms low into high entropy. ... The entropy law rules supreme over the evolution of the economic process.[332]

Georgescu-Roegen demonstrated that the common-sense view of resources as limited, expressed earlier by Boulding, has theoretical support in physics: resources are limited because their use necessarily entails their dissipation. Georgescu-Roegen points out that, rather than being an isolated exchange loop capable of perpetual expansion, the economic process is fixed to a base of materials subject to identifiable constraints. "It is because of these constraints that the economic process has a unidirectional irrevocable evolution." And these constraints stem from the entropy principle. *Our whole economic life feeds on low entropy*, i.e. high order: cloth, lumber, china, copper, etc. are all highly ordered structures. We may then take it as a brute fact that low entropy is a *necessary* condition for a thing to be useful.[333]

All economic activity necessarily involves a step, however small, towards the exhaustion of available energy – other than from the sun. Georgescu-Roegen, however, misses this latter point in his reasoning, and thus does not appreciate that the energy received annually from the sun can suffice to maintain human society. Despite his emphasis on systems-thinking, he seems unaware of the notion of *ecological* equilibrium, as well as the nature of modern hunter-gatherer societies, and mistakenly assumes that any human economy must be dependent on non-renewable resources.[334] Thus he says: "The crucial error [of those advocating a steady-state economy] consists in not seeing that not only growth, but also a zero-growth state, nay, even a declining state which does not converge toward annihilation, cannot exist forever in a finite environment."[335] And, "a world with a stationary population would ... be continually forced to change its technology as well as its mode of life in response to the inevitable decrease of resource accessibility." Daly follows him in this regard, assuming a human economy necessarily to use non-renewable resources, and admitting that on his view what he is speaking of is not a steady-state economy, but a *quasi*-steady-state economy.[336] Nevertheless, Georgescu-Roegen's reasoning is correct as regards our use of organic and inorganic minerals. On his conception, nuclear power, materials recycling, and any other technological 'solution' to matter-energy scarcity is a delusion.

As Boulding argues, it is from the stock that we derive satisfaction, not from the additions to it (production) or the subtractions from it (consumption). To be consumed, far from being a desideratum, is a deplorable property of the stock which necessitates the equally deplorable activities of production. The objective of economic policy should not be to maximise consumption or production, but rather to minimise it, i.e. to maintain our stock with as little consumption or production as possible.[337] Humans must be concerned with the *quality* of economic activity, particularly as it relates to the creation of enduring, efficient and useful stocks.

Schumacher calls for an 'economics of survival.'[338] He would agree with Boulding as regards minimising consumption: "[S]ince consumption is merely a means to human well-being, the aim should be to obtain the maximum of well-being with the minimum of consumption."

These lines of thought lead to resources coming to occupy a central position in economic thinking. To the extent that they are exchangeable, they become the fundamental form of *capital*, having not only a use value (Glossary) but also an exchange value. This *natural capital* is an exchangeable *stock* of natural resources that yields a *flow* of services and requires maintenance in the face of depreciation, and whose consumption cannot be counted as income.[339]

Throughput and Spaceship Earth

Boulding, the first modern proponent of a steady-state economy, minted the idea of 'Spaceship Earth,' and introduced the notion of throughput into economics. In a spaceship, one cannot ignore the results of production and consumption (nor, on Boulding's conception, can one jettison waste). On Spaceship Earth, consumption is the conversion of assets into waste; that is, it is a *process* involving a linear *throughput*. Therefore, in the spaceman economy, throughput is to be kept at a sufficiently low level to be able to maintain ecological homeostasis.

> [I]n the spaceman economy, throughput is by no means a
> desideratum, and is indeed to be regarded as something to be
> minimized rather than maximized. The essential measure of the
> success of the economy is not production and consumption at all, but
> the nature, extent, quality, and complexity of the total capital stock,
> including in this the state of the human bodies and minds included
> in the system. In the spaceman economy, what we are primarily
> concerned with is stock maintenance, and any technological change
> which results in the maintenance of a given total stock with a
> lessened throughput (that is, less production and consumption) is
> clearly a gain.[340]

From the point of view of thermodynamics, throughput is the quantity of energy and matter passing through the economic system. This energy and matter obey the conservation principles of physics and chemistry in that they do not diminish in quantity through the process; but they also follow the entropy principle in that they experience an increase in entropy.

Boulding's notion of throughput, in a more elaborated form, was integrated into economic theory by Georgescu-Roegen, who called it the 'metabolic flow' and emphasised the manifold consequences of its entropic nature.[341] As mentioned earlier, the structure and order (low entropy) of the economy is maintained by imposing a cost of disorder on the ecosystem. This disorder takes basically two forms, either that resultant upon the taking of resources, or that resultant upon the effects of waste. Resource use and pollution go hand in hand, thanks to the conservation principles of physics and chemistry. Regarding growth, as Daly says: "A policy of maximizing GNP is practically equivalent to a policy of maximizing depletion and pollution." The greater the growth, the higher the rate at which low entropy is being turned into high. As the stock and throughput of the economic system grow, the increasing disorder exported to the ecosystem will increasingly interfere with the environment's ability to provide natural services; and at some point the natural services lost will outweigh the artificial services won in this exchange.

From this point of view it is clear that we can define an optimum stock as one for which total service (the sum of services from the economy and the ecosystem) is maximised. This is the case when the increase in service arising from an addition to the stock is equal to the resultant decrease in service due to impaired ecosystem functioning.[342]

The steady-state economy

Already in 1848 John Stuart Mill said:

> I cannot … regard the stationary state of capital and wealth with the unaffected aversion so generally manifested towards it by political economists of the old school. I am inclined to believe that it would be, on the whole, a very considerable improvement on our present condition. I must confess I am not charmed with the ideal of life held out by those who think that the normal state of human beings is that of struggling to get on; that the trampling, crushing, elbowing, and treading on each other's heels which form the existing type of social life, are the most desirable lot of human kind, or anything but the disagreeable symptoms of one of the phases of industrial progress. The northern and middle states of America are a specimen of this stage of civilization in very favourable circumstances; … and all that these advantages seem to have yet done for them … is that the life of the whole of one sex is devoted to dollar-hunting, and of the other to breeding dollar-hunters.
>
> … Those who do not accept the present very early stage of human improvement as its ultimate type may be excused for being

comparatively indifferent to the kind of economical progress which excites the congratulations of ordinary politicians: the mere increase of production and accumulation. ... I know not why it should be a matter of congratulation that persons who are already richer than anyone needs to be, should have doubled their means of consuming things which give little or no pleasure except as representative of wealth. ... It is only in the backward countries of the world that increased production is still an important object: in those most advanced, what is economically needed is better distribution, of which one indispensable means is a stricter restraint on population.

There is room in the world, no doubt, and even in old countries, for a great increase in population, supposing the arts of life to go on improving, and capital to increase. But even if innocuous, I confess I see very little reason for desiring it. The density of population necessary to enable mankind to obtain, in the greatest degree, all the advantages both of cooperation and of social intercourse, has, in all the most populous countries, been attained. A population may be too crowded, though all be amply supplied with food and raiment. ... Nor is there much satisfaction in contemplating the world with nothing left to the spontaneous activity of nature; with every rood of land brought into cultivation, which is capable of growing food for human beings; every flowery waste or natural pasture ploughed up, all quadrupeds or birds which are not domesticated for man's use exterminated as his rivals for food, every hedgerow or superfluous tree rooted out, and scarcely a place left where a wild shrub or flower could grow without being eradicated as a weed in the name of improved agriculture. If the earth must lose that great portion of its pleasantness which it owes to things that the unlimited increase of wealth and population would extirpate from it, for the mere purpose of enabling it to support a larger, but not a happier or a better population, I sincerely hope, for the sake of posterity, that they will be content to be stationary, long before necessity compels them to it.

It is scarcely necessary to remark that a stationary condition of capital and population implies no stationary state of human improvement. There would be as much scope as ever for all kinds of mental culture, and moral and social progress; as much room for improving the Art of Living, and more likelihood of its being improved, when minds ceased to be engrossed by the art of getting on. Even the industrial arts might be as earnestly and as successfully cultivated, with this sole difference, that instead of serving no purpose but the increase of wealth, industrial improvements would produce their legitimate effect, that of abridging labour.[343]

According to Boulding, in the closed economy of the future there will be no unlimited reservoirs of anything, either as sources or as sinks, and humans will have to find their place in a cyclical ecological system which is capable of continuous reproduction of material form even though it cannot escape having inputs of energy.[344] The economy, which determines our niche, must not only be *sustainable*, it must be *sustained*. Given no drastic change in the systems with which it interacts, it should continue in its present state indefinitely, which implies that it only involve the use of resources in a sustainable way. Daly defines a steady-state economy, in keeping with the stricter sense of the term, as one "with constant stocks of people and artifacts, maintained at some desired, sufficient levels by low rates of maintenance 'throughput'." Georgescu-Roegen, before he denied the possibility of a steady-state economy, suggested that it may be considered to be an economy in a state of static equilibrium, such that whatever occurs can be indefinitely repeated.[345] And Boulding sees neo-Malthusian population control on the part of the total society to be a necessary aspect of a steady-state economy. We humans are to have *only* such an economy if we want to maintain ecological equilibrium.

* * *

A central aspect of the new view in anthropology is that the quality of the lives of modern hunter-gatherers is (or was) actually quite good; in the case of archaeology it is that prehistoric humans constantly eradicated their sources of high-quality food, leading to successive food crises and eventually the horticultural revolution; and in economics it is that for a society to enjoy a good life in the future its economy must depend solely on resources that are renewed from year to year.

It is suggested here that the new views in anthropology, archaeology and economics are all expressions of but *one* perspective, which may be termed the *ecological*, and which will be synthesised in this book in one coherent theory, the coherence of which is provided by the vicious circle principle.

3

Theoretical background to the vicious circle principle

The principle of population

The principle of population was put forward by Thomas Malthus in 1798. In the first edition of his *Essay on the Principle of Population* he characterises it as "The perpetual tendency in the race of man to increase beyond the means of subsistence," which he considers to be "one of the general laws of animated nature which we have no reason to expect will change."[346]

Though in discussions regarding the principle it has only been taken in its application to humans, as suggested by Malthus in the above quote, as well as more explicitly in his *Summary View of the Principle of Population*[347] published in 1830, it is clear that it should apply to any animal or plant species, in keeping with the energy-gathering (counter-entropic) nature of the populations of all organisms.

Thus the principle of population is a *biological* principle, and may in fact be seen as a corollary to the principle of evolution. It clearly had a profound influence on Darwin:

> In October 1838, that is, fifteen months after I had begun my systematic enquiry, I happened to read for amusement Malthus on *Population*, and being well prepared to appreciate the struggle for existence which everywhere goes on from long-continued observation of the habits of animals and plants, it at once struck me that under these circumstances favourable variations would tend to be preserved and unfavourable ones to be destroyed. The result of this would be the formation of new species. Here, then, I had at last got a theory by which to work.[348]

Hence, as more individuals are produced than can possibly survive, there must in every case be a struggle for existence, either one individual with another of the same species, or with the individuals of distinct species, or with the physical conditions of life. It is the doctrine of Malthus applied with manifold force to the whole animal and vegetable kingdoms.

The early progenitors of man must also have tended, like all other animals, to have increased beyond their means of subsistence; they must, therefore, occasionally [sic] have been exposed to a struggle for existence, and consequently to the rigid law of natural selection.[349]

Natural selection follows from the struggle for existence; and this from a rapid rate of increase. It is impossible not to regret bitterly, but whether wisely is another question, the rate at which man tends to increase; for this leads in barbarous tribes to infanticide and many other evils, and in civilised nations to abject poverty, celibacy and to the late marriages of the prudent. But as man suffers from the same physical evils as the lower animals, he has no right to expect an immunity from the evils consequent on the struggle for existence.

My guess is that in the last quotation Darwin, in saying, "but whether wisely is another question," means that the inclination to increase, in promoting a greater number and thus variety of candidates for being the fittest, may thereby support the survival of the species. Note too Darwin's recognition of the existence of internal population checks, at least in the case of humans.

Malthus had a similar effect on Alfred Russel Wallace, the 'co-founder' of the theory of natural selection. In his autobiography from 1905, Wallace says that:

Perhaps the most important book I read was Malthus' *Principle of Population*. … It was the first work I had yet read treating of any of the problems of philosophical biology, and its main principles remained with me as a permanent possession, and twenty years later gave me the long-sought clue to the effective agent in the evolution of organic species.[350]

In other words, the principle of population is necessary to the principle of evolution, for if populations of organisms didn't tend to increase in size, and compete over a common resource in so doing, there would be no survival of the fittest or natural selection resulting in species evolution. From a systems point of view, competition is implied by the attempt on the part not only of each species but of each organism and each group to counteract entropy in a world of limited energy resources.[351]

The principle of population however is not as straightforward as it might first appear. It can be given different formulations, not all of which are equivalent. Here I shall present four:

X. The principle of population

> 1. *There is a tendency for the human population (and that of any species) to grow until stopped by external checks.*

I shall call this the first formulation of the principle. A *second* possible formulation is:

> 2. *Populations (human and other) tend to produce more offspring than can reproduce themselves.*

A *third* formulation of the principle might be:

> 3. *If there were no internal or external checks to population size, that of the human (or any) population would tend to increase indefinitely.*

And a potential *fourth* formulation could be:

> 4. *There is a tendency for the human species, and all other species, to have as many members as possible.*

Note that the principle of population, in any of its formulations, is to apply at all times, and not only at some, perhaps future, time. Malthus says:

> [T]he pressure arising from the difficulty of procuring subsistence is not to be considered as a remote one, which will be felt only when the earth refuses to produce any more, but as one which not only actually exists at present over the greatest part of the globe, but, with few exceptions, has been almost [*sic*] constantly acting upon all the countries of which we have any account.[352]

The misunderstanding of the principle in this regard has recurred ever since the time of Malthus.

The *first* formulation of the principle is similar to Malthus' formulation cited above, and suffers from not taking into account the internal homeostatic nature of the population regulation of many species, including humans. In particular, it takes no account of internal checks to population growth (taking account of such checks constitutes a step towards *neo*-Malthusianism).[353] (This, in fact, was Darwin's interpretation of Malthus as he applied it in the theory of natural selection. Though, e.g. in his *Descent of Man* – as quoted above, and elsewhere[354] – Darwin does mention internal checks such as war, prolonged suckling, licentiousness, infanticide and the requirement of marriage, he does not present them as serving a function in the preservation of the species.) Nor does the *second* formulation take account of internal checks. However, that it is fundamentally correct is supported by the fact that there exist food chains. If the populations of each species did not over-reproduce, then food chains would be impossible, since predation would lead to the constant diminution and eventual extinction of what was preyed upon, and the predatory species itself would in turn starve.[355] Nevertheless, taking account of the potential or real role played by internal checks to growth would improve both the first and second formulations. The *third* formulation takes account of such checks, and is more in keeping with Malthus' thinking in the second and later editions of his *Principle of Population*.

The *fourth* formulation may at first glance appear to ignore checks. But here reference to the human *species* is crucial. For the species to have as many members as possible suggests that the size of its total population at any one time not be at a maximum, but at an *optimum*, in the sense expressed earlier. At such a level the species would not, for example, over-exploit its resource base, which would place its own existence in jeopardy. Employing systems-thinking, another way of expressing this could be: *the total amount of solar energy acquired by any species tends to be as great as is physically possible*. And, as in the case of the third formulation, this may be accomplished partly by the use of karyotypically determined internal checks, at least in the case of humans and other vertebrates. For species that have only external checks, the situation with regard to the fourth formulation of the population principle is straightforward: populations of a species grow as much as possible until they cannot grow any longer due to external factors.

In the case of species with internal checks the situation is more complicated. For example, should one or more of these checks be put out of play and not replaced by another, the species' population could grow beyond what its environment can support, and thereby undermine its own existence.

Different kinds of population check

All four of the formulations of the principle of population presented above imply that the human population has a tendency to increase in size, and is constrained from doing so by the operation of checks. Differences between kinds of internal check – somatic, behavioural and so on – for various species were discussed in Chapter 1. Here we might consider two further distinctions particularly relevant to humans.

Positive vs. preventive checks

Malthus distinguishes between positive and preventive checks to population growth. Positive checks are related to mortality and preventive checks
to fertility. We can say that, rather generally, for human populations positive checks concern mortality among potential child producers and rearers,
and preventive checks are related to various sorts of birth control (or, more
naturally, conception control, thereby making abortion a positive check). As
regards positive checks it may be noted that they can function even after male
or female menopause. As mentioned, senescence – a constant somatic check –
functions in this way by setting limits not only on 'fertility time,' but also on
'babysitting time.'

Internal vs. external checks

A second distinction between different kinds of check is that between
what I have called internal and external checks, internal checks being those
emanating from within a population/species, and external emanating from
without. Almost all external checks are positive, and include e.g. famine, disease, predation and so on. Internal checks, on the other hand, can be positive or
preventive. Positive internal checks in the case of humans include the checks
of abortion and infanticide mentioned by Carr-Saunders, as well as murder,
suicide and executions, while preventive checks include monogamy, contraception, delayed marriage, Carr-Saunders' check of sexual abstinence, and possibly long lactation. War, to the extent that it involves the killing of enemies
and the death of individuals for other reasons, is a check that is external to a
particular population while being internal to the total population and thereby
the species.

Other things being equal, if for some reason a particular form of internal
behavioural check were rendered inoperative, then it might be expected that
the karyotypic basis of the check would see to it that some other form of check
came to play a greater role. But shifting from one form of check to another
takes time, and population pressure is bound to arise in the interim. Further,
it is to be noted that humans' social checks are based on the weakest kind of
instinct, and are thus most susceptible to being overridden by more basic kinds.
In order to function, they presuppose a stable social setting. The employment of
internal checks places *demands* on individuals, and it requires their *acceptance* of
those demands. Cultural legitimation is given to such moral demands through
traditions. Thus the society, through its *culture*, must exert a strong influence on
the population in order for internal checks to be effective, and changes in culture will most likely be to the detriment of the checks in question.

Viewing *internal* checks from the point of view of systems, we could say that some of the free energy from the sun accumulated in the system (population) can function homeostatically, in a negative feedback loop, so as to counteract the system's tendency to expand, as the continued expression of this tendency would create dynamic disequilibrium and potentially lead to the demise of the system.

In biological terms, we should say that it is through the phenotypic manifestations of species' karyotypes in the form of internal population checks, such as that of territoriality, that the populations of vertebrates and various other species see to it that they do not grow to the point that their existence is placed in jeopardy by overpopulation. And, as intimated earlier, natural selection has of course weeded out relevant species that did not include this internal check in their karyotypes, by leading to their extinction.

It should be pointed out that Malthus, already in the first edition of his *Essay*, recognised the existence of internal preventive checks and the role they can play in reducing population pressure. The checks he considered were people's deciding not to marry due to the difficulty to be had in maintaining a family, and prostitution.[356] (Of course, to function as a check to population growth, prostitution must in some way reduce the number of infants who reach reproducing age.) Further, in the second and later editions of his book Malthus placed greater emphasis on the preventive check of delayed marriage.[357] And in the last edition, of 1830, he spoke of the "bad structure of society" and "unfavourable distribution of wealth" as checks to population growth,[358] both of which are internal.

Though Malthus' position, in our terms, was that the ultimate checks to population growth are external (viz. "the laws of nature"),[359] he still believed that humans, employing what we are calling internal checks, have "a great and most extensive influence on [the external checks'] character and mode of operation." At the same time, however, he never realised how powerful internal checks, both positive and preventive, can be and have been. Nor did he realise that human society need not be hierarchically ordered such that those on the bottom suffer from overwork or malnutrition, both of which insights he might have obtained from a study of other animal species or modern hunter-gatherers, the latter of whom he believed to suffer from population pressure.[360] Nor, it may be mentioned, did Malthus give any consideration to the relation between population size and ecological equilibrium.

In keeping with his view that all checks are ultimately external and thus positive, Malthus throughout maintained the attitude he expressed in the first edition of his work where, speaking of the check on population as "the difficulty of subsistence," he says: "This difficulty must fall somewhere and must necessarily be severely felt by a large portion of mankind."[361] He never envisaged a society which, through implementing internal checks, could avoid having to live "on a level with the means of subsistence." The reason that such societies are possible, as we shall see in the development of the vicious circle principle, is that need is related to *social stratification*, and that such stratification as results e.g. in the malnutrition of part of the population is not a necessary aspect of human society.

In these terms then, we are here suggesting not only that it should in principle be possible for a population to avoid 'severely feeling the difficulty of subsistence' in the event that it exercise checks which are both internal and preventive (this is similar to the position taken by neo-Malthusians), but that this appears to have been so in the case of modern hunter-gatherers. On the other hand, however, we must admit with Malthus that neither positive nor preventive internal checks, nor a combination of the two, have as yet managed to perform this function for humankind as a whole.

Population growth pushes technology

If we look for causal relations in an attempt to explain the development of the new views in archaeology, anthropology and economics, as we shall do more comprehensively in what follows, we see an interesting aspect of the discussion in a suggestion made by Ester Boserup. Boserup's line can be easily assimilated to the new views in these subjects, as we shall see.

Boserup's influential contribution to demographic theory is her drawing attention to the fact, previously noted by Jacques Ellul, that population growth can stimulate technological innovation, rather than merely result from it. (In Ellul's terms, "the growth of population entails a growth of needs which cannot be satisfied except by technical development.")[362] Boserup argues that in the case of agriculture – to which she limits her considerations – technological development is largely a function of population density, and that the various technologies employed by agriculturists constitute a series of responses (cf. the reaction principle) to growing population. Necessity is the mother of invention, or at least of use. In some cases such changing technology may be no more than briefer and briefer fallowing of cultivated plots, eventually leading to annual cultivation.

Boserup takes her own view to run counter to what she terms the neo-Malthusian view – though she names no neo-Malthusians – according to which population size is to be directly related to the availability of food. (Note that Boserup's notion of neo-Malthusianism is not the generally accepted one.) Here however she is apparently unaware that the idea that necessity is the mother of invention has already been expressed by numerous authors, including Malthus ("Necessity has been with great truth called the mother of invention").[363] Nevertheless, Boserup's emphasis on this notion brings out an important way in which the principle of population should apply differently to humans than to other species. For humans, external population checks can be pushed back (but never removed) through cultural change in the form of technological development.

Thus in situations where a population is experiencing food scarcity at the same time as its members are aware of technological means to alleviate that scarcity, on Boserup's view it is natural that they should use those means. But this suggests that populations that do *not* employ subsistence technologies of which they are aware ought not be experiencing scarcity. So in cases where human population growth is slow or non-existent, despite the knowledge and availability of technology that can increase food production and thus the size of the population, this low or absent population growth must be the result of something other than scarcity. Thus, the slow population growth in pre-industrial societies until recently cannot be explained by saying that it is the result of insufficient access to food due to overpopulation, and we must consider other factors in our attempt to explain population development.[364] Boserup gives no hint however as to what these other factors might be. But on the basis of our previous considerations regarding the principle of population, a prime candidate is the existence of *internal* population checks.

A related aspect of Boserup's view that is particularly relevant to the new views in archaeology, anthropology and economics is her recognition that technological development does not imply the ability to produce more with the same amount of human effort, or to produce the same with less effort, but on the contrary often if not always brings with it a decrease in productivity per working hour:

> The cultivators who subsist by the system of forest fallow are much more primitive in their whole way of life than cultivators who apply intensive methods of production. Moreover, there is no land preparation before sowing and no other agricultural tools than axes and digging sticks. It is tempting, therefore, to conclude that output per man-hour must be particularly low under this system of cultivation. But it is not so in actual fact.

A central aspect of the above interpretation of Boserup's account is the idea that new technology, even if available, is not employed until needed. As Wilkinson says:

Development is needed when a society outgrows its resource base and productive system. As the established economic system is proved inadequate and subsistence problems become more severe, societies are driven to change their methods. Development comes out of poverty, not out of plenty as many economic theories would lead one to suppose. Poverty stimulates the search for additional sources of income and makes people willing to do things they may previously have avoided. When for instance population growth and the division of land holdings makes units too small for subsistence, people are forced into towns to sell their labour, or else they take up rural crafts to eke out a living. It is the population's increasingly exploitable situation which provides the basis for the growth of capitalist institutions.[365]

Here we note reference once again to the idea that people *resist* the adoption of more intensive means of obtaining a living due to their involving more work.

Thus, though Boserup's view was developed presupposing a context involving primitive agriculture, it can be broadened to cover the whole of humankind's development. What it leads to is the idea treated above, that there have actually existed societies that have *not* been pushed to the limits of subsistence, namely those which did not feel the need to increase food production, though they knew how to do so. In terms of our previous considerations we should say Boserup's line suggests, not that Malthus' principle of population in any of its formulations does not apply, but that, for certain societies, some of the Malthusian checks by which population has been limited have been internal. This is in keeping with the third formulation of the principle and, ironically, with the neo-Malthusian view Boserup claims to be criticising. This being the case for particular populations would explain why their means of subsistence, even if intrinsically quite intensifiable, remain in an extensive state unless the internal checks are put out of play.

Ecological equilibrium, technological/economic development and economic growth

In Wilkinson's theory of economic development, touched on in the previous chapter, a distinction is first drawn between societies capable of limiting their own size and societies that have lost this ability. The former:

are societies which have stabilized well within the means of subsistence available to them and so have avoided the problems

which lead to development. Almost all living species have a choice
between developing methods of population limitation or facing
continuous starvation as their numbers are limited by the food
supply. Natural selection seems to have led a great many species to
adopt the former strategy, and human societies, with the help of
social controls, have often done likewise. Many primitive societies,
particularly before contact with Europeans disrupted their cultural
systems, prevented population growth [by infanticide, abortion, etc.]
and managed to live in equilibrium with their resources without
threat of hunger.[366]

When a human population has lost the ability to limit its size, its growth
leads to greater demands on the environment. Thus: "Within a stable society
in ecological equilibrium, population growth is the most dangerous threat to
continued stability." The growth of a population beyond what its environment
can support is an indication that the society is out of equilibrium. And, as in
Boserup's view, where population growth creates population pressure, necessity can well become the mother of invention, with new technologies being
developed that allow increased environmental exploitation, thereby making
it possible to meet the current needs of the population while at the same time
giving rise to such phenomena as economic growth. In Wilkinson's words:

Once one has the concept of a society existing in ecological
equilibrium there is no difficulty in accepting that the development
of need is the real cause of economic development. ... Development
is primarily a matter of increasing the rate of environmental
exploitation to support a growing population. ... Instead of regarding
development as a matter of 'progress' towards a 'better life' motivated
by an incurable dissatisfaction with our present lot, we see that it is a
process of solving a succession of problems which from time to time
threaten the productive system and the sufficiency of our subsistence.
In effect, human societies out of ecological equilibrium have to run to
keep up; their development does not necessarily imply any long-term
improvement in the quality of human life.[367]

Wilkinson realises that the whole issue of human survival as a species
revolves around our maintaining equilibrium with our surroundings – as is in
fact implied by the principles of population and evolution.

In the next chapter I shall present the vicious circle principle, in which the
development of humankind is placed in one comprehensive picture.

4

The vicious circle principle of the development of humankind

Presentation of the vicious circle principle

As regards the views of Malthus, Boserup and Wilkinson, we should say that each is right in the main. And taken together they can provide a coherent view of certain important aspects of the human condition. But even considering a synthesised version of their views something is missing, and it is the idea that, in the case where population is growing and ecological equilibrium is lacking, *the increased exploitation afforded by technology typically provides a surplus, which allows renewed population growth*. Once a new line of technological development has been opened it tends to be pursued, and once a resource has been tapped it tends to be exploited. In today's society, the profits made by entrepreneurial capitalists come from this surplus, and are obtained by receiving payment for providing the population with increased material benefits, benefits that may well overshoot the population's vital requirements. In this way technological innovation can have the effect of increasing the potential for a particular area of land to support human habitation beyond the needs of its contemporary population, thereby constituting a major factor in that population's losing its incentive to control its own numbers. This loss of incentive may be manifest e.g. in a cultural shift condoning earlier marriages, or in increasing the convenience of having larger families. And, given the surplus, which weakens internal checks, there is nothing to stop the population from once again becoming too large relative to what it is able to extract from its resource base, until external checks come into play. The way that this eventuality has in some cases been mitigated or avoided has been through the introduction of yet more efficient technology, allowing even more to be extracted from the resource base. In this way a *vicious circle* is created, in which increased consumption is made possible only by further technological development which in turn further degrades the environment (increases its entropy) while promoting the growth of the population.

All of the principles presented in Chapter 1, as well as the principle of population, apply just as well to other species as to humans. Is there a particular principle applying only to humans – a principle of *human* ecology? The vicious circle principle, which presupposes at least the third and fourth formulations of the principle of population, as well as the principle of evolution and all of the other principles, is a candidate.[368]

XI. The vicious circle principle

> *Humankind's development consists in an accelerating movement from situations of scarcity, to technological innovation, to increased resource availability, to increased consumption, to population growth, to resource depletion, to scarcity once again, and so on.*

The vicious circle principle (VCP) is both easy to understand and in keeping not only with modern science but also with common sense. Briefly put, it says that in the case of humans *the experience of need, resulting e.g. from changed environmental conditions, sometimes leads to technological innovation, which becomes widely employed, allowing more to be taken from the environment, thereby promoting population growth, which leads back to a situation of need.* Or, seeing as it is a matter of a *circle*, it could for example be expressed as: *increasing population size leads to technological innovation, which allows more to be taken from the environment, thereby promoting further population growth*; or as: *technological innovation allows more to be taken from the environment, the increase promoting population growth, which in turn creates a demand for further technological innovation.*

Note that the vicious circle principle is not a truism – for example it is not obvious that technological innovation need lead to more being taken from the environment, or that increasing population size need lead to technological innovation. Once understood however, the principle may appear to be self-evident, as any principle should.[369]

To better understand the vicious circle phenomenon we might compare the operation of the VCP, which only pertains to humans, with what is the case for other species. (Note that the present chapter constitutes primarily a *presentation* of the VCP; the bulk of its support comes in the next chapter.) Any species, including the human, can become extinct due to a change in its environment; but in those conditions where a species is not imminently endangered, it is represented by some optimum range of numbers of individuals, as suggested earlier. If there are too few the species may be subject e.g. to problems related to interbreeding; and if there are too many its source of sustenance may be eliminated, resulting in either case in disequilibrium and the possible extinction of the species. This is very much in keeping with, if it does not directly follow from, the various formulations of the principle of population.

But what distinguishes humans from other life forms in this regard is our development and use of *technology*. Unlike other species, humans have invented and employed such devices as the hand-axe, fire, clothing, the bowl, spears, boats, the bow and arrow, the hoe, the plough, irrigation, watermills and windmills, sailboats, various engines, and electricity generators operated by nuclear power. And this technology, paradigmatically, has had the effect of pushing back the limits to population size, a phenomenon we do not see in other species.

Humans' development of technology has been *exponential*, and has led to a corresponding exponential increase in our total resource consumption as well as in the size of our population – right from when we first came into existence as a species. Most notable in this regard are our harnessing of fire some 1.5 million years ago, the horticultural revolution of 10,000 years ago, the beginning of the mining of metals 6000 years ago, and the industrial revolution of 250 years ago. But this is a process that is going on all the time (like the operation of the principle of population), with such apparently minor technological innovations as that of the stirrup or horseshoe, or ball-bearing or adjustable wrench, each contributing to the end result of increasing the number of humans that can occupy a given area of land.

Here is the expression of a sophisticated form of the vicious circle principle, or of the form the principle might take in being applied to a relatively complex situation:

> A situation of *scarcity* leads to the experience of *need*
> which creates a demand for *new* or *previously unused technology*
> which in certain cases is *developed* and then *widely employed*
> making the population *genetically dependent* on developing
>   technology
> and giving rise to other unintended *side-effects*
> but which allows the exploitation of previously *inaccessible resources* –
>   renewable, non-renewable or both
> an exploitation which presupposes the *existence* of those resources
> and sometimes makes it possible for human populations to *expand*
>   to areas where the new technology is necessary for their
>   existence
> the taking of resources *reducing the quantity* remaining

and producing *increasing quantities of polluting waste*

as well as leading to the *extinctions* of various species of plants and
animals

while at the same time creating *new needs*

technological development itself becoming *self-perpetuating*

while typically producing a *surplus* of consumables

which allows an *increase in resource consumption*

the consumables however normally or often being of *lower quality*
than those they are replacing

while at the same time the availability of the surplus weakens *internal
population checks*

allowing *population growth*

which gives rise to *population pressure*

and underlies *migration for economic reasons*, first to areas where the
new technology is being used to produce the surplus, resulting in *centralisation* and *urbanisation*, then, when possible, to
areas where it is *not*, taking it along

the new technology most often being *more complex* than the old

and requiring *specialisation* for its use

which gives rise to a *division of labour* and an *increase* in the *complexity of
society* as a whole

thereby promoting *social stratification*

and an *unequal distribution* of the surplus

which promotes an increase in the *property* and thereby *power* of the
upper strata

while the lower strata experience an *increase in work* and *illness*, and a
general worsening of their *quality of life*

such social discrepancy being maintained by *laws*

a reaction to which, and to the power of the upper strata more generally, may be conflict between the weak and the powerful in
the form of *revolt*, *terrorism* and/or *revolution*

while the surplus in the hands of the upper strata leads to conflict in
the form of *war* amongst themselves

which gives rise to *migration due to conflict*

the surplus at the same time allowing the *consumption of luxury goods*
amongst the upper strata, which can be produced thanks to
technological development

as well as providing them with *leisure*

some of which is devoted to *cultural development*: the arts, architecture,
philosophy, science and medicine

while the presence of the surplus also leads to *increased trade* amongst the upper strata

which contributes to reducing the *self-sufficiency* and thereby the *security* of society

while the population grows so as to *overshoot* the surplus, i.e. to *over-exploit its resources*, such that the surplus begins to dwindle and can no longer maintain the population's quality of life at the same level

the excess population combined with the reduction in available resources meaning *diminishing returns* to the use of the current technology

the results of the employment of the technology thereby *undermining its own usefulness*, and, since people have become genetically dependent on it, potentially *undermining their existence* as well

which leads to *economic decline*

eventuating once again in *scarcity* and *need*; and possible *population reduction*.

Looking over these factors, we see that some of them play more of a causal role while others are more *results* of the operation of the principle. These latter should include e.g. the creation of new needs, increasing pollution, species extinctions, the weakening of internal population checks, increasing complexity, social stratification, cultural development and economic growth.

In terms of systems, the vicious circle principle represents an instance of a positive feedback loop of the increasing kind. If the elements in such a circle are not physical, the circle need not be vicious – for example inflation and increasing nominal wages, or the production of fractals using a computer program, could in principle continue indefinitely. But if the circle produces elements that are physical, its continued turning will lead to disequilibrium in the system of which it is a part, either internal and/or in relation to the other systems with which it interacts. The size or immediacy of the threat to the system will depend on the extent to which vital elements in it are involved in the circle. In the case of the vicious circle principle of the development of humankind, this threat couldn't be greater, as it involves the very biological core of the human species.

Explication of the vicious circle principle

Here follows a rather detailed depiction of the various elements that can go into the manifestation of the VCP, as given above. It is to be kept in mind that various of these elements can and do work in concert, and that as a consequence the order in which they are presented here is to some extent arbitrary.

Needs and scarcity
Vital vs. non-vital needs

In considering needs in biological contexts one must take into account the sort of biological entity being considered. In the present case we should thus distinguish (at least) between the needs of individual organisms and the needs of the species to which they belong; and we should further distinguish between those needs which must be met for the organism or species to survive, and those which must be met for it to function in an adequate way according to some criteria or other.

Needs in order to survive I shall call *vital* – though they might perhaps just as well be called *basic*, or *absolute*, or *subsistence* – and in the case of virtually all animal organisms they include oxygen, water, food, shelter, etc. The vital needs of animal *species* include the organisms' vital needs (though of course not all organisms of the species must have their vital needs met in order for the species to have *its* vital needs met) as well as the need to procreate, which includes the need for breeding sites. The breeding sites of individuals will be in their or their mates' territories, as will, in the case of many species, their and their families' source of food. In the event that the vital needs of a population go from being met for the whole population to not being met for at least part of it, it may be expected that there will be a reduction in the size of the population. As regards humans, we have essentially the same vital needs as do other large mammals.

But it is only humans that have *non*-vital needs, i.e. needs individuals and/or the species have, the lacking of which will cause disruption (static disequilibrium) but not the demise of the individual or species. In the case of individuals, such needs are often needs in order to accomplish some end, the attaining of which will or may improve the person's situation in some way or other. Such needs might include the felt need for a holiday, what a woman experiences as the need for a particular kind of cosmetic, or an academic's need for e-mail facilities.

Real, imagined and experienced needs

Vital needs are also *real* needs, while non-vital needs may or may not be real. As is understandable, in many cases it may be difficult to determine whether a particular non-vital need is real or not. If a need is not real, it may be termed *imagined*.

When a need is difficult or impossible to meet, I shall call it an *experienced* need. Thus experienced needs may be real or imagined; and they may be vital or non-vital. In all cases, however, experienced needs result from a scarcity of whatever it is that is needed. So, for example, it may be the case that vegetable foods have become relatively scarce, such that the human population as a whole has an experienced need for more food – an experienced real need which is also a vital need. Or, due to the general scarcity of money, a capitalist will have an experienced need for more of it – a need which is non-vital and may well be only imagined.

In the case of other species, as taken up in Chapter 1, an experienced vital need is typically brought on by changes the populations of the species have not themselves influenced, such as a change in the climate. Experienced need can also be induced in humans in this way. But in our case we typically bring about such changes ourselves, through the operation of the VCP. For example, over the past century the human population has itself had a profound effect upon the earth's climate, a consequence of which will quite likely be an increase in the experienced vital needs we have in the future.

Furthermore, for other species experienced vital need is something that simply must be lived through, and, if prolonged, normally leads to a demographic transition in which the size of the population is reduced. This, however, need not be the case for humans, for reasons mentioned above and which will be considered more thoroughly below.

The needs of the powerful vs. the needs of the weak

Experienced vital need was the only sort of need that contributed to the turning of the vicious circle before the advent of horticulture. The vicious circle of humankind's development has taken us, however, from this situation to one in which the number and variety of humans' experienced needs are greater, both as regards vital and non-vital needs, and both for the powerful and for the weak. But for the weak, experienced needs are still of the vital sort, while for the powerful they are non-vital, and often only imagined (the powerful and the weak might even be distinguished in terms of whether their experienced needs are vital or non-vital): no matter how much some wealthy persons have, they still experience the need for more, either to maintain their position, or to improve it.

Two dominant forms taken by the non-vital needs of the powerful are political rulers' experienced (though perhaps only imagined) need to expand their territory (through military action) and capitalists' experienced (though perhaps only imagined) need to increase their wealth (through economic action).

But where, through the operation of the VCP, the more than sufficient meeting of both vital and non-vital needs leads to such phenomena as increasing resource depletion and production of waste, it is the more than sufficient meeting of vital rather than non-vital needs that tends to lead to population increase – e.g. a surplus of food and housing among the poor will tend to give rise to an increase in population, while a 'surplus' of mobile phones and pleasure boats among the rich need not. It is the meeting of vital needs and not non-vital needs that provides the preconditions for the survival and procreation of the species, and for the turning of the vicious circle.

Technological innovation

In periods of widespread experience of vital needs, the size of the human population tends to diminish (as does that of other species) as a result of the operation of internal or external checks or a combination of the two. But in this regard – and this is a key aspect of the VCP – the human species is ecologically unique in being able to meet some instances of experienced vital need not by reducing the size of its population but by taking more resources from the environment, though there may be population diminution even in such cases. Thus it may be said that where other species are genetically modified by their environments while they themselves exert only minor influence on those environments, humans modify their environment, and their genetic change is a result of their so doing.

As suggested by Malthus and Boserup, as noted earlier, as well as by Schumacher, and as implied by Wilkinson, in situations of scarcity necessity can become the mother of invention.[370] In the case of humans, scarcity leading to experienced need (whether it be vital or non-vital) may be overcome via *technological innovation*. (Such innovation need not imply a change in technological devices; but if the devices remain the same, then technological change implies using them in a different way or on a different scale.) This does not mean that every instance of experienced need will lead to innovation or the use of previously unused technology – all that is required for the vicious circle to operate is that *every now and then* a technological solution be found to a problem of scarcity. Thus technological change is here seen as *paradigmatically* being a defensive move on the part of humans in reaction to a worsening life-situation resulting from an increase in the number of people living on a particular area of land. Here we have a broad application of the reaction principle.

Though the discovery of new technological solutions may be rare, once made they are remembered, and their use spreads to other cultures. Technological solutions are easily transmitted largely because of their objective nature;[371] and knowledge of how to employ innovations is eventually transmitted to all areas of the world where they can be of use.

Technology only used where it reduces experienced need

The application of a new technology is like the birth of a new species: once its viability has been established its use spreads. The dissemination of technology is often via trade routes, and extends to all areas in which it can reduce experienced need.[372] With the turning of the vicious circle, the role played by commerce in the dissemination of new technology has constantly increased.

Our species' internal homeostatic mechanism regulating the size of our population weakened successively as larger and larger domains were opened up by new technology. But when the homeostatic element was still relatively strong, new technology was not employed until its use could relieve *vital* experienced needs, the reason being that it involved more *work*. (Cf. Boserup.) Thus what may be noted here, particularly against the background of the idea that necessity is the mother of invention, is that, as argued by Wilkinson[373] and as is supported by the VCP, after an innovation has been made, necessity has more been the mother of the *employment* of invention.

Here we have a manifestation of the pioneering principle in a context particular to humans. The new technology can lead to population growth by, for example, opening up new geographical areas containing food and breeding sites. The difference however is that the growth is also affected by the fact that the new technology is not employed unless need is experienced. Note that the pioneering principle may lead to the *development* of technology, as when it leads people to find a way over a body of water. Need may in fact be evident in both cases, as may the necessity to do more work.

Tools, weapons

When we modern humans came into existence with the particular karyotype we have that gave the basic direction to our behaviour, that behaviour, compared with later, involved relatively little use of tools. Concerning the niche into which our species evolved, but which we long ago left, our relentless use of new technology was not part of the picture (though its basis was already being laid by earlier pre-human and human species). And it may be asked to what extent we can or could develop technology and remain viable as a species.

The first form of human technological development resulted in the production of *tools*, which could almost be defined as any means of obtaining *more* from the environment. In keeping with the view of Mumford,[374] we can see tools as being of essentially two types: containers and piercers, paralleling the woman/man, defence/offence and yin/yang dichotomies.

Military technology

It may be noted that new technology may also be employed to increase available resources for one population of humans by taking them from another.

Thus advances in military technology giving one group an advantage over another are of this sort. Here a distinction along the lines of that between piercers and containers can be made between military (political) and economic power: military power depends on *pointed* weapons, while economic power depends on *accumulated* capital. And we see that it is precisely the development of *weapons* that has always been at the forefront of technological development – first in the form of weapons for hunting; later, weapons for war.

Language

Though the development of tools was the first manifestation of technological development incumbent on the turning of the vicious circle, many other innovations have resulted as well, such as that in the form of language.

Rather than merely being a tool, language can be seen as itself constituting a technology; and it has played an important part in the turning of the vicious circle. For one thing, it has been an aid to other technological innovation through allowing the exchange of ideas regarding how to solve particular problems of scarcity. Language has also aided the dissemination of new technology through putting the concepts required for its manufacture or realisation into a form that can easily be used by others. And language's making possible the giving of orders or commands, and the following of chains of command, has subsequently meant that the implementation of technology on a massive scale can originate with a single person. The most important development in the use of language has been *writing*, which has reinforced such uses as those mentioned above.

Domestication of plants and animals

After the use of fire, which includes today's burning of fossil fuels, the domestication of plants and animals for human purposes constitutes humankind's greatest harnessing of solar energy.[375] This domestication was a 'one time event' like the invention of any tool or machine, one which, however, unlike the invention of artefacts (which become obsolete), constitutes a development on which we have become increasingly dependent.

The process of domestication is of interest not only with respect to how wild organisms have been tamed, but also with regard to how they have been genetically modified. In this regard domestication could in fact be defined as the genetic adaptation of various species to technology,[376] which, when that technology is used to support them, leads to their becoming genetically *dependent* on the relevant technology.

Domestication of the weak

Of the animals domesticated by humans, one can count humans themselves. Just as the ancestors of domesticated animals roamed freely in the wild, while their domesticated descendants have to remain stationary and close to a shelter, the same may be said of humans, particularly after we became sedentary. And just as species of domesticated animals experienced genetic changes as a result, so did we.

Since the weak constitute the vast majority of the total population, and it is primarily they who have been required to adapt to the operation of technology, the effect has been the domestication of the species as a whole, which has manifest itself in various ways. One of these may be its having led to a general increase in human intelligence (ability to innovate, and to operate sophisticated machines). Another has been pointed out by Darwin: "Civilized races can certainly resist changes of all kinds far better than savages; and in this respect they resemble domesticated animals."[377] And just as domestic animals would perish if the support of technology were removed, we could expect the same of modern humans. We are both karyotypically and genetically adapted to a situation of increasing technological dependence.

Unintended side-effects

Generally we see that the employment of virtually any technology gives rise to unwanted – and often unforeseen – side-effects. As expressed by Aldous Huxley, the advantages accruing from recent technological advances are generally accompanied by corresponding disadvantages; gains in one direction entail losses in other directions; and we never get something except for something.[378] As suggested by Commoner, our most celebrated technological achievements – the car, the jet aircraft, the power plant, industry in general, and indeed the modern city itself – are ecological failures.[379] But where Huxley and Commoner limit their claims to *recent* technological development, on the VCP what they say applies to *all* technological development.

Technology, by its very nature, is employed as a means to accomplish a certain end. But, as emphasised by Garrett Hardin, its effects are always more wide-ranging than the accomplishment of the particular end it is intended to achieve.[380] The very solution of problems by technological means gives rise to new problems.[381] And as further suggested by Ellul, history shows that every technical application gives rise to unforeseeable side-effects which are much more disastrous than what would have been the case if the situation had just been left alone.[382] And we see that this must be so, since the more we counter entropy by bringing technological order to particular parts of the physical world, the more entropy we create elsewhere, such that there is an increase in the entropy of the system as a whole. As Commoner emphasises, the new problems are not the consequences of incidental failure but of technological success.[383]

Here humans are like other animals in that we follow the reaction principle and focus on accomplishing things in the here and now, and tend to ignore what does not impinge on us in a striking fashion. This is an orientation that has served well in the evolution of all species; but in a species with a developing technology that gives rise to unwanted side-effects, it is bound to lead to problems. In a way, the vicious aspect of the vicious circle consists precisely in these unwanted side-effects: population growth, resource depletion, increasing pollution, social stratification and so on.

Further, as pointed out by Jay Forrester, a series of actions all aimed at short-term improvement can eventually burden a system with long-term depressants so severe that even heroic short-run measures no longer suffice. The perceived short-term needs are more visible and more compelling, and speak loudly for immediate attention. (As Boulding wryly notes, "It seems to be very hard to organize a long-run crisis.")[384] Policies which produce long-term improvement on the other hand may initially depress the behaviour of the system in which they are implemented before the improvement is manifest.[385]

Economic development

Economics is the means by which we humans maintain ourselves as individuals, as groups, and as a species. Economic *development* is essentially the same thing as applied technological development, consisting as it does in the new use of technology to meet needs. It is not to be confused with economic *growth* (to be treated later), which implies an increase in available consumables, and which may be said to be the *aim* of economic development. And these two should be distinguished from economic *efficiency*, which concerns the amount of energy required to obtain a product.

As suggested by Wilkinson, a society's being out of ecological equilibrium due to overpopulation necessitates economic development, which, if successful, will increase the extent to which the environment can be exploited. "As the level of environmental exploitation increases, more and more of the production and processing of raw materials is dependent on the work of man [and his machines] rather than on purely natural processes."[386] On the whole this results in declining economic efficiency. Whenever the constraints which maintain a society in ecological equilibrium break down, the society will try to find ways of developing its technology to increase the yield from the environment.

As Wilkinson also says, the appearance of subsistence problems makes people willing to accept changes which previously seemed to require too much work, or which suffered from other prohibitive disadvantages. Most of the changes are accepted because they represent improvements in the supply of subsistence materials, not because they represent increases in efficiency for societies that already have an adequate living. Economic development is primarily the result of attempts to increase the output from the environment rather than produce a given output more efficiently, its main features including changes in the resource base, the division of labour, the development of trade and industry, increasingly intensive agricultural methods and many other aspects of a society's changing productive activity, all of which are responses to the growth of need.[387]

Increase in available resources

Before considering the nature of resource availability and depletion it is important to distinguish between *resources*, *reserves* and *stocks*. With the implementation of a new technology, both reserves and stocks may be increased, and it is the quantity of these, not resources, that affects the immediate economic value of such substances. So the implementation of a new oil-drilling technology, for example, may increase oil reserves and lower the price of oil, while the quantity of oil as a resource is constantly decreasing.

The use to which new technology is put typically involves increasing the amount taken from the environment by making available resources that were previously inaccessible, the paradigmatic such resource being food. Though the employment of new technology will mean an increase in the available resources at the time of its employment, nothing says in advance that the overall quantity of available resources will be greater than that made available by the older technology. (For example, in Palaeolithic times the employment of the bow and arrow produced less meat than did the spear.) Thus we cannot be sure exactly what our future reserves will be, since there may exist unknown resources that future technological innovations will succeed in making available. Uranium, always a resource, was not appreciated as such – not turned into a reserve – until the advent of nuclear technology. It has in fact often been the case that some substance not known to be a resource or that cannot be treated as a resource at one point in time can later be so treated thanks to technological development, such as in the case of uranium, or e.g. the seeds of wild grasses that could not be digested before the invention of cooking, or coal and petroleum that could not be extracted or refined before the requisite technology was in place. On the other hand, it may also turn out that something assumed to be a resource is not, as is the case with uranium for example, the energy obtained from it undoubtedly being less than the energy required to acquire and use it and properly get rid of the waste.

Prerequisite of the existence of resources

That we use non-renewable resources is to the detriment of the species, first because such resources are *finite* and there's a great risk that we become dependent on them, and second because after they have been used they become harmful *waste*. It has only been possible for us to get caught up in the vicious circle, as we have been since we first came into existence, because the increasing quantities of resources it requires – renewable and non-renewable – have been there to be exploited. However, that such resources have existed at all, or whether other resources will exist in the future to replace them when they disappear, is a matter of chance. And the more resources we use up, the smaller that chance becomes. As regards non-renewable resources, though technological innovation has in many cases made them available, it has not created them. To take but one important example, all technology employed in the extraction and use of fossil fuels would not have produced a surplus of anything if fossil fuels had not existed.

However, up until the present, nature has provided us not only with the materials necessary to produce such complex technology as is involved e.g. in the use of nuclear reactors, but the energy sources required to operate them, both of these conditions having to be met for such complex technology to be able to fill a need. In the case of specialisation, human development parallels the evolution of species, for the survival of new species depends precisely on the increased specialisation necessary to find an unoccupied niche or oust some other species from its niche. Typically, each new species, when it first comes into existence, should occupy the same niche as, or a very slightly different one from, the species from which it evolved. But in the human case, thanks to the turning of the vicious circle, we have constantly been adopting ever larger niches.

Each use of a non-renewable resource diminishes the total quantity remaining, and moves us closer to the point at which the only resources that may be left apart from water and air are plants, animals and stones. Thus, if not for some other reason, sooner or later the vicious circle will stop turning due to the lack of resources to fuel it. From the point of view of our species' survival, the role played in the vicious circle by non-renewable resources and the unsustainable use of resources is therefore particularly noteworthy, for it implies a dependence of the species on a state of affairs which, by the very nature of the case, cannot continue.

Increased energy use

The implementation of a new technology will quite generally mean an increase in a society's use of *energy*. Historically, the first non-human source of such energy was wood used in fires, and later domesticated animals such as the ox and the horse, and then water and wind, and since the industrial revolution mainly fossil fuels. While usable energy is itself a non-renewable resource, its *source* may be renewable.

All use of technology demands energy; when the technology is used in production it is generally the case that the more sophisticated it is, the more energy it requires per unit produced, unless or until benefits of scale appear. At the same time, the energy required to obtain non-renewable resources constantly increases as they become less accessible. Similarly with trade, the greater the distance between trading partners, the greater the energy required to trade a particular entity. From the other end, increase in the availability of energy promotes economic expansion, thanks to both the increasing number of products technological development makes available and the ease of transportation it makes possible. And this economic expansion implies an increase in the extent to which we counteract entropy within society while increasing it without.

Geographic expansion made possible by technological change

Throughout the existence of humans, technological development has meant the ability of the human population to spread to new areas. Originally this meant expansion to areas unoccupied by humans (as in the spread of *Homo erectus* and to some extent *Homo sapiens*), and then later the spread to areas occupied by karyotypically more primitive humans (e.g. the occupation of Europe by modern humans despite the presence of Neanderthals), and lastly the occupation by more technologically developed humans of areas occupied by those who are less developed (e.g. the occupation of the New World by Europeans). In each case we have a manifestation of the pioneering principle.

Resource depletion

As regards the ecological equilibrium of the human species when it comes to resource use, what is of primary importance for the turning of the vicious circle is not our use of non-renewable resources *per se*, but our use of resources *in an unsustainable way*. Thus some resources, such as particular types of stone, may be non-renewable but nevertheless exist in such quantities that we could never use them up, in which case their use would be sustainable. And other resources, which are renewable, may be used in such a way that they cannot renew themselves.

The differences between such states as the vicious circle's turning normally, slowly, quickly or with greater momentum, and its accelerating or stopping, are directly related to the rate at which resources are being used. The *normal* turning of the vicious circle requires constantly *accelerating* quantities of resources, and it will continue as long as they are provided. The turning of the vicious circle, like the growth of population on the population principle, has no internal limiting mechanism, and under ideal conditions grows physically at an exponential rate.

Our increasing unsustainable use of resources is itself a precondition for the turning of the vicious circle. If all potentially renewable resources were renewed or allowed to renew themselves, as in the case of other species, and if at the same time we hadn't become dependent on the use of non-renewables, then it would have been possible to avoid getting caught up in the vicious circle, as many modern hunter-gatherer communities had not before the intrusion of other cultures. But, as noted by Wilkinson, as population increases, the aggregate subsistence demand for the particular resources on which the cultural system is based come to exceed the supply the environment can provide, and the system loses its (quasi-)equilibrium. The scarcity of vital resources encountered at this point forces the society to alter the way it gains its living from the environment, and, through technological change, it ends up occupying a different ecological niche.[388] Our unsustainable use of resources has led to a situation in which technological innovation, and with it the turning of the vicious circle, becomes a *must* (cf. technology perpetuating itself, above) in order to extract a replacement when the acquisition of a particular resource becomes uneconomical.

From one point of view, our non-sustainable use of resources is, on the VCP, primarily due to our inability to foresee, and/or disinclination to consider, the consequences of the application of technology; that is, it is a manifestation of the reaction principle. Here again we act as other animals would: given that a particular resource is available, we use it, without considering the long-term effects of so doing. That we share this predilection/disposition with other animals suggests that it is *instinctual* – that it stems from our karyotype and not simply from our genes. Here we have an instance of a karyotypical aspect of our natures that is not pre-adapted to our using and developing technology – an aspect deeply rooted in our nature as a species.

Waste and pollution

Any use of resources will result in waste; even the mere consumption of food involves at least body waste. Wastes resulting from the sustainable use of resources are biodegradable, and if properly dealt with need not be of concern to humans. On the other hand, the production of wastes from the use of organic and inorganic minerals (increasing entropy outside the social system) constitutes a negative and unintended side-effect of technological development. Such wastes can not only lead to a worsening of humans' life-situation, but can undermine the conditions for the functioning of technology itself. Rivers may become so mired that their water can no longer be used either for drinking or as engine coolants; waste deposited at sea can kill fish and thereby undermine the fishing industry; non-biodegradable garbage can swamp agricultural land making it useless. Such wastes as can have this effect may be termed *pollution*. (We might define pollution as any substance deposited in an ecosystem that tends to disrupt its equilibrium.) Thus we see that technological development, and with it the vicious circle of the development of humankind, can come to a standstill not only through the unsustainable use of resources, but also from the effects of the pollution that their use gives rise to.

Increasing extinctions

Thanks to the turning of the vicious circle, *Homo sapiens* is the only species to exterminate other species *systematically*, a process which began with the extinctions of large mammals during the Pleistocene. In that case the species directly affected were the targets of human predation, their elimination being made possible by technological development, while at the same time many of the non-human predators of such prey also became extinct for lack of food. To this may be added the effects of foreign organisms imported by humans, such as rats and the micro-organisms they carry, as we spread over the world and increased in numbers.

Apart from eradication through hunting, the extinctions of various plant and animal species can be seen to result from the operation of the VCP in the form of the constantly accelerating growth of the human population, together with its increasing per capita use of technology, and the greater pollution to which this use gives rise. Humans' constantly increasing exploitation of the physical and biological environment has eradicated the habitats of the relevant populations, and thereby the preconditions for the continuing existence of their species. Not only has the turning of the vicious circle led to the extinctions of various species, but with its constantly increasing momentum it has also led to an *acceleration* in their extinction, such that more species are becoming extinct per unit time at present than ever before in the past 65 million years.

New needs

Once a technological innovation meeting a vital need has been adopted, the pioneering principle, as noted above, has the effect that the population tends to expand to the size that the resources made available by the new technology allow. And as also intimated earlier, once this happens the population becomes dependent on these resources, and, moreover, on the technology required to obtain them. In other words, these resources and technology become *new needs* for the population.

Typically, other new needs also arise with the adoption of new technology, such as the needs to service the technology and provide adequate conditions for its use. Thus, just as we might say that necessity is the mother of invention, we might also say that invention can be the mother of necessity, necessity which in turn requires further invention. Once a new invention has been adopted in a particular society, it tends to create pressures within the society to make other innovative changes. In this way we humans are the only species that has come to create new needs for itself.

The application of new technology creates not only the needs associated with its own maintenance, but also those associated with its dissemination. Once telephones come into use, telephone cable technology and switchboard technology have to be developed, and the results manufactured and put to use. The development of the car not only made roads and service stations necessary for the first cars, but for all of an increasing number of cars as they filled the experienced non-vital needs of an increasing number of people.

Such needs are not removed until or unless the technology in question is usurped by a newer technology; and the newer technology will, of course, bring with it yet newer needs related to *its* maintenance and application. And new needs, created by the impact of the employment of new technology on people's lifestyles, provide the stimulus to higher levels of consumption. Also, technological development leads to the need of a broadened resource base (larger niche) – more different kinds of metals, for example. This is related to the new technology's only providing *part* of what the old technology provided, and to constantly increasing complexity/division of labour in society. In effect, as suggested by Wilkinson, the real cost of living is increased by technological development. And in the modern era this growth of need makes new activities profitable to capitalists.[389]

Technological development self-perpetuating

In terms of systems, technological development, unlike the development of individual organisms and certain species, but like the development of the biosphere as a whole, has no internal homeostatic mechanisms or checks to its own growth. It is the key part of the turning of the vicious circle, which itself has no internal check. As expressed by Schumacher: "Technology recognises no self-limiting principle – in terms, for instance, of size, speed or violence. It therefore does not possess the virtues of being self-balancing, self-adjusting, and self-cleansing."[390]

Though technological development could in principle be limited by human intervention, this has so far not occurred on any significant scale; and if it involves the non-sustainable use of resources, it will prevent humans from coming into equilibrium with their surroundings. Left to itself technology will continue to develop, and grow in physical size, until the resources on which its development or growth depend have become exhausted – unless the wastes it produces prevent it from continuing before that. And, as Schumacher says, "In the subtle system of nature, technology, and in particular the super-technology of the modern world, acts like a foreign body, and there are now numerous signs of rejection." Thus technology and technological development tend to undermine their own existence, as well as that of any species dependent on or otherwise affected by them. And when technological development comes to a standstill, the vicious circle of the development of humankind will stop turning as well.

Due to the need acquired for a particular technology, the use of that technology becomes – perhaps genetically – ensconced in the population or society, and cannot be given up but only replaced. As in the case of specialisation, this replacement parallels evolution in that just as new species tend to be more complex than their predecessors, so do new forms of technology. Following the innovation–speciation analogy further, we should say that each group of people consisting of all specialists of a particular sort is comparable to a species.

New techniques tend to be applied to a smaller range of resources than those they are replacing.[391] Thus each technique that is replaced is often replaced by a *number* of new techniques to cover all of the necessary resources. Due to technology's tending to eliminate its own resource base, it is normally only a matter of time before its replacement itself becomes a necessity. In this way a positive feedback loop of the increasing kind is created in which technological development is a driving force behind further technological development.[392] And the population becomes dependent not only on technology, but on the *development* of technology. This is part of the expression of the VCP, and constitutes a vicious circle in itself.

Creation of a surplus and increased consumption

Technological change originally employed to counteract need has often *overshot the mark*, giving rise to a surplus, i.e. to more resources, reserves and/or stocks than is necessary to meet the needs of the current population. The acquisition of such a surplus is a precondition for the turning of the vicious circle. It is to be noted however that the existence of a stored surplus itself implies a relative scarcity of what is stored.[393]

Here we might distinguish between cache, store and surplus. Both a *cache* and a *store* are paradigmatically ways of preserving food, and it may be suggested that where a cache is typically of hunter-gatherers' dried meat, a store is typically of agrarians' grain. More generally, caches and stores may consist of resources, reserves and/or stocks, as well as entities that have only exchange value. Money, for example, is stored in a bank. A *surplus*, on the other hand, is essentially that portion of a physical cache or store that exceeds the vital needs of the extant population.

Note that a surplus is essentially of *vital* resources, i.e. of resources, reserves and/or stocks meeting vital needs. This is necessary for population growth and the turning of the vicious circle. In a situation where such a surplus is lacking, all needs are vital needs. Given such a surplus, however, *non-vital* needs may develop. These needs may be met by non-vital resources, of which there may or may not be a surplus. Thus we can speak of surpluses of both vital and/or non-vital resources, the existence of a surplus of non-vital resources presupposing that of vital.

It is the constant presence of surpluses of vital resources that has allowed the vicious circle to turn. Most importantly in this regard, it has allowed an increase in the meeting of vital needs, including through the production of an excess of food as well as the construction of a greater number of shelters or homes in which families can be raised (an increase in perceived territory for the masses). As dictated by Liebig's law, an increase in both food and breeding sites is necessary for population growth, and in the case of humans, for the vicious circle to turn. What this implies, among other things, is that a surplus of available resources, i.e. of *reserves*, must further be converted into *stocks* if it is to be used. For example, reserves of oil must be *refined*; land amenable to agriculture must produce *food*; and so on.

A surplus of a resource (or resources) can take a variety of forms; and it may be of a resource different from but nevertheless able to meet the same needs as the resource that became scarce. What is typically the case is that the scarce resource is at least partly *replaced* by another resource that meets the same need(s). At the same time, however, it may be the case that a newly acquired resource have an application that is *broader* than that of the resource it is replacing, though this is not the norm. The presence of more resources meeting vital needs will mean an increase in the consumption of these resources (taking both breeding sites and food into account), in keeping with the pioneering principle.

Inferior substitutes

The implementation of new technology may result in the acquiring of products that are superior to those they are replacing. However, this is not always the case, and, as has been implied by Wilkinson,[394] more often has *not* been the case. On the other hand, however, replacements have generally been greater in *quantity* than what they are replacing. When it comes to food, for example, the quantity of food available to the whole of humankind over the ages has constantly been increasing, while at the same time its quality has on the whole been worsening over the past 25,000 years. A modern example of inferior substitutes given by Wilkinson is plastics as compared with leather in shoes and handbags.[395]

As regards resources more generally, the trend towards substitutes of lower quality is the same. We must recognise however that the inferiority of replacements is not a priori, unless such a view is taken as that given our evolutionary nature, those resources that are nearest to hand are those of the highest quality, i.e. best supportive of the continued existence of our species.

The idea that substitutes will tend to be inferior receives support from looking at ourselves as rational agents: given the choice we will first take what is best. The more difficult a resource is to acquire, the lower its use value for that very reason. As expressed by Daly, we do not satisfy ends in any arbitrary sequence but seek rationally to satisfy our most pressing needs first. Likewise, we do not use resources in any order but first exploit the most accessible ones known to us. The former fact gives rise to the law of diminishing marginal benefits, the latter to the law of increasing marginal costs.[396]

Population checks

The survival of the human species is to be ensured by the replication of its karyotype by fit organisms successfully reproducing neither too many nor too few reproducing offspring. As treated in Chapter 1, if too many are produced, other things being equal, homeostatic mechanisms checking growth should come into play, so as to keep the species in equilibrium with its surroundings. It might here be kept in mind that the modern human ideal of no mortality until old age suggests that through artificial means the human species should differ from all others in eliminating infant mortality. (Cf. the second formulation of the principle of population.) The implications of this ideal for potential population growth are clear.

Positive vs. preventive checks

As regards checks to growth then, we should perhaps begin by noting Malthus' *positive* and *preventive* checks. As remarked earlier in discussing the principle of population, positive checks are causes of premature death (i.e. death prior to the loss of the capacity to produce or rear children); and preventive checks are checks on birth rate (or rate of conception). Note that population control, when positive, means higher mortality and lower life expectancy.

Internal vs. external checks

The distinction between internal and external checks has also been dealt with earlier. Where the positive checks to human population growth are typically external (e.g. predation, disease, starvation), preventive checks – to the extent they have been operative – are normally internal.

In keeping with what was suggested earlier, the weakening of *internal* checks – both positive and preventive – may be seen on the VCP to be the result of technology continually providing humans with a surplus of vital resources, including both food and the materials requisite for the building of dwellings. Had we only been provided with surpluses of food, but not an increase in breeding sites, or vice versa, then external checks would alone have been able to limit the size of the population, in keeping with Liebig's law. We have reacted to these continual increases in the quantity of food and breeding sites as virtually all species of animal would, and as accords with the pioneering and reaction principles – by increasing our population.

Constant vs. transitory (stress-provoked) checks

All transitory internal checks are stimulated by crowding. Such internal checks may appear only in case of stress; or they may exist independently of stress and be intensified by it.

Modern human population density, in order not to lead to stress, should be about the same as (or at most slightly greater than) that of the species humans are directly descended from. Thus due to constant human population growth, stress is de facto endemic to virtually all human populations, though it of course increases and decreases with increasing and decreasing population pressure.

Territoriality

The more complex the species, the more its basic instincts are supervened upon by its less basic, and the greater its behavioural adaptability. Complex species are more dependent on instinctually less-basic behaviour, the highest form of which is *intelligent* behaviour. In the present context this is manifest in humans' greater reliance on learned (cultural) behavioural checks. In our case, all behavioural checks are variable manifestations of territoriality, taking the form they do as the result of the mediation of other factors, particularly culture on the macro level and genes on the micro.

Territoriality tends to keep the population size in check *before* there is a scarcity of food. Here it is important to note how the territory in question is perceived. Though there may in fact be sufficient territory to support population growth, if it is not *perceived* as sufficient no such growth will occur (cf. Wynne-Edwards' smallholder cocks). Similarly, when the territory available for potentially mating pairs is perceived as being too small, even if there should exist sufficient food, action will normally be taken to reduce the size of the reproducing group (as in the guppies example). It is this flexibility that makes it possible for territoriality (stemming from the sexual instincts) rather than food availability (the survival instincts) to function as an *internal* population check.

As taken up in Chapter 1, in virtually all species this excluding of others from an individual's or group's territory involves aggression, which, particularly in the case of mammals, may or may not result in death(s). At least since we first became sedentary, the fit man's aggressive dominance over nearby men has ensured him his territory; and it is the similar dominance of one group over another that assures the group its territory.

For non-human species when sedentary, as well as for Neolithic humans, the size of any particular male's territory is limited by his ability to roam and defend it. And the same can be said of groups. We might also expect that in the absence of other fit males in the vicinity, both individuals' and groups' territories would tend to be larger than otherwise (the pioneering principle). Increasingly for humans this 'territory' of the average person is not patrolled by the individual owning or ruling over it, but consists rather of those areas where the person and his family live, plus those areas where the food and other resources they consume are produced or obtained.

In many sophisticated species other males must be allowed to mate in group territory ruled over by a leader. And, like the leader, other males may have a number of mates – in the case of humans, as in that of many other mammals, polygyny being evident in the marked differences in males' and females' body size and strength. This was so with our primate ancestors and it is so with us. (Among other things, this suggests that human monogamy is a *cultural* population check.)

It may also be kept in mind that, as noted earlier, while territoriality promotes aggression between males, once individual territories are decided on it brings about peace, and in fact group territoriality leads to social cohesion amongst those occupying each of the individual territories in question. As in the case of other social animals, a ranking system is developed with the leader (fittest individual) on top. And, also as in the case of other animals, the peace continues until the next occasion for conflict arises with the challenging of the leader's position. While the positions of the members of the hierarchy are only temporary, the form of the hierarchy itself is permanent, stemming directly from the species' karyotype.

Thus the coherent unit constituting a tribe, with its leader, fights other tribes occupying other territories, it being *morally obligatory*[397] – and in fact unquestioned – for fit males to do so: cf. the Murngin above and Darwin's bees, below. In this way territoriality can account for both divisive (individual territoriality) and cohesive (group territoriality) forces in society. And when societies become more complex and kin/tribal relations are obliterated, the social instincts operate so as to form similar relations in groups that are more artificial, such as nations, religious communities, platoons of soldiers, ethnic groups and social classes.

Thus in the case of humans, dominant males have been able to become the leaders of huge populations and/or populations occupying huge territories. By being able to gain control over a significant proportion of the population's *weapons*, the leader can see to it that the territory he rules over is protected for him by others. Being the leader of his nation, his ruling over the group's territory through his controlling its members gives him his status and social position. But the *motor* behind this development, viz. the dominating behaviour of males, is present in all territorial species; it's just that in the case of humans many of the normal checks to particular individuals' increasing the size of their territory have been removed.

In the case of kings, each must be able, using his *army*, to defend his (group's) territory from peoples led by other kings, and in the best of cases increase its size by taking territory from other peoples and kings. (That he have an army at his disposal is largely thanks to the social instinct of all fit males in the territory he rules over to fight for their group.) It is thus to the leader's advantage that he have large numbers of followers when it comes to conflict with other groups.

As expressed by Charles Galton Darwin, any nation that should limit the size of its population would be forced off its land (territory) by some other nation or nations that had not done likewise.[398] But having large numbers is also to the advantage of the group as a whole, in that in the case of war it increases the likelihood not only of personal survival but a sharing of spoils. While on the one hand infanticide and killing in war, both resulting from social instincts supporting group territoriality, tend to reduce the size of the population, the social instinct for males to defend and enlarge group territory, which supports population growth, is the stronger influence, since it is reinforced by the survival of those groups that evince it. So we see that *where individual territoriality tends to limit population size, group territoriality tends to increase it*. This tendency is counteracted by inter-group killing (among other things), a population check that does not exist in non-social species, the result being that there is a balance between group territoriality's promoting population growth and inter-band killing's repressing it. Because of the population-increasing effect of group territoriality, infanticide is only employed when increasing the size of the population is of detriment to the group; and then it is mainly the females that are killed, who cannot be employed in defending or expanding the group's territory. In other species the expansive influence of group territoriality is sufficiently countered by inter-group competition, resulting in an inter-group homeostasis and relatively constant population sizes. In the case of humans, on the other hand, the influence of group territoriality in increasing the population has *not* (to date) been sufficiently countered for population sizes to level off.

To ensure that his army be as powerful as possible, the king supports the inclination of the group to grow, so that the number of males in his group's territory able to bear arms be as large as possible (which also provides him with more income from taxes). Despite this, however, even in the case of post-horticultural people, the actual fight over group territory tends to function as a population check. This is due both to soldiers' and civilians' succumbing directly to enemy weapons, and even more to war's leading to starvation and the spread of disease; it is only that this check is overridden by the experienced need of the powerful male to have as many followers as possible, together with the presence of a surplus and the natural tendency of the social group to increase its population so as to be able to defend its territory. Regarding the surplus, as Grahame Clark observed, war is "directly limited by the basis of subsistence, since the conduct of any sustained conflict presupposes a surplus of goods and manpower."

Thus not only the presence of a surplus, but also that of males constantly driven to acquire more territory in the form of land or capital, supports a relaxing of population checks in society. And human leaders' desire to have large populations, given the constantly recurring surpluses in our past, can explain why they have consistently turned their backs on problems of overpopulation.

The adaptability of members of the human species – to a large extent manifest in our intelligence – may be important not only with regard to our innovative ability, which has increased the actual number of humans who can be raised, but also with regard to our perception of territory. Though the members of many non-human K-selected species are resistant to a certain amount of crowding, none appear to be so to the extent that humans are – though still with increased violence as a result.

Looking over the whole of the development of our species, we see that the average space available to each pair for breeding, i.e. the male's territory, has constantly been shrinking due to population growth, at the same time as the space *needed* for successful breeding has also been shrinking, thanks to the results of technological innovation (the building of many-storied dwellings, etc.). That smaller breeding sites are perceived as sufficient and will in fact suffice does not eliminate however the karyotypically based *psychological* effect of reduced space – the effects of *crowding* – from expressing itself in various forms of violence of man against man. We note however that, as in the case of other animals, conflicts amongst humans become less intense after social adjustments to crowding have been made.[399]

In any case, given crowding, at some point territorial animals can be expected instinctively to begin positively or preventively reducing their own numbers. We humans, on the other hand, though our territoriality is also instinctual, thanks to our intelligence can nevertheless adapt to such situations, as long as our vital needs are met. And a higher density can be more easily tolerated if boundaries of ownership are clearly marked, a house for example having clearer boundaries than a flat.[400] Human adaptability to crowding is a necessary aspect of the weakening of our initial population checks.

Somatic vs. behavioural checks

Somatic checks of particular interest are those, such as reduced ovulation, which are variable and can come into operation through stress due to crowding.

Behavioural checks, also induced by crowding, are notably either *preventive*, such as coitus interruptus or failure to mate; or *positive*: negligent maternal behaviour, infanticide, abortion, initiation rites. They can vary from species to species, and in humans from culture to culture and/or individual to individual.

Cultural (learned) vs. instinctual checks

In the case of humans internal checks take various forms, and it is as regards *behavioural* checks, all of which are internal, that humans evince great variety. This variety may be seen as stemming from greater human intelligence/ adaptability/learning ability.

Cultural checks are a more sophisticated form of learned check, cultures themselves existing *only* in human populations, and in *all* of them, and being essentially determined by the population's technology. Such checks are a manifestation of the *social* instincts, as exist in all social animals. Like all learned checks they are *variable* and are *mediated* on the macro level by learning, and on the micro by our genes. Further, they may be either constant or transitory (requiring an external stimulus).

As Divale says, unlike other animals, man adapts primarily through culture.[401] Thus, as Wilkinson says more particularly, the most important (internal, behavioural, social) mechanisms for limiting *human* populations are cultural. As with most aspects of human behaviour, the physiological and other invariable mechanisms for homeostatically controlling reproduction are inadequate on their own: they can serve only as fall-back systems when cultural checks fail. Human populations are only adequately checked and starvation avoided in cultural systems which are sufficiently well adapted to have developed their own homeostatic controls. It is variations in the cultural system, not in man's physiology, which decide whether starvation occurs in human populations[402] – though we note that in modern humans cultural change both influences and is influenced by genotypic change.

Cultural population checks may have been sufficient to keep the populations of many hunter-gatherer societies, in which technological development was non-existent (and into and out of which there was no migration – to be taken up below), from pressing against their environmental limits for thousands of years at a time. But when technological development related to vital needs takes place internal checks can be overridden. This is particularly so at the time of major revolutions in the turning of the vicious circle, with the cultural upheavals they entail. Though it may be expected that after such revolutions internal checks gradually build up again, they should never do so completely so long as there is a surplus, i.e. as long as the vicious circle continues to turn.

Morality

What we term *altruism* in the case of social animals quite generally may be called *morality* in the case of humans, i.e. the *cultural* requirement to act in such a way as potentially or actually reduces one's own genetic fitness (reduces one's fecundity or fertility) while supporting the fitness of one's community and thereby one's species.[403] (If we consider the nature of human values, we could say that they are of essentially two kinds: egotistical – supporting the individual's gene line, and altruistic – supporting the group's gene line. The latter, when they are to the real or potential detriment of the individual's gene line, are morals.) However, morality is of greatest relevance to the species' survival in situations of overpopulation, due to our species' inordinate tendency to grow in numbers, its function in such cases being to check population growth; and the weakening of population checks in the presence of a constant surplus is expressed in moral terms in an increase in moral laxity.

Thus when a mother feels obliged to kill her infant during a period of scarcity, we can speak of a cultural check to population growth. In this regard, Darwin has pointed out that:

> The murder of infants has prevailed on the largest scale throughout the world, and has met with no reproach; but infanticide, especially of females, has been thought to be good for the tribe, or at least not injurious. Suicide during former times was not generally considered as a crime, but rather, from the courage displayed, as an honorable act; and it is still practised by some semi-civilized and savage nations without reproach, for it does not obviously concern others of the tribe.[404]

And he elsewhere suggests that:

> If, for instance, to take an extreme case, men were reared under precisely the same conditions as bees, there can hardly be a doubt that our unmarried females would, like the worker-bees, think it a sacred duty to kill their brothers, and mothers would strive to kill their fertile daughters; and no one would think of interfering.

All species are karyotypically inclined to produce more offspring than can be expected to survive, with the *fittest* of them being those that do survive and themselves have offspring. On the average each individual of any sexually reproducing species can only have *one* reproducing offspring; thus it is not natural for any species, including the human, that all infants live to reproducing age. This means that if we assume the fertility of the average human female to be, say, five children, then three of those children must die before themselves reproducing. The death of half or more of the children born in a human society is the natural state of affairs. And if these children do not die due to external checks, then they must be killed through the internal checks of infanticide, etc.

As suggested by Baschetti, we may say that the selective advantages of animal aggregation explain why many species, including humans and other primates, live in social groups. In ancestral times early communities of apes, to enhance their chances of survival, had to evolve selectively advantageous social behaviours, which constitute precisely the essence of morality.[405]

By means of selective pressures evolution has favoured morality over immorality in human groups, rewarding socially beneficial behaviour by favouring the survival of those groups that evinced it to the appropriate degree. Darwin wrote: "At all times throughout the world tribes have supplanted other tribes; and as morality is one element in their success, the standard of morality and the number of well-endowed men will thus everywhere tend to rise and increase."[406] (But we might add: until the 'well-endowed men' started getting away with pretending that what is best for them is best for society.) And Mayr has said:

> [C]ultural group selection may reward altruism and any other virtues that strengthen the group. … It is easy to imagine how a particular value system within a culture might lead to the prosperity and numerical increase of the group, which might, in turn, lead to genocidal warfare against neighbors, with the victor taking over the territory of the defeated. Any divisive tendencies within a group would weaken it and in due time lead to its extinction. Thus, the ethical system of each social group or tribe would be modified continuously by trial and error, success and failure.

Note that *cooperation*, while based on the social instincts, is not in itself an expression of morality, though morality can foster cooperation.

Religion and myth

Religions are *cultural traditions* the most important function of which is to reinforce instincts benefiting the community as a whole, and this they do by giving higher authority to and making explicit certain rules of behaviour the members of the community are to follow. These instincts are *social*, and benefit from religion's reinforcement particularly due to their being the weakest of the instincts; and the explicit rules authorised are *moral*. Note further that religions more particularly support group territoriality and its pushing of population growth – a point to be returned to in Chapter 5.

As E. J. Mishan suggests, no moral law, no matter how enlightened, will command the allegiance of men if it is known to be founded explicitly on considerations of social expediency. Socially instinctive submission to its precepts is ensured only if they are engraved on stone on the inner layers of the conscience, distinct and inerasable, resistant alike to exemptions and concessions. All past religions have been of divine origin.

Not only the great monotheistic religions, but all the supernatural beliefs that guided and influenced the behaviours of societies large and small, imparted stability and cohesion to those societies. And these beliefs include not only the sacred myths but also the secular ones – those sustaining beliefs held by a tribe, a folk, a race, a nation, about its heroes, and about its heroic origins and its heroic achievements.[407]

Conscious vs. cultural checks

Another distinction may be made between kinds of check, namely between cultural and conscious checks. Thus, for example, some people may use contraceptives in societies where their use is not condoned. But even conscious checks, which depend, say, on reason, have a karyotypic basis, just as reason itself has such a basis. Here it may be noted, however, as has been pointed out by C. G. Darwin, that purely voluntary population control selects for its own failure.[408] The gene lines of those families that have few children will become fewer, while those who have many will increase.[409] Thus we see, for example, the impracticability of the (neo-Malthusian) suggestion of Russell and Russell,[410] that the regulation of human populations by voluntary birth-control would be the most important first step towards eliminating human violence.

Population growth

As suggested by Allen W. Johnson and Timothy Earle, there can be no population growth beyond a certain limit without technological changes permitting more food to be provided per given unit of land. Population and technology have a feedback relationship: population growth provides the push, technological change the pull. But it is fundamentally population growth (or its concomitant population pressure) that propels the evolution of the economy.[411]

Humans must *eat* to survive, so an increase in the size of the population will mean an increase in its food requirement. Due to the presence of a surplus of food and breeding sites thanks to technological development, we humans, as would other animals in a similar situation, and as is suggested by the pioneering principle and the second formulation of the principle of population, tend to have more than a replacement number of offspring. And as suggested by the reaction principle, the inclination of the members of *any* species is naturally to react to their immediate situation. Though our *reason* may tell us that an alternative mode of action is appropriate, in the main we follow our instincts, including our social instincts. And, again thanks to technological innovation and the surplus it provides, since at least some of the extra children we produce are not eliminated by internal or external checks, the result is the constant growth of the human population.

While the populations of modern hunter-gatherers have apparently not had a tendency to grow, according to Cohen, as intimated earlier, this phenomenon is anomalous. Thus "the concept of carrying capacity as a fixed ceiling to which population responds, although applicable to specific populations under particular conditions, has little general validity for human history."[412]

As has been emphasised by Virginia Abernethy, throughout human history periods of surplus have as a matter of fact been followed by periods of population growth. As noted at the beginning of this chapter, such seemingly minor innovations as the adjustable wrench can have the ultimate effect of providing or increasing a surplus. The increase in the amount of resources that can be extracted from the environment is then taken as permanent, and what Abernethy terms a 'euphoria effect' (or what we might term a 'pioneering effect') takes hold, leading people to have larger families.[413] Without such a surplus, population growth would be impossible. Thus human adaptability, both with regard to our ability to find technological solutions in situations of need, and to reproduce in situations of crowding, together with the surplus that technology has provided, leads to population growth. All of these factors are necessary to the normal turning of the vicious circle.

In terms of systems, technological development undermines the homeostasis of the human species; where there is no technological development, and resources are being used sustainably, homeostasis tends to assert or reassert itself. In terms of our thermostat analogy, the constant presence of a surplus of food and the diminution in the area necessary to raise a family lead to higher settings at which the thermostat controlling population growth clicks in.

According to the VCP, technology's role in the growth of the human population is central, such that one may say that without technological development its size would be minuscule as compared to what it is today. And technological development itself, together with the existence of resources to which it can be applied, constitutes the most important aspect of the vicious circle.

Closed vs. open populations

In *closed* populations of K-selected species, i.e. populations in which there is no immigration or emigration (such as those of modern hunter-gatherers and virtually all non-human social mammals), population homeostasis is maintained by internal checks. A population may or may not be closed for a number of reasons, including geographical location, social mores and politics.

The operation of population checks in *open* populations (such as those of Palaeolithic hunter-gatherers and today's various human populations) is more difficult, however. In human groups that allow immigration and emigration, it is harder to establish the social cohesion necessary for the maintenance of cultural traditions checking population growth. The situation is similar to that of Wynne-Edwards' red grouse in unstable or transitory environments, where their internal population-controlling mechanism is destabilised, resulting in overpopulation. In such populations, it is also possible for people to acquire knowledge about innovations made elsewhere, as well as to emigrate in the case of overpopulation. An open population also makes *trade* possible, without which the resources for technological development are more limited.

The fact that in a closed population excess members have nowhere to emigrate means that if there were to be such an excess, those in it must die prematurely – but this being the case inclines the society not to produce too many people in the first place, so that ecological disequilibrium due to overpopulation is avoided. Thus the likelihood of closed societies' attaining or maintaining equilibrium is greater than that of open societies – the smaller the society, the greater the likelihood. In such a society technological development may come to a standstill, and the population tend to remain at a sustainable size – as happened in the case of various tribes of modern hunter-gatherers.[414]

Population pressure and crowding

Population pressure occurs when the size of a population is too great relative to the carrying capacity of its habitat. According to the VCP, increasing population pressure is endemic to *Homo sapiens'* development.

In some cases population pressure, while it exists, may not be *experienced*, due e.g. to the presence of a surplus of non-renewables. When experienced, population pressure is manifest in *stress* – which may be psychological and/or somatic – due to a diminution in consumables and/or land (territory). Stress due to a diminution in the amount of land is *crowding*; and due to a diminution in consumables takes the form e.g. of malnutrition. Consumables can be considered in terms of whether they are reserves or resources, *experienced* population pressure existing only in regard to reserves. Non-experienced population pressure is related to *resources* – and of course both sorts of pressure can be manifest at the same time. All non-experienced population pressure eventually becomes experienced, unless a renewable alternative is found in the meantime. The use of non-renewable resources to relieve experienced population pressure exacerbates the problem humanity faces as regards non-experienced population pressure and our total resources. Both sorts of population pressure result from per capita resource diminution due to population growth and/or total resource diminution.

When it comes to population pressure and violence, one is inclined to say that if a certain level and form of mortality is well ensconced in the culture of the society – i.e. if the size of the population has been regulated through self-imposed mortality over a long period of time (cf. Darwin's bees), then population pressure has been avoided. In this case continual warfare, rather than being a sign of population pressure, could be seen as a means of eliminating or lessening it – though it could be both. If, on the other hand, there is a sudden *increase* in mortality through e.g. infanticide or war, or external checks such as starvation or disease, we should say that this is a sign of population pressure.

Migration and centralisation

Migration occurs in the populations of virtually all species of plants and animals. As suggested by the principle of population, and more particularly by the pioneering principle, if new territory becomes available adjacent to that already occupied, plants will tend to move into it through the spreading of their seeds or roots, and animals will do the same by themselves moving, in both cases increasing the likelihood of the population's having reproducing offspring. This also applies to humans.

Population pressure functions so as to strengthen the pioneering phenomenon. This was the case both when fire was first controlled and people began moving into colder environments, and when the land bridge opened to America and people crossed it in pursuit of prey. And population pressure can also play a role in the migration of animals other than humans.

In the case of humans in particular, the reasons for migrating become more complex due to the operation of the VCP. Thus human migration can be to places where changed employment of technology has led to an increase in available resources such that e.g. more people can be fed there than before. In this way we are the only species that has expressed the pioneering principle geographically *inwards*. When reserves or stocks are stored, the places where they are stored will function as magnets to those who want to partake of them, particularly if paying jobs dealing with them can be found. The result is a *centralisation* not only of stocks but of people, with urbanisation being the result if the population is sufficiently large. Thus, as expressed by Ellul in this context: "The idea of effecting decentralization while maintaining technical progress is purely utopian."[415] This may be seen as being partly due to the fact that the cost of distributing the stored consumables – a cost borne by the consumer – increases with distance. (It may be noted that the centralisation of people works against the natural inclination of humans to associate only with their, from birth, 'significant others,' as in crowded urban settings they have mainly to deal with strangers.) Quite generally, any amassing of social power at one point will tend to attract people, either competitors, beggars or something in between. The greater the store a group can obtain, the greater will be the centralisation of power and institutions dependent upon power (as all are) in that group.

The operation of the VCP can also lead to migration to or expansion into places where a particular form of technology has not yet been implemented, with the migrants intending to implement it upon arrival, as in the case of colonisation. Another form of migration incumbent upon the turning of the vicious circle is migration due to weaponed conflict, most notably when those not engaged in the conflict flee the area where it is taking place.

Increasing complexity

Of technology

As mentioned earlier, it is natural in the case of biological evolution that more complex organisms evolve from simpler, such that organisms as complex as humans might result, and that the biosphere itself constantly become more complex until some biophysical limit is reached at which increased complexity no longer provides survival benefits. In keeping with Darwin as cited in Chapter 1, this increase in complexity results from more-complex species being able to occupy niches 'between' those of already existing species, or to usurp the niches of those species, which suggests their being able, or better able, to acquire the relevant resources. Following Wilkinson, we can say that substituting one resource base for another implies changing from one niche to another.[416] Such a resource base includes the species' vital resources – both food and breeding sites.

There is also increased energy use in the case of evolution, generally the more complex the species, the higher it is on the food chain, i.e. the more solar energy it requires relative to body weight. In systems terms, as expressed by Boulding, evolution moves the world towards less probable and more complicated arrangements in both society and the biological world. Order is created within each of them at the cost of creating a greater quantity of disorder without.[417] Each more complex species (higher on the food chain) that is introduced into the biosphere means an increase in the conversion of usable solar energy per unit biomass into unusable energy such as heat.

As regards social evolution or development, the more-complex technologies can, like successful new species, acquire resources unavailable to simpler technologies. Also, social change may be seen as resulting from the 'mutation' of earlier technology, just as new species come into existence through the mutation of earlier species. The simplest technology capable of acquiring particular resources is usually developed first, as it is normally easiest (most natural) both to create and to use. More complex technology – or more complex *tools* – may be employed before simpler however (e.g. the bow and arrow was widely used before the hoe), if their use provides at least the same quantity of consumables while requiring less work or energy.

However, just as there is a limit to the complexity of the biosphere, there is also a limit to the complexity of technology. In the case of technology, its limit depends, among other things, on the nature of physical reality and its potential for 'moulding.' In this regard the quantity of non-renewable resources amenable to technological development is constantly dwindling and will eventually disappear. Our renewable resources are also constantly dwindling through use. Even if they were not, however, in that they themselves are limited there must be a limit to what technology, no matter how sophisticated, can obtain from them (particularly given that increased sophistication suggests increased energy use).

So, like the biosphere, the technosphere has a tendency to develop towards greater complexity as long as there exist resources that can be employed in this development, and resources to which it can be applied, plus a sink to receive waste. The force behind this process is the human drive to meet experienced needs, needs which are constantly present largely due, in particular, to the growth of the human population that technological development has itself made possible, and due more generally to the turning of the vicious circle.

Of society: specialisation, or the division of labour

The human population as a whole, like the populations of other species and as is suggested by the reaction principle, is karyotypically adapted to the immediate exploitation of its available resources which, given its size and the nature of human intelligence, has resulted in specialisation and order. This order increases with increases in energy use, population size, and the specialisation required for handling the new technology.

Specialisation and the division of labour are the same phenomenon seen from different points of view. Division of labour occurs *in a group* through the specialisation of *individuals*. (According to John Ruskin, the division of labour was misnamed. It was not the labour that was divided but the men, "into mere segments of men – broken into small fragments and crumbs of life." Men were now condemned to forms of labour that made them 'less than men' in their own eyes.)[418]

Specialisation and increased societal complexity were originally the result of the increasing effort required for a society to obtain the resources requisite for survival; and it may be that people with particular skills which most other people lack come to concentrate on employing those skills. Note that while this phenomenon implies an increase in the complexity of society, it need not mean an increase in the complexity of the tasks performed by individuals, but quite the reverse. Thus the division of labour between the making of weapons and using them to hunt, while it increases the complexity of society, makes individuals' tasks in the food quest simpler. In fact specialisation, or the division of labour, resultant upon the turning of the vicious circle has meant an increase in the simplicity, and thus monotony, of the lives of most individuals.

Specialisation in human society is necessary not only for the design of new technology, but also for the acquisition of the resources needed to construct and fuel it. Unlike in the case of specialisation through biological evolution however, this specialisation is not directly the result of karyotypic change, nor even of genetic change – though there may well be interaction between human specialisation and either of them. What it involves rather is *cultural* change, i.e. learning and the acquisition of skills. In this regard, like culture itself, it is uniquely human.

People's having to perform ever more specialised tasks of course means a general movement away from individual self-sufficiency and flexibility as regards the filling of needs, and towards greater dependence on society. And as regards society as a whole, it means a constantly increasing dependence on both technology and its development. Included in this is the necessity for individuals to devote years of their lives to learning a trade, during which time they have to be housed and fed by others.

Territory, property and commerce

Where individual or family territory and its contents are *owned* by individual persons, group territory may be *ruled over* by individuals, but may be considered to be owned – i.e. be the property of – the group. As Daly recognises, both individual and collective property-holding are manifestations amongst humans as well as in the animal kingdom of the territorial instinct.[419] The nature of the ownership becomes more attenuated moving from individual to family to group. In the event that the property can also be *traded* it constitutes *capital*.

It has earlier been suggested that access to food and territory constitute the most fundamental population checks for all territorial species. Food is required for the survival of the individual; territory for the survival of offspring. Following this distinction further, we can say that what consists solely of food for the lower vertebrates, in the case of humans consists not only of food but more generally of *consumables*. (Stocks, reserves and resources are all consumables – and potentially capital – in various states of refinement.) In other words, where other animals consume only food, as noted in discussing the food chain, we 'consume' many other things as well, such as manufactured products; that is, we convert the low entropy in many other things than just food into high entropy. Consumables have a *use value*; if they have an *exchange value* as well, they are also *commodities*. *Debt*, including in the form of *money*, has *only* an exchange value. (Note the general transition from use value to exchange value as the vicious circle turns.) Consumables and commodities may be the *property* of *individuals*, which suggests that their ownership is primarily an expression of the acquisitive aspect of the survival instinct.

The behavioural manifestation of ownership of places and objects is very highly developed in humans, but this predilection is not peculiar to modern man. It is almost universally present in terrestrial vertebrates, on either a permanent or seasonal basis.[420] Thus, as pointed out by Darwin, a dog's bone, a monkey's stone or a bird's nest might also be considered their property.[421] And *trading* has its basis in the fact that consumables can be *stored*, the trading of consumables, i.e. making commodities of them, constituting an *economic system*.

Now, for humans, and potentially for other animals as well, social power consists in an individual's ability to control others, and may be manifest among other ways in his owning or ruling over territory. Power over other people can take the form of either owning them or of otherwise controlling them, both of which imply a controlling of their access to vital resources, as will be returned to in Chapter 5. As implied in Chapter 1, economic power lies in being able to control other people through owning individual territory or what it produces, while political power lies in being able to control other people through military strength.

Social stratification and class societies: the powerful vs. the weak

Social stratification is a manifestation of Darwinian *intraspecies* survival of the fittest, and exists in the populations of virtually all social animals, which includes most of those that are medium to large size. The basis of social stratification lies in inter-organismic relations of two types, both stemming from the sexual instincts: the power of the male over the female, and the power of the male who rules over territory (or owns property) over the one who does not.

As regards social stratification generally, it would appear that the territorial dominance of males over other males plays a greater role than does the dominance of males over females; and this form of dominance behaviour increases with the complexity of the species. In the case of modern humans, it increases with the complexity of *society*. Thus society becomes stratified such that the application of new technology can be directed by the powerful and the work performed by the weak, while the status of the female generally rises or falls with that of her mate.

In terms of our division among the survival, sexual and social instincts, class societies develop on the basis of social instincts, strongly influenced by the sexual instincts. They concern the *group*. Furthermore they are more particularly *cultural* (variable, learned) rather than invariable.

The larger and more complex the society, the greater the number of levels in the hierarchy and the more complicated their interrelations. Moreover, since the time of the horticultural revolution it has virtually always been the case that there has not been sufficient breeding space for the weakest of the weak. Recognising this, we obtain three classes of people: the small class of the powerful, the large class of the weak who can breed, and the variable class of the weak who cannot.

Further classes can also exist in various societies, some of which are hereditary, such as in the case of caste systems favouring the powerful. As regards classes, we note in particular the existence of the *middle class* in Western societies – stronger (richer) than the weak, and weaker (poorer) than the powerful.

Unequal distribution of the surplus

The existence of class societies is also particular to humans, and is made possible by technological development and the redistribution of resources that it gives rise to; that is, it is dependent on the turning of the vicious circle. While the amount of resources going to the weak is normally only sufficient to allow them to raise children, that which remains – including virtually the whole of the surplus – goes to the powerful, a small portion of this going to the middle class, if such a class exists. The higher one is in the hierarchy, the greater the amount of resources at one's disposal.

Note that the mere increase in a surplus, even if evenly distributed, does not in itself guarantee an improvement in standard of living. It must be possible to *use* that surplus in a way that such an improvement results.

Property and power

Political and economic power, the possession of the one reinforcing that of the other, each consists in being able to control the behaviour of other people. Both sorts of power are thus forms of *social control*. Thanks to such control, the politically powerful individual *rules over* group territory, i.e. he *sets the rules* for those living in the territory. Ruling over territory does not mean owning it – for example the ruler cannot sell it. One result of the foregoing is that, given our social instincts, political power can be used to have others defend or increase the territory one rules over. As is in keeping with the nature of group territoriality amongst primates and early humans (*social* animals), a person's political power does not rest in his personal ownership of land (territory). It rests rather in his being the leader of a *group* (nation) that occupies land (a state), and which he thereby controls due to his position as leader (access to soldiers and arms). (Of course if the form of leadership is tyrannical, the leader as much as owns the group's land – if not the people themselves – in which case it is his property.) Political power can in this way be seen as ultimately stemming from the necessity for social species to ensure the existence of food and breeding sites through their groups' possessing territories.

Where political power concerns group territory, *economic* power concerns individual territory, and consists in individuals' *ownership* of property in the context of an economic system, i.e. consists in their ownership of *capital*. It is through the ownership of capital in the form of vital resources, reserves and/or stocks that the economically powerful are able to control others. The seeking of economic power stems ultimately from the necessity of the individuals of all species to acquire energy in the form of food in order to counteract entropy.

As regards the drive to attain political power, it is natural in all social species for males to be inclined to attempt to lead their group and take prime responsibility for the group's territory. This drive can be seen to be based on our social instincts as evolved from our sexual (fighting) instincts. The drive to obtain economic power, on the other hand, is a more recent development, and rather than be based on our sexual and social instincts, is based on our survival and sexual instincts. Here the survival instincts play a greater role, for what economics basically concerns is not territory *per se*, but the products of the territory: originally food; later capital. It concerns what is needed in order to survive, not reproduce. Nevertheless it is part of the male hominid's role to provide food (meat) and living space for his family, and so the sexual instincts are also involved. In any case, the drive to obtain economic power hasn't to do with group territory, but with individual territory. Individual human territoriality is further removed from the territoriality of other animal species in that, with the turning of the vicious circle, it has constantly become more abstract, the most important development in this regard being capital's increasingly taking over the role of individual territory from land.

Here we must turn to a human trait, touched upon earlier, which was genetically strengthened through the turning of the vicious circle, namely a latent tendency to *hoard*. Though hunter-gatherers began to cache dried meat towards the end of the Palaeolithic, actual hoarding on the part of individuals began with sedentism, when it was no longer necessary to be able to carry all one's belongings. And once hoarding became possible it also became a potential route to power, i.e. a means of controlling other people,[422] as intimated above.

We can say that the fact that there exist people with economic and/or political power is a result of the turning of the vicious circle. In both cases the circle's technological innovation of language has been necessary to the having of such power, since without language commands cannot be given; and in the case of political (military) power, the development of weapons-technology incumbent on the turning of the circle has made possible the existence of power of enormous physical magnitude. In the case of economic power it is rather the vicious circle's provision of a surplus of consumables that has been key, though here too the hugeness of that surplus to date has given tremendous power to capitalists in relation to society's weak.

Both political and economic power provide *social status*, though social status can also be derived e.g. from fame. Politicians and businessmen have power and status; movie stars *qua* movie stars have only status. The middle class phenomenon of 'keeping up with the Joneses'[423] – already taken up by Mill as quoted in Chapter 2 – has to do with social status, but not power. Thus ordinary people's ownership of property beyond what is required for comfortable survival is not acquired for the sake of having the property (e.g. in horticultural times a man and his family could only eat so much of the grain they owned), but for the sake of maintaining or improving their status.

Social position is determined by political power, economic power and/or social status. Both social status and social position are partly determined by the way an individual is *perceived*; social *power*, on the other hand, is determined by what the individual can actually *do*.

Kings with vast territories will war with other kings in order to increase the importance of their nations and thereby improve their own social position, the ultimate reward being to rule the world – even if it's a world of destroyed cities and starving people. And virtually any means available will be used to do this. Where, for example, a capitalist will never take on more workers than he has to, a general will never spare his soldiers' lives if doing so would interfere with his winning a battle. Capitalists and generals have the power they do *because* they behave in this way. If they did not, they would be ousted by others who did.

Given the nature of both political and economic power, the more power a person has, the more he can get. When it comes to competition over power, it's an advantage to be more powerful. In this regard, as noted by Ellul, "Competition is thus an incitement to such technical progress as will bring victory over the competitor. This means that competition tends to destroy liberalism."[424] Quite generally, an individual's or group's having power, in whatever form, is part of a positive feedback loop of the increasing kind which generates more power for the individual or group.

Social stratification as a population check

The function of ranking (resulting from the operation of the sexual instincts) as a population check among higher animals in general is described by Wynne-Edwards:

> The hierarchy [ranking] … produces the same kind of result as a territorial system in that it admits a limited quota of individuals to share the food resources and excludes the extras. [I]t can operate in exactly the same way with respect to reproduction.[425]

In the case of humans, however, where the stratification is *social*, it works in quite the opposite direction, supporting population growth due to the needs of leaders for large populations.

Social Darwinism

Note that the above discussion does not support Social Darwinism, taking Social Darwinism to be the idea that the dominance of the powerful over the weak is to the advantage of the species. The reason that Social Darwinism is not supported is again due to the operation of the VCP. It is the activities of the socially 'more fit' (the powerful) to a greater extent than those who are socially 'less fit' (the weak) that are paramount in turning the vicious circle and thereby undermining the preconditions for the survival of the species.

Furthermore, given their constant desire to acquire more property, and the subordination of a large population being an aid to doing so in times of surplus, throughout history the powerful have tended to support the reproduction of the weak to a greater extent than reproduction amongst themselves. This means that their gene lines are not as strongly represented in later populations, which distinguishes them from the socially powerful in hunter-gatherer societies. As expressed by Lorenz: "It is fortunate that the accumulation of riches and power does not necessarily lead to large families – rather the opposite – or else the future of mankind would look even darker than it does."[426] On the other hand, however, were the 'more fit' suddenly to be eliminated, it may be expected that the 'less fit,' since they are driven by the same instincts and operate in the same milieu, would soon come to act in the same or a similar way, à la Orwell's *Animal Farm*. (We might term this the *Animal Farm* syndrome.)

Lower quality of life for the weak

According to the VCP, the nature of human males' territoriality, given a sufficiently large population together with the appropriate means, should lead to a disproportionate part of that population living lives only slightly above the level of survival. Meanwhile the small minority with power reap the benefits of the surplus that the turning of the vicious circle has provided, as is the case in the world today.

Increased work

As Wilkinson points out, under the impact of ecological problems the productive workload tends to grow throughout society.[427] The need to employ new technology to acquire a particular resource typically means that more energy is required to obtain the resource, which in turn means more work for the weak. And during periods when there is no surplus, it is the weak who suffer first and most, often with massive death.[428] Furthermore, the population growth in the labour force promoted by the powerful widens the gulf between the powerful and the weak through, among other things, its reducing the value of a person's labour.[429]

Labour, as a concept, may be compared to the broader concept of *work*. To labour is to do physical work for someone of a higher economic class, the products of one's labour belonging to the person one is working for; thus we should say that the middle and upper classes don't perform labour. Work, on the other hand, may be simply defined as the creation or use of technology, or the provision of the preconditions for its use.

Poorer diet

That the weak have less leisure and do more work means a lowering in the quality of their lives, particularly for those drawn into the extraction of resources or the production of the goods or services (consumables) resulting from the implementation of the new technology. The extra labour on the part of the weak, together with their much poorer diet, is manifest in increased mortality. Generally, however, as the use of the new technology becomes an integral part of society, this effect tends to lessen – until a peak is reached and returns begin to decline.[430]

Disease and other causes of mortality

Even more important than heavy workloads and poor diets, however, though abetted by both, is the influence of disease. It is an effect of the principle of population that the populations of all species tend to grow to be as large as possible while not over-exploiting their habitats. This also applies to micro-organisms.

Unlike in the case of other animals, throughout history humans' infectious diseases have accounted for the vast majority of mortalities prior to male and female menopause. As regards other animals, in the case of wild birds, for example, it seems unlikely that disease is an important factor in regulating their numbers; and in the case of North American deer, the influence of disease seems to be of secondary importance.[431] Of course all populations of all wild animals are generally healthy thanks to natural selection.

As Wilkinson says, disease in general tends to act as a homeostatic population regulator, taking a higher toll from human populations living in bad conditions, particularly if people are crowded or their resistance is weakened by malnutrition.[432] That this scourge of humankind is the result of the turning of the vicious circle can be seen from the fact that the vast majority of the infectious diseases from which humans suffer have arisen from the technological innovation of animal domestication, and their spread has been incumbent upon the constantly increasing size of the human population made possible by technological development.

Stress, aggression and conflict

As treated in Chapter 1, intraspecific territorial conflict amongst animals was seen to have two main effects: the survival and reproduction of the fittest organisms, and, by allowing only a portion of the population to reproduce, the checking of its size such that it doesn't tend to outgrow its resource base. These 'functions' of conflict, existing in virtually all animal species, also constitute the basis of conflict amongst humans (men). Note that this means that human extra-familial conflict is related to the relative scarcity of females and/or breeding sites, and thus to the procreation of the species, and not to a relative scarcity of food, i.e. the survival of the individual – though food, as a form of property, can nevertheless play a role in inciting conflict.

According to Russell and Russell, violence in human societies is not the result of an innate propensity towards aggression irrespective of conditions, but a response to *stress*. In an overly dense population cooperation and parental behaviour are replaced by competition, dominance and violence, violence being part of a complex of responses evolved to achieve a drastic reduction in the size of a population that is in danger of outgrowing its resources. Recurrent population crises produce what Russell and Russell call a *stress culture*, consisting of behavioural aberrations transmitted through the generations,[433] which include increased violence and greater emphasis on maintaining or improving one's place in the pecking order rather than sharing; in other words, a stress culture leads to the social instincts coming increasingly to be usurped by the survival and sexual instincts, and morality thereby coming to be replaced by immorality.

The difference is to be noted between aggressive behaviour in a population that is in equilibrium with its surroundings, and aggressive behaviour in situations of crowding. Aggressive behaviour functions in virtually all animal species as an instinctual population check; but it can vary in intensity, becoming more strongly manifest in situations of crowding. Thus aggressive behaviour is favoured by natural selection, since its manifestation as a population check enhances the likelihood of the survival of the species in question.[434]

There are essentially three kinds of conflict amongst humans: conflict between the weak and the powerful, conflict amongst the powerful, and conflict amongst the weak. The two classic types are first the powerful against the powerful, which can take the form of commercial competition, but which, when weapons are employed, can become *war*; and second, the weak with breeding sites against the powerful, which can be manifest economically e.g. in trade unionism or, when weapons are involved, in *social revolt*. Conflict between the weak and the powerful can also involve *terrorism* on the part of the weak. Conflict amongst the weak, on the other hand, consists in competition with one's neighbours over jobs and property, or fighting strangers in war.

The greater the power gap between the strong and the weak, and the greater the population pressure, the greater the likelihood of all three kinds of conflict. Population pressure clearly increases conflict between the weak and the powerful, and is easily understood to play a role in conflict amongst the weak in the event that the property allowed them by the powerful is too small for healthy procreation, or is perceived as such. In the case of conflict amongst the powerful, the greater a man's power, the greater the likelihood that he will attack somebody else to further increase it, since already being powerful he may believe his chances of winning are good. But the experience of population pressure in his own nation will give him an added incentive to act. Also, having an advantage in weaponry almost always leads those seeking military power to attack weaker parties.

The population pressure that exacerbates conflict may be caused e.g. by a diminishing surplus, as may be manifest in economic decline. In the case of war, even if the potentially warring group lacks a surplus, the powerful can often still scrape together what is required to outfit and feed an army by taking (more) resources from the weak, e.g. by increasing taxes – of course in such a case the army would not be as large or strong. Decreasing per capita property in a population leads individuals to attempt to avoid losing their own property and/or to obtain property from others, the result being conflict – a conflict in which the powerful will have an advantage over the weak. More generally, however, all *increases* in mortal conflict amongst humans may be seen as being the result of population pressure combined with the availability of weapons, and will tend to check population size – whether effectively or not. Decreases in mortal conflict, on the other hand, may be seen as resulting from the greater order required for political leaders to have soldiers, and capitalists to have cheap labour.

According to Abernethy, though the underlying cause of much domestic and international conflict is rapid population growth, violent upheavals are often reported as class, ethnic and religious conflicts because regions and societies fracture along these lines, while reports of political and social stresses and associated individual pathology often omit mention of the causal role of rapid population growth.[435] To this it may be added that these different conflicting groups stem from what were originally different tribes.

Note however that, as is implied above, population growth, economic decline and so on, while increasing the likelihood or extent of human conflict, are not *necessary* to it; such conflict should be expected even in the case of a population in equilibrium with its surroundings, it being an expression of our territorial instinct, which, other things being equal, should check the size of the population *before* overpopulation develops.

Laws; crime

According to Jean-Jacques Rousseau, whose view here is partly in keeping with the VCP, once men begin to claim possessions, the inequality of their talents and skills leads to an inequality of fortunes (the rich get richer, and the poor ...). Wealth enables some men to enslave others; and the very idea of possession excites men's passions and provokes conflict. This, according to Rousseau, leads in turn to a demand for a system of *law* to impose order and tranquillity (countering entropy). The rich (note: not primarily the militarily powerful, who have *weapons* to defend their property) especially voice this demand, for while the state of violence threatens everyone's life, it is worse for the rich because it threatens their possessions as well. Hence, in keeping with Hobbes, the agreement among men to live under a political system.

> Such was, or may have been, the origin of civil society and laws,
> which gave new fetters to the poor, and new powers to the rich;
> which destroyed natural liberty for ever, fixed for all time the law of
> property and inequality, transformed shrewd usurpation into settled
> right, and, to benefit a few ambitious persons, subjected the whole of
> the human race thenceforth to labour, servitude and wretchedness.[436]

Further according to Rousseau, the effect of the establishment of political societies is to both institutionalise and increase inequalities; and the establishment of such things as property rights and titles of nobility sets the seal of law on inequality.[437]

Note here the relation between laws and the individual/group territory distinction. Laws are maintained *within* the group's territory – they are *intra*tribal or *domestic* – and are fundamentally an economic and not a political concern. (There are no laws against waging war.) They pertain to the property of *individuals*, and serve to strengthen the hierarchy within the community.

A view similar to Rousseau's was expressed already in ancient Greece, e.g. by Plato, where he has Thrasymachus say: "I affirm that the just is nothing other than the advantage of the stronger,"[438] a position Socrates tries to undermine. And in Hesiod's *Theogony* Zeus overcomes his elders and rivals by sheer force, which Hesiod for his part clearly considers to be quite acceptable.

While we here accept these views as adequate characterisations of the nature of social organisation, I suggest that in a complex society laws might very well be to the advantage of the weak as well, at least in the short term. Since it is to the benefit of the powerful that there exist numerous weak to labour and fight for them, laws protecting the position of the powerful will also see to it that the position of the weak, including their ability to raise offspring, is also protected. Thus there are two sides to the coin when it comes to the role of laws in society. (Cf. *Pax Romana*.)

Also when it comes to social organisation, in a small society of the hunter-gatherer-type moral action, i.e. action benefiting the group at the expense of the individual and/or his or her family, can be directly reinforced, due to everyone's being intimately acquainted with everyone else. But in larger societies, where the members do not know most of their fellows, the shame, guilt and pride felt vis-à-vis others becomes greatly lessened. The feelings one has for the other members of society who are strangers moves in the direction of the feelings one has towards strangers in hunter-gatherer societies, where they belong to other tribes, which places them immediately under suspicion. And this feeling is strengthened in the event that the stranger can be *seen* to be of another culture. Thus laws enforced by paid police provide the function in society of maintaining peace where there otherwise would be a return to tribal groupings and armed fighting among them. Society would still be organised, but into smaller states with or without an overordinate political power. It is to be noted that *economics* requires the existence of such a power (cf. Hobbes).

Nevertheless, the basic function of the laws is to maintain the status quo, and most particularly the positions of the powerful. Thus in a broader perspective it may well be the case that the laws themselves are unjust. As Jeffrey Reiman says, a criminal justice system is a means to protect the social order, and it can be no more just than the order it protects. A law against theft may be enforced with an even and just hand. But if it protects an unjust distribution of property, the result is *injustice evenly enforced*.[439]

Revolt, terrorism and revolution

Given humans' territorial instincts, it is natural that there be conflict not only between classes, but within them. As regards conflict between classes, the almost total lack of power on the part of the weakest of the weak means that their access to significant weapons will be minimal, as will therefore the military threat they pose to the higher class or classes.

The basis of the conflict between the poor and the rich is the rich's taking more from the poor than the poor can tolerate, i.e. the creation of a situation amongst them in which a scarcity of vital resources is the norm ("Let them eat cake"). And this, as noted above, is more likely to occur in times of economic recession. In such a situation, as emphasised by Joseph Tainter, the marginal returns to energy expended will be constantly decreasing, and population pressure will be more severely felt. Thus, when the marginal cost of participating in a complex society becomes too high, productive units across the economic spectrum passively or actively increase their resistance to the demands of the hierarchy, or overtly attempt to break away.[440]

Through the weak's experience of increasing desperation, what was originally a relatively benign form of conflict can thus escalate into one of rebellion. As regards social revolt, as long as the weak experience their lives as worth living, in that they have food, shelter, and can raise a family, they are not likely to rebel. When families of the weak are threatened, however, as in the case of famine, their dissatisfaction with their lot increases. The size of the threat this constitutes to the powerful minority is not dependent so much on the numbers of the weak as on their access to meaningful weapons, which also plays a role in whether they actually go so far as to rebel. When the weapons at the disposal of the rebels are less effective, then of course their number and the number of people who bear them become more relevant. In keeping with what has been noted by Tainter, the willingness of strong groups other than the one being threatened to provide the weak with arms is often vital in this context. Such weapons will not be provided by other strong parties for humanitarian or egalitarian reasons, however, but only if it is thought that their use will improve their own political position – or that their sale will provide a profit.

Terrorism can be seen as an extension of the above. Since in modern times the weapons available to the weak are not such that they could win an all-out military conflict with the powerful – with or without the support of other powerful groups – they can still inflict notable or even grievous harm by making isolated attacks on particular places. The likelihood of this sort of revolt increases as the vicious circle turns, its development being largely due to improvements in military technology. It has meant that highly lethal weapons can be operated by but a few, unlike in the days of the Bastille. Note, however, that each terrorist group has a leader who, in having control of the group, is himself powerful. Thus in some cases terrorism may be seen more as a conflict amongst the powerful, the leader of the terrorists being a sort of Robin Hood.

Another reason terrorism becomes more likely as the vicious circle turns is that, as intimated above, constant technological development leads to a centralisation of services, such as defence, business and energy transformation. This makes the society as a whole, and the powerful who own these services, more vulnerable to attack, since one well-placed blow could paralyse them; and terrorists could take advantage of this fact. Also, at the same time as the wealth-gap between the rich and the poor constantly increases, the population of the poor grows both in absolute terms and relative to that of the rich, which means a constant reduction in the relative size of the territory available to them.

Revolt or rebellion is an attempt on the part of the weak to wrest property (territory) from the powerful, with the use of whatever military technology they can get their hands on. If the weapons employed by the weak are on the whole superior to those of the powerful, the revolt may end in a *revolution*. While revolution may lead to a more equitable distribution of power and wealth, the effect will only be temporary, since the sexual territorial instinct increasing the distance between the powerful and the weak is always operative. Thus some of the weak who benefit from the revolution eventually become as powerful as those the revolution deposed, while the power of others does not increase at all – the *Animal Farm* syndrome – and thus we have the 'necessity' of continual revolution.

As suggested by Georgescu-Roegen,[441] Marx and Engels admitted that all social movements up until the time of their writing (1848) had been accomplished by minorities for the benefit of minorities. They, of course, believed and preached that the Communist revolution would be an exception to this rule. By now, we know that it is not: a new privileged class crystallises under every communist regime. As Milovan Djilas says regarding the Communist system: "The new class may be said to be made up of those who have special privileges and economic preference because of the administrative monopoly they hold."[442] Georgescu-Roegen adds:

> For a few glaring examples from some countries leaning heavily
> toward socialism: in Indonesia scores of luxurious villas have been
> built in the most attractive spots for the use of the *president*, who
> cannot visit them all during one year; in Bombay, scarce though the
> medical resources are all over India, the best-equipped clinic has
> been earmarked by … law for the *exclusive* use of the families of the
> members of the local government and legislature.[443]

[T]here is absolutely nothing in the constitution of the average man that could make him not wish to be the king. And the question is why he should be a rickshaw man and not the king.

[I]n the future as in the past, human society will pass from the control of one elite to another and … each elite will have to influence not the genotypes of people, but their beliefs, with the aid of a seemingly different, yet basically homologous, mythology.

And he continues, that only in the late twilight of the human species, when human society will very likely disintegrate into small packs of people, will the social factors which produce the circulation of elites fade away too. Class conflict, therefore, will not be choked forever if one of its phases – say, that where the captains of industry, commerce and banking claim their income in the name of private property – is dissolved. Nor is there any reason to justify the belief that social and political evolution will come to an end with the next system, whatever that system may be.

War

War, like virtually any form of male–male violence in human society, stems from the instinctual basis of human territoriality.[444] To this it may be added however that war is in particular a manifestation of *group* territoriality, i.e. stems from the *social* instincts. As succinctly put by Keeley, war is a method, derived directly from hunting, for getting from one group what another group lacks and cannot peacefully obtain.[445] That the basis of war lies in group territoriality means that it is fostered not only by the powerful few – the group leaders – but also by the fit weak, whose social instincts include protecting their group. As a result of these social instincts, as described by Diamond, humans have always practised a dual standard of behaviour: strong inhibitions about killing one of 'us,' and a green light to killing one of 'them.'[446] Note too how war itself reinforces group solidarity, as evinced by the big cheer that often goes up on the part of the populace when its nation goes to war.

That war has a karyotypic basis means that the impulses that give rise to it may be modified, but not eradicated.[447] But this leaves open the question as to the *extent* they may be modified. If we follow Russell and Russell in seeing human violence as directly related to crowding-induced stress incumbent on overpopulation (or population pressure; note that the idea that overpopulation leads to war is at least as old as Plato),[448] we could imagine that territorial instinct that gives rise to war might manifest itself in a non-lethal form of conflict, such as when nations compete at sporting events. Such thoughts are highly speculative, however, for the general trend in our species' development has always been towards increasing population pressure; and as regards our past we have little to go on apart from our knowledge of modern hunter-gatherers and the skeletal remains of Palaeolithic peoples (which tell us, among other things, that a lot more killing went on amongst them).

Distinct from social revolt, not only does war involve conflict between different *states* or *nations* (cf. Rousseau: "War, then, is not a relation between men, but between states; in war individuals are enemies wholly by chance, not as men, not even as citizens, but only as soldiers; not as members of their country, but only as its defenders. In a word, a state can have as an enemy only another state, not men."),[449] but it also typically involves conflict amongst the *powerful*, rather than between the powerful and the weak, or the weak and the weak. (Cf. Quincy Wright: War implies a struggle between equals.)[450] According to Schumacher, the wealth of the rich in the modern world depends on making inordinately large demands on limited world resources, and thus puts them on an unavoidable collision course – not primarily with the poor (who are weak and defenceless) but with others who are rich.[451]

Part of what makes a rebellion a rebellion, and a revolution a revolution, is that the weak occupy the same geographical territory as do the powerful against whom they are rebelling. War, on the other hand, typically takes place in situations where one powerful individual (or individuals) rules over a group territory different from that ruled over by another individual, and at the same time believes himself to possess a military strength superior to that of the other. The winning of a war, then, will decrease the territory and thus power of the loser, and increase that of the winner, including his wealth, if the war hasn't been too costly.

The attacks of the Mongols and Turkic peoples against sedentary peoples, while constituting war in being between different states (or nations and states) as well as between powerful men, are nevertheless similar to revolts in being uprisings on the part of the poor against the rich.

Further by way of distinguishing war and revolt or rebellion, it may be said that war involves *armies* in a situation in which the leaders are powerful and the soldiers weak, and in which there must be *provisions* for *campaigns*. Thus the waging of war, as noted earlier, is dependent on the existence of a surplus, to feed, clothe and arm the non-productive soldiers.

Apart from having a larger army, as mentioned earlier, a second factor that will improve a ruler's odds when it comes to winning a war is his army's having weapons superior to those of his opponent's army. We thus see that the territoriality of rulers also spurs innovation in weapon technology. A smaller army can defeat a larger one if it has superior weapons. More important than its particular weapons, however, is their potential total effect, which would normally depend on their number and the existence of soldiers to wield them. But as weapon technology advances and weapons become more destructive, it may be expected that the need for large armies will decline, just as, with the development of machinery, the need for a large labour force in industry should decline. This development however will be to a disadvantage in the case of guerrilla warfare.

Feuding and protowar

Major conflicts between groups of humans that involve killing but not armies may be termed *protowars*, and may be considered to involve *skirmishes* between armed *warriors* of different *tribes* in *protoarmies* the leaders of which are *chiefs*, rather than *battles* between armed *soldiers* of different *nations/states* in *armies* the leaders of which are *generals*. Protowars are typically related to the relieving of *vital* rather than non-vital needs. Where many warriors of a particular tribe are closely genetically related or have known each other all their lives, thereby constituting something like an 'army of lovers,'[452] soldiers are almost always (to begin with) strangers to their fellows. And it may be noted that already in the case of protowar, success is supported by population growth in the community, as is, due to population pressure, the extent to which the group engages in such conflict.

We should also here make explicit the distinction between protowar and *feuding*, i.e. prolonged *intra*societal conflict between families, bands or clans. Feuding, protowar and war all have their phylogenetic basis in the sexual instincts and the fight over females and/or individual territory, the connection being strongest in the case of feuding and weakest in the case of war. Feuding occurs mainly in primitive groups and involves blood-revenge; and, as intimated in Chapter 2, the fighting that sets it off is often over women (adultery, rape, wife-stealing), which may be seen as being at least partly a result of their relative scarcity due to female infanticide.[453]

Armed conflict as a population check

Stanislav Andreski depicts civil disturbance and war as 'alternative releases of population pressure, as they are alternative methods of organising emigration to the hereafter.' Russell and Russell concur, suggesting that war is a response to experienced population pressure in each of the contending countries. According to them, the reason violence exists both at the level of the individual and the whole society is ultimately as a means for reducing the size of the population when confronted with population crises. (And they also agree with Andreski that the limiting of population growth, and a determined attempt to bring the majority of the population of the world out of its present condition of misery, offer the best hope of abolishing war.)[454]

Other things being equal, the use of weapons by humans should reinforce intraspecific conflict as a form of population check, particularly in a situation of experienced population pressure.[455] But other things are not equal. The need of the powerful to have large armies – and in the case of business, a large labour market – together with the continual presence of a surplus, has meant that the human population has to date continued to grow despite armed conflict.

As regards different sorts of conflict functioning as population checks, *armed revolution* constitutes a check which is internal to a society; *terrorism* may or may not do so, but generally takes relatively few lives; and *war*, which is *between* societies, tends to check the sizes of the populations of both. All three, particularly indirectly through their effect on female fertility, are cultural checks to the growth of the total human population.

In Chapter 2 *warfare* was considered as a possible check to the growth of human populations, not only through the direct killing of men on the battlefield, but also through the female infanticide that can be practised by warring societies. (The population-curbing effect of female infanticide is greater than the death of men in battle, since, as taken up earlier, any number of females can be impregnated by the same male.) And a further population check associated with war is the aforementioned starvation and disease that usually accompany it. Normally, the crowded and unsanitary conditions of warfare breed infectious diseases which kill many more people, civilians as well as soldiers, than does actual fighting.[456] And the almost invariable period of starvation in the state that has just lost the conflict further increases mortality amongst women as well as men. According to Moran, warfare is the cultural price that must be paid to keep these mechanisms of population control operative.[457]

Not only does political leaders' drive to acquire power lead to population growth, but that growth itself becomes an excuse for acquiring more power. As expressed by Harold Cox:

> As soon as a population grows big, its leaders say: 'Our people are so numerous we must fight for more space.' As soon as war has taken place, the leaders invert this appeal, and say: 'We must breed more people in preparation for the next war.'

And as Cox further says, human beings would never hesitate to kill one another when, as a result of population pressure, they find that war is the only alternative to starvation.[458]

As regards the function of armed conflict in its capacity to counter the VCP, it should be pointed out that in the broader perspective it actually speeds up its operation. Not only does war, more than anything else, promote technological innovation, but the destruction it wreaks requires taking more from the environment to replace what has been destroyed. As Russell and Russell say regarding human violence generally, it can no longer serve the fundamental function for which it was evolved in animals – the conservation of the natural resources of a species. On the contrary, it does the exact opposite, destroying resources on which the species must depend in the future. This means too that selection for power favours those who exploit nature and discards those who revere it. In keeping with C. G. Darwin's conception, a society that exploits its resources quickly accrues more power than a similarly based society that husbands its resources and protects its environment. The resource-exploitative society may then overpower the more nature-protective society and seize its resources for additional quick exploitation.[459]

Cultural development

As expressed by Torsten Malmberg, human beings can become adapted to almost anything – polluted air, treeless avenues, starless skies, the rat race of overly competitive societies, even life in concentration camps.[460] This adaptation is accompanied by genetic change. The fundamental determinants of human behaviour, on the other hand, have their biological basis in the human karyotype, not in individual genes. Thus human adaptability, whether it be on the part of groups or individuals, takes place within the limits set by our instincts.[461] In fact, since the human tendency to quick adaptation has its basis in our karyotype, it may itself be considered instinctual. It is to be kept in mind that since we modern humans came into existence some 200,000 years ago, our karyotype has remained the same, while our individual genotypes and cultures have changed.

The adaptability of human *groups* – not as pronounced as that of individuals but having a greater influence on the species – is manifest in changes in their *social organisation* (changes of social *system*), particular populations adopting particular cultures ultimately because those cultures in some way tend to support the continuing short-term existence of the population.[462] As expressed by Lorenz, natural selection determines the evolution of cultures in the same manner as it does that of species.[463] Though our species' karyotype remains constant, it allows for genetic change capable of producing a wide spectrum of behaviour. Note that this is not the same as simply allowing a great deal of genetic change; dogs, for example, are capable of more genetic change than we are – and might therefore be more adaptable as a species in the long term. But we are more intelligent, i.e. adaptable in the short term. In having culture humans have *exosomatic* means of adaptation, means which are being called upon to a constantly greater extent as our impact on our surroundings increases.[464]

The most important of cultural changes are those related to the economy, for it is only with a functioning economy that individuals can survive and reproduce and the species thereby continue to exist. As expressed by Wilkinson: "If the demands which basic production imposes on society change, then the rest of the cultural system will have to change to meet them." And, as he continues: "Most fundamental within the adaptive context is a society's technology." Thus in the case of humans we have a situation where the survival of the species is dependent on the adaptability of human culture, where cultural and genetic change are dynamically related.[465] And where on the micro level the *continuity* of human group behaviour stems from the constancy of the human karyotype and the transmission of gene types, on the macro level it stems from human culture, or, more particularly, from the maintenance of *traditions*. As expressed by Georgescu-Roegen, it is the role of tradition to transmit knowledge as well as propensities from one generation to another.[466]

As Wilkinson points out, our capacity to produce culture provides us with a means – much greater than that available to other animals – of exploiting different ecological niches, suggesting that in biological terms cultural change is nothing other than the adoption of a new niche. When people change the source of food and raw materials they depend on, when they find ways of increasing their resistance to diseases and parasites, or develop forms of protection against inclement climatic and geographic conditions, then they may be said to have changed their ecological niche. The adaptive function of culture is its *raison d'être*.[467] Just as living creatures have evolved over time to fill particular environmental niches, so have human societies.

While it is true that culture is a particularly efficient adaptive mechanism, it is nevertheless too slow to react to the changes in the environment it itself produced the previous time it made an adaptive move. Each cultural (economic) adaptation has repercussions necessitating a new adaptation, and this at an accelerating rate. Cultural change generally, like technological development specifically, is an adaptive mechanism that undermines the preconditions for its own functioning.

Luxury goods and leisure; art and architecture; philosophy and science

The increasing complexity of society means that the needs of the powerful also increase in complexity, and come to include not just vital needs but non-vital ones as well. Such needs may be filled by converting vital resources into non-vital, or there may be a 'mining' of non-vital resources directly. Thus given a sufficient surplus of vital resources there may arise the production of *luxury* goods – meeting non-vital and often only imagined needs – over and above the goods necessary for the survival of the population. Such goods may themselves be of a technical nature, as are e.g. the aforementioned mobile phones and pleasure boats, and thereby constitute instances of non-typical technological change, i.e. technological change that does not support the maintenance or growth of the population.

Leisure too, or the potential for leisure, will also be greater the greater one's power. Where in hunter-gatherer societies everyone had ample leisure, with the turning of the vicious circle and the more marked stratification of society, leisure becomes available only to those of the upper strata. This leisure had by the powerful, in combination with the human tendency to innovate, has given rise to the arts and philosophy. And, in combination with constantly increasing technological know-how, it has allowed for their expression and development e.g. in the form of monumental architecture and science. Thus, on the VCP, with its strong biological-ecological orientation, the arts and sciences are an 'emergent property' of the basic dynamics of our species' development – they could be seen as a side-effect.[468] The existence of both is dependent on the existence of a surplus. In the case of modern science, however, as expected already by Francis Bacon, some of the results of the search for knowledge and understanding of the physical world have been channelled back into the productive effort – particularly that of the military – thereby speeding up the course of the vicious circle. (Cf. Ellul: "Science is becoming more and more subordinate to the search for technical application.")

Medicine

From the point of view of the VCP, medicine is a typical technological development which has the potential to improve the life-situation of the existing generation while worsening that of posterity – for example by making it possible for biologically less fit individuals to procreate, thereby weakening the human strain and making humans dependent on medical technology for their existence. (Cf. the state of health of the !Kung and the Hadza.) It is only in recent years however that the development of medicine has meant an increase in what is taken from the environment, through the building of hospitals and the development of complex diagnostic and treatment apparatus. And where originally medicine was available to everyone, as the vicious circle turns and medicine becomes more sophisticated it increasingly becomes available only to the powerful.

Economic growth

The surplus of consumables resulting from the new use of technology may be, and in modern times virtually always is, put on a market, thereby giving rise to trade, or, in the case of ongoing trade, giving rise to an increase in trade, i.e. exchange of capital. The notion of economic growth is often thought to apply to such an increase, but we shall adopt another common line which has been assumed earlier and is here more to the point, namely that of seeing economic growth as consisting in *an increase in the quantity of consumables*. Thus, for example, the invention of the bow and arrow resulted in economic growth in this sense, for its use originally led to an increase in the quantity of meat obtained – meat being a consumable but not a *commodity* in the non-trading economies of hunter-gatherers.

Note that economic growth in the context of trading is dependent on the existence of a surplus that can be traded,[469] and furthermore that economic growth speeds up the diminution of that surplus, the greater the rate of growth, the more quickly the surplus is used up. If the surplus is sufficiently great, and of the right sort, economic growth will not only involve an increase in the production of goods, but an increase in other activities as well, such as those related to distributing the goods. The *social* phenomenon of *economic growth* is, thanks to the principle of the conservation of matter, nothing other than the *physical* phenomenon of *increasing resource depletion*. And here, rather than in our prehistory, do we have the expression of the entrepreneurial spirit: to obtain, by means of business, as much of the present surplus as possible. This spirit will lead to such enterprises as the improvement or building of roads and other transportation systems, and the building of dwellings, all of these things constituting consumables/commodities, and all of them providing capitalists with a profit.

As Schumacher points out, though economic growth as seen in the context of economics, physics, chemistry and technology has no discernible limit, it must run into bottlenecks when viewed from the point of view of the environmental sciences,[470] and thus must do so in reality. Economic growth, like technological development and the turning of the vicious circle more generally, will stop when the surplus on which it is dependent no longer exists, or the waste it produces can no longer be disposed of. Seen as a system, it lacks internal checks to its own expansion. In this way it is similar not only to technology but to the biosphere. But unlike the biosphere, which simply stops growing, at the end of a period of economic growth there tends to be a period of economic collapse, due to the over-exploitation of resources supporting the growth.

The orientation to economics which suggests a positive value in economic growth (but which does not take account of its requiring a surplus) goes back at least to Hobbes' social contract theory. Hobbes' basic orientation involved the application of Greek atomistic thinking and its notion of perpetual motion to social phenomena.[471] Having constructed a system which would purportedly explain humans' motions relative to one another, his aim was then to deduce what kind of government they would have to have to enable them to *maximise* that motion.[472]

And Adam Smith, for his part, suggests in a similar vein that:

> It is in the progressive state, while the society is advancing to
> the further acquisition, rather than when it has acquired its full
> complement of riches, that the condition of the labouring poor, of
> the great body of people, seems to be the happiest and the most
> comfortable. It is hard in the stationary, and miserable in the
> declining state. The progressive state is in reality the cheerful and the
> hearty state to all the different orders of the society. The stationary is
> dull; the declining melancholy.

Of course technological development may mean, and has meant in the past, the creation of a new surplus; but the amount that can be taken from the environment is finite, and growth will end when it can no longer be taken in increasing quantities. Furthermore, the dependence of growth on non-renewables means that its end will be more abrupt than otherwise. For an economic system to be sustainable, it must involve as an integral element only the use of renewable resources that in fact are being renewed. In any case, economic growth cannot continue indefinitely, and, whether or not it involves the use of non-renewables, it directly detracts from a society's becoming sustainable.[473]

Decreased security

Increased centralisation and trade, and the destructive power of weapons, all lead to decreased security. In the case of trade, trade routes are created on which the society becomes dependent, and with economic growth become more numerous and longer, thus opening the society to attack at many points. As pointed out by James Bonar, dependence on other nations for the first necessity of life is a source of political insecurity to the nation so depending; and, though the dependence is mutual, identity of commercial interests seldom prevents interdependent nations from going to war with one another,[474] economic relations in fact *increasing* the chance of war.[475] Security is further decreased both for the individual and for society with the turning of the vicious circle due to increasing specialisation and technological dependence.

Over-exploitation of resources and population overshoot

Not only does population grow when provided with a surplus, but it grows beyond what the surplus – which is itself dwindling faster than otherwise – can support at the same subsistence level.[476] And, as intimated earlier, culture, as an adaptive mechanism, is too slow to react to such changes. Here we see the impact of the overshoot principle, where in a pioneering situation populations expand beyond the carrying capacity of their environment. This overshoot manifests itself, among other ways, as a lowering of the quality of people's lives, and may well result in a population crisis, with an increase in mortality, and depopulation. During such crises large political units may disintegrate into many smaller ones, at the same time as there are massive famines and the society opens itself to foreign invasion, as was the case e.g. on various occasions in China. Resources are wasted, and previous concerns with art, nature and the past all disappear.[477]

The new, larger, population, in order to avoid this eventuality, thus employs technology it would not otherwise have needed, thereby taking the vicious circle further round its path. So while the quality of people's lives thus tends to decline at this stage in the turning of the circle, an increase in mortality may be reduced or avoided.

Population overshoot may thus be seen to be potentially of two levels of intensity. On the first level there is a decline in the quality of life; on the second, which may or may not be avoided thanks to economic development, there is an increase in mortality. It is also to be noted that, as the notion is being employed here, overshoot, whether it be of population or the provision of resources, is relative to *short-term* carrying capacity (Glossary).

Diminishing returns and the undermining of technology and human existence

The phenomenon of diminishing returns is the expression of a negative feedback loop of the decreasing kind, and is taken up elsewhere in this book with regard to predation, increasing evolutionary complexity, and economics. When resources are being used in an unsustainable way, as suggested by Carr-Saunders, if there is no technological improvement (if new needs are not met), the returns to the same doses of capital and labour will diminish.[478] However, if there *is* technological improvement, returns can increase. As mentioned earlier, however, more energy is normally required for the operation of new technology than was needed for the operation of old, and in the beginning it provides little more by way of output. Its output then increases as it becomes properly operative – during which process energy expenditure per unit produced may well drop and increasing returns *for the new technology* result – reaching a peak at some point in time. After the peak, diminishing returns set in, mainly because of the increasing energy that must be expended both to obtain the resources on which the technology is dependent as well as – from society's point of view – to get rid of them when they become waste. Eventually the returns become so low that a yet newer form of technology may be introduced, and the cycle repeat itself.

This is the notion of diminishing returns as applied to one turn of the vicious circle. But the whole process involving the turning of the vicious circle can be seen on the same pattern. Technology itself, rather than a new instance of technology, moves through a phase of bringing increasing returns, at the end of which it peaks, and after which there are diminishing returns. I would suggest that in the industrialised countries this peak was during the 1950s and 1960s (see Chapter 6). Note that this is prior to the peak in resource use.

The quality of life of the ordinary person follows this process. It drops when the new technology is first introduced, due to the extra energy (work) needed to operate it. Then, once the new technology is functioning, it tends to rise, only to fall again when returns begin diminishing; and then with the introduction of yet newer technology to drop even further. In the long run the size of the population increases due to the increase in consumables made available by the new application of technology, while at the same time the diminishing returns from technology produce a situation of experienced population pressure.

Note how this process is dependent on the turning of the vicious circle, with its technological development and non-sustainable use of resources. If resources are used sustainably, the employment of technology needn't mean diminishing returns. We can imagine, for example, the technology of fire being used in a sustainable way, such that it doesn't involve the turning of the vicious circle.

The employment of new technology to obtain a needed resource will first be devoted to obtaining that which is the most easily acquired, and as efforts come to be focused on resources that are more difficult to obtain, it will be found that greater energy is required to do so, with an *overall* decline in returns *and* increasing complexity *as results*. But in this way an increase in the complexity of technology itself can be seen as part of technological (economic) development as a whole, in that more complex forms of technology, which require more energy to develop and employ, will be required to obtain the resources that are beyond the reach of the simpler technology. Thus while the constantly increasing energy required to obtain ever less-accessible resources will in itself mean diminishing returns in relation to energy input, the constantly increasing complexity of the process required to obtain them will exacerbate the problem, and shorten the time before the point of decreasing returns is reached for the currently employed technology. From a systems point of view, however, *all* processes leading to decreasing returns involve an increase in energy expended vis-à-vis the result obtained.

Economic decline

Just as the life of the ordinary person normally follows the rise and fall of returns, so does the growth and shrinking of the economy. Imagine a situation where the employment of a particular technological innovation results in the obtaining of a surplus, either by turning previously unusable resources into reserves, or by increasing the quantity of products given extant reserves. Due at least to the growing human population, that surplus is constantly being eroded, however, and at an increasing rate. As resource consumption increases, a point is reached at which the amount of reserves per capita starts to decline, and returns to economic activity start to diminish. The surplus becomes a deficit, and economic decline sets in. In modern society this may be manifest in rising prices. This depression lasts until and unless the use of new technology begins to create a new surplus (to give rise to increasing returns). Note that such depressions need not be worldwide. In fact, they are occurring all the time in every industry, depending on the particular resource(s) used by that industry.

Here we have an example of the turning of the vicious circle on a small scale, giving rise to the idea of there being a hierarchy of vicious circles in encompassing and overlapping relations.

Such depressions are naturally more severely felt by the poor than the rich, their experienced needs being *vital* (though the absolute and even perhaps proportional decrease in wealth may be greater for the rich). This is particularly so when the lost surplus is that of food. Further, in such times of economic decline, the strong take from the weak whatever they perceive themselves as needing.

Scarcity, need and population reduction

There is both a *momentum* and a *time-delay* in social systems – manifestations of the overshoot principle. The momentum consists in the slowly changing mores inclining people to have the same number of children as their parents, even when it is no longer ecologically reasonable to do so; and the time-delay consists in the fact that, due to the increasing consumption and the consequent reduction in resources, the surplus that existed when the children were conceived may no longer exist when they reach adulthood. In any case, the end result, if new technology is not employed earlier, is the return to a situation of *scarcity*, and possible population reduction.

Conclusion

The theory of *Homo sapiens'* development based on the vicious circle principle, which itself presupposes the fundamental principles of modern science, unifies and makes understandable the new views in anthropology, archaeology and economics presented in Chapter 2. It suggests technological development to be essentially a reaction to population pressure, as is in keeping with Cohen's view of the horticultural revolution and Wilkinson's theory of economic development. And on the vicious circle principle population growth can just as well push technological development as be the result of it, the former idea being advocated by Boserup and Wilkinson.

Thus on the present theory the essence of technological change, looked upon as *progressive* on the traditional, Western perspective, is seen as actually being *regressive* when it comes to the long-term existence of the human species, since its employment undermines the preconditions for our survival – as is partly exemplified in the phenomenon of prehistoric overkill.

In the next chapter the correctness of our theory will be demonstrated through its application to the whole of the development of humankind. Given its correctness, this application will also serve to reveal the dependence of that development on the operation of the vicious circle principle.

5

The development of humankind

The application of the vicious circle principle to humankind's development is intended to *explain* that development. To do this it should indicate its *cause(s)*. On the VCP, the one key cause (which is also an effect) distinguishing our development from what is the case with other animals is technological innovation. Other causes/effects necessary to the operation of the principle include population growth, resource depletion and so on. Together, all of these phenomena constitute the vicious circle, the turning of which itself constitutes the development of humankind.

In order to demonstrate the applicability and thus explanatory power of the VCP, in this chapter each of the major episodes of humankind's development will be described in some detail and then analysed in terms of the principle.

Apes and protohominids *7 million* BP

After the demise of the dinosaurs 65 million years ago, mammals gradually became the most dominant life form on earth, the most intelligent of the mammals being the primates, and the most intelligent of the primates, the apes. Due to the separation of the South American, Eurasian and African continents beginning some 150 million years ago, the primates in these three areas afterwards evolved separately. In Africa, apes diverged from monkeys about 23 million years ago,[479] with humans descending from these African apes, our branch splitting off from the ancestors of chimpanzees perhaps as recently as five and a half million years ago,[480] and no earlier than eight million years ago.[481]

Our nearest living relatives are the chimpanzees, who are genetically more similar to us than they are to the gorillas, but karyotypically more similar to the gorillas, their and our next nearest kin. Where humans have 23 pairs of chromosomes, chimps and gorillas each have 24, the second human chromosome being a fusion of what in an earlier species were two separate chromosomes. We ourselves may still be considered a type of ape, one going under the name of *hominid*.

The genetic, as distinct from karyotypic, distance between humans and chimpanzees is well within the range found for sibling species of other organisms. This shows the importance of karyotype with regard to phenotype, since, despite this genetic similarity, chimps and modern humans differ far more than sibling species in anatomy and way of life. Although humans and chimps are rather similar in the structure of the thorax and arms, they differ substantially not only in brain size but also in the anatomy of the pelvis, foot and jaws, as well as in relative lengths of limbs and digits. They also differ significantly in many other anatomical respects, to the extent that nearly every bone in the body of a chimpanzee is readily distinguishable in shape or size from its human counterpart. Associated with these anatomical differences are major differences in posture, mode of locomotion, methods of procuring food, and means of communication. Because these differences in anatomy and way of life are so great, biologists place the two species not just in separate genera but in separate families.[482]

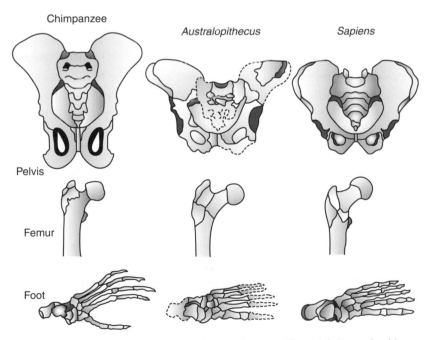

Skeletal parts of the chimp, *Australopithecus* and *sapiens*. The pelvis is much wider and shorter in hominids than in the ape.

As mentioned earlier, organisms tend to evolve from small to large, and from less to more complex. Ecosystems also increase in complexity, because of the increased complexity of the organisms constituting them as well as the increased number of species. As part of the general evolution towards the greater size and complexity of organisms, primates and a number of other mammalian orders have shown a strong evolutionary trend towards large brains. This development has been most pronounced among primates however, giving them the largest brains relative to body size of all mammals (and all other animals). And as regards primates, where non-human apes have twice the brain:body ratio of monkeys, humans have twice that of non-human apes.

Unlike humans, other apes, due to their smaller brains, are unable to plan ahead for more than a few minutes, e.g. abandoning tools when they have finished using them;[483] and they are very limited in their ability to visualise and think about relationships between objects when those objects are not in sight.[484] And, of course, they can't speak. On the other hand, however, some non-human primates hunt game, share food, cooperate in carrying objects, have lifelong kinship ties, practise work division by sex, have incest prohibitions, make tools, exhibit linguistic ability, possess long-term memory, and practise infanticide and cannibalism.[485]

Since all primates are social and thus defend group territories, and since modern hunter-gatherers are known to do the same, it may safely be assumed that all of our pre-human ancestors also defended group territories. This trait would have been a karyotypically-based characteristic retained from a primate or pre-primate ancestor, a trait structured around the family or extended family (band). A significant aspect of this characteristic is the maintenance of dominance–subordinance relationships, such relationships, based on the sexual instincts, being important in the social behaviour of vertebrates quite generally. Learning behaviour also plays a role here, it being a particularly well-developed factor in determining population density in the case of humans.[486]

There is no seasonal sexual periodism in either humans or the other great apes, which leads to sexual ties constituting a sustained and continuing bond that provides a biological basis for the long-surviving family unit. As has frequently been pointed out, this is a central element in human sociality (which is also positively correlated with biological complexity). Also, there is a long period of growth and maturation in both humans and the other great apes, which is related to the unique dependence for survival on learning in the higher primates.[487]

The change in the karyotype of an ape some five to eight million years ago that reinforced its leaving the trees was the first of the many karyotypic changes since then that have resulted in modern humans. It would seem that the evolution of protohominids from moving about on four legs to moving about on two was at least partly a response to a climate change in Africa resulting in less rainfall. Thus as the forest area shrank and the savannah expanded, walking and running rather than swinging from branches became an advantage in certain areas. For the forest dwellers, stress increased as the forests shrank, and many ape species became extinct.[488] Adapting to the grassland, on the other hand, would have supported a genetically- or karyotypically-based increase in intelligence due to the greater and differing demands made by the new environment when it came to acquiring food. This increase in intelligence was also abetted by the need to avoid being taken by other predators; and it would have afforded an increased ability to create and deal with complexity. The situation is importantly similar to that of modern hunter-gatherers living in tropical areas who have had little incentive to develop technologically, such an incentive promoting genetic increases in intelligence.

A primate group living on the ground, the members of which cannot climb to safety in the trees, is much more vulnerable to attack by predators (except other primates) than are their arboreal cousins.[489] Thus the use of large canines in defending the group against predators is more pronounced in ground-living primates such as baboons and gorillas. But the canines of the protohominids, who, like the baboons, lived in open plains country, were small. Thus the shift from teeth to tools occurred early in the evolution of pre-humans, long before the appearance of the forms that have been directly associated with tools,[490] and millions of years before the evolution of larger brains.

With our ancestors' move out of the trees vegetable foods became less available to them, and they had to survive on what the growing savannah could offer. Thus instead of fruit they had to eat roots that they could get at with their digging sticks. (It is suggestive in this context that the chimpanzee, not known to be a tool-user in its forest environment, makes use of sticks for various purposes as soon as captivity forces it to spend most of its time on the ground.)[491] And they had to eat more meat. (Whenever there is a dry period, even if only temporary, one can expect an increase in meat consumption on the part of omnivorous animals – including omnivorous arboreal primates.) Early humans and their ape precursors probably ate meat when they could get it, but in the case of pre-humans it was only when they became bipedal that it was energetically feasible for them to search for meat actively.[492] The animals the protohominids hunted were unlike the prey of their arboreal forebears but like that of baboons, often being the faster and/or larger mammals of the savannah, including hares, and young antelopes and gazelles. The very first bipedal protohominids, like chimpanzees and baboons, would have cooperated to hunt[493] (social instincts) – which is not to deny that they also scavenged.[494] Almost all carnivores, including lions and hyenas, scavenge as well as hunt,[495] as do some African hunter-gatherers,[496] including e.g. the Hadza, and as did people in the Italian Palaeolithic.[497] Scavenging does not constitute an intermediate stage between apes and humans,[498] and there was no particular 'scavenger phase' in human evolution.[499]

As with other vertebrates, the territoriality of protohominids and early hominids had a significant influence on their behaviour.[500] In the case of primates quite generally, the relevant territory is that of the band as a whole, i.e. the *group's* territory. (In this regard it may be noted that pre-sedentary human individuals can at best only have temporary territories, as their bands are constantly on the move; cf. other primates as alluded to in Chapter 1.) With primitive humans, this band territory becomes part of the larger group territory of the tribe. Among other things this means that, as in most social species, the leader is not the only male in the band to mate, and the defence of the territory is taken on by all of the fit members (males) in the group. However, while both protohominids and their arboreal forebears were group-territorial, the nature of their territories differed. When the ancestors of humans became hunters, their group territories went from being relatively small and vegetation-dense to being large and open,[501] encompassing the whole of the area in which they migrated throughout the year. And it may be expected that this difference, together with their high mobility, meant not only their manifesting their territoriality differently from other primates,[502] but, combined with their sharing of meat, also their being more social.

As with other primates, each protohominid band probably had a dominant male who, while he didn't prevent the other males from mating, nevertheless had first choice in such activity. According to the Darwinian conception of the intraspecific selection of individuals, one consequence of this is that the gene lines of the leader should have been relatively well represented in the group's future genetic constitution. Leaders would have the *fittest* mates and, where relevant, the *most* mates. In the present case this would mean a positive correlation between males' hunting prowess and genetic fitness.

The first hominids: *Australopithecus 4 million* BP

Protohominids became extinct about four million BP. At this time, or somewhat earlier, the first hominids, of the genus *Australopithecus* (southern ape), appeared. They existed for over two million years; from four to one million years ago there existed at various times seven or more early hominid, i.e. *Australopithecus*, species in eastern and southern Africa, some contemporary with one another, and all comparable to 'bipedal chimps.' Their brain size was still small and, though they probably used tools more than the protohominids, their similarity to modern humans has not been considered sufficient to include them in the genus *Homo* (human).

Australopithecus

There were essentially two kinds of *Australopithecus*, one *gracile* and one *robust*. One gracile species, to which the remains of the famous Lucy (107 cm tall, discovered in Ethiopia in 1974) belong, existed from three to two million years ago and constituted the main hominid line during that period. The gracile aspect of the members of this line is mainly to be found in the shape of their heads, the bodies of both sorts of *Australopithecus* being very similar.

The gracile type, though at one time dominant, died out earlier than the robust, but not before mutating into at least one other species. The robust line existed up until 2–1.25 million years ago; but it is the gracile, of which there may have been as many as eight species, who are our ancestors, while the robust are not.

The size of emerging humans' prey increased dramatically sometime after stone tools entered the fossil record, and with larger prey came the greater importance of meat in the diet.[503] The first tools must have been weapons: natural stones which were thrown, and then clubs in the form of the thigh bones of hoofed mammals,[504] which could also be thrown. Our progenitors, like chimpanzees today, were clearly able to hurl objects,[505] the hand of *Australopithecus afarensis* being adapted to throw with precision and force.[506] (As regards clubs, it may be noted that G/wi hunters throw clubs at small mammals and birds,[507] as did e.g. Tasmanians, who could throw them over 40 m.[508])

Bipedalism

The development of bipedalism was a critical feature in the structural development of the hominid line, changes of this kind having occurred among mammals only rarely in the last 30 million years.[509] We are the only mammals to walk and run on two legs (rather than hop, as do kangaroos and salutatory rodents). Early hominids' use of stones and bone clubs to kill animals, and stones as knives to cut their skin and flesh, must be the original technological steps giving immediate survival value to bipedalism and free forelimbs.[510]

The 1976 discovery of the fossilised footprints at Laetoli in Tanzania shows at least some hominids to have been fully bipedal already 3.6 million years ago, the original specialisation of the human leg probably taking place with *Australopithecus* and somewhat earlier, related species. Of all the unique hominid features, bipedalism was the first to appear.[511] (Though australopithecines were fully bipedal, they exhibited skeletal traits of a knuckle-walking ancestor.) By about two to three million years ago, hominids already had a leg and foot structure almost identical to our own.[512]

Laetoli footprints[513]

The above-described open prairie or savannah setting, where continuous running and leaping were possible, was conducive to running specialisation in all animals, and generally resulted not only in leg lengthening but in an overall slender build.[514] Longer limbs create a more rapid walking speed during natural stride, and are more energy efficient in long-distance running.[515] V. L. Smith suggests further that bipedalism would have reduced heat stress by its exposing less body surface to direct sunlight, which would at least have partly compensated for protohominids' increased exposure.[516] Bipedalism would also have raised the body from the ground, so it could benefit from cooling breezes.[517] As Berndt Heinrich points out, the human brain has a special network of veins that acts as a heat radiator to dissipate extra heat. Vein tracks on fossil skulls indicate that the gracile australopithecines already had this type of circulatory system, which indicates that they had experienced strong selective pressure to prevent overheating. In this regard it may also be noted that we humans are unique in having bushy head hair that protects both head and shoulders from the sun's rays.[518] Standing upright would also have improved australopithecines' ability to see over obstructions, grass and shrubs,[519] which would have given them an advantage in both predation and defence.[520]

Teeth and head

The function of large teeth – particularly canines – for primates is mainly for fighting. Primates use their large canines not only in fighting over territory and mates, but also in the defence of the group against predators; and together with incisors, canines are also important in removing the skin of prey, and in dividing the flesh.

The skull of *Australopithecus*, on the other hand, is that of an ape that lacks the ability to fight or seize and pull with its teeth. As expressed by Darwin,

> as man gradually became erect, and continually used his hands and arms for fighting with sticks and stones, as well as for the other purposes of life, he would have used his jaws and teeth less and less.[521]

> The free use of the arms and hands, partly the cause and partly the result of man's erect position, appears to have led in an indirect manner to other modifications of structure. The early male forefathers of man were … probably furnished with great canine teeth; but as they gradually acquired the habit of using stones, clubs, or other weapons, for fighting with their enemies or rivals, they would use their jaws and teeth less and less. In this case, the jaws, together with the teeth, would become reduced in size.

The reduction in canine size, which was more striking than that of the incisors, strongly suggests that australopithecines not only used tools to kill and butcher prey and defend their communities against predators, but to fight with one another over territory and/or females.[522] Small canines and incisors are biological marks of a changed way of life, their primitive functions being replaced by hands and tools.[523]

The actual change in protohominid dentition not only involved a reduction in the size of the canines and incisors, but also the moving together of the other teeth, which thereby functioned better in grinding food, as would be necessary in the eating of roots rather than fruit. The result of these various changes was that *Australopithecus'* dentition was unique among the large carnivorous mammals in having reduced canines and incisors and non-sectorial premolars and molars.[524]

Gorilla and *sapiens* dentition.[525] Canine teeth shaded. In the man-apes the molars were very large, larger than in either apes or humans.

With the reduction in the size of the teeth, there occurred throughout the development of hominids a change in facial structure from the more wolf-like face of a baboon towards the flat face of modern humans. Thus, as noted in the previous chapter, the modern-human head is the result of a form of 'self-domestication' similar to what has occurred to the heads of other animals when domesticated.

Arms, hands and the use of tools

Though human bipedalism affords particular advantages, at the same time, as it developed in humans, it is relatively ineffective when compared with the locomotion of quadrupeds as regards speed over short distances. (This differs from the development of bipedalism in other animals: e.g. the basilisk and some other lizards achieve their full running speed only by rearing up onto their hind legs. All of the bipedal animals that run fast do so by a rapid succession of long leaps, either alternating between legs or kicking off with both legs at the same time.)[526] That humans survived and flourished means that this loss of speed was more than compensated by the survival value of the freeing of the forelimbs.

As the various primate species evolved before pre-humans left the trees, there was already a movement towards forelimbs taking on functions previously performed by the teeth. Primates constitute the only major order of mammal which is characteristically arboreal (squirrels belonging to the order Rodentia), and pre-human primates' swinging from branches gave dexterity to their arms and hands. Thus we can see hominid and human tool-use as an extension of this early tendency, with tools largely being substitutes for teeth.[527]

We can imagine quadrupedal protohominids using tools in a sitting position, as apes do. But the more important tool-use becomes, the more the forelimbs have to adapt to it, and the less effective they become for locomotion. In this way bipedalism is the result of an interactive process in which it both makes tool-use easier and is further developed genetically or karyotypically as a result of tool-use.[528]

This dependence on the learned use of tools indicates a movement into a previously unexploited dimension of behaviour for life on earth, a movement which accompanied the advent of protohominid bipedalism. Though tools had been used earlier, it is with the assumption of erect posture that the regular use of tools became necessary to the continuing existence (and successful mutation) of pre-human species.[529]

Along with being replacements for teeth, tools are also *detachable extensions of the limbs*. What makes them detachable is the *hand*, with its long, versatile, powerful, skilful and opposable *thumb* (the last characteristic also being found on other primates), all of which qualities taken together support hominids' and humans' physical adaptation to tool-use better than that of other apes. Though the hands of other apes are like ours in a general way, their thumbs are less versatile, being much shorter. A human can trace a circle with the end of his thumb, move it to the middle of his palm, and touch each separate fingertip. This affords us various 'precision grips,' i.e. grips that use the thumb and one or more fingers, with or without the palm serving as a passive prop. Humans are unique in using three very powerful precision grips in tool-making. These oppose the thumb with the side of the index finger, or the index and one other finger (baseball grip; pen grip), or one of the finger pads.[530] Other apes are only capable of a much less precise 'power grip,' such as we would use when clasping a branch.[531] As functional maps of the cortex of the brain show, the human sensory-motor cortex is not just an enlargement of that of other apes. The areas for the hand, especially the thumb, are disproportionately large in humans, and this is an integral part of what makes the skilful use of the hand possible.[532] This dexterity is crucial to our development of tools, in the context of which the brain, hand and eye co-evolved.[533]

Brain size

In all apes and monkeys, the baby clings to the mother; and to be able to do so it must be born with its central nervous system in a sufficiently developed state. At the same time however the brain of the foetus must be small enough for the infant's head to pass through the mother's bony birth-canal. (Cf. the marsupial solution to this sort of problem.) In the case of hominids this created a biological quandary, for on the one hand adaptation to bipedalism decreased the size of the birth-canal, while on the other tool-use was selecting for larger brains. This problem was solved by the delivery of the foetus at a much earlier stage of development. Though this meant that the infant was unable to cling to the mother (which it would not have been able to do in any case, since it couldn't grasp with its feet)[534] this difficulty was met by the freeing of the mother's forelimbs that bipedalism made possible, so that she could hold and carry the helpless infant. (The early, small-brained hominids probably developed in the uterus to the same extent as non-human apes do.) Thus bipedalism, tool-use and selection for large brains can be seen as having slowed our species' development (though it occurred rapidly all the same), and at the same time led to far greater maternal responsibility.[535] Among other things, the prolongation of infancy must thus have intensified the bond between mother and infant.[536]

Body size

In the evolution of humans we see a continuation of primates' mutating so as to produce successive species with larger and more complex members. In the case of the original bipedal tool-using mammals, the conditions of terrestrial life virtually demanded that they be large, i.e. at least in the 25 to 50 kg range, large size providing important biological advantages without which protohominids might well not have survived. Their size alone would have kept them from being the prey of carnivorous birds and reptiles, and of all carnivorous mammals except the big cats and pack-hunting wolves and dogs.[537]

Hunting, scavenging and long-distance running

An important aspect of the freeing of forelimbs is that it not only makes it easier to use tools, but also to *carry* them,[538] as well as to carry food. The carrying of tools did not begin with hominids or even protohominids, however, chimpanzees being able to carry e.g. termite-collecting tools for several hundred metres or more when there is motivation to do so.[539] Imagine however hominids developing as intelligent quadrupeds able to use their forelimbs as arms, but also needing them as legs. As compared with the actual development, this would have meant great awkwardness in taking tools to places where they could be used, and it would have made it impossible to use them while walking or on the run – this being of particular relevance when it comes to weapons. Washburn argues that the new-found hominid ability to carry weapons was the cause of (increased) meat eating, male dominance and cooperative male hunting, and that these in turn together were responsible for the differentiation of early hominids from other primates, and for hominid survival.[540]

Australopithecines were slower than baboons, a disadvantage partly counterbalanced however by their ability to use weapons, thanks both to their larger cerebral cortices and free forelimbs. To this must be added that greater running endurance than other large predators of the savannah was beginning to evolve with them. Humans are unique in having a sweating response, combined with reduced body hair which increases convection rates,[541] that makes sustained running in the heat possible, even under direct sunlight. One cost of human sweating, however, is that we are particularly dependent on water.[542] Other features conducive to endurance running are humans' expanded joint surfaces in the spine, hip and legs,[543] as well as our tendency to breathe through the mouth (but not pant) during strenuous activity.[544]

Meat was in abundance on the plains for those who could catch it, or take it from such other carnivores as leopards, cheetahs and lions, or compete for it against hyenas, jackals and vultures. Travelling fast and long would have been a great premium for getting to a predator-killed carcass before competitors consumed it. Travelling in groups and coming upon such a carcass, early hominids might have chased off the owner with sticks and rocks. The hominids' mobility in the heat could also have been employed in getting their own fresh meat, by hunting. Once hominids or early humans were fast enough, they could potentially run down such weaker prey as young, old or injured antelope. It generally pays prey animals, since they are usually not pursued for very long, to sprint fast but only over short distances, a behavioural trait that the human predator could exploit. Pre-humans and humans who capitalised on this weakness – thanks to their ability to anticipate due to their larger cerebral cortices – became superpredators.

As regards modern humans, the Khoikhoi and Bushmen of southern Africa were able to run down swift prey, including steenboks, gemsboks, wildebeests and zebras, *provided* they could hunt in the heat of the day.[545] American Indians can run down horses and deer by pacing them, and by taking advantage of their tendency to move in an arc, by traversing the chord,[546] the members of many North American Indian tribes having chased down animals on foot.[547] Australian Aborigines chase down kangaroos by forcing them to reach lethal body temperatures. Early hominids, given what is known of them, must also have had this ability. Those of our predecessors who didn't have the taste for the *long* hunt would very seldom have been successful, and would have left fewer descendants.[548]

The fact that early hominids were able to carry their tools and use them on the run was of tremendous immediate survival value to the hominid life form. The hunting success of *Australopithecus* depended on technical efficiency, and since efficiency is advantageous in the short-term workings of evolution, the ability to use weapons, and to employ technology more generally, were selected for amongst hominids.[549]

Thus we have, newly arrived on the scene in the form of *Australopithecus*, a large, relatively intelligent omnivore moving about (often running) with tools in its hands with which to hunt other animals, cut them open and remove their skin, as well as dig for roots. Their prey was similar to that of their protohominid forebears, generally being small[550] but including e.g. baboons and other australopithecines.[551] (Primates may long have been the chief prey of other primates.)[552] Since baboons, who travel in large groups, were a significant item among the australopithecine sources of meat, it seems likely that the australopithecines hunted in bands. A single individual, even armed with a club, would not have been a serious threat to a band of baboons.[553]

This change from primitive primate to hominid enormously broadened our ancestors' potential terrestrial sources of food. Without imputing to the australopithecines any technical abilities beyond the use of a digging stick, at least the following types of vegetable food would have been available to them: berries, fruits, nuts, buds and shoots, shallow-growing roots and tubers, and the fruiting bodies of fungi. (The digging stick is also used by modern great apes;[554] but no non-human primates, including australopithecines, are/were able to obtain tubers buried deep in the ground.[555]) But humans and their ancestors have always been meat eaters to some extent. It was with the change from forest living to living in the open that the diet of protohominids became more varied, and changed from consisting almost solely of plants and the fruits of plants to consisting to a large extent of meat.[556] As suggested by Kenneth Oakley, all the main uses of stone tools by hominids and humans were connected in the first place with carnivorous habits. Not only do we lack the teeth evolved by true carnivores, but we have the long large intestine associated with a herbivorous diet.[557] (With a transition to meat eating the small intestine becomes longer while the large intestine becomes shorter.)[558]

The food that was most nutritious was meat; and the larger the prey, the greater the reward for effort spent. In coming from life forms higher on the food chain, meat is generally a more concentrated form of energy than are plants, carnivores thus needing smaller quantities of food by weight than do herbivores. Meat contains all of the amino acids needed to build body proteins.

It also contains fats essential for brain growth during childhood, iron needed for oxygen transport, the B vitamins necessary for oxidative metabolism, and vitamins A, D, E and K.[559] The hominid diet – thanks to weapon innovation – increasingly consisted of meat. And instead of eating almost continuously like their forest-living ancestors, pre-human and early human males ate more sporadically, which they could do due to the greater nutritional density of their food. Thus hunting eventually became *necessary* for hominid survival (which it is not for today's humans), and genetic configurations supporting hunting became favoured through the increase in meat consumption.[560]

Meat sharing, male dominance and the strengthening of the social bond

In the new situation of living on the ground there was also an increase in the division of labour. Not only did females have to suckle their young, as before, but they had to devote more time and energy to their development due to their greater immaturity when born; and the males had to hunt. The sexual division of labour present in all known hunting peoples was already evident at this early stage of hominid evolution.[561] Hunting must have been a basic part of pre-humans' cultural adaptation, even if it provided only a relatively modest proportion of their food.[562]

Early hominids spent relatively much of their time in hunting, which led to increased interdependence, both in hunting itself and in sharing the spoils. When the males obtained meat they would first eat their fill, and then return to camp with the remains, which they would share with those not taking part in the hunt, including the females and young, thereby encouraging sexual and family bonding,[563] as well as reinforcing male dominance. In the case of monkeys and apes an individual simply eats what it needs; and after an infant is weaned, it is on its own and is not dependent on adults. This means that adult males never have economic responsibility for any other member of the group, and adult females do so only when nursing. Thus hunting makes females and young de facto dependent on the success of male skills,[564] and sharing becomes necessary to the survival of the group. According to Washburn and Lancaster, the human family is the result of the reciprocity of hunting, the addition of a male to the mother-plus-young social group of monkeys and non-human apes.[565] And though females also shared with males the vegetable products they gathered while the males were hunting, this food consisted of staples, and the males could and often did procure it themselves. Thus *meat*, not only due to its growing place in the pre-human diet, but also because of the necessity of its being shared while vegetable products could be acquired by everyone, became an object of even more special value to hominids as compared with their forebears, as it is with modern hunter-gatherers.

Though we have spoken of hunters' *sharing* meat with the rest of their band, the situation was not always quite so egalitarian as this might suggest. When men in virtually any culture gain control over resources that women need, they, like chimpanzees, use those resources to control and coerce them.[566] Thus, apart from being used to enhance a man's status or to show his benefi-cence, some of the meat brought to the group is traded for sex[567] (cf. 'the oldest profession'), thereby improving the genetic fitness of the best hunters. Sexual attractiveness and kinship ties play an important role in non-hunters' obtain-ing meat already amongst chimps[568] and undoubtedly amongst other primates as well. The strategy of supplying protein to secure mating privileges is wide-spread in the animal kingdom, being the rule for many male birds, spiders and insects. Male mantids, for example, provide their own bodies as food. Similarly, women of the Aché hunter-gatherers of Paraguay prefer successful hunters, the meat providers.[569] As with chimps, meat was a prized form of social currency among hunter-gatherers and their predecessors.

Territoriality

Given the highly social nature of hominids and protohominids there were basically two sorts of competition, connected to individual as versus group territoriality – as taken up in Chapter 1 with regard to social species generally. The one took the form of a fight between individual males in the group, and included the obtaining of the fittest females as mates, who would generally be won by the most proficient hunter (a change was to come in this regard how-ever with hominids' increasing intelligence, and when tribes rather than just bands fought); and the other was a fight between *groups*. Fights between males of the same band or tribe ultimately concerned who was to be the leader of the group – the pre-human leader having access to all the females,[570] but also leading other males in the defence of the group – this latter point inclining the winners not to kill the vanquished, since they were needed to help protect the group. Fights between groups, on the other hand, concerned each group's maintaining control over its territory, or obtaining control over another group's territory. In social species, intragroup fights are often over females, and result in one individual's taking command of the group. And in the case of our ances-tors both sorts of fight, perhaps more so in the case of inter-group conflicts, probably often led to fatalities (and cannibalism), as is attested to by the known frequency of violent death among early men caused by other men.[571]

Like protohominids, as intimated above, in the case of inter-group conflict over territory the physically fit males in the hominid group had to be pre-pared to fight to defend their group's territory against rival groups. This was an expression of group cohesion supported by social instincts,[572] the engaging in which itself reinforced social solidarity,[573] as it does with modern humans. And such conflicts may have occurred more amongst hominids than amongst their forebears due to such factors as their greater mobility and the fact that, because of their biological prowess, they were constantly increasing in numbers and thus experiencing greater population pressure. Furthermore, it may be surmised that it is in such cases that what has developed into human morality comes into play, where an individual male must be willing to give up his life in defending the territory of the group.[574]

As regards social instincts, as pointed out by Lorenz, if it were not the case that humans were richly endowed with them, we could never have developed the power of speech, cultural traditions or moral responsibility. These specifically human faculties could only have evolved in a being which lived in well-organised communities.[575] Humans are the most social of all animals.

The first humans 2.5 million BP

Significant and accelerating karyotypic mutation took place in the hominid line from its inception some four million BP up to about 200,000 years ago, with the arrival of our own species in central East Africa; and on the average each successive hominid species existed for a shorter time than its predecessor. Also, the number of co-existing hominid species tended to decrease over time, being down to only one during the *Homo erectus* period and at present. The step in this process considered to mark the arrival of humans (genus *Homo*) occurred some 2.5 million years ago, at a time when there was a five to ten degree drop in temperature in Africa.[576] The ability to make and employ new technology was the karyotypic response of a hominid species to these worsening climatic conditions, just as entering the grassland was our arboreal forefathers' karyotypic response to shrinking forests.

With the arrival of this new genus began the Stone Age, with the *manufacturing* of the first recognisable stone tools some 2.3 million years ago. From three to one million years ago there always existed at least two hominid species simultaneously; and two million years ago some three to six *Homo* species lived at the same time.

Homo habilis and the making of tools

The earliest named species of the new, human, genus, *Homo habilis* (handy human), lived in Africa from about 2 to 1.5 million years ago, and was clearly engaged in the making of stone tools.

While the brain volume of *Australopithecus* averaged around 400 cm³, well within the range of gorillas and chimpanzees, the part of the brain responsible for mental dexterity, the cerebral cortex, was well developed in comparison. By three million years ago, hominids already had expanded frontal lobes more human-like than that of any other non-human primate. But it is on the basis of brain size that scientists include species in the genus *Homo*, the minimum being 600 cm³.

Though apes use tools, and, as noted, chimps have been reported to make them in captivity, and pre-humans may have made the occasional tool, what distinguishes all of them from early humans is that it was the latter who, mainly through species succession, began to deal with the raw material from which the tools were made in a variety of ways.[577] It is essentially through the evolution of successively more sophisticated species, beginning with *habilis*, that our ancestors first engaged in *technological development*; and the first technology, involving a technique of chipping stones to create a chopping or cutting edge, was the *Oldowan* (Glossary).

Here we note that there may be said to be four stages in human evolution, all related to changes in our or our predecessors' relation to tools, as manifest in (i) protohominids and hominids, who are bipedal, have small brains, and *use* tools; (ii) the first humans, who have larger brains, and *make* tools; (iii) humans through *karyotypic* change up until *sapiens*, who, with increasing brain size, *develop* tools; and (iv) *sapiens*, possibly through *genetic* change, who, having essentially the largest brain, develop tools within the one species.[578] We have by far the largest brains relative to our body size of all animals, and are the only ones to use machines or develop technology, all of these characteristics being evolutionarily related.

Through the development of tools, particularly with the arrival of each new *Homo* species, our various ancestor species became karyotypically dependent on tools for their very existence. (Here it may be noted that undoubtedly right from the beginning it has always been men and not women who have made and developed tools.) Among other things, this implies an evolving life form able to acquire increasing control over its environment, a state of affairs which is not only unique but, in evolutionary terms, has not existed for very long. Noting that Darwinian natural selection promotes what is immediately useful to a species even if it may lead to its extinction in the long run, this aberrant life form had yet to prove its staying-power by coming into equilibrium with its environment.

Homo erectus

The next major step in hominid karyotypic development was the evolution of the larger, more robust, and bigger-brained *Homo erectus* (the erect human), presumably from *Homo habilis*, some 1.8 million years ago.[579] *Homo erectus* began producing bifacial hand-axes as cleavers – beginning the *Acheulean* tradition (Glossary) – about 1.5 million years ago, a practice which continued for more than a million years with very little change;[580] and at around the same time *erectus* learned to control, though not create, fire, which he used in northern, temperate areas.[581]

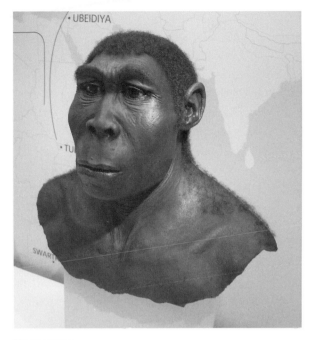

Homo erectus

Fire

As mentioned earlier, humans' mastering of fire constituted our greatest harnessing of the world's natural energy. According to C. G. Darwin it was the first revolution in our development, making possible the consumption of a much wider variety of food as well as the heating of dwellings, the latter allowing geographic expansion northwards, and both together supporting a huge increase in the human population.[582] Originally, fire was used by humans for warmth rather than for cooking,[583] though *erectus* presumably used it also for cooking tubers.[584]

As regards food preparation, protohominids and early hominids of course ate their food raw;[585] and *Homo habilis* and the Acheulean hunters of Olorgesailie (1,000,000–750,000 BP) apparently did not use fire. Fire thus allowed humans to expand their range of edibles, enabling them to eat foods that would otherwise have been too hard to digest, or have been poisonous or caused nausea.[586]

As concerns northward expansion, fire created a microclimate that resembled the climate to which humans had karyotypically adapted, permitting

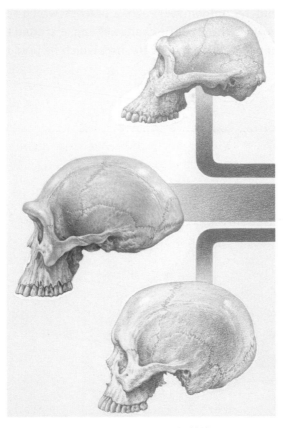

Australopithecus, erectus and *sapiens* skulls[587]

them to live in areas into which they could only have physiologically evolved after a process resulting in the reacquisition of fur, an increase in body volume, and/or the development of fat deposits under the skin. The evolution of new bodily equipment in response to environmental change normally requires hundreds of thousands, if not millions, of years; but thanks to fire and the invention of exosomatic equipment – such as weapons and other tools, shelter and clothing – which could be discarded or changed as circumstance dictated, humans became more adaptable than any other animal, and free to spread into every climatic zone.[588] As expressed by Mumford, the hearth lay at the centre of Palaeolithic man's life.[589] Humans are the only animals to control fire; and fire has been used in all modern-human cultures.

Geographic expansion

Intraspecific territorial behaviour would have tended to spread out the population of Pleistocene humans in a way similar to that in which it spreads out populations of other mammals.[590] In the case of early humans, due to their social nature such behaviour was primarily related to group rather than individual territories however, such that the spreading out was of tribes, and within them bands, rather than of individuals. In the case of *Homo erectus* this involved the occupation of a larger area than was occupied by earlier hominids thanks to *erectus'* greater mastery of nature.

Group territoriality maintains order amongst bands and tribes, while individual territoriality, involving dominance behaviour, maintains order in the group. Like individual territories, group territories have fairly definite boundaries that are not transgressed by the members of other groups – in the case of bands from other tribes, at least not in the presence of members of the band occupying the particular area,[591] and by members of other bands of the same tribe probably only with permission. The degree and nature of the violence that could be involved is evident from the Zhoukoudien *erectus* site in China, where people's brains were eaten.[592]

Erectus' population started to expand some 1.5 million years ago, at which time members of the species began moving into dry regions of Africa that were previously unoccupied by humans, and, thanks to the development of the use of fire, began migrating to northern Africa and from there to even colder regions. (Where in tropical Africa head hair served to keep the head cool by keeping off the rays of the sun, in the north it served as insulation to reduce body heat *loss*.)[593] Up until this time, just after the beginning of the Pleistocene with its various ice ages, all protohominids and hominids had lived in sub-Saharan Africa. This expansion from the tropics to the temperate zone not only meant a movement into a colder climate, but into an area where plant foods were scarcer and meat more plentiful, thereby increasing the importance of hunting for the migrating groups. Already while still in Africa, *Homo erectus* may have hunted elephants, horses, rhinoceroses and giant baboons – though some argue that the hunting of big game did not occur until some 110,000 years ago.[594]

Outside of Africa, *erectus* first migrated to the Near East, and from there as far as to China and Java around one million years ago,[595] and to Mediterranean Europe between one million and 700,000 years ago. From 700,000 to 300,000 BP these early humans moved further north and then east, colonising the northerly temperate areas of Eurasia, and becoming ever more reliant on a meat diet.[596] All of these migrations were made possible by the use of fire.

Homo erectus and the lion, leopard, hyena and wolf all reached temperate Eurasia at about the same time.[597] No other primate species had occupied more than a fraction of this area. Each *erectus* band probably had some 20 to 50 members[598] (about the size of a band of Kalahari Bushmen). This set a low limit on the number of adult males who might cooperate in hunting or protowar, as well as on the kinds of social organisation that were possible.[599] By one million years ago *erectus* was the only living hominid, *Australopithecus*, *Homo habilis*, and all other species of hominid having by that time become extinct.

Archaic Homo

Mitochondrial evidence suggests that at least two human life forms evolved from *Homo erectus*. One of them, constituting the archaic *Homo* line, led to and included the Neanderthals, a *Eurasian* descendant of *erectus*. The other, the modern *Homo* line, led to and included *Homo sapiens*, an *African* descendant of *erectus*. This separation of the archaic and modern *Homo* lines took place about 500,000 years ago[600] or earlier, though the first archaic *Homo* may have evolved as early as 650,000 years ago, while most lived after 300,000 BP.

As mentioned earlier, bipedal locomotion was fully developed in our ancestors long before they had brains noticeably larger than those of chimps. Bipedalism, with its freeing of the forelimbs, brought with it constantly increasing techno-logical complexity, including stone-tool assemblages; and interactive with this, of course, increasing brain size. In the hands of early humans the number of tool-types increased dramatically, as did the standardisation of their manufac-ture. Though these people continued to make use of hand-axes, mainly through species change they also developed the wooden lance, as well as wooden con-tainers and canoes. With the development of the lance, the importance of hunt-ing over scavenging increased greatly, much larger prey now being available; and it may perhaps be said that humans who had access to the lance no longer had much to fear from predators. Between 500,000 and 100,000 years ago there were four or more distinct archaic *Homo* species, *Homo neanderthalensis* being the last of them to survive.

The lance

The lance is used in a repeated thrusting motion, unlike the javelin which is thrown. It was perhaps introduced somewhat gradually, which allowed the medium- to large-size mammals (primarily in Africa) to adapt genetically and exhibit a flight reaction to the presence of humans. The lack of flight response on the part of animals people met when they moved into new areas would have made the lance much more lethal there. There is little evidence of the extinc-tion of larger species until about 100,000 years ago, when it began in Africa.

Language

Speech, like other forms of technology, is dependent on humans' abil-ity to conceptualise.[601] It is not clear when language first appeared, since speech leaves no direct archaeological record, but its development must have been gradual. In any case, it is generally believed that archaic *Homo* species were capable of speech some 300,000 years ago. There is some debate however as to how this skill was acquired. Here we might consider Darwin's view:

> I cannot doubt that language owes its origin to the imitation and modification of various natural sounds, the voices of other animals, and man's own instinctive cries, aided by signs and gestures. When we treat of sexual selection we shall see that primeval man, or rather some early progenitor of man, probably first used his voice in producing true musical cadences, that is in singing, as do some of the gibbon-apes at the present day; and we may conclude from a widely-spread analogy, that this power would have been especially exerted during the courtship of the sexes – would have expressed various emotions, such as love, jealousy, triumph – and would have served as a challenge to rivals. It is, therefore, probable that the imitation of musical cries by articulate sounds may have given rise to words expressive of various complex emotions. The strong tendency in our nearest allies, the monkeys, in microcephalous idiots, and in the barbarous races of mankind, to imitate whatever they hear deserves notice, as bearing on the subject of imitation. Since monkeys certainly understand much that is said to them by man, and when wild, utter signal-cries of danger to their fellows; and since fowls give distinct warnings for danger on the ground, or in the sky from hawks (both, as well as a third cry, intelligible to dogs), may not some unusually wise ape-like animal have imitated the growl of a beast of prey, and thus told his fellow-monkeys the nature of the expected danger? This would have been a first step in the formation of a language.

It has also been suggested that the repetition of sounds involved in *ritual* – the sharing of hunted meat – and *magic* (quite possibly originally related to the hunt) could have played an important role in the first acquisition of language.[602] But, however acquired, speech must have proved valuable when it came to modern humans' systematic *development* of tools.[603] And if we do accept the role of ritual in language learning, we can see a self-reinforcing movement (positive feedback loop) from ritual, to weapon innovation, to hunting, to meat sharing and back to ritual.

As regards the *making* of primitive tools, the relevance of language might not have been so great. As suggested by Mumford, "tool-making as such calls forth little verbal facility, and most of the knowledge needed can be handed on without verbal instruction, by example, as one best learns to tie a knot;" and he suggests that language may rather have originated with big-game hunting: "there is reason, indeed, to suspect that one extremely primitive mode of communication then came into existence: verbs of command."[604] On the other hand, however, group hunting does not of itself imply a high level of communication, such as speech, for, apart from the fact that in stalking one must be silent in any case, hunting in groups is characteristic of a number of non-primate carnivorous vertebrates, including many canids (e.g. wolves), some fish-eating birds, and killer-whales.[605]

Modern Homo

The *Homo* line leading to and including modern humans is the result of a mutation either in *Homo erectus* or in a descendant of *erectus*, *before* the arrival of the first archaic *Homo* some 500,000 or more years ago.

We can surmise that in comparison with archaics, who were robust, pre-*sapiens* modern *Homo* species must have been gracile. We can also suppose that the whole of the modern *Homo* line (unlike the archaic) evolved in Africa, since both *erectus* and *sapiens* originated there.

The Neanderthals *230,000* BP

The archaic line included a number of species, some of whom lived simultaneously, the Neanderthals (230,000–28,000 BP) probably being the last to come into existence. Physiologically, Neanderthals clearly culminated the archaic line in being stronger[606] and more robust than all other *Homo* species, and, together with *sapiens*, also the heaviest. Though their brains were as large as *sapiens'*, they were not as intelligent, and they can rather generally be seen as constituting a comparatively regressive evolutionary step back towards quadrupedalism.

Neanderthals obtained meat predominantly by either confrontational scavenging or cooperative ambush hunting,[607] most meat coming from hunting;[608] and cannibalism was perhaps practised.[609] Neanderthals were clearly temperate and cold-climate adapted Eurasians, but they did not live in all habitats, preferring hilly or semi-mountainous areas. Their total population of less than 100,000[610] may have been the smallest of all humans.[611]

In this regard it is interesting to note Clutton-Brock's considerations regarding the fact that, though man is a primate and the wolf belongs to the quite separate order of Carnivora, the life of Palaeolithic hunters in the northern hemisphere was much more similar to that of wolves than it was to that of other members of the primate order. Both wolves and humans evolved as social hunters, and, in the case of *sapiens*, during the Late Pleistocene they had the same ubiquitous distribution and preyed on the same herds of large mammals. Highly sophisticated social behaviour evolved, including altruism, in both

Neanderthalensis[612]

humans and wolves in response to the harsh conditions of life on the tundra.[613] To exist there, they had to function as a unified group – which meant a further development of social skills.

The Neanderthals had fire, and probably needed some form of clothing to hunt in winter; this clothing would have been very crude, however, since they lacked needles. Meat for the Neanderthals was plentiful nearly year round, while vegetable foods were in scant supply except in the spring. But where Neanderthals *evolved* into this colder climate some 230,000 years ago, both modern humans and the wolf *migrated* there. Nevertheless, the wolf has been an established member of the European and Asian ecosystems for nearly a million years (having moved north at the same time as *erectus*), while *Homo sapiens* arrived there only 50,000–40,000 years ago. Neanderthals' having evolved further north than *sapiens* means that they would have been physiologically better adapted to northern latitudes, though their degree of acclimatisation was not so great that they could do without fire. But due to their physiological adaptation they did not have to be as technologically resourceful as we did in order to survive there. Here we have an instance of marginalised individuals – the Cro-Magnons – because of their more advanced technology, succeeding better than their other-species competitors in what is originally for them a marginal area, and coming to experience it as an area of plenty.

Humans' size increased during the post-*erectus* period, as did that of many of the ice age mammals; but in humans there was a difference in that the relative size of the brain also increased, a change which was probably conjoined with a great expansion in intelligence as tools became more sophisticated. Both humans and wolves were highly intelligent; perhaps next to humans, wolves are the most intelligent (and most social) of all non-arboreal land animals. When it came to hunting they could seek out their prey, and keep it in sight while they followed it for up to three kilometers.[614] The canid strategy generally is to work in teams and to identify and then chase prey with weaknesses,[615] and finally kill it in a co-ordinated group effort. We see another manifestation of wolves' intelligence in the learning ability of dogs.

The adult wolf, like other canids, swallows more meat at the site of a kill than it requires itself, returns to its den, and regurgitates it for its young pups and their mother – if she had given birth too recently to join in the hunt. In early human societies the same end was achieved in a way that evolved from the combination of the primate strategy of gathering food with the hands, and the new carnivorous diet. Humans carried the food acquired in the hunt in their hands, and what was more important for the evolution of social behaviour, like earlier hominids they carried their children around with them as well.[616] From this, one can imagine that if wolves had been as intelligent as us so as to have developed technology, their being so would also have involved their becoming bipedal so as to be able to carry their weapons. The reason this ability evolved in humans rather than wolves was largely because of primates' use of their forelimbs for swinging from branches – which developed interactively with an increase in their intelligence.

Neanderthals obtained most of their sustenance from meat: medium-sized and large animals, including reindeer. Though they had clubs/bludgeons which, like modern primitive hunters, they could probably throw, they had no projectiles proper, typically capturing game with traps or surrounds and then killing them with their clubs[617] or lances. And they were not usually nomadic, occupying the same caves and valleys year-round, unlike the Cro-Magnons, who migrated (and had denser populations). Neanderthals are associated most particularly with a *Mousterian* technology – which involves fashioning stones into tools by chipping many small flakes from them; and they were not inventive, not using bone in their tool-making, for example, as did the contemporaneous Cro-Magnons.[618]

Franz Weidenreich observed that *most* of the fossil remains of prehistoric humans clearly indicate violent death, while Ludwik Krzywicki drew the same conclusion regarding modern hunter-gatherers, the most frequent causes of death amongst them being infanticide, protowar and headhunting.[619] One study determined that 40 per cent of Neanderthal skulls in a particular sample had suffered head injuries;[620] and cut marks and other signs of post-mortem processing possibly associated with cannibalism have been reported in several collections of Neanderthal remains. (The evidence for Neanderthal cannibalism has been greatly strengthened recently through studies of the human and animal remains from Moula-Guercy, a French cave site.)[621] Though some Neanderthals and members of at least one other archaic *Homo* species attained ages of 60–70 years, on average these humans only had a life expectancy (including infant mortality) of about 15 to 18 years.[622] In general, it seems that Neanderthals were more prone to violent death than were Cro-Magnons.

You could say that there have been two karyotypic types of human, corresponding to the robust/gracile distinction. These are manifest in the Neanderthals, who made much use of the bludgeon and developed a muscular physique to go with it, and *Homo sapiens*, who invented the javelin and developed a more linear build. The Neanderthals had curved thigh bones[623] and, compared with us, short lower legs and forearms; and while their average height was only around 162 cm, their weight would have been about 10 kg more than that of a modern person of that height, and this excess was mostly in the form of muscle. The Neanderthals' heads were disproportionately large,[624] with an elongated skull and sloping forehead, jutting brow, protruding nose, long distance between nose and mouth,[625] receding chin, strong facial muscles and big teeth, and thick cranial and neck bones.

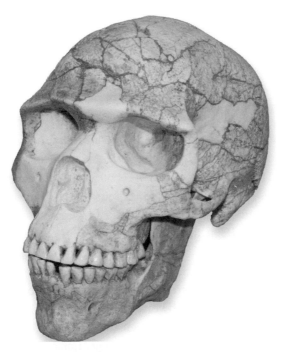

Neanderthal skull[626]

The Neanderthals populated mainly western Europe and the Near East. Their more northerly populations occupied southern Britain, northern Germany, Kiev and the area around the Caspian Sea. In the Levant they lived for a time with another archaic species, about which much less is known, but which disappeared around 100,000 BP.[627]

Erectus and Neanderthal distribution

Homo sapiens 100,000 BP

While the notion of a 'missing link' may have originated with regard to the relation between us and apes, it has today come to mean the relation between us and *erectus*. Thus there are some 500,000 years of species-change between *erectus* and *sapiens* that we do not know about, apart from our having developed outside the archaic line. The one or more unknown species filling this gap belong to the modern *Homo* line.

Geographic expansion

As already mentioned, our species originated in Africa some 200,000 years ago. From Africa we wandered northwards during an interglacial period, arriving in the Levant about 100,000 BP, at which time, as mentioned, it was already populated by at least two species of archaic *Homo*, including the Neanderthals, who had moved into this area from Europe some 20,000 years

"WELL, IF IT ISN'T THE MISSING LINK!"

The Missing Link[628]

previously. There was no technological difference between modern humans and the Neanderthals at this time, both peoples employing rather typical Mousterian-like industries.[629] Nevertheless, it was at the time of our arrival that the other archaic species became extinct, and one might speculate as to whether there was a connection between the two events. At that time our total population numbered only about 10,000;[630] and shortly after that we were forced back south to Africa, perhaps by the advancing ice age.[631]

Our second wave northward occurred around 60,000 BP. The first new areas we migrated to out of Africa beginning ca. 55,000 years ago were already inhabited by other humans, and in those areas we had to compete to obtain a niche. (In such areas, as may be expected, the weaponry of modern humans had a smaller impact on wildlife than in pristine environments.) These include the Levant, where we co-existed with Neanderthals for some 15,000 years, the southern parts of Asia, and the Far East, including New Guinea and Australia, as well as Europe – all of this migration being prompted at least in part by population growth.[632]

As regards competition with the archaics, and with the Neanderthals in particular, we were much better at planning and organising than they were, which would have given us a distinct advantage. Thus our superior technological and/or biological (karyotypic) adaptability allowed us to expand beyond the areas inhabited by the archaics – in the Old World, eventually to the Arctic regions.[633] We were better runners (and hunters) than the Neanderthals, due to the karyotypic enhancement on the African savannah of the bipedal skill already manifest in *Australopithecus*.[634] This development also involved the raising and shortening of the cranial vault and the tucking under of the face.[635] We expanded into areas uninhabited by other humans in populating the New World, and later, Micronesia and Polynesia. (Here it might be kept in mind that most animal species – including, as mentioned, other primates – are distributed over only a small part of the earth's surface.) Once entered, these areas must have become filled rapidly, so that subsequent immigrants were faced for the most part with the problem of replacing established populations.[636]

Mitochondrial geneticists have determined that all living people have a common African origin. Thus when *Eve's* (the first female *sapiens'*) descendants spread throughout the world, the indigenous non-*sapiens* human populations were replaced without admixture[637] – most notably without incorporating women of the supplanted natives (archaics) into the populations of the intruders (moderns). This can perhaps be partly explained by the idea that in natural selection similar species tend to avoid one another: a species of bird that ousts a similar species from an area doesn't interbreed with the ousted species. In addition it may well have been the case that at least the Cro-Magnons looked upon their human relatives primarily as food rather than as competitors.

The first *Homo sapiens* were karyotypically identical with modern humans, though they were genetically different in being more robust. The Cro-Magnons – Palaeolithic *sapiens* living in Europe – had a cranial capacity of up to 1590 cm³, which is large even in comparison with people of today; and the height of the Cro-Magnon male was from 162 to 183 cm.

Where Neanderthal archaeological sites are concentrated in areas that had varied vegetation, Cro-Magnon sites are located in or near what were wide steppe-tundra plains. The Neanderthals' resources, while more limited, were also more constant and required less foresight in their exploitation than did the Cro-Magnons'. And where the Neanderthals were sedentary or semi-sedentary, hunting within a small area in which their tools were also made, the Cro-Magnons followed herds of large mammals that migrated great distances.[638] *Sapiens* and *neanderthalensis* both lived in the Levant between ca. 55,000 and 40,000 BP, and in Europe from about 35,000 to 28,000 BP, *sapiens* arriving e.g. in France around 32,000 years ago.[639] By 28,000 BP the Neanderthals were extinct, having constantly been driven into rougher and rougher terrain, with the last bands living in the mountains of Spain and the former Yugoslavia.

A view over the whole of the development of the hominid line up to and including modern humans shows technological and other cultural change to have occurred at an accelerating pace from *Australopithecus* some four million years ago right up to the present. Modern humans' technological development took a particular spurt shortly after our first arrival in the Levant. It was then that we began to create purely aesthetic objects, include bone in tools, make tools with built-in handles, and probably use skin clothing. About 80,000 years ago bolas – stones attached by strings used to entangle animals – were first employed.[640] If we define a machine as a tool with moving parts, bolas were perhaps the first machines.

As mentioned, it was first with *sapiens* that *one* species began to develop technology. Thus to the extent that technological development has been dependent on micro-biological change, for the last 200,000 years this change has been genetic, not karyotypic, and has constituted biological development, not evolution. Apart from slight bodily changes and changes in intelligence, the phenotypic manifestation of these changes has been *cultural*.

Internal population checks

Though internal population checks were operative throughout hominid development, they become of greater interest as they manifest themselves in our own species. In this case, due to our knowledge of modern hunter-gatherers, we also have a better idea of how they functioned.

A key aspect of early *sapiens'* checking the size of their own populations was their need to be mobile in the pursuit of their migrating prey. Among other things, this meant that infants had to be carried by their mothers on such treks, so that a child could not be reared until its next older sibling was able to walk long distances. Australian data indicate that the inability of a mother to carry more than one child at a time together with her baggage imposes an insurmountable barrier to her having a large number of children.[641] It also means that infirm persons unable to keep up on such journeys would have to be left to die, whether or not they were beyond reproducing age.

There existed both preventive and positive internal checks in this regard. The preventive checks included lengthy periods of sexual continence, possibly sustained by long lactation. The long lactation may have been supported by infants' difficulty digesting the available foods. And there may also have existed some homosexuality.

The positive checks included infanticide – as is common among monkeys and apes[642] – and, to the extent relevant, senilicide, while cannibalism existed throughout the early *sapiens* period;[643] initiation rites also performed the function of reducing the population by eliminating the weak. The circumstances demanding infanticide are those such as when the infant is malformed, or the community is experiencing resource scarcity, or the child is female and the community requires more males – typically as warriors, as taken up in Chapter 2. Furthermore, a combination of positive and preventive checks may also have existed in various forms of male and female mutilation at puberty. And fighting between tribes played a clear role in restraining the size of the population.[644]

Social stratification: hunting ability; men vs. women

Though not pronounced in comparison with what was to come, social stratification was already evident in early *sapiens* society, first in the relation between men (hunters) and women (non-hunters). Not only did the men provide the group with meat, but through sexual selection they were bigger, stronger, faster and better co-ordinated than women (as well as being, according to Darwin,[645] more intelligent), and women had the encumbrance of having to carry and nurse their infants. It may be assumed that most if not all such groups had a leader, and that that leader was a man. Adult sex roles differ far more in humans than they do in non-human primates,[646] due to the greater division of labour in human populations – which can be related to humans' greater intelligence – itself underlying a greater degree of sociality and increased social complexity.

The sharing of meat – particularly relatively large-game meat – became perhaps the first, and in any case the most important ritual among *sapiens* hunter-gatherers.[647] (Cf. religious ritual.) In a way, meat was like capital (and, like capital, took on more than merely a use value), and the fittest males had it to disburse, and received status and influence from their ability to do so.

In this way meat became an 'essential luxury' for *sapiens* hunter-gatherers, the availability of which tended to regulate their population size at levels that protected them from environmental fluctuations.[648] The scarcity of meat would thus function as an epideictic phenomenon, triggering the implementation of more stringent population checks. To fill this role, meat had also to be highly valued by all members of the community (as it was), and, whenever available, to be shared evenly among them, so any shortage was felt by all – which implies the existence of little social stratification.[649] Everybody, except for infants, the sick, and the old, joined in the search for food. Thus there could be no kings, no full-time professionals, and no class of social parasites to grow fat on food seized from others.[650]

The social discrepancy that did exist was probably manifest e.g. in females' perhaps having less leisure than men, as well as – as in the case of other species – the widespread custom of certain males' having more than one mate. In a study of modern hunter-gatherers, only ten per cent of the groups investigated were monogamous. It was a mark of status for the headman to have more than one mate, which from a biological point of view of course implies that head-men had greater genetic fitness than other males. The wife of each man would have had her husband's relative status for a female, and the headman generally fared better than the others.[651]

As in the case of other species, such groups were stratified with respect to male status, if not power. Compared many agriculturally based societies of the last 10,000 years, the sophistication of hunter-gatherer tribes' political organ-isation was slight, however. As in the case of modern simple societies, the headman was probably not an overall leader, nor was there a council. Rather, forceful, able men, and later, *older* men – which would have distinguished us from other apes – acted as semi-formal headmen of bands, subject to consider-able pressure of opinion from other adult members of the group. Leadership was probably awarded to men who, although they were often expected to be or have been brave and skilled in war, were more proficient in the arts of peace – oratory, food acquisition, generosity, negotiation and ritual knowledge.[652] Intratribal inter-band affairs were probably regulated by ad hoc negotiations dominated but not controlled by the headmen.

The Upper Palaeolithic in Europe *40,000* BP

In our second expansion wave northward, which was during a warm period in the middle of the last ice age, we arrived first in the Near East (undoubt-edly pursuing prey that moved northwards with the warming climate) and then moved eastwards through India and arrived in Indonesia around 55,000 BP. At that time, like *erectus*, we also crossed to New Guinea, making a sea voyage which at its shortest would have been less than 100 km,[653] and continued on into Australia, which at that time – still around 55,000 BP or perhaps somewhat later – was connected to New Guinea by a land bridge. We arrived in Europe around 35,000 BP and in Asia north of the Indian Ocean and in Japan around 30,000 BP, America 11,500 BP, the West Indies 4000 BP, Hawaii 1700 BP, Easter Island between 1600 and 1100 BP, and New Zealand and Madagascar 1000 BP.

Some 40,000 years ago there was a shift in our behaviour – from archaic to modern. This shift constitutes the transition from the Middle to the Upper Palaeolithic, or from Mousterian flake technology to blade technology (Glossary), which in Africa is the transition from the Middle to the Late Stone Age. Innovation now came to occur every few thousand years, not after hun-dreds of millennia as in earlier eras.

To judge from modern hunter-gatherers, three overlapping levels of social organisation characterised Upper Palaeolithic societies: the family, the band,

and the collection of bands the members of which intermarry, speak a common language, and have a common set of myths and rituals – i.e. the tribe. Members of a tribe generally maintain relatively peaceful relations with each other, and cooperate in subsistence, defence and other activities. Nothing like the relatively peaceful, cooperative relationship among hunting and gathering bands of the same tribe is known from any other animal species. Relations between tribes vary, much like relations between nations, from close alliance to traditional enmity. Partly through considering modern hunter-gatherer groups we can say that the whole linguistic/cultural group constituting the tribe in Palaeolithic times consisted of a few hundred to a few thousand people, which may be contrasted with modern ethnic groups that range up to many millions. As regards the eradication of various tribes through protowarfare, we might say that this did not just happen every now and again, but that it was a constantly recurring phenomenon. Modern nation states are not the result of the amalgamation of different tribes so much as the successive eradication (cultural if not biological) of one tribe by another, in which the eradicating tribe grows in size, eventually to become a nation.

The shift from the Middle to the Upper Palaeolithic meant a spurt in human cultural development, as well as an increase in the size of *sapiens'* total population,[654] and at the same time a decrease in its members' robusticity. On the technical level the shift to a blade technology was accompanied by the manufacture of sharpened bone, ivory and antler points and standardised, hafted compound tools, as well as awls and punches.[655] About 35,000 BP, just after we arrived in western Europe, there was a sudden emergence of body ornamentation there, as well as an explosion in artistic expression – both in mobile and in cave art – some 5000 years later.[656] Clay figures were fashioned, weapons ornamented, and graves supplied with accessories for the dead. Huge leaps also occurred in the complexity of *sapiens'* social organisation, symbolic abilities, calendrical knowledge, information acquisition and exchange, transgenerational transmission of knowledge, and so on,[657] all these being ingredients in our culture.

Our arrival in Europe was coincident with another intraglacial warm period that began at that time. When we arrived, the Neanderthals had already been established there for at least 100,000 years – during the whole of the last ice age. Their presence, together with the previous lower temperatures – for which they were physiologically better adapted than *sapiens* – may have been responsible for our not entering Europe earlier.

Distance from the equator

The further north or south of the equator, other things being equal, the less lush and the drier the vegetation, and the harsher the living conditions for

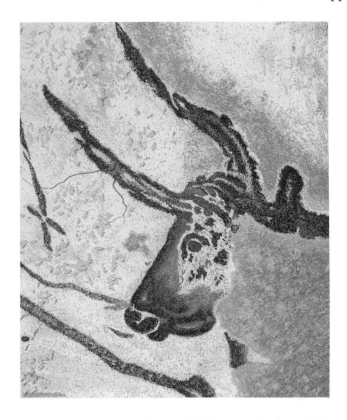

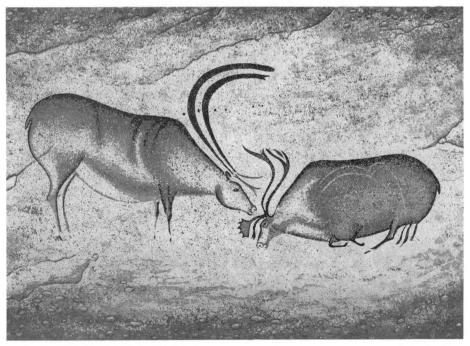

Cave art

a species that evolved in the tropics. The importance of hunting over gathering increases with distance from the equator,[658] and appears as the dominant mode of subsistence at latitudes of 60 or more degrees. This of course is understandable due not only to the decrease in vegetation, but also to the increase in the numbers of herding animals.[659] Note too that it meant a superior diet. As regards modern hunter-gatherers, hunting is primary in the Arctic, fishing in the temperate zones, and gathering in the tropics. With a single exception, *all* such societies at all latitudes derive at least 20 per cent of their diet from the hunting of mammals.[660]

The further tropically-bred humans are from the equator, the greater the effort they must expend creating a warmer environment around themselves, both as regards their clothing and their habitation. The technological superiority required for Cro-Magnons to live far from their original habitat, manifest in their greater hunting prowess, could well have been the main factor leading to the extinction of the Neanderthals. The difference between Neanderthals and Cro-Magnons might be compared to that between wolves and apes. Both are highly social, but in the inter-specific survival of the fittest the Cro-Magnons probably won due to their superior intelligence, which at this stage in human evolution can be equated with their better ability to fashion and use weapons for hunting (blade technology), and to organise themselves in their employment.

We see an example of this phenomenon of technological development in a harsh climate in the case of the Neo-Eskimos, their technology being amongst the most sophisticated of those of modern hunter-gatherers, due to the scarcity of materials with which to fashion it. Living the furthest from the equator of all humans, hunting constitutes almost their only means of acquiring food, while at the same time the resources available to them for the manufacture of weapons are the scantiest. In order to survive in the Arctic they had to improvise. Thus, unlike peoples living in wooded areas, they could not simply cut down a sapling to make a bow, but had to construct a compound bow by piecing together a variety of materials, such as driftwood, horn and antler.[661]

The sophisticated hunting weapons of Neo-Eskimos include their harpoons, which were of both the lance and the javelin variety. They further developed the Palaeolithic float-harpoon, using either an arrow shaft with a feathered butt (widely recognised as a masterpiece of primitive technique) or an inflated sealskin as the float. The prey would drag the float until it tired, which allowed hunters to harpoon larger animals than otherwise. And they invented the most elaborate of primitive harpoons, the toggle harpoon, where a line is attached to both the middle of the shaft and the middle of the harpoon head. After the animal has been pierced, tugging the line both pulls away the shaft from the head (so that the animal has to drag the crossways-turned shaft) and makes the head pivot (toggle) in the wound, embedding it more securely.[662] A similar device is the Eskimo gorge, a small rod of bone sharply pointed at each end. Used in fishing, it is baited and attached to a line. It enters the fish lengthwise, but, when the line is jerked, it turns sideways athwart the gullet, holding its victim as firmly as a hook.[663] To this it may be added that Eskimo winter houses are the best that could be devised under the circumstances to meet the rigours of an arctic climate.[664]

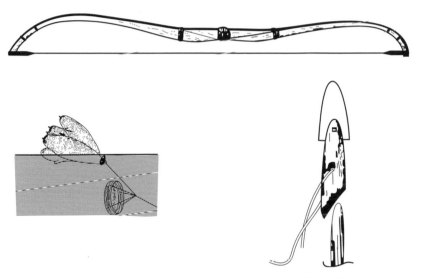

Eskimo bow, sealskin floats and drogue, and toggle-harpoon head. The bow is made of three pieces of reindeer antler, riveted together and strengthened by plaited sinews.

Regarding hunting, as expressed by Sollas:

> The Eskimo hunter, while possessing much in common with the Indian, is distinguished by greater aptitude and by special methods of his own. He represents the triumph of human adaptation to the changing conditions of a rigorous climate; by the variety and ingenuity of his implements, weapons, and devices he has brought the art of hunting to its very highest state of differentiation, and in the exercise of this art he stands supreme among all the hunting races of the world.

Applying this reasoning concerning distance from the equator to early *sapiens*, what we see in their culture is an increasing need of technological support for survival the higher the latitude at which they live, due in part to the increasingly inclement climate and in part to the need to hunt more for food. Thus virtually all of the many innovations of the Upper Palaeolithic were made in Europe, not in Africa or other warm areas where modern humans also lived.

Quality of life

If we compare the quality of life of humans living at this time with that of modern hunter-gatherers – who themselves appear to be well nourished in both qualitative and quantitative terms – we should expect a much better state of affairs to have existed for Cro-Magnons 30,000 years ago. Rather than live in areas where both meat and vegetable foods require some effort to locate and acquire, they lived in a choice environment where large game was plentiful; and the food available to them quite generally must have been both greater in quantity and superior in quality.[665]

On the other hand, however, what we know of the high rates of violent death throughout this period suggests that mortal raids by one band or tribe on another, along with female infanticide, were a part of life. Otherwise, however, it may be suggested that the Upper Palaeolithic (40,000–25,000 BP) constitutes the high point in the human way of life to date. In keeping with Sahlins' 'affluent society' line (or, taking a precondition for a society's being affluent that *all* its members have adequate diets and sufficiently large breeding sites), it can fairly be said that we never had it so good before, and we've never had it as good since. Though average longevity was short by modern Western standards, those who survived infanticide and death related to protowar lived to an advanced age, 60 to 70 being quite possible.[666] And on the purely positive side, unlike today, thanks to unimpeded natural selection no one suffered from any sort of malnutrition; there were very few diseases;[667] and men never really had to work but were engaged rather in the highly meaningful activities of hunting or protowar, or simply lazing about. (As expressed by Washburn and Lancaster, and as intimated in Chapter 2, men enjoy hunting and killing other animals, and these activities are continued as sports even when they are no longer economically necessary.)[668] Women worked only a few hours a day gathering food and wood, and, though they had lower social status than men, they could still make their voices heard; social stratification was at its lowest, with no difference in the lifestyles of the most and least powerful; neither men nor women were alienated from the products of their labour; and all personal relations of both sexes were highly meaningful, being with individuals people not only knew but often were related to, and not with strangers. On the other hand, however, our understanding of the physical and biological world in which we live was little at that time as compared with today; and the art we enjoyed was less sophisticated than that of today – though at that time we at least produced it ourselves and weren't simply passive consumers.

Protowar

Humans' development of weapons meant not only increased hunting but increased killing of other humans, where originally people from other tribes were food. The accessibility of tools (weapons) also led to discrepancies first between and then within human societies, depending on such factors as the number and quality of the weapons available to the relevant parties, the effect of which, among other things, was to check the growth of the population of the area as a whole.

Protowars between bands or tribes involved small groups of people, not armies; and the fighting men might be considered warriors, but not soldiers. Due to the lack of storable provisions, these engagements could not have been protracted (though they were often repeated) nor could they have extended over any large area; and due to the relatively small population, the fighting units could not have been large. Furthermore the discipline and chain-of-command one sees in an army would have been lacking. Such factors lead us to say that we here have to do not with war, but with protowar.

Several of the rare burials of Cro-Magnons in central and western Europe, dating from 34,000 to 24,000 years ago, show evidence of violent death.[669] These deaths invariably resulted from protowarfare between tribes. Though the weapons were primitive, protowars nevertheless took a larger proportion of lives in the population than do modern wars. Key to the nature of such conflicts is that most non-state societies did not produce the surplus food, ammunition or population necessary for prolonged episodes of combat. On top of this, even if they had done so, they had no means of transporting food or ammunition any distance, and so could not engage in far-reaching campaigns. Though in New Guinea groups of modern hunter-gatherers have been known to fight battles lasting for several days or even weeks, this was possible only because the fighting was so local that the warriors could retire each evening to their own camps.

The pattern of abandoning territory out of fear in order to widen a buffer zone, followed by gradually intensified use of the zone by the victors, illustrates the most common mechanism by which primitive warfare expanded and contracted the domains of pre-state societies. The scale of such territorial gains and losses was from about five to ten per cent per generation in some instances involving modern hunter-gatherers.

It is of interest to cite Lorenz in this regard, who has received support from the archaeological findings:

> [Assume] that the men of such a palaeolithic tribe did indeed have the same natural inclinations, the same endowment with social instincts as we have ourselves; let us imagine a life, lived dangerously in the exclusive company of a dozen or so close friends and their wives and children. There would be some friction, some jealousy about girls, or rank order, but on the whole I think that this kind of rivalry would come second to the continuous necessity for mutual defence against hostile neighbouring tribes.[670]

And as regards what we here term protoarmies:

> The men would have fought side by side ever since they could remember; they would have saved each other's lives many times; all would have ample opportunity to discharge intra-specific aggression against their enemies, none would feel the urge to injure a member of his own community. In short, the sociological situation must have been, in very many respects, comparable to that of the soldiers of a small fighting unit on a particularly dangerous and independent assignment. … Loving your neighbour like yourself and risking your life in trying to save his is a matter of course if he is your best friend and has saved yours a number of times; you do it without even thinking. The situation is entirely different if the man for whose life you are expected to risk your own, or for whom you are supposed to make other sacrifices, is an anonymous contemporary on whom you have never set eyes.

But, of course, as Lorenz intimates himself, it is not the case that modern soldiers have never set eyes on those whom they fight beside. They both live and fight together, experiencing the same perils while engaging with the same enemies; and they experience a sense of identity wearing the same uniform, which differs from the enemy's. These factors, I suggest, in combination with the plasticity of human nature, rather quickly lead to their becoming 'buddies,' and being willing to risk their lives for one another.

Here we might note that Hobbes has in fact been partly vindicated, namely in his belief that in humankind's natural condition there is continual danger of violent death. And he was also correct in his belief that through smaller political units being incorporated into nations, the relative amount of killing would decline. But the situation in humans' natural state was not one of war of every individual against every other, as Hobbes believed, but rather of every *society* against *some* other. In fact, on Hobbes' bourgeois conception of the state, and of conflict as being *within* the state, competition of everyone against everyone most naturally takes the form of *economic* competition, which only arose once society had *left* the natural state.

Population growth

As already mentioned, at the same time as such technological and other cultural developments took place starting about 40,000 years ago, the modern-human population began to increase noticeably, where, in comparison, the populations of modern hunter-gatherers, lacking technological innovation and constantly being encroached upon by other groups, have shrunk. The availability of big game, each carcass of which could feed many people, suggests further that the size of these Cro-Magnon groups was also markedly larger than those of modern hunter-gatherers. And this in turn implies that Cro-Magnon society was also more complex than that of today's hunters and gatherers.[671] This population growth was abetted both by their expansion into new territories and by the effectiveness of their hunting in these territories, particularly as regards big game. Not only this, but there was apparently a dramatic increase in *Homo sapiens'* longevity around 34,000 BP, the number of people surviving to an older age almost quadrupling.[672]

The latter half of the Upper Palaeolithic in Europe *25,000 BP*

Technological development

The whole of the Upper Palaeolithic, from 40,000 to 12,000 BP, was sprinkled with innovations of increasing sophistication. Ceramics appeared around 28,000 years ago, and after 20,000 BP the artefacts produced included tools with stones inserted in antlers, leisters (three-pronged fishing spears), and a wide variety of clothing.[673] At around 20,000 BP it was also learned how to *create* fire with the twisting stick, and later the bow drill. The twisting stick and bow drill were each machines, both employing rotary motion.[674] Towards the end of this period (around 20,000–15,000 BP) such tools as large bone and ivory needles (bodkins) and belt fasteners (both machines), lamps, spoons, pestles and buttons were being made,[675] and the spear-thrower (another machine) was invented.[676] And in North America grooved axes – polished stone axes with a groove for a shaft – appeared around 11,500 BP.[677]

More significant however is our development of weapon technology, not only during this period, but right from the time of *Homo sapiens'* first migration northward around 100,000 BP. In fact, as already noted, throughout the evolution of hominid hunter-gatherers meat was increasingly necessary to

Grooved axe head[678]

survival, and the tools used to procure it were often the ones most ardently developed. This was certainly the case among *Homo sapiens*, in whose hands weapon technology accelerated noticeably. In this regard it is of interest to cite Lorenz again:

> When a man invents, let us say, the bow and arrow, not only his progeny but his entire community will inherit the knowledge and the use of these tools and possess them just as surely as organs grown in the body. Nor is their loss any more likely than the rudimentation of an organ of equal survival value. Thus, within one or two generations, a process of ecological adaptation can be achieved which, in normal phylogeny and without the interference of conceptual thought, would have taken a time of an altogether different, much greater order of magnitude. Small wonder indeed if the evolution of social instincts, and what is even more important, social inhibitions, could not keep pace with the rapid development forced on human society by the growth of traditional culture, particularly material culture.
>
> Obviously, instinctive behavioural mechanisms failed to cope with the new circumstances which culture unavoidably produced even at its very dawn. There is evidence that the first inventors of stone tools, the African Australopithecines, promptly used their new weapons to kill not only game, but fellow-members of their own species as well. Peking Man, the Prometheus who learned to preserve fire, used it to roast his brothers.[679]
>
> When man, by virtue of his weapons and other tools, of his clothing and of fire, had more or less mastered the inimical forces of

his extra-specific environment, a state of affairs must have prevailed in which the counter-pressures of the hostile neighbouring hordes had become the chief selecting factor determining the next steps of human evolution. It is hardly surprising if it produced a dangerous excess of what has been termed the 'warrior virtues' of man.

Note however that Lorenz doesn't take up either stress or the population-controlling function of conflict; nor the high killing rate among e.g. gorillas and chimps.

Here we can expect that, as in the case of modern hunter-gatherers[680] and ourselves, it was the men rather than the women who made and used the greater number of the more complex tools, weapons, and implements, and who used more complex techniques in making and using them. The more complex the technology, the more the men participate, and the simpler the technology, the more the women.

Weapons

The development of humans' basic weapons to the end of the hunter-gatherer phase went something as follows: from the club (*Australopithecus*), to the wooden lance (archaic *Homo*), to the lance with a stone tip, to the javelin, to the harpoon (later with a detachable head connected to a line), to the javelin and the harpoon with barbed bone points, each used with a spearthrower, and on to the bow and arrow and the blowgun, each with poison-tipped projectiles.

Most of this development – from the javelin on – occurred with *Homo sapiens* after 40,000 BP. The lance with a stone point may have been developed earlier, but the development of pointed weapons that could be hurled began in the first half of the Upper Palaeolithic,[681] the ability to throw stones evolving into the ability to hurl javelins.

As Alice Brues points out, as soon as a sharp point is placed on a projectile weapon, the user is committed to a technique of accurate aim and maximum penetration. Using a pointed weapon, it is not only possible to operate at a distance, but better to do so, since if actual bodily contact with the victim occurs, aim is difficult and the withdrawal of the weapon becomes dangerous. This is probably why the pike has never been popular as a weapon except for purposes such as boar-hunting, where danger is construed as sport.[682]

As regards weapons that are projected, it is of some interest to note that they must be symmetrical. More generally, as suggested by Washburn and Lancaster, while this is not the case with tools that are moved slowly, balance becomes important when an object is swung rapidly or thrown with speed, irregularities otherwise leading to deviations in the course of the blow or the trajectory of the flight.[683]

The first pointed projected weapon was the javelin, originally a sapling with a sharpened end, later with a stone head shaped to penetrate deeply rather that to stun. After 25,000 BP, the points were multiplied for hunting elusive targets, such as birds and fish. Barbs were used on spearheads to keep them in the wound; and spears were incorporated in pit, trip-line, and spring traps[684] – the latter two also being instances of machines.

Lances, harpoons and arrows were developed which had detachable heads that would remain in the wound if the shaft were knocked or drawn away (the most sophisticated of which being the toggle harpoon, treated above), and later such heads were devised for use with poison. In the case of the harpoon, which as intimated above could also take the form of an arrow, already during the latter half of the Upper Palaeolithic the detachable head was attached to a line, which could be paid out or tied to a float (float-harpoon, above), being in this way similar to a fishing line, the advantage afforded by the float harpoon being its preventing aquatic animals from diving beyond recovery.[685]

The first clear representation of the bow is from northern Africa, at a period between 30,000 and 15,000 BP,[686] it being first employed at about the same time as snares, traps and nets.[687] Shortly afterwards it was used in Europe, while its use spread more slowly to the rest of the inhabited world. The effective wounding range of the bow and arrow is roughly four times that of the javelin and twice that of the javelin thrown with a spearthrower. Also, an arrow travels up to three times faster than a javelin; and the hunter need not move to deliver it, which greatly improves his aim.[688]

All of humankind's first major machines appeared shortly after the latter half of the Upper Palaeolithic: the twisting stick, the harpoon, the spear and spearthrower, the bow drill, the bow and arrow, and various kinds of traps.

Where clubbing and jabbing with a lance require overall strength, favouring a robust body type, as considered above, javelin-throwing is better performed by a person with long limbs and muscles, favouring a gracile body type. As Brues comments in this regard, the possibility that the spearthrower represents a compromise with body build finds confirmation in the fact that the very linear javelin-users of Africa generally throw the javelin with the bare hand; apparently the spearthrower has little to offer a physique with maximum built-in speed leverages.

The shooting of the bow, on the other hand, requires power leverage in the arm, favoured by short limb segments and short, thick muscles. Each of these body types arose genetically in association with the use of these weapons, and each may be seen as part of the domestication of humans. Brues has further suggested that even if javelins with or without spearthrowers had been available to the Neanderthals, they would have been unable to use them to advantage due to their heavy build and adaptation to power rather than speed.[689]

The use of the sling (another machine) may also have developed from stone-throwing around this time (20,000–15,000 BP), though it involves a different kind of motion. Noting that the ability to hit objects with a thrown or slung stone would have been genetically favoured in the population at least from the time of *Australopithecus*, we can appreciate that a David armed with such a weapon would have had a distinct advantage over a Goliath without one.

Conflict amongst modern humans leading to death was evident throughout this period, but particularly at around 20,000 BP, after the development of the bow and arrow. In Egyptian Nubia, 14,000 to 12,000 years ago, protowarfare was common and particularly brutal. Of the 59 skeletons discovered in one cemetery, over 40 per cent had stone projectile points intimately associated with or embedded in them.[690]

Brues also points out that it has been a general trend in the development of projectile weapons, from javelin to bullet, that the size of the projectile decrease and its velocity increase. In the case of the light but rapidly moving projectile, the influence of gravity is far less in proportion to the momentum carrying it in the direction it is aimed. Hence the range and accuracy of a projectile can be greatly increased by trading weight for speed, while its destructive power remains the same. As a result, the history of projectile weapons is a succession of inventions for increasing the speed at which the projectile flies,[691] and, to attain this, a reduction in its size.

Hunting

Looking at hominid hunting as a whole, we see a constant increase in the size of prey taken up into the Upper Palaeolithic. At one site a thousand mammoth were killed at one time.[692] While on the one hand this suggests an abundance of big game, on the other, due to the scale of the kill, it suggests that when it occurred people had to acquire a *cache* of meat, which would have been eaten dried. Thus starting about 25,000 BP, due to the previous elimination of the largest game, successively smaller prey began to be taken. It was then that Cro-Magnons began conducting the regular, planned, efficient slaughter of large numbers of such herding ungulates as caribou, reindeer, horses, red deer and ibex (large-horned goats)[693] through the organising of stampedes; and it was then that they began taking elusive and dangerous game.[694] Reindeer came to replace mammoth as *sapiens'* preferred quarry in much of western Europe, while mammoth still predominated in central Europe (becoming extinct in Eurasia ca. 13,000 BP, as noted earlier); in central Russia it was reindeer and horse, and in southern Russia bison.[695]

In southern latitudes, this form of hunting was combined with the exploitation of divers other animal food resources. These included fish, birds, lagomorphs (such as rabbits and hares), mustelids (e.g. weasels, otters, badgers) and foxes. During this period, improvement in weapon technology – such technology being essentially for hunting – was required only *after* a new area had been populated, when the big game began to become scarce.

Note how hunting is related to inter-tribal fighting. Hunting involves cooperative group effort using weapons to kill animals, and other humans are animals, which originally were treated as food. It is only natural that more sophisticated humans would use their weapons on one another when it's a question of territory rather than food (sexual/social vs. survival/sexual instincts). In the case of West African horticulturist tribes, there is a very close connection between hunting- and war-parties, "and the search for beasts may develop into a hunt for men."[696]

Longevity and possible population reduction

Average human life expectancy during this period has been estimated as being slightly over 30 years for men and slightly under 30 for women[697] – an estimation which probably does not take account of infant mortality. (The lifespans of archaic *Homo* species, Neanderthals and Cro-Magnons were all about the same.)[698] While in Africa the human population seems to have grown more or less continuously throughout the Late Pleistocene,[699] i.e. during the last ice age, that of Europe grew until around 25,000 BP, after which, just before the greatest glacial coverage at 18,000 BP, it apparently began to shrink.[700]

The Palaeolithic–Mesolithic transformation *12,000* BP

Population growth and possible decline

The modern-human population in Upper Palaeolithic Europe has been estimated to have been between 150,000 and 200,000. During the Mesolithic in Europe, however, prior to sedentism, there appears to have been an economic crisis and major reduction in population,[701] suggesting that at this time humans living there suffered a major population crisis,[702] with a decrease in life expectancy.[703] This is further attested to by Mesolithic societies' leaving an impression of extreme poverty.[704] At the same time there was a widespread transformation from the specialised hunting of particular large mammals to the more scattered hunting of both large and small animals, and an intensification in the taking of fish and shellfish.[705]

The Mesolithic societies inhabiting the wooded plain extending from the central English Pennine chain to the Urals seem to be the first to have devised equipment for dealing with timber – and the forest was the outstanding factor differentiating the Holocene environment from that of the Late Pleistocene. Beginning with splitting tools – wedges – of antler, such as had already been employed earlier in south-eastern Europe (Romania and Hungary), the Mesolithic forest-folk provided their wedges with blades of flint or stone, sharpened like antler tools by grinding. Eventually they thus created a regular kit of carpenter's tools of axes, adzes and gouges with which they could make, among other things, sledges for transporting goods and people over snow and ice. (Runners of sledges found embedded in Mesolithic peat in Finnish bogs are probably the oldest remnants of vehicles.)[706]

The transformation in America

The people who migrated to America over the Beringian land bridge had earlier, probably around 20,000 BP, migrated from China into eastern Mongolia.[707] To expand from the sub-arctic conditions of Late Pleistocene China into the Siberian Arctic required, along with fire, three key inventions: windproof fur clothing, sleds, and dogs to pull the sleds. On the sleds could be packed tents, bedding, cooking utensils and other gear necessary for a nomadic arctic hunting way of life, this being the first context in which people were able to accumulate more than they could carry. These 'mobile homes' afforded a form of 'mobile sedentism,' which allowed expansion into still colder regions.

Clothing is evident in the Siberian Arctic from 18,000 BP, a bone needle from that time being recovered from excavations there; and dogs are evident from 11,000 BP.[708] The most reasonable assumption is that the increased competition from other populations, prompted by the disappearance of the mammoths in Eurasia together with population growth,[709] forced the people then living in Siberia to move even further north into the north-eastern part of the continent, pursuing prey which themselves moved northwards with the warming climate. Hunting groups might thus gradually have expanded across the Beringian land mass – part of the Mammoth Steppe – with no alteration in their livelihood.[710] The earliest this land bridge would have existed around this time was ca. 18,000 BP, which would have been at the height of the glacial period. During previous glaciations the Beringian land mass was for long periods not covered by ice, and thus capable of supporting life, thanks to a paucity of snowfall due to the nature of the prevailing winds.[711] However, practically all modelling experiments generate ice sheets around the Arctic at the last glacial maximum, including a Beringian ice sheet. In addition, some Alaskan archaeologists find the Beringian ice sheet more consistent with their evidence that human migration across the land bridge was delayed until 12,000 years ago, even though a hazardous marine crossing to Australia had been accomplished some 34,000 years earlier. At such times as when the sea-level dropped 100 m, the land uncovered between Asia and North America constituted a plain up to

1600 km wide. These Asian people would have moved east only at the rate at which competition with other humans behind them and the thinning of prey before them made the occupation of new territory attractive,[712] and only given the absence of the Beringian ice sheet.

Around 11,500 BP some 200 of what were probably the first and only Palaeo-Indian people from Asia arrived in North America, about 70 of them leaving their genetic print in modern descendants.[713] That they be followed by others was made impossible some 500 years later by the re-formation of the Bering Strait as a result of the warming climate. This warming climate also opened up an ice-free corridor along the eastern side of the Rocky Mountains – between the eastern edge of the Cordilleran ice sheet and the western edge of the Laurentide ice sheet – allowing the human arrivals to move south through the Americas.[714] These people brought with them lances, the lances having large flaked and fluted points which may have been developed in North America – the essential part of the Clovis toolkit referred to in Chapter 2; and they probably had javelins and spearthrowers as well. Remnants of the Clovis toolkit have been found throughout North America, and points similar to the Clovis have been found over the whole of South America.[715]

Thus the original inhabitants of the Americas in 11,500 BP are descendants of the Mongoloid peoples of Asia,[716] as are the Eskimos, who crossed the Bering Sea some seven thousand years later. Note the hardy, nomadic, meat-eating, pioneering nature of this group. As discussed in Chapter 2, as these people travelled south from Alaska via the ice-free corridor they hunted such K-selected animals as the mammoth, mastodon, giant deer, camel, horse, musk-ox and different species of large bison, eliminating about two-thirds of these species over a one thousand year period.

With the disappearance of the mammoths so too disappeared the Clovis toolkit,[717] and there followed a population crash.[718] Thus at around 10,500 BP there was a transition in the Americas to the Archaic period. The central element in the Folsom toolkit that replaced the Clovis at this time, which was used in the hunting of smaller animals, was the flaked-stone point attached to a spear,[719] a point which was noticeably smaller than the corresponding Clovis point. This spear was used particularly in the killing of members of the now-extinct *Bison antiguus* and *Bison occidentalis* species, both of which were smaller than the *Bison latifrons* which disappeared during the Clovis period,[720] but larger than the presently existing *Bison bison*, which had evolved from *Bison occidentalis* by about 6000 BP. This hunting of bison was well established by 10,000 BP, and continued into historic times with the hunting of *Bison bison*. Driving bison over bluffs or sharp inclines into corrals, sinkholes or deep-sided gullies was a well established strategy by 5000 BP.[721]

Clovis points

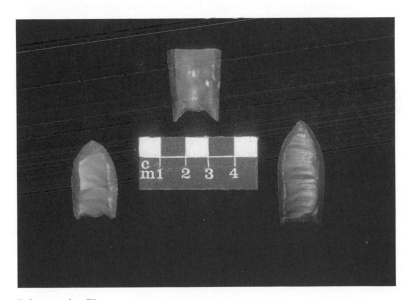

Folsom points[722]

Innovation

As regards technology, the transition to the Mesolithic (Archaic) era consisted essentially in a change from the use of weapons/tools suitable for the obtaining of large game to the use of those suitable for acquiring smaller game and aquatic fauna. Thus there was a general shift from the use of lances and javelins with simple points to the use of javelins with barbed points, bows and arrows, and various kinds of nets and traps. In the case of bows and arrows – as well as blowguns – this involved the development and use of *microlith* technology (Glossary), involving the production of smaller missile points.[723]

Quality of life

With the transition to the Mesolithic, the quality of human life once again declined, to a lower level than, but in a way similar to, its decline at the beginning of and probably throughout the latter half of the Upper Palaeolithic. This decline took the form of a worsening of people's nutrition and health, as manifest e.g. in a 5 cm reduction in the height of adults,[724] and a marked increase in homicide.[725] One Mesolithic site provides the first clear evidence of the mass killing of humans[726] – with women and children predominating[727] – and the collecting of trophy skulls.[728] Here we have mass killing *before* we became sedentary. In such societies fear of being killed by other humans was a fact of everyday life.[729]

Increasing population pressure and the beginnings of sedentism

As intimated above, through the Late Pleistocene and into the post-Pleistocene, as the sea level rose and forest vegetation became re-established, the disappearance of the megafauna in Europe was followed first by greater dependence on ungulates, which migrate seasonally over considerable distances, and subsequently by the taking of smaller terrestrial animals, birds, fish, sea mammals, invertebrates and plant foods such as hazelnuts.[730] Here we have a situation of positive feedback of the decreasing kind (diminishing returns), in which the numbers of migratory ungulates decline, with or without human population growth, and population pressure results.

With the adoption of village life in the form of *fisher-forage sedentism*, population grew through an increase in fertility: the age at first pregnancy was lower, the time between births was less, and the tendency towards infanticide diminished.[731]

As regards population growth associated with non-agricultural sedentism, Binford studied an Eskimo group whose population doubled in the decade from 1950 to 1960, during which time they became sedentary. The direct cause of this increase was partly a decrease in miscarriages, but chiefly the closer spacing of births, which was itself related to both reduced female mobility and less-prolonged male absence on hunting trips.[732]

One factor that may have contributed to the formation of sedentary cultures was rising sea levels at the end of the Pleistocene,[733] the increased population pressure experienced by Mesolithic people forcing them to concentrate along seashores, where aquatic food was available. The rising sea levels also created numerous drowned valleys, where potential aquatic food flourished.

Where local resources were and are adequate, settled villages can develop, and have done so, without agriculture. Three early examples of non-agricultural sedentary cultures are those of the Natufians, who inhabited caves in the Levant, the coastal peoples of northern Peru, and the Tolowa of the north-west coast of the United States. Of these, the Natufians eventually developed a form of agriculture, the coastal Peruvians finally adopted agriculture about which they had known but which they had been avoiding for a thousand years or more, and the Tolowa retained their traditional lifestyle until it was destroyed by European invaders.[734]

Non-agricultural sedentism and large group aggregations occurred in many parts of the world, and often without any evidence of significant new technology. In fact the Mesolithic use of aquatic resources was for the most part *not* based on new technology. Mesolithic peoples added one or two new techniques to the quest for aquatic foods, but for the most part they simply placed increasing emphasis on resources which had long been readily available but which had previously been ignored or exploited only minimally. Similarly, seal hunting, traceable back to the Late Pleistocene, gradually assumed greater importance in Europe during the Mesolithic, and this trend continued after the appearance of Neolithic farming economies.[735]

According to Binford and as intimated above, the first instances of sedentism were the consequence of the increased dependence on aquatic resources during the terminal Pleistocene and Mesolithic periods. Not all portions of rivers and shorelines favour the harvesting of fish, molluscs and migratory fowl; and it is with the dependence on just these resources that we find a higher degree of sedentism in both the Archaic period in the New World and the Upper Palaeolithic and Mesolithic periods in the Old World. This hypothesis is lent strong support by the fact that it is also in the Mesolithic and Archaic that we find, once sedentism began to increase, evidence for marked population growth and the development of food-storage techniques, the latter being functionally linked to the highly seasonal nature of the migratory fowl and fish exploited as food.[736] Thus where fisher-forager sedentism occurred emphasis came to be placed on the preparation and storage of food.[737] Here we have the beginnings of private property and trading, where those who had possession of the most stored food not only had the highest status, but were also the most powerful, due to their ability to control others' access to vital resources. Also, according to Wayne Suttles, storing food to trade with others was recognised as an important method of acquiring wealth; hence the interest the Tolowa fisher-foragers showed in acquiring women, the processors of food.[738]

VCP analysis of the hunter-gatherer era

As regards pre-humans' leaving the forest for the savannah, it may have been the case that, in keeping with e.g. Burt's analysis of vertebrate territoriality cited in Chapter 1, it was a matter of the weaker (less fit for forest living) apes being pushed to the periphery. There, they would have had to adapt behaviourally; or, if unable to do this, to produce viable mutations, or leave no descendants at all. That they did leave descendants, and in fact thrived in the new environment, was dependent on their ability to use tools.

Though one might be inclined to see the invention of early stone tools, such as the hand-axe, as a response to population pressure, we lack sufficient information regarding the size and living conditions of the human population at that time to warrant the making of such a connection. Nevertheless we see the VCP at work right from our beginnings as tool-using hominids. Once australopithecines began to employ technology they became dependent on it, and changed genetically and karyotypically (evolving into new species) in adaptation to that dependence. And as regards *Homo sapiens*, as has been implied by Darwin, already at the earliest stage of our development we had become "the most dominant animal that has ever appeared on this earth."[739]

One key aspect of the development of humans is our tools' originally being extensions of our limbs/hands functioning as substitutes for our teeth. It was as though we had grown various kinds of arms, of greater length or strength than our actual ones, which we could use and discard at will. This of course gave us a great 'advantage' over other species – too great an advantage for our own good, as this book tries to show.

Fire

According to Greek myth, Prometheus, watching humans as they shivered in the cold winter nights, felt sorry for them and gave them fire, previously only available to the gods. Later, however, in retaliation for a trick Prometheus had played on him, Zeus took the fire away from them, so that they had once again to live in cold and darkness, and could not cook their meat.[740] But then Prometheus stole the fire from Zeus and gave it back to humans. In Hesiod, Zeus then says to Prometheus: "Son of Iapetos, clever above all others, you are pleased at having stolen fire and outwitted me – a great calamity both for yourself and for men to come. To set against the fire I shall give them an affliction in which they will all delight as they embrace their own misfortune." He then had a woman, Pandora, created from clay, and sent her as a gift to Prometheus who, being suspicious, would have nothing to do with her. She was then sent on to Prometheus' brother, who, despite Prometheus' warning, married her. Pandora brought with her a box (a jar) containing evil, disease, poverty, war, and other troubles. When she opened the box, she released these sorrows into the world of men, and Zeus gained his revenge.

Apart from the recognition of the importance of fire to humankind, from the point of view of the VCP the Prometheus myth is of some interest in that it links this particular form of technological development to subsequent ills that befell humankind.

The fundamental factor that made the spread of *erectus* and *sapiens* possible was the use of fire; and fire and the later use of clothing are, Cohen suggests, explainable in terms of population pressure – combined with the pioneering principle. Their use constituted adaptations necessitated by enforced migrations into colder climates. Evidence of pre-*sapiens* use of fire is found only in the northern, temperate portions of the occupied world, not being evident in most mid-Pleistocene sites in Africa, for example.[741] (It is of some interest that modern hunter-gatherer Pygmies – living in tropical Africa – did not know how to make a fire.)[742] The fact that the earliest evidence of fire clusters so strikingly along the northern margins of human expansion suggests that the mastery of fire was not a recent 'discovery,' but was rather a previously known technique which was necessary for the survival of the populations that penetrated these new geographical zones.[743] When fire did begin to be used in Africa, its first use was primarily to expand the range of edible foods. The very uneven geographical distribution of fire in the Pleistocene suggests that the use of fire did not diffuse, but rather emerged as a specific response to local need.[744]

The role played by humans' mastery of fire constitutes a clear and important instance for the application of the VCP. Acquisition of the technology of fire made it possible for *erectus* to leave Africa and inhabit colder parts of the world (and one can imagine that the *erectus* population would have expanded further if the people had had clothing). Thanks to the pioneering principle, pushed by population pressure, the increase primarily in breeding sites but also in food meant the spread and growth of the human population. Here again, however, as e.g. in the case of pre-humans leaving the forest and moving onto the savannah, the situation would not have been one of certain adventurous families heading off to realise new vistas, but rather one where the weaker in the community were continually being pushed into non-optimum environments – in this case into colder environments – with the result over the years that the population expanded northwards beyond the tropics. And this they were only able to do thanks to their mastery of fire.

Fire was used only where it was *needed* – consider the constant vigilance and effort required always to keep a fire burning in a mobile community – whether for warmth or for the preparation of food. If, for whatever reason, wood – the resource then necessary for the use of fire technology – should have been removed, the population of early humans would at least have experienced a marked decline in size, no longer existing in such northerly latitudes; and it may be expected that much suffering would have ensued, depending on how quickly the decline occurred.

The ability to utilise fire thus went from being an advantage to a small population to being a need for a large one. At least in the case of *sapiens*, wherever fire was used the amount of available food would have increased thanks to its now existing in the form of e.g. grass seeds; and, given sufficient breeding sites, it may be expected that this increase in available food would have resulted in population growth. And as regards populations that had expanded north beyond the tropics in particular, their very existence was dependent on fire to keep them warm. In fact, *all* populations of early humans who used fire were or became dependent on it. Furthermore this would have had a genetic influence: individuals who would not have survived without fire, for whatever reason, would now be able to do so and to procreate, and their children procreate and so on, such that whatever genes as required fire for their phenotypic manifestation would have spread through the whole of the fire-using population.

In sum, in the case of fire in the hands of early humans we see the vicious circle turning from need, to innovation in learning to master fire, to its use, to an increase in available resources, to expansion into new areas, to new needs, to possible genetic dependence, to the creation of a (non- or poorly-storable) surplus, to increased consumption – including increased energy use – and on to population growth and possible overshoot.

Homo sapiens

Modern humans, i.e. *Homo sapiens*, with their own unique karyotype, came into existence in a situation that included the use of fire. One can imagine that if that situation had not included fire our karyotype might not have afforded sufficient of an advantage for our species to have gained a foothold. Given this, our dependence on the various sorts of technology up to and including fire may be said to be not just genetically, but karyotypically based. In fact this reasoning applies to the various species of modern *Homo* that led up to the arrival of *sapiens* as well, each step in the evolutionary chain from them to us reinforcing humans' dependence on earlier forms of technology for their existence. And it may be assumed that our more general dependence on technology has been karyotypically injected in successively stronger doses into our evolution right from when our ancestors first left the trees.

As in the case of *erectus*, our use of fire, through making it possible to expand out of the tropics and to eat what would otherwise have been inedible, allowed our population to grow to a size that could only be sustained so long as fire was available to warm dwellings and cook the relevant food. In fact, for those of us who migrated north, not only did we have a need for fire, but we also had to contend with other new needs, most particularly the need for clothing and shelter. Here we have an instance of 'invention becoming a breeder of necessities'[745] – a state of affairs which, on the VCP, is endemic to technological change.

Quite generally as regards the development of *Homo sapiens* from our inception up until our second major expansion-wave northward around 60,000 BP, we see a turning of the vicious circle from the innovation and use of improved weapons, clothing, etc., to demographic expansion, to the acquisition of a surplus, this process becoming self-perpetuating and reinforcing our genetic/karyotypic dependence on developing technology.

Though the climate in those areas where the migrants were pushed was sub-optimal, on the whole the new resources provided a better diet; and the increased ease in acquiring prey and the surplus of food this meant may have increased leisure – not only for the most powerful in the band or tribe, but for everyone. This, together with a general increase in technological innovativeness, may have resulted in the non-technological cultural development (e.g. cave art) that occurred in Europe around this time. Our innovative ability also made possible increased consumption of both food and energy (the burning of wood), leading to population growth, while at the same time the increased food resources also provided a surplus. In this case the food resources that became available were superior to the old not only as regards ease of capture but also size of prey. This plethora of vital resources meant a pushing back of external population checks, and undoubtedly a relaxing of internal checks as well, which in turn prompted further population growth. This growth was of course abetted by the fact that humans now occupied a larger area.

Here we may note that the fact that human resources were almost solely biological (and thus biodegradable) and existed on a relatively small scale meant that such side-effects of the VCP as pollution and increasing social stratification had yet to make themselves felt; though already at this time humans' overexploiting their environment in the form of hunting was beginning to lead to the extinction of other species.

Geographic expansion and the Upper Palaeolithic

One can see three main factors enabling us to expand northwards as far as we did in the middle of the Upper Palaeolithic. The first, shared by many other life forms, was the warming climate; the second was our *use* of the technology of fire and the lance; and the third was our *invention* of clothing and of weapons more formidable than the lance, in particular the javelin. It was our possession of these weapons that made us as successful as we were in hunting large prey. Our hunting resulted in, among other things, such 'domesticating' genetic changes in ourselves as those manifest in a decrease in robusticity due to our increased employment of tools, and those manifest in our dentition and digestive systems more generally as a result of our becoming more carnivorous.

The type of ecosystem in which early modern humans participated, and, by inference, the type of climate to which they are ideally suited and would have preferred to live, was tropical. Since the climate to the north was less hospitable, we can say that our northward migration was a result of population pressure[746] – in conjunction with the operation of the pioneering principle.

Distance from the equator

On the VCP, increased need of technology spurs technological innovation, and what we have seen throughout *Homo sapiens*' development is a greater technological innovativeness among people living further from the equator – particularly towards the north, where the greater land mass affords the preconditions for larger populations. Thus high-latitude populations have more tool types, and their tools are more complex and have more components.[747] As noted by Binford, northern adaptations in the Pleistocene required a more elaborate material basis, including fixed facilities such as fish weirs and game fences; and they also required more sophisticated toolkits, more substantial dwellings and warmer clothing.[748]

Note how in the extreme case of the Eskimos the lack of resources (such as wood, arable land) led to technological sophistication in order to overcome the lack. Note also however that this same lack of resources – both for the making of tools and to use them upon – meant that the Eskimos were more limited than other people in the sorts of tools they could make or profitably use.

In this way we can understand the surge in technological development both when *sapiens* moved north to the Levant 100,000 years ago, as well as when they moved further north to Europe, some 35,000 years ago. And we can also understand why Europeans developed technology further than the Africans, and why the Chinese developed it further than the Indians.

Genetic changes occasioned by the changed living conditions made possible by technological development can themselves be such that individuals cannot live under the same conditions as those under which their ancestors lived. White people and Eskimos[749] are dependent on the technologies of fire and clothing not only to survive in a cold climate but perhaps to survive at all. As regards the existence of the resources requisite for technological development, as implied above, they concern both those necessary to make various devices, and those to which the devices are to be applied. In this latter regard, there would be no point, for example, in developing a plough if the soil in the area in which you live would not provide increased regular harvests as a result of its use. Thus we can further understand why e.g. the Eskimos and the Mongols, despite living in inhospitable climates, did not develop technology so as to become sedentary, namely that horticultural technology was useless in the environments in which they lived, while the nature of their major food source (seals and horses) supported nomadism.

Morals, protowar and infanticide

As Darwin suggests, the basis of morals is the survival of the *tribe* in particular:

> [A]ctions are regarded by savages, and were probably so regarded by primeval man, as good or bad, solely as they obviously affect the welfare of the tribe …. This conclusion agrees well with the belief that the so-called moral sense is aboriginally derived from the social instincts. [These] social instincts … no doubt are acquired by man as by the lower animals for the good of the community, [and] will from the first have given him some wish to aid his fellows, some feeling of sympathy, and have compelled him to regard their approbation and disapprobation. Such impulses will have served him at a very early period as a rude rule of right and wrong.[750]

In micro terms, what Darwin is saying is that the protection of the gene line of the tribe must take precedence over that of the individual. Spread throughout the total population, this means the protection of the species' karyotype over the genotypes of individuals.[751] Morality is a cultural expedient for the preservation of *Homo sapiens'* karyotype through supporting the genotypes of *groups* (populations).

The maintenance of the tribe had to take precedence over everything else, and actions which threatened that maintenance were deemed immoral.[752] Thus the tribal territory became sacrosanct, and the taking of the territory of other tribes – the obtaining of which would further support the continuing existence of one's own tribe – met with the highest praise. As suggested in Chapter 4, when it comes to the survival of the human species, morals are particularly important in their population-checking effect. There are two basic ways in which this effect is manifest. One is through their supporting *infanticide*, where a mother feels the moral obligation to kill her infant; and the other is in the form of *protowar*, where a young man feels the moral obligation to participate in inter-tribal killing, and possibly get killed himself. Both of these *cultural* expressions of morals are *biological* expressions of population checks.

It is actions that potentially reduce one's own genetic fitness while improving that of the tribe that are moral. Thus having many male offspring, while it would support the tribe in its protowars, would not decrease but rather increase one's own genetic fitness; and the same may be said of e.g. making a big kill, which would help meet the band's food requirement. In such cases there would be no 'conflict of interest' between the sexual and social instincts, and so no moral action.

It may be expected that in the case of killing other humans the power of the social bond operated in two directions, one suppressing intratribal killing (as might occur e.g. in feuding or fights directly over mates or individual territory), and the other licensing inter-tribal killing.[753] As noted by Derek Freeman, almost every human society has regarded the killing of members of particular other human societies as desirable.[754] And, as pointed out by Washburn and Lancaster, such killing has been a major factor in humans' view of the world, every folklore containing tales of heroes whose fame is based on the enemies they eradicated.[755]

As regards protowars, Lorenz emphasises the *genetic change* they occasion, where he suggests that:

> When man had reached the stage of having weapons, clothing and social organization, so overcoming the dangers of starving, freezing and being eaten by wild animals, and these dangers ceased to be the essential factors influencing selection, an evil intra-specific selection must have set in. The factor influencing selection was now the [proto] wars waged between hostile neighbouring tribes. These must have evolved into an extreme form of all those so-called 'warring virtues' which unfortunately many people still regard as desirable ideals.

And he further cites Sydney Margolin, who suggests that during the few centuries when Prairie Indians led a life consisting almost entirely of protowar and raids, there must have been a selection pressure at work breeding extreme aggressiveness.[756]

The extinction of the Neanderthals

Whether the demise of the Neanderthals was the result of such conflicts is open to debate. However, mitochondrial evidence suggests that none of the Neanderthal females were taken as mates, which implies the possibility that the Neanderthals were treated mainly as food, in which case their extinction might be seen as yet another instance of *Homo sapiens* overkill. On the other hand, no unequivocal co-occurrence of archaic and modern remains has been found, the closest being the discovery of a stone tool tradition in western Europe about 35,000 BP that employed a mix of both archaic and modern tools.[757]

The latter half of the Upper Palaeolithic
Overkill

The transformation from the first to second halves of the Upper Palaeolithic is perhaps clearest in the Americas, where it occurred over the shortest period of time. The ecologically most important aspect of this transition was its resulting from overkill.

Prehistoric overkill is not an exceptional phenomenon in the development of humankind; it is just a particularly clear example of what we humans have always been doing – at a constantly accelerating rate – since we first appeared on the face of the earth.

In keeping with Martin's view, Pleistocene overkill in America may be broadly understood against the background of the VCP in the following terms: Newly arriving humans equipped with stone-pointed spears (the Clovis toolkit), eradicate a major portion of the largest mammals on the newly entered continent. Medium-sized and smaller mammals, partly because they constitute less desirable game, survive this onslaught. Once the easily accessible larger prey are gone – the result of diminishing returns from the use of the Clovis toolkit – smaller mammals are hunted. With this we have the transition from the Clovis to the Folsom toolkit. As Brues points out in keeping with this, any improvement in hunting technique thins out the supply of game and increases its wariness, so that the less well-equipped group or individual can no longer survive with a technique that was formerly adequate.[758] This of course relates to the idea taken up in Chapter 1, that any predator group's overdevelopment of hunting prowess would eventually eliminate its source of food.

Looking at the overkill phenomenon generally in terms of the VCP, we see that prior to its occurrence technological development in the form of e.g. fire and improved weapons led to population growth and pressure, the latter being partially and temporarily alleviated through geographic expansion (modern humans moving into e.g. Europe and America). At this time there was a marked increase in available resources due to the employment of the technologies of fire and weapons in areas in which they had not been employed before, with the result that the size of the total human population grew as more and more people were having their vital needs met. That the technology of fire was necessary to survival in some of these new environments meant that it was self-perpetuating there. It is to be noted that the existence of resources in the form of megafauna and wood in these areas was a matter of chance, and that the situation could have been such that they had not existed. In that case the human expansion would not have taken place, at least not at that time, and the vicious circle of population growth and technological innovation that ensued might not have occurred, or might have occurred later and have taken a different form.

In sum, the continued hunting of megafauna on the part of an ever-growing population, and the lack of new areas from which to obtain resources, led to the over-exploitation and consequent depletion of potentially renewable resources through the extinction of the most accessible megafaunal species and genera. That the population continued to grow while the food resources diminished was sooner or later manifest in population overshoot, i.e. in the manifestation of the overshoot principle. Thus we have an instance of the over-use of technology (javelins) not only undermining its own usefulness, but the preconditions of human existence as well. The diminishing returns from the employment of the original technology meant economic decline, leading to a situation of scarcity and need, and an increase in the influence of external population checks manifest in increased mortality. The response to this population crisis was to develop new technology suitable for the hunting of small game, which marks the beginning of the Mesolithic.

As intimated in Chapter 4, you could say that we were from the beginning not biologically equipped as a species to handle developing technology, which our eradicating a huge proportion of the genera of the world's large animals when we were still in our hunter-gatherer stage makes clear. If technological development were truly an aid to the survival of the human species, it would not have led to the elimination of a significant part of the population's resource base (albeit a part which technological development had itself made available) only to replace it with an inferior substitute. Something is wrong with technological development in the hands of humans – but then, you could say, something would be wrong with technological development in the 'hands' of any organism.

The end of the road, hunting, and quality of life

Sapiens geographic expansion was *prompted* by an increase in our numbers, and *resulted* in a further increase in our numbers. But this momentum of population growth ran into a wall when, around 25,000 years ago, there no longer existed significantly large new areas we could inhabit (the Beringian land bridge not yet having been established) while at the same time, again thanks to our technological ability, we had eradicated most of our large game. And the result was, among other things, an intensification of economic activity.[759]

As suggested by Cohen, there are a number of reasons for believing that the specialised hunting patterns of the Upper Palaeolithic with their massive kills were not so much the enjoyment of plenty as an adjustment to the increasing difficulty of feeding dense populations. Among modern hunter-gatherers, for example, hunting focused on individual species does not occur in game-rich environments, but is confined rather to regions where the particular species hunted is the only source of meat existing in significant quantity. Most important however is simply that the hunting of herding animals would have involved the considerable difficulty of driving them into swamps etc., which would only have been undertaken for lack of alternatives.[760] Such large-scale cooperation is likely to have emerged only as the available resources became increasingly insufficient. Furthermore, in order for the kill to have value for more than a few days, it would be necessary to preserve it,[761] which would not only require a surplus relative to daily requirements but would involve more work and the eating of less-desirable food. (Woodburn notes the Hadzas' dislike of preserving meat, "which is tedious, time-consuming, and renders the meat less palatable.")

Looking at reindeer in particular, their exploitation, whether by hunting or herd management, would seem to involve very significant costs in terms of work. Neither the specialisation on reindeer nor the pattern of occasional interception of their herds and storage of the meat seems an attractive strategy, and it is plausible that this pattern would not have been adopted except in the absence of other reliable prey. And looking at musk-oxen as another example, as pointed out by P. F. Wilkinson their increasing exploitation in the later stages of the Upper Palaeolithic suggests that other vital resources were scarce.[762]

The diminution in available food incumbent upon the eradication of most of our large game led to increased mortality. And one can expect that this diminution in food, combined with the impossibility of expanding into uninhabited territories[763] where there would be fresh game, would have meant an increase in the manifestation of such internal population checks as stricter mating requirements, infanticide and protowar.

Since the remains of infants – particularly infants that had been killed – are less frequently preserved in graves, and infant bones decompose faster than adult bones, it is impossible to obtain an accurate idea of human longevity for any period before the industrial revolution.[764] The estimated average longevity of 30 years for Cro-Magnons would have been lower if infants had been equally represented. Birdsell for example estimates that between 15 and 50 per cent of all live births were eliminated by systematic infanticide during the Pleistocene as part of the cultural machinery for creating equilibrium between population and resources.[765] (Here it might be remembered that infanticide is common among chimps and gorillas.) Fekri Hassan has suggested that an abortion or infanticide rate on the order of 25–35 per cent would have been necessary to account for the difference between potential and real growth rates during the Pleistocene.[766] And Divale has suggested that a whole complex of population control mechanisms, including systematic female infanticide and feuding, may have been universal to human populations for at least the last 70,000 years.[767] Here we have a clear instance of cultural checks on population growth.[768]

What archaeological evidence indicates even more clearly, however, is that, beginning at this time and continuing during the whole of the latter half of the Upper Palaeolithic, there was a significant increase in the development of more sophisticated weapons capable of taking prey that were previously inaccessible (and unnecessary).[769]

Population growth and possible decline

The reason for the difference in human demography between Europe and Africa, where in the former the population may have declined after 25,000 BP while in the latter it grew continuously, may well have to do with the fact that Europe had only recently been inhabited by *Homo sapiens* with their javelins. (During the last glaciation the human population of Africa was many times that of Europe.)[770] Thus the prey in Europe at first provided a source of food that the weaponry of modern humans could cull more effectively than could that of the already-resident Neanderthals, thereby leading to the prey's being over-exploited and human population pressure ensuing. (Note that *not* to have population pressure presupposes ecological equilibrium, which the Cro-Magnons never possessed.) As regards Africa, on the other hand, we modern humans had been living there for over 100,000 years, and, as suggested in Chapter 2, during this period some of the African animals had had time to adapt to our constantly increasing predatory skills. Furthermore, we did not use javelins there. Such a reason for possible population growth and decline has been put forward by Martin with regard to the Americas, where he hypothesises (as in our figure in Chapter 2) that the population of the front of invading humans as it moved south was some ten times denser than the population behind it.[771]

Cro-Magnon groups that allowed their populations to grow, and responded to need by devoting extra work to the food-gathering process, were able to compete successfully for space with groups that remained small. Thus, even if only relatively few populations grew, over a period of time they would have replaced the more conservative groups in all but the most marginal environments, thereby, among other things, spreading genes that favoured increased fecundity. In keeping with the population-increasing effect of group territoriality, awareness of the need to win inter-tribal conflicts may very well have been a factor in many peoples' *not* limiting the size of their populations.[772]

On the VCP virtually every technological advance during the latter half of the Upper Palaeolithic, due to its helping to meet *vital* needs, can be seen as directly or indirectly increasing the number of humans who could occupy a particular area of land. In this case such developments as the invention of needles, or even bowls and spoons, each in its own way contributed to creating an environment which could support a greater number of people. As regards *containers* in the case of hunter-gatherers, Washburn and Lancaster suggest that where meat can easily be carried away from the site of a kill, the development of receptacles for carrying vegetable products was an important advance in human cultural evolution already in hunter-gatherer times. Without a means of carrying, the advantages of a large area are greatly reduced; and sharing implies that a person carry much more than he or she can use.[773]

Among the needs incumbent on the turning of the vicious circle during the latter half of the Upper Palaeolithic was the need to develop technology. In Europe the technology employed in the form of javelins led to the eradication of our major food source, and resulted in our own possible depopulation there at the beginning of this period. At the same time, it addicted us to *developing* technology; in this way the self-perpetuation of developing technology becomes part of the vicious circle.

During the early *sapiens* period we can see two major turnings of the vicious circle, the first during the Upper Palaeolithic and involving the over-hunting of megafauna, and the second, its aftermath, constituting the Mesolithic.

The Mesolithic

Whether or not the population grew or shrank at the beginning of the latter half of the Upper Palaeolithic, it is clear that population pressure constantly increased. One indication of this is the aforementioned evidence from Egypt. The Nubian tribe sites along the Nile around 14,000 BP appear to have formed a tightly-knit mosaic, each culture maintaining its individuality despite close proximity to others and the opportunity for easy communication. This distribution is indicative of carefully maintained territorial divisions, suggesting intense competition for resources. That there was fighting over territory in Nubia is suggested by the earlier-mentioned high proportion of skeletons excavated from this area that display traumata from projectile points; and burials in California suggest the same there at around the same time. The scarcer resources are, the more tribes will compete over retaining or acquiring them.[774]

A general indication of the existence of increasing population pressure and the experience of economic crisis during the Mesolithic was the transformation from specialised to eclectic hunting, as manifest e.g. in the Italian Mesolithic in the replacement of big-game hunting with shellfish collecting and the hunting of small game. And the possible reduction in the number of European sites inhabited in the Mesolithic prior to sedentism suggests that during this period the population in Europe was small, until it was increased by the arrival of horticulturists from the east. As Cohen says, when a group increases its concentration on water-based resources – which previously were more difficult to acquire than land-based ones – this shift may be viewed as resulting from demographic necessity, i.e. population pressure, rather than choice. In our terms, it is a manifestation of a more general aspect of the VCP, which says that the using of greater effort to obtain a particular resource is a result of necessity – in the present case, vital necessity.

Sedentism, population growth and population pressure

As Cohen suggests, sedentism in most cases occurs not because of newly discovered resources which *permit* year-round residence in a single location, but rather because of the decline of resources associated with other parts of the traditional annual cycle, or because of territorial impingement by other groups.[775] Sedentism is itself a form of technology resulting from population pressure operating inwards. And as regards technological change in this context, as Rolf Sieferle says, it was only worthwhile to produce fireproof, fired pots in a sedentary setting; and it was only then that those plants became edible that required a higher temperature.[776] And much of the food had to be stored. Amongst the Tolowa of the Pacific north-west, women performed the time-consuming work of cleaning and drying salmon, which could be caught only during brief annual runs but were a staple food year-round.[777] In any case, sedentism, with its fired pots, enlarged the range of foods, and pushed population growth.[778]

Sedentism originally came into existence because of the requirements set by aquatic resources becoming a necessary part of the human diet. In a broader perspective, fisher-forager sedentism can be seen as the result of turnings of the vicious circle that took us from the hunting of megafauna, round to the hunting of ungulates, and from that round to the hunting of small mammals and the consumption of marine fauna, all the while drying and storing a greater proportion of the food we acquired.

As regards the transition from the Palaeolithic to the Mesolithic eras around 12,000 BP, Cohen points out that, as mentioned above, the Mesolithic use of aquatic resources was for the most part *not* based on new technology. Mesolithic humans added one or two new techniques to the quest for aquatic foods, but for the most part they simply placed increasing emphasis on resources which had long been readily available, but which had previously been ignored or exploited only minimally.[779] Humans were capable of sedentism long before they adopted it as a way of life, as is clear from the Neanderthals being relatively sedentary. Thus not only did it not come into existence until it was needed, but it is not inextricably linked to, or dependent upon, the domestication of plants or animals.[780]

Mothers not needing to carry their children on long treks, and increased contact between the sexes, gave rise to demographic stress or population pressure in sedentary situations during the Mesolithic, and triggered population increases which cultural checks failed to counter. This increased contact was largely the result of a shift to a broader spectrum of wild foods, which led to reductions in mobility due to their location and the means by which they were acquired and/or stored. Thus sedentism was both an effect and a cause of population pressure. The population pressure that arose resulted in an intensification of labour input into the food quest, and led, via improved seasonal scheduling of the procurement of wild foods, to the increasingly specialised exploitation of particular resources[781] – some of which would not have been eaten previously.

During the Mesolithic, new, potentially renewable, resources became depleted (though not eradicated) at least in part due to the growing need for food incumbent upon continually increasing population pressure. With this shift to a new form of food acquisition new skills became genetically favoured. While the quantity of food consumed increased due to the increasing population pressure, its quality became lower; i.e. it contained less meat. Though internal population checks probably increased, they did not do so sufficiently to avoid the over-exploitation of the relatively newly available food resources.

At the same time technology became more complex, while the quality of life became lower for everyone, particularly as work increased in the food-acquisition process, with men having to spend more time hunting and women more time gathering due to the constantly increasing distances to be traversed. And hunting continually brought less meat, with the result that everyone's diet became poorer. The constantly increasing work meant steadily decreasing leisure for everyone, and probably a decline in non-subsistence cultural development.

As time went by each man-hour spent on the task of acquiring food gave diminishing returns, and economic decline set in as the population overshot its resources. The new fisher-forager technology at least to some extent undermined its own operation through its sedentism supporting population growth; and it thereby also undermined the existence of those populations that had become dependent on the technology. The result was a new period of scarcity, with an increase in the effect of external population checks, leading to the need to domesticate plants and animals. Whereas from 25,000 BP to and including the hunter-gatherer part of the Mesolithic there was constant *population pressure*, with or without population growth or shrinking, beginning with *sedentism* population pressure and population growth pretty much went hand in hand.

As has already been noted, the genetic effects of the movement to domestication began even before we became sedentary. As Mumford says: "Paleolithic man began to domesticate himself before he domesticated either plants or animals."[782] Thus we find, for example, that the skull structure of the Neanderthals bears the same relation to that of *Homo sapiens* as do the skulls of wild animals to tame animals in the case of e.g. the rat, cat, dog, pig, horse and cow; in each case the wild form has the larger face and muscular ridges. And our temperament and glands have also developed in the same way as those of other domesticated animals.[783]

Why should such a change have begun to occur long before humans became sedentary? The answer again lies in the operation of the VCP. The VCP begins to apply as soon as technological dependency begins, and this began shortly after our ancestors started using tools. Through our ancestors' dependence on such things as tools to kill and butcher prey and fight off rivals and enemies, and fires for warmth and cooking, their karyotypes adopted forms where, among other things, the strength of body and jaws was selected against.

With regard to the application of the VCP to the development of humankind, the fact that it applies so clearly already in prehistoric times is noteworthy. It is not the case that our environmentally destructive ways are a relatively recent occurrence, starting, say, with agriculture or, as some believe, with the industrial revolution. The fact that the turning of the vicious circle began even before modern humans came into existence, and that we are karyotypically and genetically adapted to it, indicates just how hard it would be for us to break out of.

Note too that through the external check of food scarcity leading to increased mortality, and the ability of small animals to avoid hunters and breed more rapidly than large, the people living at the end of the Mesolithic might have established something akin to an equilibrium. On the VCP, however, as we shall see below, it is precisely the fact that there was further technological development, and a resource to which it could be applied, that ecological equilibrium was *not* attained, the result being rather the horticultural revolution.

VCP models of increasing complexity

While the vicious circle principle is to apply to the whole of *Homo sapiens'* development, this development has constantly been increasing in complexity due to the operation of the VCP itself. For this reason the form taken by the VCP in its application also becomes successively more complex. Thus we can say that *the vicious circle principle adopts more complex forms as its impact on society increases*. To capture this increasing complexity, different models for the application of the VCP can be devised, starting with a model to be applied to humans in our simplest guise – i.e. as prehistoric hunter-gatherers – and on to human society today.

In what follows I shall present four successively more complex models: the *hunter-gatherer*, the *horticultural*, the *agrarian* and the *industrial*. Each model corresponds to a major revolution in human social, technological and more general cultural development. Nothing is lost of any more-primitive model in a more-complex one: the process is one of new factors coming into the picture and the existing factors becoming more complex. So all lower-order models are applicable to higher-order situations, though their applicability becomes less clear-cut the more sophisticated the situation.

For the sake of brevity, each model is presented simply as a list of ingredients in the VCP, it being expected that the reader can see the links between these ingredients on the basis of the explication of the VCP given in the previous chapter. Remember that many of the factors presented in these models may occur simultaneously, overlap temporally, or occur in a slightly different order than that given here, depending on the circumstances of the particular case. And keep in mind that, as implied earlier, any one of the elements in the model might have topped the list, with the others following in the same order.

In that it contains the central aspects of the VCP, which can be found in all subsequent models, the hunter-gatherer model constitutes the *core* model of the VCP.

The hunter-gatherer model

Scarcity and need
Innovation (particularly as regards weapons)
Economic development
Increase in available resources
Prerequisite of the existence of resources
Increased energy use: fire
Expansion to new areas made possible by technological change
Resource depletion
Species extinctions
New needs: fire, new weapons, clothing

Technological development self-perpetuating
Karyotypic and genetic dependence on developing technology
Surplus: non-storable (except for dried meat)
Varying quality of substitutes (first larger game, then smaller)
Population checks: internal (the spacing of births) and external
Morals
Population growth (up to 25,000 BP)
Population pressure (after 25,000 BP)
Increasing complexity of tools, particularly weapons
Social stratification: in terms of hunting ability; men vs. women
Lowering of the quality of life of everyone: increased work and poorer
       diet beginning at 25,000 BP
Conflict: feuding and protowar
Leisure: declines during the Mesolithic
Cultural development: cave art etc.
Decreased (economic) security (particularly during the Mesolithic)
Population overshoot
Diminishing returns
Undermining of technology and human existence
Economic decline (starting at 25,000 BP)
Scarcity and need
Possible population reduction

The horticultural (domestication) revolution *10,000* BP

Perhaps along with the taming of fire, the horticultural and industrial revolutions constitute the greatest changes in the development of humankind. The horticultural revolution came about at the end of the last ice age, when the tundra, which was ideal for human hunting, was replaced by forests. By 10,000 years ago the climate was almost as warm as today, and the increase in temperature brought with it an increase in the distribution of edible plants,[784] particularly wild grasses; and by this time the density of game animals had diminished, many of their species having become extinct. This reduction in the size of open habitats such as savannah and tundra, combined with habitat loss in the coastal zone as sea levels rose and big-game species became extinct, meant for humans a greater dependence on vegetable products.[785] In this respect it constituted a major reinforcement of the backward move towards our more vegetarian ancestry that began with the hunting of smaller game 15,000 years earlier.

The horticultural revolution was essentially a shift to *domestication*: the direct manipulation of plants and animals for the purpose of obtaining food, and with it a change in their genetic constitutions. Domestication constitutes humankind's second greatest harnessing, after fire, of the world's natural energy,[786] and includes the continuing genetic modification of humans resultant upon our employment of technology. Technologically, but not necessarily chronologically, the first step towards domesticating other species was our herding of animals.

Herding

The process of domesticating animals can be seen as in principle going through the stages of herding, in which the animals are essentially wild but are rounded-up for slaughter; taming, in which a small number of young wild animals are raised by humans; and proper domestication, in which the animals are born in captivity and have undergone genetic change. The keeping of domesticated animals normally means their being fed and sheltered, and may include e.g. their being milked, having their blood let, or being sheared.

The first animals to be herded include goats and sheep, 9000 years ago (in the Near East). Later additions include swine and oxen (8000 BP, also in the Near East), llamas (7000 BP, South America), horses (6000 BP, central Asia), and camels (5000 BP, Arabian Peninsula). Reindeer were (and still are – using helicopters) also herded (Scandinavia, date unknown), but, unlike the other animals mentioned here, never came to be domesticated. Herding greatly increases the temporary carrying capacity of the relevant area for humans,[787] though not to the same extent as horticulture in more fertile areas.

Herding was normally if not always directly preceded by the progressive elimination through hunting of species of the more vulnerable slow-breeding large mammals of the area in question, such areas being drier than those where cultivation first occurred, as was the case in the Sahara and northern Africa.[788] Milking and presumably the blood-letting of cattle began around 8000 BP, if not earlier, in the Sahara,[789] which was an extensive desert from 18,000 to 13,000 BP, but was covered with lakes surrounded by savannah from 9000 to 5000 BP.[790]

Herders first appeared after humans had spread over the globe, and there were no new areas left to occupy. Like hunter-gatherers, herders are nomads, following the migration of their herds, much as Cro-Magnons did during the Upper Palaeolithic. Despite herders' constantly being on the move, the phenomenon of individuals owning property, as evinced already in fisher-forager cultures, reappears with them. The animals they herded – the ownership of which could have been indicated by branding – became their property. Their territories, i.e. the lands on which they herded, on the other hand, were more 'owned' by the group as a whole,[791] while the animals in the herd constituted both capital and a medium of exchange for individuals. Thus something's formally being considered the property of a particular person concerned animals before it concerned non-biotic commodities or land, the ownership of land presupposing a sedentary setting; and it is with the herders, more so than with the fisher-foragers, whose exchangeable property consisted essentially of dried fish, that we see the birth of the prerequisite for commerce.

Domestication

About 10,000 years ago the horticultural lifestyle started to provide the primary means of human subsistence, with hunting and gathering playing a constantly more peripheral role. At that time the earth's total human population was around five million (still less than that of other primates such as baboons), after which it began to increase markedly.

Plant domestication

One of the environmental changes in parts of south-western Asia at the end of the Pleistocene was an increase in grasslands; this had the effect that the people living in, or moving into, the area increased their consumption of grass seeds, particularly those of wheat and barley,[792] the practice of harvesting wild grains extending back to before 15,000 BP.[793]

Here we note, as has been emphasised by Cohen, that the transition to horticulture occurred at more or less the same time over the whole world – in this way it was similar to the adoption of fire – at the beginning of the Holocene. The generally accepted view today, and the view that has been argued for by Cohen, is that agriculture was not a phenomenon that spread from a founder community to other communities so as eventually to cover the whole world, but once the climate changed it developed independently in widely different types of environment. Thus it involved the planting of e.g. root crops in the tropics and cereals in semi-arid areas,[794] and it is from these areas that it afterwards spread.

As has been pointed out by Cohen, domestication in e.g. both North America and the Near East appears not to have involved choice resources, but low priority foods, which were used extensively only after the range of the diet had been broadened to include a number of other items.[795] The domestication of crops, in particular grains,[796] began in areas adjacent to relatively food-rich aquatic pre-horticultural settlements along the Tigris and Euphrates. In the tropics such plants as yams, potatoes, taro and cassava were grown. Tree-borne crops such as bananas, coconuts, olives and various kinds of nut also began to be cultivated around this time. Meanwhile herding, with its relative freedom from work, was more prestigious than horticulture.[797]

Horticulture (which may have slightly preceded the domestication of animals) reduced space requirements from the ca. 25 km²/person during the Upper Palaeolithic and Mesolithic to ca. 1 km²/person; and with the domestication of animals this figure fell to 0.5 km²/person.[798]

As regards plant domestication, the first and subsequently predominant form of horticulture was swidden, or slash-and-burn, cultivation. (The word *swidden* is from the Old English term for a burned clearing.) First, an area of forest was cut down by men using polished-stone axes, and then burned. Then, if grass seeds were sown, the land was probably prepared, if at all, with a digging stick.[799] The sowing of seeds was of particular importance to both the horticultural and agrarian periods, grass seeds constituting the basic staple of both in non-tropical areas.[800] If tubers were planted, women would stick them in holes they had made with a digging stick. Harvesting, in the case of cereals, was performed using a flint-edged sickle.

Much of the efficacy of the swidden system consisted in the fact that the surrounding forest vegetation does not include the low-growing plants which quickly invade cleared ground, while at the same time the burning that the swidden method involves destroys whatever weed seeds there may be, so there is little or no need to do any weeding. Nor is it necessary to spread fertiliser or grow rotation crops to keep up soil fertility, as this is provided for a time by the forest ashes.

The same plot would be used for a number of years, until its productivity declined to the point where it was more worthwhile to chop down and burn a new area of forest and start afresh. The original patch would be left fallow to eventually regain its forest character, which would take some 20 or 30 years, after which it could again be used for crops. With increasing population density, however, the length of time a swidden plot was used before being left fallow increased, and the length of time it was fallow shortened. Longer use of the plot increased the amount of weeds, and a shorter fallow period did not allow the re-growth of forest but only bush. Because the ground was no longer covered by forest, sunlight reached it, which resulted in more weeds in the form of unwanted grasses and other plants. Thus the cultivators had to clear land of bush and rough scrub, which became the predominant vegetation in the area. This increase in weeds was responsible for the transition within horticulture from the use of the digging stick in forest-fallowing to the hoe in bush-fallowing, the first hoes being essentially forked digging sticks, with the handle longer than the head. Hoes were used in a manner similar to the axe, thus constituting a tool essentially different from the digging stick.

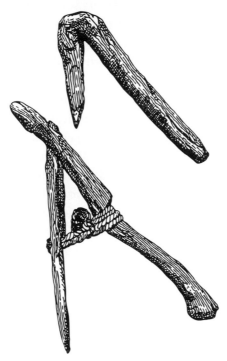

Primitive hoes.[801] A primitive Egyptian hoe cut from a forked branch; and a more developed form with hafted wooden blade. Both Middle Kingdom (2030–1640 BC)

By the bush-fallowing stage the soil had also become more compact,[802] which meant that the forked-stick hoe had eventually to be replaced by one with a stone head, which required much greater effort to wield. The stone heads were of *polished* stone, the grinding of which also required extra work – all on the part of *men* – as did the grinding of other stone-headed implements related to this form of horticulture, such as axes. By the final stage of the swidden process, described as 'short fallow,' the soil had become infested with perennial weeds, which could not be controlled by hoeing without exceptional effort.[803] Eventually bush-fallowing in sufficiently fertile places was replaced by *gardening*, where the same plot was used continuously, and crops were rotated to support productivity. Gardening, compared with swiddening, gives an even poorer caloric return to labour, but given greater labour, can provide greater quantities of food. Gardening can absorb ten times more labour input per unit land than can swiddening, and constitutes the most intensive form of cultivation in rainforests today.[804] This specialised food production meant a lack of flexibility however, which led to famine when crops failed e.g. as a result of drought.

Animal domestication

As mentioned above, to become domesticated, animals must *breed* in captivity. At the same time we note that, in contrast to the domestication of humans, while breeding has led to genetic changes in plants and animals, it has not led to karyotypic changes; and we thus speak of *breeds* rather than species or subspecies of domestic animals.[805]

The first domesticated animal was the dog, domesticated from the wolf some 14,000 or more years ago. Like the horse, the dog was originally used mainly as food, and like the pig, it is able to feed on human waste products, such animals being in this respect pre-adapted for domestication.[806] As in the case of fire, all human settlements have dogs. They may act as free-roaming carnivores which perform such functions as keeping down the vermin, warning of approaching animal and human predators, and helping in the hunt. In the north they could pull sleds, sled-pulling dogs being among the human migrants to America some 11,500 years ago.

Later animal domestication was mainly of herbivores, such domestication commencing shortly after the herding of the animals in question, their generally being domesticated around the same time as plants in the area. Thus we see a transition where hunter-gatherers, lacking sufficient game, begin to herd their prey to be assured of obtaining meat, and eventually domesticate it to obtain as much meat as possible, as well as a greater proportion of fresh meat.

Sieferle outlines a conjectural history excluding overkill in which hunters constantly follow a large animal herd, which gradually becomes used to them. These hunters would have driven away other predators and taken care not to kill any gestating or lactating animals; and, e.g. by preferentially killing aggressive males, they would have encouraged certain characteristics in the herd, in this way eventually domesticating it.[807]

The first animals to be domesticated after dogs were sheep and goats 9000 or more[808] years ago. For the first 4000 years after domestication sheep were used only for meat, wool sheep appearing after about 5400 BP.[809] And cats were first domesticated at about the same time – like sheep and goats, in the Near East, cats probably in association with rats and mice being attracted to stores of grain; then cattle (aurochs: *Bos primigenius*) around 8500 BP in south-eastern Europe;[810] or perhaps somewhat earlier in the Sahara;[811] pigs and bees; and then the horse in the Proto-Indo-European, Black Sea–Caspian Sea steppe region, 5900–5800 years ago;[812] and the donkey some 5000 years ago[813] in east Africa.[814]

As Neolithic farmers became increasingly dependent on cattle as draught animals they at the same time found that they were quite troublesome to keep. During the day they had to be kept away from water supplies and growing crops, while at night they had to be protected from the predators they attracted, such as lions and wolves.[815]

Being raised by people living on the margins of subsistence, the first domestic animals probably had to find their food themselves – as is the case even today with primitive people's livestock – or at best were fed waste. As a result they did not obtain a proper diet, and generally suffered from poor health, a situation in which the smaller animals tended to survive.[816] This, possibly together with crowding (à la the Norway rats), as well as their actually having been bred for small size, meant that domesticated animals were noticeably smaller than their wild counterparts, such that by the Iron Age cattle were being bred that would have been considered dwarfs by today's standards.[817] Since 5000 BP no new species of plant or animal for any use whatever – whether it be for food, clothing, shelter or transportation – has been domesticated.[818]

Pastoralism

Though horticulturists did not herd, they were often shepherds, i.e. they tended *flocks* of domesticated animals that were let roam in the wild to obtain food, but were either daily or seasonally returned to a fenced-in area. In such societies, as in herding societies, it was the men who were responsible for the animals. Those people whose subsistence depends wholly or mainly on shepherding are *pastoralists*.

The idea of ownership of course applies in the case of pastoral animals at least as much as it does in the case of herded animals. The property constituted by flocks was probably kept 'in the family' by some form of inheritance such as lineal descent.[819] Note the human, cultural nature of this; there's no inheritance among other animals. Note too how it weakens the species, allowing individuals dependent on cultural inheritance to propagate.

Areas where horticulture practised

Horticulture was first practised in areas *adjacent* to those occupied by the already sedentary fisher-foragers.[820] These areas provided good conditions for the domestication of both plants and animals, particularly if they lay between major ecosystems and along the edges of woodlands.[821] In being drier than riverine locations they both contained more potential animal domesticates as well as provided better conditions for the growing of grasses.

As regards grasses and plants more generally, the seeds of those adapted to hot and dry climates normally contain less moisture, which means that they are not as subject to rot and can thus be stored for longer periods (the storability of seeds such as those of wheat and barley was often enhanced by charring them over a fire),[822] which in turn means that less of the year must be devoted to producing food. Since a certain portion of these seeds had to be set aside for planting the next year, and the rest constituted food to be consumed year-round, their storability weighed heavily in their favour. Dense stands of such plants are typical in semi-arid regions, and already existed there before the influx of humans.[823] In fact, they existed in such quantities that it was possible to store grain for year-round use on a massive scale – a prerequisite for *agrarian* sedentism.[824] Though individual families owned their own plots, in the case of both horticulturists and pastoralists the area they occupied as a whole belonged to, and was defended by, the tribe.[825]

As population increased, the villages that horticultural sedentism gave rise to eventually became towns. Among the first were Jericho in the Levant and Çatal Hüyük in south-central Anatolia. Jericho originally, i.e. about 9000 years ago, contained some 150 to 250 people. Their staple food consisted of grains, which were supplemented by seasonal gazelle hunting and, later, domesticated gazelles. As evidenced by rich child burials, there also existed a system of social ranking in Jericho during the horticultural period. After some time, but still during the Neolithic, there was a shift from the domestication of gazelles to sheep and goats. By around 6000 BP the population had increased to about 2000, which at the time was unusually many to be in one place, and grain storage bins were in use. At this time Jericho had walls and a tower, which were probably for defence, and were the first ever built – the fact of their being built itself suggesting a social hierarchy and division of labour.

Though horticulture was originally adopted almost simultaneously at many different points in the world, it advanced across Europe from the Near East at barely 1 km per year. From its origins there around 10,000 BP, agriculture crept north-westwards to reach Greece around 8000 BP, and Britain and Scandinavia 2500 years later.

Innovation

While domestication constituted the fundamental innovation of the horticultural revolution, humans had been aware of horticultural techniques for a long time, the primary tools associated with the beginnings of horticulture being already well known and occasionally employed by Palaeolithic peoples,[826] which can partly account for how horticulture could appear simultaneously all over the world. By 30,000 to 20,000 years ago, apart from the digging stick, people already had knives (sickles) that could be used for reaping grain, polished-stone tools, particularly axes (first used in Japan ca. 30,000 years ago),[827] mortars and pestles, and grindstones for crushing paint pigments and cracking nuts, which were later adapted to the milling of grain.[828] Here we might also note that already in Palaeolithic times flour was milled by strewing grass seeds over the surface of a stone and pounding them with another stone; in time the bottom stone became hollowed out, and the one used for pounding became rounded – resulting in the first mortar and pestle. Other adaptations, just after the horticultural revolution, include the sliding of a perforated stone to near the bottom of the digging stick to weight it (a similar method is employed by the Bushmen),[829] and the digging stick's eventual development into the hoe.

The hoe with a stone head is a form of bludgeon, and its use requires neither the attaining of great speed nor momentum, but rather a constant, back-breaking effort. The wood-cutting polished-stone axe of this period is also a bludgeon as regards how it is used and its demands on the physique of the user. Thus the optimum food-obtainer of the horticultural economy was not the long-limbed spearman nor the broad-shouldered archer, but a sturdy peasant of medium build, a stature which then began to be genetically favoured in the population.[830]

The fact that humans no longer needed to move about – or were *unable* to move about – meant that many more possessions could be and were accumulated. Given sedentism, such heavy articles as pottery, first produced some 9000 years ago (though, as mentioned, the first ceramics were much earlier), could be used, and these articles began to be produced in number.

Technological innovations during the horticultural era include cloth, baskets, sailboats, fishnets, fishhooks, ice picks and combs. The innovations from 6000 to 5000 BP, at the end of the horticultural period and presaging the agrarian, include the plough (the first instance of which was a simple hoe, i.e. a forked digging stick, in traction),[831] the use of animals to pull ploughs and function as pack-animals (the oldest pack-animal would seem to have been the donkey), and a yoke and harness to enable them to do this, as well as the wheel and its application to both wagons and the manufacture of pottery (in Mesopotamia and, to wagons, on the steppe between the Black and Caspian Seas, both ca. 5500 BP),[832] the invention of writing and numerical notation, the use of seals to mark property, and the invention of the calendar.[833]

Social stratification: chiefs and workers/warriors

A horticultural tribe consists of a collection of bands that unite for protowar. While tribal leaders may be called big men or chiefs, they are not formal full-time political officials, and, like hunter-gatherer headmen, they usually exercise influence rather than what we would call power. *Chiefdoms* are organisations that unite many thousands or tens of thousands of people under formal, full-time political leadership. The populace of a chiefdom is usually divided into hereditary ranks or incipient social classes, often consisting of no more than a small chiefly or noble class and a large body of commoners. The term *tribal societies* usually encompasses bands, tribes and weak chiefdoms but excludes strong chiefdoms and states.[834] Social stratification becomes more evident the larger the social unit considered.

The size of any particular group depended essentially on the size of the surplus the herding and horticultural lifestyles could amass for it. This surplus constituted capital, in the case of herders in the form of livestock, and in the case of horticulturists in the form of seeds. Beginning with the horticulturists this economic surplus became *regular*, and with the agrarians it became the property of but a few.

Thus during the horticultural era social inequality increased. Slavery began in western Asia, and for the first time there existed both affluence and poverty in the same society, and crime and human sacrifice became widespread.[835] At the same time government began to develop.

The inequality of men and women already manifest in hunter-gatherer societies (as well as amongst pre-humans) increased in the horticultural era. While women did the hard work and the carrying, the men, at least at the beginning of the era, sat around and talked, except for seasonable bursts of activity,[836] some of which involved making tools and fighting. With the advent of horticulture, women often became beasts of burden and at the same time experienced more frequent pregnancies, and as a consequence suffered poorer health.[837]

Quality of life

The quality of horticulturists' lives was markedly lower than that of hunter-gatherers. Judging from the lifestyles of modern hunter-gatherers, horticulturists must have worked much more than their predecessors, in both making and using stone tools. (Note that the work required of modern horticulturists is lessened by their having tools of steel, which they obtain through barter, rather than of stone, which they make.) While women worked planting and tending crops, a task involving a good deal of drudgery itself, as the horticultural era continued men found themselves grinding, polishing and boring holes in stone tools for use in the horticultural effort.

As regards horticulturists' food, not only did these people starve when their crops occasionally failed, but their diets, based to such a large extent on vegetables and dairy products, tended to be less nutritious than foragers' diets, which were higher in proteins and lower in fats and carbohydrates. As a result of this worsened diet, combined with genetic selection for compactness, there was a five to ten per cent reduction in body size relative to the people of the Mesolithic,[838] who, as mentioned, were already smaller than their Palaeolithic predecessors. In Greece and Turkey average heights of 1.78 m (men) and 1.68 m (women) were found among postglacial hunters. In agricultural societies that settled the same area around 6000 BP the heights were only 1.60 and 1.54 m. At the same time there were signs of a social differentiation in the diet: members of the horticultural upper class were larger than the peasants.[839] Bronze Age (Late Neolithic) burials reveal an aristocratic world with richly developed upper-class life based on organised luxury trade and the labour of the lower classes.[840]

Of greater importance for the whole of the subsequent development of humankind however was the contraction of new diseases from animals during the horticultural period. The majority of infectious diseases that have plagued humankind since that time have been those acquired from domesticated animals. These include smallpox, brucellosis, malignant boils (anthrax), diphtheria and tuberculosis, all contracted from cattle, influenza contracted from pigs, leprosy from water buffalo, the common cold from horses, and measles, rabies, hydatid cysts and tapeworm from dogs.[841] (Note the prevalence of disease in animals that in the wild constitute herds or packs, due to their higher population density.) As a result of the domestication of animals, humans now share 65 diseases with dogs, 50 with cattle, 46 with sheep and goats, and 42 with pigs.[842] And deforestation for agriculture at this time promoted falciparum malaria, the most devastating of all tropical diseases, which led to genetic change in the form of sickle-cell anaemia in human populations.[843] The effect of contagious diseases constantly increased with the increase in population density[844] and constantly growing trade.

The effect of the contraction of new diseases from domesticated animals, and the more general spread of parasites, including bacteria, viruses, worms, etc., was exacerbated by the horticulturists' crowded, unsanitary, sedentary existence. These factors, including contaminated stored food, would sometimes have caused widespread illness.[845] The infant mortality rate soared relative to hunter-gatherers as parasites passed through the horticultural population.[846] Here we note that the number and variety of the parasites preying on humans, and the seriousness of their effect, increases steadily as one moves from polar regions to the tropics.[847]

As a result of these various influences, human life expectancy at birth may well have fallen – reports differ in this regard. But if mortality rose, fertility climbed even more. Such factors as mothers no longer having to carry their infants long distances, being able to provide them with softer (boiled) foods, thus influencing lactation, and the different sexes generally being in closer proximity to one another – all this against the background of the horticultural surplus – led to a shorter spacing of births[848] and the survival of a greater number of infants.

Protowar and feuding

Interaction between different societies increased drastically with the horticultural revolution. This interaction took the form of commerce, protowarfare and the spread of infectious diseases, all of which increased with time.

Warfare, unlike protowarfare, was virtually impossible for hunter-gatherers, not only due to their small numbers, but also because of their not being able to accumulate sufficient food to see them through such engagements. It was first with the horticultural era, when food from domesticated sources could be stored as grain or livestock, that humans began moving towards the true warfare involving large armies and lengthier campaigns that blossomed in the agrarian era. Furthermore, this grain and livestock, in constituting property, was also an immediate *cause* of such conflict, as well as being a prerequisite for commerce. In the case of herding peoples, apart from their often fighting for other reasons, the nature of their capital meant they had to contend with thieves, which itself was a constant inducement to conflict.[849] As the size of populations increased, so did stress related to resource acquisition, and with it, violence.[850]

As has been emphasised by Keeley,[851] warfare has been more frequent and deadly in primitive societies than in state societies; and it has been worse among horticulturists than among hunter-gatherers, while at the same time the cannibalism evident among hunter-gatherers continued among horticulturists.[852] Following Keeley, we can divide the forms of battle used by non-state peoples into formal battles, small ambush raids, and large raids or massacres. For most primitive groups, small raids have been the most and massacres the least frequent form of combat. It is becoming increasingly certain that many prehistoric cases of intensive protowarfare in various regions corresponded with hard times created by ecological and climatic changes. As a result of these altercations, the victors acquired more territory or choice resources with striking regularity, territorial change being a very common outcome of protowars.

When one of the contending parties in pre-state warfare was routed, the subsequent rampage by the victors through the losers' territory often claimed the lives of many women and children as well as men. In several cases, formal battles with controlled casualties were restricted to fighting within a tribe or linguistic group. As is the case in all societies, however bellicose or violent, the members of primitive groups used social and cultural devices to preserve havens of peace and cooperation within the group; and in armed fights quite generally, blood-kin or in-laws would try to avoid harming one another. When the adversary was truly 'foreign,' fighting was more relentless, ruthless and uncontrolled, and often involved cannibalism. Thus the 'rules of war' applied only to others within the tribe, while unrestricted killing, without rules and aimed at annihilation, was practised against outsiders.

According to Divale, the presence of protowarfare as versus feuding is a function of the population densities of the groups involved. When the ecological conditions are such that the members of a population or populations are widely distributed, usually only feuding occurs; but as population density increases protowarfare also takes place. Among primitive societies with dense populations, feuding is usually suppressed and one finds only protowarfare. Densely populated horticulturists such as the Kapauku of highland New Guinea (four per km²) suppress feuding with social controls and only engage in protowarfare.[853] (This of course suggests that the denser the population the larger the particular conflict.)

The extinguishing of subpopulations – similarly to the instances involving the bands of chimpanzees discussed earlier, and probably stemming from our arboreal ancestors – was clearly evinced in the horticultural period, warring bands and tribes often completely eliminating their neighbours.[854] Before any possible contact with civilisation, the tribesmen of Neolithic Europe, like those of the prehistoric United States, were wiping out whole settlements, with raids killing a higher proportion of women than even the routs that followed battles. Everyone was constantly engaged in guerrilla warfare. Armed or unarmed, adult males were killed without hesitation in raids, battles or subsequent routs in the great majority of primitive societies. Surrender was not a practical option for adult tribesmen because survival after capture was unthinkable. (Early horticulturists had little use for slaves, who could easily escape and return to their tribes or otherwise survive.) The proportion of war casualties in primitive societies almost always exceeds that suffered by even the most war-torn modern states.

And protowarfare produces an even higher rate of female fatalities than does modern warfare. Primitive social units could be reduced to a famine footing by the consequences of a few days of raiding or even of a single surprise attack. Not only massacres could decimate a small clan or tribe, but so could the cumulative effects of all these forms of violence undertaken a few days a month but sustained over time. Instances of tribes or subtribes being driven to extinction by persistent tribal warfare have been recorded from several areas of the world. Social extinction in tribal societies seems normally not to have entailed the killing of every person in the victimised group; rather, after a significant portion of the group (including most of its adult men) was killed, the surviving remnants were incorporated into the societies of the victors or into friendly groups with whom they sought refuge. Thus a social or linguistic entity was destroyed, if not necessarily the whole of the population that composed it.

Horticulturists not only fought with each other, but also with the remaining hunter-gatherers occupying the same region, as well as with herders invading from other regions. Hunter-gatherers in such conflicts were invariably out-numbered and out-armed, and they were forced to relinquish the areas they occupied that could be used for horticulture. The result of conflicts amongst horticulturists themselves may have largely been determined by the relative sizes of the fighting communities, as well as by the quality of the respective parties' weapons, the use of bronze weapons, developed towards the end of the horticultural era, providing a distinct advantage.

Population checks and growth; morals

Apart from the direct loss of life resulting from inter-tribal conflicts, protowarfare functioned as an even greater population check by reducing the number of potential mothers through its indirectly increasing female infanti-cide, as mentioned earlier, allowing the group to devote more of its resources to the nurture of its male warriors-to-be. It is to be noted that not only was such behaviour condoned by the group, it was morally obligatory.

Another cultural population check of horticulturists was their continuation of the practice of initiation rites performed already by hunter-gatherers. These rites involved such surgical operations as the aforementioned circumcision, as well as subincision (the splitting of the underside of the penis from the urethral opening to the base – both types of operation being practised by Australian Aborigines, for example)[855] and superficial mutilation. They were a preparation for a life of guerrilla warfare. The male West African horticulturist "was not supposed to have attained to the full dignity of manhood until he had killed someone."[856] As in the case of hunter-gatherers, eliminating those of delicate health in this way not only reduced population pressure,[857] but bred the hardi-est warriors.[858]

Formalised marriages with weddings, in which the tribe determined the couple to be paired,[859] were another population check employed and perhaps initiated by horticulturists. In spite of these checks, however, including a lower-ing in people's quality of life, during the 5000 years of the horticultural period the human population grew from some five million to around 80 million.

VCP analysis of the horticultural era

Herding and the beginnings of class society

The move from hunter-gatherer to herder is a movement of the vicious circle, being a technological response to the inability of a particular lifestyle to provide sufficient food for a population, and involving a more thorough and systematic exploitation of the environment. Thus in herding, the source of food (meat) is brought under human control (though not domesticated) in order to increase its availability.

According to Adam Smith, where private property must be relatively insignificant for hunter-gatherers, it is first with herders, where it can be *accumulated*, that inequality becomes significant. (We note however that, as mentioned earlier, the accumulation of private property began with fisher-forager sedentism.) Here then, the greater the number of animals, the higher the status of their owner. In this context it is of interest to look at the laws and rituals of hunting in medieval Europe, where the beasts of the chase were treated more like ranched than like wild animals in that they could be hunted only by the elite of society, and were protected by royalty.[860]

According to Smith, ambition on the part of the rich, and envy and hatred of labour on the part of the poor, both stemming from personal ownership, give rise to a new source of social tension in human society. These passions are, he thought, "steady in their operation," and fairly "universal in their influence." And according to him it is this state of affairs, in which property at least exceeding the value of two or three days' labour can be acquired by individuals, that necessitated the establishment of civil government.[861]

With herding we have what might be termed humans' *third* major step away from the general behaviour patterns of other animals, the first being our use and development of tools – particularly weapons – that in effect extended the range of what we could do with our bodies/teeth, and the second our harnessing of fire, which extended the areas in which we could live. The third step is the development of social hierarchies of the sort described by Smith. It is with herding that *class* societies develop. To begin with, herding is performed by men, and the herded animals are their property, thus making society more male-centred and strengthening or increasing the social distance between men and women. And further, the fact that some men own more or fewer animals than others increases the gap between the haves and the have-nots, such that distinct social statuses arise.[862]

Here we should distinguish between violent and non-violent means of acquiring power, i.e. between political and economic means. Regarding political power, there are two forms of aggression in social species. One, based on the sexual instincts, concerns the acquisition and defence of individual territory, and the other, based on the social instincts, the acquisition and defence of group territory. (The *social* instinct driving males to act aggressively in defence of group territory, like that leading to competition for group leadership, has of course evolved from the *sexual* instinct promoting individual territoriality.) Power consists in controlling people's access to vital resources, and the stronger you are, the easier it is to do so. But with sedentism comes *hoarding* as a route to power. Hoarding is based on the survival and sexual instincts. What is hoarded becomes the property (individual territory) of the hoarder, with the sexual instincts dominating once the hoarder's vital needs have been met (e.g. the male trades food for sex). This is the basis of economics. Hoarding presupposes a stable intratribal setting; and in that setting it provides power in the same way as does physical strength, i.e. by controlling people's access to vital resources.

The position a person has in the economic hierarchy is not determined solely by the physiological characteristics of the individuals involved. In other words, such hierarchies as are based on the individual ownership of property are not dependent simply on the innate expressions of instincts, with the physiologically better-adapted individuals occupying positions above the physiologically less-well adapted; they are also to some degree dependent on the plasticity of the human make-up, i.e. on intelligence or learning ability combined with life-experience, as well as on cultural inheritance. In this way human property-based hierarchies, starting with the herders if not earlier, may be said to be *culturally* rather that directly physiologically based. Our species becomes *dependent* on cultural (technological) adaptation. Where the economic hierarchy is *intra*tribal – property cannot be owned outside the tribe's boundaries – and stems from individual territoriality, the political ranking system is both intra- and inter-tribal. The intratribal part consists in individual males' ability to defend or extend *individual* territory by force, and the inter-tribal part by leaders' ability to defend or extend *group* territory by force. Thus in the intratribal part, the political question of who is to be the tribal leader is to a large part determined by the sexual instincts as manifest in individual territoriality. Of course the maintenance of peace within the tribe means that the defence of individual territory can't be too violent too often, and we see how physical strength might be usurped by economic power the more sophisticated the society becomes, with the wealthy having laws created to protect their property. The *inter*-tribal part of the political ranking system is determined by the relative armed strength of the leaders' tribes.

Whether civil government and the formal laws it involves came into being with the herders or with the horticulturists is not of great moment to the VCP; what *is* important to it, and in keeping with it, is Smith's conception of the key role played by property owned by individuals in the development of government and laws. Thus government and laws pertain to *intra*tribal (and eventually *intra*national) relations, supporting the status quo with respect to the ownership of property, and discouraging feuding and intratribal killing generally. Government is a manifestation of the sexual and social instincts combined, though the order-preserving role played by internal territoriality may perhaps be said to be greater than the cohesive influence of group territoriality. It is fundamentally the sexual instincts that give rise to the human pecking order, that order being supported in the group however by the social instincts, which see to it that there exist leaders for inter-tribal fights. Thus government requires the operation of social instincts.

More generally, one might suggest that tribal leaders are more oriented to group conflict than are band leaders. For the sake of the tribe, and more so in horticultural than in hunter-gatherer societies, as suggested by Johnson and Earle, leadership "becomes a necessity for defence and alliance formation. In this light, population growth and a chain reaction of economic and social changes underlie cultural evolution."[863]

Why did the horticultural revolution occur?

The Garden of Eden allegory

A consideration of the parable of the Garden of Eden in *Genesis* is of relevance to the horticultural revolution. In the Old Testament (ca. 1450 BC), after God has forbidden Adam and Eve to eat the fruit of the tree of the knowledge of good and evil, the serpent says to Eve:

> But God knows that the day you eat of the fruit your eyes will be opened, and you will become as gods, with knowledge of good and evil. [Eve] took of the fruit and ate. She also gave to her husband, who was with her, and he ate. And the eyes of both of them were opened.

And God learned of this:

> To the woman [God] said, I will greatly multiply your sorrow and your conception; in sorrow you shall bring forth children; and your desire shall be to your husband .... And to Adam he said, Because you listened to the voice of your wife, and have eaten of the tree of which I commanded you, saying, You shall not eat of it: cursed is the ground for your sake; in sorrow shall you eat of it all the days of your life; thorns and thistles shall it bring forth to you. You shall eat of the plants on the ground; you shall toil for your bread by the sweat of your brow until you return to the earth. ...
>
> [And] the Lord God sent him forth from the garden of Eden, to till the ground from whence he was taken.[864]

As relates to our considerations, Wilkinson has suggested that the Garden of Eden symbolises the hunting and gathering way of life, where food is there for the taking. The one condition for maintaining this ideal state of affairs is to practise sexual restraint or some other means of population limitation. As soon as Adam and Eve break the sexual taboos, they are cast out of the garden and thenceforth have to till the ground.[865]

On Wilkinson's interpretation then, as is in keeping with a key aspect of the VCP, the allegory in the story of the Garden of Eden concerns the weakening of internal population checks in hunter and gatherer societies, the resultant population growth requiring people to work tilling the soil in order to increase food production. We might alter Wilkinson's interpretation slightly, suggesting that the knowledge provided by the fruit was technological rather than carnal, and that an increase in carnal knowledge resulted from this technological knowledge. It was through learning to control the environment that the internal checks were weakened, and Adam was thrown into the vicious circle.

The warming climate

On the traditional perspective, the reason the horticultural revolution occurred just when it did was because it was then that the climate began to get warmer, which resulted in an increase in vegetation, a state of affairs humans took advantage of by developing horticulture. On the VCP, on the other hand, as intimated above and as is in keeping with the new views in anthropology and archaeology, humans' adoption of this form of food acquisition was rather a response to a worsening economic situation. The horticultural revolution began in the Near East, and, as argued by R. J. Braidwood, there is no evidence that when it occurred there were climatic changes of sufficient magnitude in that area to warrant it.[866] Rather, with the demise of our favourite prey of large mammals due to overkill during the Palaeolithic, we managed to overcome scarcity by developing such hunting technologies as those involving the use of the bow and arrow; and when this technology was no longer able to meet the vital needs of the population due to an increasing per capita reduction in available game throughout the Mesolithic, we overcame scarcity by developing horticultural technology. This change meant a decline not only in the quality of humans' lives, but also in the robusticity of the ecosystems they inhabited. At the same time, however, technological change once again made possible increased consumption leading to population growth.

In Chapter 2 a number of reasons were presented for believing that the adoption of horticultural technology was not simply the opportunistic response of a clever species to a warming climate. Here we shall look at them more closely.

Homo sapiens long acquainted with horticultural technology

By the time of the horticultural revolution humans were already aware of horticultural tools and techniques from earlier activities resembling cultivation, their experience in this regard going a good way back into the Palaeolithic. As a matter of fact, as regards tools, those of the hunter-gatherers required more skill in the making and using than those of the horticulturists.[867] Though most people probably knew enough about botany to understand how to reproduce plants using seeds and tuber cuttings long before the time of the beginnings of agriculture, this knowledge was not by itself sufficient to induce gatherers to begin planting.[868] As expressed by Sieferle, the fact that hunter-gatherer societies possessed all the components of the knowledge that agriculturists employed speaks against the view that the 'invention' of agriculture was the result of an autonomous improvement in technological knowledge.[869]

What was unique about the horticultural revolution was neither the particular tools used nor the individual activities their use involved, but the *comprehensive* employment of those tools in such a way as to constitute horticultural *technology*. As expressed by Cohen, none of these behaviours alone constitutes agriculture (horticulture), but taken together they *are* agriculture. Thus the beginnings of agriculture should be conceived more as a de facto accumulation of new habits than as the result of a conceptual breakthrough.[870]

Given this, as regards the traditional perspective one might then ask why, considering that humans had long known how to bring about such a change, it did not take place earlier in those parts of the world where it could have. As expressed by Reed: "Anatomically modern man was present over the Eastern Hemisphere for some 30,000 years before the end of the Pleistocene but is not known to have had agriculture, even though many parts of the world were quite warm enough." Braidwood sees the reason for this, as is in keeping with the VCP, as being that it was simply the case that people did not experience the transition to horticultural technology as being necessary at that time; and that if they had, they would have made the change then.[871]

Resistance

The above implies that even when they knew of horticultural technology hunter-gatherers consistently resisted employing it. As Wilkinson suggests, "primitive societies of hunters and gatherers no doubt thought that much more work was involved in cultivating the crops they needed than in gathering those that grew naturally."[872] As an example of work avoidance, the West African horticulturist male will not work unless there is a great necessity, such as hunger or a superior force; "there is no working for the love of it, there is no such thing as the dignity of labour." And as noted earlier, the use of more sophisticated technology (while the horticultural tools are simpler, the technology is more complex) to meet vital needs is generally avoided by primitive peoples, while anthropological and archaeological findings further suggest this to be the case even when they are aware of changes that could increase yield (cf. Boserup). For example, there is good evidence of there having been a significant delay in the acceptance of such technology in certain areas in Peru, in the eastern United States, and in much of Europe, even when a diffusion of horticultural techniques from nearby regions was quite possible.[873] The Cape York Peninsula people of Australia had constant communication with the New Guinea horticulturalists for a long time, but did not adopt horticulture.[874] Cohen notes that in the case of one particular society, agriculture developed as much as 2000 years later than it did in a neighbouring group, despite interaction between the groups. And we see the same phenomenon today in the case of modern hunter-gatherers, who often live next to farmers without assuming their 'progressive' lifestyle.[875] (Of course here it should be kept in mind that the hunter-gatherers' land is invariably of poorer quality for horticulture than is the land already occupied by the horticulturists, and that if the hunter-gatherers were to begin employing horticultural methods it would have to be on marginal land beside the resident horticulturists.)

In fact, one may say rather generally that horticulturists would gladly return to a nomadic way of life if they could. With the introduction of the Spanish mustang on the American prairie, many Plains Indian tribes, such as the Cheyenne and Arapahoe, abandoned their villages, agriculture and pottery and reverted to the bison hunt as a permanent way of life.[876] And, quite generally amongst primitive groups with mixed economies, there is a persistence of a 'hunting mentality' due to the fact that many activities related to hunting and gathering are apparently perceived by primitive people as being less arduous and, like herding, more prestigious than agricultural work.[877] It was not ignorance that prevented human populations from becoming agricultural sooner than they did, but their desire to avoid work. Or, from the other end, it was not new technological knowledge that led human populations to become agricultural when they did, but the fact that they had to do so in order to survive. Once again, what was required for horticulture to take place was not new knowledge so much as motivation or incentive prompted by need.[878]

Thus, to cite Cohen:

> Agriculture, in other words, appears to have been adopted not when first exposure might have occurred, but rather at the point when the absence of wild resources necessitated cultivation. Domestic crops first appeared in a hunting and gathering context where they supplemented and supported the old economy in a time of increasing shortage. Only later did any revolutionary effects emerge in the form of stable settlement and accelerated growth of population.[879]

And Reed:

> People in general did not seek to become agriculturalists, probably preferring hunting/gathering, but were forced eventually into agriculture by a series of events which they had not planned and the consequences of which they did not foresee.

Began in marginal areas

Sedentism began before horticulture, and was engaged in by those who could acquire the food necessary for survival throughout the year in one particular area. In practice, the food included both vegetable products and meat, obtained in such areas as the Nile flood basin and the Tigris–Euphrates valley, where a year-round supply of meat was assured mainly by fishing. Horticulture, on the other hand, did not begin in such areas, but in areas *adjacent* to these.[880]

As regards the question of whether horticulture began in marginal zones or not, the VCP suggests that it did. Thus it supports Binford's 'population-pressure' model, while being antithetical to Braidwood's 'hilly flanks' hypothesis. According to Braidwood, horticulture began rather in a *favoured* zone. As technology and knowledge about the physical environment grew more sophisticated, human populations colonising and 'settling into' the Hilly Flanks of the Zagros Mountains realised the potential inherent in the local flora and fauna, and exploited that potential by domesticating the appropriate species. (More-arid areas are better for both pastoralism and the growing of grains.) Thus, "[a]round 8000 BC the inhabitants of the hills around the fertile crescent had come to know their habitat so well that they were beginning to domesticate the plants and animals they had been collecting and hunting." This is where wild ancestors of wheat, barley, legumes, sheep, goats and cattle would have been found;[881] and it is where the earliest food-producing villages such as Jarmo, Karim Shahir and Banahilk developed.

But Braidwood here misses the fact that what may be optimal from one economic perspective can be marginal from another. From the point of view of the people who found themselves in the hills with only their fisher-forager toolkits, the situation would have looked pretty marginal. On the other hand, these areas, in being largely covered with wild stands of wheat and barley, *were* ideal from the point of view of horticulture. (Cf. e.g. the advantages provided to technology-developing pre-humans on the savannah in relation to their kin still living in the forest.)

Thus, as expressed by Clutton-Brock, increased pressure on the available food supplies in many parts of the world during the early Holocene meant that new methods of obtaining food had to be resorted to.[882] And, as has been noted earlier, horticulture was originally the result of an effort on the part of the *weaker* members of society, i.e. the 'defeated' males (and their mates). These individuals were forced out of the riverine group territories of their bands or tribes – an instance of marginalisation as found in all territorial animals – and had to expend greater energy on horticulture than their contemporaries did in continuing their fisher-forager ways.

As suggested more generally by Divale, and in keeping with the idea of conflict avoidance by group fission taken up in Chapter 2, it appears that expansion into new regions by primitive groups was accomplished through such a fissioning process. As a result of an internal dispute between members of the same band or village, the smaller of the dissident factions, either a family in the band or a lineage in a village, would move to a new location. This fissioning process also has a domino effect and results in people in villages on the tribal periphery being pushed into new territory.[883]

Thus we see on the VCP a situation where, e.g. in the Fertile Crescent, socially weaker people – those at the bottom of the social hierarchy – were forced out into marginal areas, i.e. into the Hilly Flanks. These areas were naturally more arid than sites closer to the river, and may have responded to planted grasses better than to tubers. Note also that while drier areas are poor from the point of view of growing vegetables, they should be relatively rich from the point of view of non-aquatic hunting, since they contain more game (present-day hunter-gatherers are largely confined to arid grasslands); this would have made the domestication of animals both easier and more natural. Grasslands would even be superior from an agrarian point of view in that they allow the growing of grasses without requiring the removal of trees. Perhaps the main difference in the diet of these emigrants from that of the cultures from which they came was in their greater consumption of red meat as compared with fish and shellfish – meat provided not only by hunted land animals but also by domesticated ones.

Not only does the fact that horticulture began in peripheral areas support the view that the people most likely to adopt a new subsistence strategy are those having difficulty following their traditional one, but it supports the VCP more generally, since it too suggests that horticulture was a response to need, rather than being an entrepreneurial endeavour.

In a broader biological perspective, the phenomenon of being pushed out to the periphery is of course an expression of vertebrate territoriality. From the point of view of the group and of the species, the most important effect of the forcing of individuals out into marginal areas is the preservation of the group's resources. Thus the marginalisation of humans at the time of the horticultural revolution has essential similarities to the marginalisation of Wynne-Edwards' less fit red grouse, African pre-humans, and those early *Homo sapiens* forced to move into new areas. The less fit grouse were pushed from the individual territories of others out into marginal areas, where they died; the pre-humans were forced from their group's territory in the forest out onto the savannah, where they had to employ new methods of obtaining food in order to survive; and the marginalised early *Homo sapiens* were obliged to move into colder areas, where they had to employ technologies of fire and clothing. In all four cases the marginalised individuals were those most poorly fit for living in the habitat the populations of their species then occupied: the particular red grouse because they were weak; protohominids because they were more poorly adapted to arboreal life; certain early *Homo sapiens* because of their being poorly culturally adapted to their groups; and the original horticulturists because of their low position in the cultural pecking order of their fisher-forager communities. (Here one might speculate that this form of behaviour is paradigmatic in the case of *speciation*. It is radically new environments that call for karyotypic adaptations; and it is such environments that are typically inhabited by the *less fit* of the species.)

But where the excess red grouse did not adapt to their marginal situation so as to produce offspring, at least some of the pre-humans and modern humans did – ultimately those who were either karyotypically or genetically better adapted to the new situation. As suggested by the VCP, the difference of course has to do with the application of technology in the marginal areas. Thus it was through such a cultural (technological), and perhaps genetic, change that some humans were able to meet the challenge of their new surroundings at the time of the horticultural revolution.

As regards potential advantages of drier as versus wetter localities, it may be pointed out that grain as compared with most other forms of plant food has the particular advantage of being *storable*. As noted earlier, the seeds of plants adapted to hot and dry climates store better than those adapted to wet climates; and in semi-arid regions such plants tend to constitute dense monocultures. Prior to the employment of money the fact that grain was storable, and that grasses grew better in more arid regions, may have given a certain economic advantage to humans living there. Thus the areas that were *marginal* from the point of view of providing a well-rounded potentially sustainable diet for a hunter-gatherer population may have been *central* for the production of the greatest *quantities* of food, and moreover food in a form which could be used as *capital*.

Population pressure

One can imagine a scenario where the first sedentary fisher-foragers obtained a large and regular part of their protein in the form of fish and shellfish from the Tigris or Euphrates. In this situation the population would have begun to increase due to the influence of sedentism on internal population checks combined with an increase in the quantity of food acquired once aquatic resources began to be reaped regularly. Despite this increase in population, according to Birdsell these sedentary cultures were in equilibrium with their environments, excess people being sent off *before* this equilibrium was jeopardised.[884]

However, the fact that fisher-forager cultures sent out emigrant horticulturists-to-be suggests that they were nevertheless experiencing population pressure, i.e. that their sustainability, like that of Wynne-Edwards' red grouse, or of Norwegian lemmings, *depended* on their getting rid of excess individuals. (Cf. open vs. closed populations.) But where in the case of the red grouse population pressure is an annual occurrence, and in the case of the lemmings it is also a recurrent phenomenon, one can imagine that it was relatively constant for fisher-forager communities. And since horticulture and the domestication of plants and animals was begun by the emigrants from such cultures, we can affirm, as suggested by Leslie White,[885] that domestication was a result of population pressure. As argued by Cohen,

> population growth (and population pressure) was a more ubiquitous and more significant trend among pre-agricultural peoples than is usually recognized. [H]uman population has been growing throughout its history, and ... such growth is the cause, rather than simply the result, of much human 'progress' or technological change, particularly in the subsistence sphere. While hunting and gathering is an extremely successful mode of adaptation for small human groups, it is not [successful when it comes] to the support of large or dense human populations. I suggest therefore that the development of agriculture was an adjustment which human populations were forced to make in response to their own increasing numbers.[886]

Thus, in keeping with the VCP, both nomadic *and* sedentary hunter-gatherers experienced population pressure. In the case of the nomads, this pressure resulted from the reduction in the carrying capacity of their land due to an over-use of hunting technology. In the case of the sedentists, as noted by David Harris, it resulted from an increase in population due to the use of a sedentary technology, i.e. a technology required to maintain a non-migratory population. Both cases constitute instances of the turning of the vicious circle: from the employment of new technology (javelin-hunting, sedentism), to, respectively, resource depletion and population growth. And in both cases a contributing factor was the fact that humans had recently occupied virtually all potential habitats on the planet, so that virgin areas into which growing populations might expand had become non-existent.

As suggested by Cohen:

> It seems likely, therefore, that agriculture [horticulture] would only have been adopted under conditions in which the demand for calories was increasing, or at least where the demand was out of balance with the productive potential of the existing economy. [O]nly one possible explanation – actual population growth [or population pressure] – could account for increasing demand or ecological imbalance which appeared to span a wide range of ecological zones and to embrace a large number of different cultures. [T]he nearly simultaneous adoption of horticultural economies throughout the world could only be accounted for by assuming that hunting and gathering populations had saturated the world approximately 10,000 years ago and had exhausted all possible (or palatable) strategies for increasing their food supply within the constraints of the hunting-gathering life-style. The only possible reaction to further growth in population, world-wide, was to begin artificial augmentation of the food supply.

The VCP also receives support from the fact, mentioned in the above quote, that horticulture did not originate at a centre from which it spread, but rather occurred more or less simultaneously in various parts of the world (parts each of which could arguably be said to be experiencing population pressure).

Cohen makes the further suggestion, also in keeping with the VCP, that since vegetable foods provide the main source of calories under conditions of intense population pressure, once humans have become confined to a particular area it is these resources which will increasingly dictate their behaviour. The domestication of animals, he suggests, is primarily a means of subjecting them to the requirements of a system increasingly geared to intense exploitation of spatially limited vegetable resources.[887]

Rather generally then as regards the transition to horticulture, we have such facts as that: humans had just recently completely occupied virtually all areas of the globe that could support a hunter-gatherer lifestyle; we were long acquainted with horticultural techniques prior to the horticultural revolution; historical hunter-gatherers consistently resisted engaging in horticulture; there is no evidence of major climatic changes where the horticultural revolution first occurred; it was not the result of the dissemination of technical knowledge from a point or points, but occurred more or less simultaneously around the globe; it occurred in marginal areas; and with the employment of horticultural technology the quality of individuals' lives declined vis-à-vis that of hunter-gatherers. These facts suggest, as does the VCP, and as is in keeping with the new views in anthropology and archaeology as expressed by Clutton-Brock, Wilkinson, Cohen, Hassan, Binford, Reed and Harris, that the horticultural revolution did not represent an opportunistic step forward for humankind, but was rather a response to the need resulting from constantly increasing population pressure.[888]

Hassan, assuming this to be the case, suggests further that it is more likely that the horticultural revolution came about through the use of those innovations which supported year-round residence in one locality by relatively large groups, and by a selection of those practices which ensured a more reliable yield from year to year.[889] Thus where in the case of hunter-gatherers the biological complexity of the supply ensured its reliability, the horticulturists' artificial biotope could be ravaged by the elements,[890] which meant that a back-up supply was necessary in case this should happen – as it often did. The importance of a reliable supply, i.e. a surplus that could be used in bad times, thus came to characterise not only horticultural but *all* agricultural societies.

As a result of these considerations then we may say that, while it is true that the horticultural revolution was supported by the increase in vegetation concomitant with the warming climate at the end of the last ice age, its occurrence was nevertheless a response to *need* on the part of the human population – as would be expected on the vicious circle principle.

What were its effects?

As suggested by C. G. Darwin, we might say that for a transition to be considered an actual revolution in *Homo sapiens'* development, what is key is that it occasion an irreversible change in our way of life.[891] Here we shall look at such changes in the case of the horticultural revolution.

Work

We have already seen that a major reason we humans did not adopt horticulture earlier than we did was that it would involve more work. Continuing to gather wild plants, if possible, would have been much easier than beginning to cultivate and harvest crops; and it would have been much easier to obtain meat by hunting than to see to the survival of livestock through e.g. building fences and otherwise tending to the animals.[892]

That increased work should play a role in the adoption of horticulture is evidenced by the behaviour of the horticulturist Kuikuru of Brazil who, as described by Robert Carneiro, were capable of producing several times more food than they actually did. Though a small increment in the amount of time devoted to planting and harvesting would have brought about a marked increase in the size of their crops, the Kuikuru chose to spend this time in other ways. Enough food was produced to meet the demand, and once it was met work stopped. According to Carneiro equilibrium had been reached, and neither population nor production increased.[893]

According to the VCP, the fact that a particular resource can only be significantly exploited thanks to the introduction of new technology suggests that the acquisition of that resource should require greater inputs of energy – in the present case human energy – than did that of the original resource before it became scarce. Thus, as has been pointed out by Boserup and others, we see that the employment of horticultural technology in the Neolithic era, involving e.g. the construction and use of stone gardening implements, requires more effort per unit food acquired than did the earlier hunter-gatherer technology. And we note that the amount of work constantly increased as we moved through the forest- and bush-fallowing phases and on to gardening. Given present knowledge of work/leisure preferences among primitive peoples, even more support is given to the view of the VCP, according to which food production would not have begun, and the level of technological development it required would not have been exercised, had it not been for need.

Poorer diet

Taking the human population to have been five million at 10,000 BP, the existence of the surplus provided by horticultural technology supported its growing some *sixteen*fold from the end of the Mesolithic up to the agrarian revolution, despite people's having a worsened diet. As compared to the amount of food that could be acquired using hunting and gathering techniques, horticulture is reported as providing about 20 times as many edible calories per unit land. But the grass seeds eaten by horticulturists contain little protein in comparison with the plants eaten by hunter-gatherers. Furthermore they are only available for at most a few weeks of the year and must be stored. And more effort is required to make them edible: though in the case of some seeds, grinding alone may have sufficed, the vast majority could not be eaten without soaking, grinding and boiling.[894]

We might compare the advent of horticultural technology with previous major changes incumbent upon technological development. In the case of humans' first use of the *lance* in the Palaeolithic, the new food source, meat, was superior to (higher in protein content than) what it replaced. The later use of the *javelin* provided a diet superior to that afforded by the lance. If we then look at the Mesolithic transition and take as paradigmatic the widespread use of the *bow and arrow*, we see a decrease in the quantity of meat in the diet as compared with when the javelin was most effective, due in part to hunting's resulting in smaller kills. In the case of e.g. the *nets* used by fisher-foragers, we see a further increase in effort expended per calorie obtained. And in the case of the horticultural revolution and the widespread use of the *hoe*, high-protein meat was largely though not wholly replaced with low-protein and high-carbo-hydrate vegetable products.

Thus just as the Mesolithic diet is inferior to that of the earlier Upper Palaeolithic, the small mammals hunted with the bow and arrow being an infe-rior substitute for the larger mammals hunted with the javelin, the horticul-tural diet is inferior to the Mesolithic. Once again we see that an immediate increase in the *quantity* of a resource of a particular type, thanks to a new use of technology, often goes hand in hand with a worsening of its *quality*. Of course new technological applications always mean an improvement over the situa-tion just prior to their use – otherwise they would not be used. The question then is whether they provide an improvement in the long run – which they almost always do not.

Poorer health

As has also been noted earlier, with the change to the horticultural life-style humans' health did not improve, but rather worsened. This was in part a result of our poorer diet, but was also affected by the increase in disease brought on by the domestication of animals, sedentism, and increasing population dens-ity. Horticulturists' worsened diet was probably the main cause of the decrease in their stature as relative to people at the end of the Cro-Magnon era.

Diamond provides an example of the worsening health that attended the horticultural revolution. Corn, first domesticated in Central America thou-sands of years ago, became the basis of intensive farming in the Illinois and Ohio river valleys around 1000 AD. Until then, Indian hunter-gatherers had skeletons 'so healthy it is somewhat discouraging to work with them,' as one palaeopathologist complained. But with the arrival of corn (carbohydrates), the number of cavities in an average adult's mouth jumped from fewer than one to nearly seven, and tooth loss and abscesses became rampant. Enamel defects in children's milk teeth imply that pregnant and nursing mothers were severely undernourished. Anaemia quadrupled in frequency; tuberculosis became established as an epidemic disease; half the population suffered from yaws or syphilis; and two-thirds suffered from osteoarthritis and other degenerative diseases. Mortality rates at every age increased, with the result that only one per cent of the population survived beyond the age of 50, as compared with five per cent in the golden days before corn. Almost one-fifth of the whole popula-tion died between the ages of one and four, probably because weaned toddlers succumbed to malnutrition and infectious diseases.[895]

Also, for the first time in our development, during the horticultural era our production of waste became a serious problem. Where hunter-gatherers can simply walk away from the waste they produce, sedentary horticulturists cannot, a situation which was constantly being worsened by their growing populations. This increasing waste and population density combined to support an increase in parasite and disease infection. Tuberculosis, leprosy and cholera came with the rise of horticulture, while smallpox, bubonic plague and measles appeared only in the past few thousand years with increasingly dense populations in cities.[896]

Division of labour

Throughout the development of the horticultural era there was an increase in specialisation or the division of labour. In most cases this specialisation involved the repetitive performance of particular tasks. As suggested by Mumford, making holes in stone is *boring* work[897] (the daily *grind*), particularly so if it is all that you do, i.e. if it's your *job*. The vicious circle promotes and requires specialisation – which does not in itself imply expertise – as it increases a group's ability to exploit its environment. But one price paid for this increased exploitation on the part of a society is an increase in the monotony of the tasks to be performed.

Social stratification and the redistribution of resources

Specialisation is a manifestation of increasing societal complexity; and while a group that engages in specialisation may be able to exploit the environment more thoroughly, greater order must be maintained in the society (which implies a greater *disorder* outside the society) for the economic interdependence this implies to be possible. This order, due to the territorial instincts of human males, has always taken a hierarchical form, with the society's most powerful individuals (typically, males) on top.

Already in hunter-gatherer groups the food from the hunt is distributed by the dominant, hunting males, as in the case of australopithecines, in this case to other males in the group, upon which it is partially *re*distributed by them, e.g. to women and children, as in the case of the !Kung.[898] Normally, however, a *redistributional system* is thought of as one involving *trade*; and, according to Alan Barnard and James Woodburn, such systems exacerbate the inequality between males and females. In the context of redistributional systems women are treated as minors, which is seldom the case in the more direct systems of hunters and gatherers.[899] More generally, according to Sahlins, given a redistributional system, the greater the surplus, the greater the degree of production-related stratification.[900] Thus we have a direct connection between the existence of surpluses and social stratification in societies sufficiently complex to involve a redistributional system.

Increasing inter-tribal conflict and the domestication of humans

Constantly increasing surpluses together with the need for warriors supported the further domestication of humans, i.e. supported a change in the human genetic make-up resultant upon the technological requirements of a horticultural as versus a hunter-gatherer lifestyle. Apart from the genetic favouring of a short, stocky build for the wielding of horticultural implements, somatic characteristics supporting survival in armed confrontations using the particular weapons of the time would also have been favoured.

Politics vs. economics

By looking back to our simpler past we can see why the exercise of political power fundamentally involves the use of weapons, while the exercise of economic power does not. Intraspecific competition over group territory amongst pre-civilised *sapiens* often involved weaponed fights between bands from different tribes or between whole tribes. Within one's own tribe (amongst one's relatives and potential relatives), however, such fights were often not allowed (warriors could not be wasted, as they were needed to protect the group's territory), though a certain amount of killing e.g. in the form of feuding nevertheless took place. Intragroup fighting was discouraged not only in hunter-gatherer societies but also in herding and sedentary societies. And it is with the latter that economic power came into existence, i.e. *within* herding and sedentary societies, where mortal combat was proscribed. Originally consumables were *shared* (or traded for sex) within one's band (extended family), but as the quantity of storable consumables, both edible and manufactured, began to increase, they came to be *owned* by individuals and families. Here we have what may be seen as a weakening of morals, abetted by the presence of a surplus. Ownership (a manifestation of survival and sexual instincts) began 10,000 years ago and, in the case of sedentary societies, has been increasing ever since. Apart from the accumulation made possible by sedentism, this is also due to both the ability to transport goods and the increased autonomy of individuals in such societies. In our increasingly relating to strangers, individual ownership has successively taken over from group sharing within the tribe (nation).

The disarming of the environment (shrinking of the resource base)

No matter what period of our development we consider, the key indicator of the health of the ecosystem in which we live is the number and variety of species it contains and/or its total biomass – the two being positively correlated to each other and to the degree to which a climax ecosystem is being approached. In this regard, throughout the period humans have existed on earth, the number and variety of other species in ecosystems affected by us have constantly dwindled. In other words, ecosystems around the world have constantly been moving further away from the climax nature they had prior to their being impacted upon by human technology. Thus the series of species extinctions that began during the Palaeolithic era continued on into the horti-cultural; for example the species *Myotragus* (like a small deer) was exterminated by human hunting during the Neolithic,[901] and the small horse, *Equus hydrunti-nus*, was hunted to extinction between 6000 and 5000 BP.[902]

One manifestation of species depauperisation was the desertification that began in the horticultural era and has been increasing ever since, contributing factors being both pastoralism and overly intensive swidden agriculture. As regards pastoralism, in much of the Mediterranean region for example, the nat-ural vegetation of which was a mixed evergreen and deciduous forest of oaks, beech, pines and cedars,[903] goat-herding led to the desertification of agricul-tural lands, and through its continuation is preventing their recovery. A simi-lar phenomenon occurred in Late-Neolithic Egypt, where the huge supplies of charcoal needed for smelting metals led to the clearing of large wooded areas. The newly-cleared land was then heavily exploited for the successive grazing of cattle, sheep, goats and camels.[904]

The *need* to acquire a surplus (of food)

With the advent of horticultural technology we see once again – as in the case of the large-scale slaughter of ungulates during the Upper Palaeolithic – the need to *cache* food. Food is not available at all times; in the Upper Palaeolithic this meant preserving some of the meat from big kills, and during the horticultural era it meant preserving almost all of the harvest. However, in the horticultural era the food was both more easily stored – vegetable products, in particular grain, as versus meat – and provided a much greater surplus. As expressed by Johnson and Earle with reference to Sahlins, we have to understand surplus pro-duction at the time of the horticultural revolution not as a sign of affluence, but as a form of insurance against bad weather and/or other contingencies that can cut production drastically and threaten people with starvation.[905]

Both systems were open to disruption, however. An unsuccessful big-kill hunt could mean starvation, as could an unsuccessful harvest. Thus it was necessary in both cultures to acquire a cache or store to be reverted to when food was scarce; and that store had to include more food than was normally necessary to see the group through lean periods. It was this extra food that con-stituted the surplus, not the store itself. In other words, the surplus went *beyond* what would normally be used, such that a major failure in the food acquisition or storing process would not lead to starvation. And this requirement was not always met.

The ability to provide this surplus depended on the viability of the more intensive application of technology, which in the case of horticulture in turn depended on the productivity of the land. As mentioned above, it has been suggested that horticultural technology can provide something like 20 times as many calories per unit arable land as compared with hunter-gatherer technology, so it is clear that the potential for increased resource exploitation and subsequent/consequent population growth was great. Note however that, in keeping with the idea that alternative forms of technology are employed only when needed, increasing technological exploitation in the horticultural era did not mean increasing technological innovation, the reason on the VCP being that the potential of horticultural technology to increase supply was so great that innovation was not required for quite some time. Rather generally one might say that it is when the flow of resources begins to *dwindle* that innovation or the employment of previously unused technology becomes necessary.

Sieferle, like Sahlins, suggests that this surplus fuelled social stratification, it increasingly being employed to meet the non-vital needs of the powerful rather than to lessen the vital needs of the weak. By the end of the horticultural period there was a division of labour that freed entire groups of people from work; and this was only possible if more food was produced than the producers consumed. And the production of such a surplus gave rise to conflicts over distribution, and to power struggles within and between societies.[906]

Technological intensity, population growth, and sustainability

Considering that virtually all of the resources used by the early horticulturists were renewable – metals not being mined until about 4000 years later – one can imagine a scenario in which their populations first experienced overshoot and then reduction, after which they might have, through their sustainable use of resources, plateaued so as to maintain ecological equilibrium. Horticultural populations' succeeding in this, however, would have meant, as is in keeping with the VCP, that they would not have increased the intensity of their use of technology for the fulfilment of vital needs – or at least not beyond what the environment could sustain. But, as we know, throughout the horticultural era horticultural technology was constantly used more intensively, and the population of horticulturists constantly grew, contributing to the turning of the vicious circle.

Internal checks: morality; infanticide and protowar

Population checks, whether internal or external, or positive or preventive, operate so as to counteract the turning of the vicious circle. What we see with the expression of the VCP through the whole of our species' development is a steady weakening of these checks at the same time as the constantly altering conditions and increasing complexity of human society lead to their taking new forms, or to certain forms manifesting themselves to a greater or lesser extent than earlier.

What the more or less constant presence of a surplus during the horticultural period meant in practice, as it would have for almost any species under similar conditions, was that the human population increased to the size that its vital resources could maintain in the short term. Given the availability of breeding sites thanks to the building of dwellings, it may be expected – in keeping with the pioneering principle – that a surplus of food should lead to a growth in the human population.

Looking at the transitions involved in the horticultural period as a whole, we note that as human populations became more acclimatised to the presence of a surplus, and the powerful experienced an increasing need for muscle-power, whatever internal (cultural) population checks their societies may have had would have been successively weakened; and this weakening in turn suggests that they would have been strongest at the beginning of the period.

Morals were already evident in hunter-gatherer societies in the case of such phenomena as e.g. intragroup sharing, infanticide and inter-tribal fights, and these phenomena continued on into the horticultural era. C. P. Kottak provides an example involving infanticide:

> Among the [horticultural] Tapirapé Indians of Brazil, couples could raise two children of one sex, three in all. Tapirapé culture banned parents from raising more, because additional mouths would siphon resources needed by other families. The Tapirapé considered it selfish and immoral to try to keep a surplus baby. The death of the infant, who was not defined as human, was considered morally necessary for other members of the group to survive.[907]

As expressed more generally by R. A. Fisher regarding early agriculturists:

> The act of infanticide, however offensive to civilized feelings, has, in these circumstances, a certain natural propriety, and it is not surprising that it should often be regarded as a moral action and be, to some extent, compulsory. … In the pre-Islamic period we hear of reluctant parents being urged to kill their baby daughters for the sake of their tribe.

The population checks that became weakened during this period were primarily those related to the spacing of children. Where for hunter-gatherers there were some four or five years between births, this period shortened with the sedentism and surplus of the horticulturists.

As in the hunter-gatherer period, armed group conflict was one way in which the population was restrained in its rate of growth. Thus inter-tribal fighting early in the horticultural period also constitutes a population check, one whose main direct impact however would not have been through the deaths of many males, but of many females, when enemy villages were attacked.

Inter-population conflict in the case of hunter-gatherers and horticulturists was a matter of tribe against tribe. The population check constituted by such conflict was constantly worn away however by humans' ability continually to take more from the environment. Nevertheless, constantly increasing population pressure meant that as the horticultural period advanced, so did the relative amount of protowar.

Given the division of labour and the larger population, the complexity of society increased, and a class society formed in which chiefs were at the top. Inter-tribal fights, supported by the surplus provided by domestication, became larger and more frequent, the stored food/capital itself coming to be submitted to the threat of destruction by invading groups. And this, together with threats to the stores from the weather and various organisms, as well as such phenomena as famines, decreased social security.

External checks: disease and parasites; mortality

The horticultural revolution could just as well be called the era of the birth of infectious disease, thanks to the fact that the vast majority of such diseases came from the animals we domesticated. (As regards disease, it may be mentioned that early *sapiens'* long lactation period may not only have meant a greater spacing of births, but may also have tended to increase resistance to infection, for *both* mothers and children.)[908] Not only did contact with these animals give rise to disease in humans, but the maintenance of the animals in crowded, dirty conditions increased the incidence of such diseases amongst them, and thus amongst humans as well.

Population growth and overshoot

Thus population grew during the horticultural period as internal checks weakened and a surplus continued to be provided. In the end this surplus could not be maintained, however, with the result being population overshoot and the agrarian revolution.

* * *

From the point of view of the VCP, a key aspect of the horticultural period's contribution to the turning of the vicious circle was its involving population growth. This growth would not have been possible had the horticultural period not produced a surplus of consumables meeting vital needs. This suggests that at least part of the surplus that might have gone to relieving hunger in bad times was going to increasing the size of the population, resulting in severe dieoff when there was e.g. drought or flooding. Horticulture constitutes a less secure lifestyle than does hunting and gathering.

Regarding the idea that technological advance leads to an improvement in the quality of life through raising the standard of living, not only was there no such advance at the time of the horticultural revolution, but people's standard of living declined. This was manifest in increased energy expenditure (in the form of human work), a poorer diet, a decrease in size, more killing of other humans through armed conflict and infanticide, a tremendous increase in infectious disease and parasites, and a presumable increase in mortality.

Other central aspects of this major turn of the vicious circle during the horticultural period were a general disarming of the environment, the virtually constant presence of a surplus of (lower-quality) food, and the further population growth the surplus abetted.

The horticultural model

Scarcity and need
Innovation: **domestication; widespread use of the hoe**
Economic development
Increase in available resources
Prerequisite of the existence of resources
Increased energy use: fire
Resource depletion
Continuing species extinctions
New needs: **domestication; hoe, axe**
Technological development self-perpetuating
Genetic dependence on developing technology
Environmental damage: **waste and the beginnings of desertification**
Surplus: **storable grains = capital**
Increase in total consumption
Inferior substitutes: **vegetable products partly replace meat**
Population checks weakened
Morals: **hoarding rather than sharing**
Population growth
Population pressure
Increasing social complexity: **division of labour**
Social stratification: **chiefs vs. workers/warriors**
Lower quality of life for virtually everyone: **increased work** (started
      to increase already before the Mesolithic), **poorer diet,**
    famines, increase in diseases and mortality

Conflict: **protowar develops towards war**
Less leisure
Little cultural development
Decreased security
Population overshoot
Diminishing returns
Undermining of technology and human existence
Economic decline
Scarcity and need

Differences between the horticultural and hunter-gatherer VCP models consist in the addition of new factors and the increasing in intensity and changing in form of factors already present. Innovation is present in all VCP models, and its most important form here is the widespread use of the hoe and the domestication of plants and animals, which together provide a greater quantity of food, but food that is of a lower quality, with vegetable products partly replacing meat. Domestication in itself also means an increase in work and a general lowering of the quality of people's lives, in the latter regard including an increase in disease and the new risk of famine. And, through its provision of a storable surplus – which becomes the property of powerful individuals and can function as capital – domestication also affects morals related to sharing and the production of children. As regards the latter, the effect of the surplus is to weaken internal population checks, thereby leading (along with sedentism) to an increase in population. This population increase in turn means an increase in the production of waste and in desertification, and supports social complexity as manifest in specialisation or the division of labour. As compared with the hunter-gatherers' varied and relatively rapid acquisition of food, horticulture came to constitute monotonous, time-consuming work for many, contributing to the general lowering in the quality of people's lives. There is a clearer distinction between leaders and others in the form of chiefs vs. workers/warriors; protowar develops towards war and there is little cultural development.

Mining metals *6000* BP

The beginning of the horticultural period coincided with that of the Neolithic era, or New Stone Age, and involved the grinding and polishing of stone tools. This phase in our development evolved into the use of metals, the first being copper about 8000 years ago, and later bronze, the *mining* of copper marking the beginning of the Late Neolithic or advanced horticultural period.

Non-renewable vs. renewable resources

Beginning with the use of metals, and more particularly the mining of them, humans for the first time began to move towards a lifestyle dependent on non-renewable resources. While *Homo sapiens* had earlier eradicated various species of animals, each species being non-renewable, it nevertheless would have been possible to harvest the populations of these animals in a way that allowed them to renew themselves. In the case of mining, on the other hand, resource renewal is impossible. Mining in its very essence involves the depletion of non-renewable resources.

Copper and bronze

People were probably first attracted to copper through their interest in unusual rocks. In its native form, copper appears as a purplish-green or greenish-black nugget which, when scratched or rubbed, shows a yellowish kernel of pure unoxidised copper. At first, copper was simply hammered cold into small tools and ornaments, such as awls, pins and hooks. The earliest use of metal yet known is in southern Turkey, where a few hammered copper objects from ca. 9000 BP were found,[909] and a few articles made by this method have also been found in Near Eastern sites dating from before 8000 BP. Later, between 7000 and 6000 BP, the technique of softening copper by *annealing* was discovered, in which the metal is held at a high temperature but below its melting point for some time, after which it is slowly cooled. Another means of softening the metal discovered around this time is *tempering*, i.e. the *re*heating of the metal. By alternately heating copper using a simple wood fire, and hammering it, it was made less brittle and thus could be used for a wider variety of purposes. Later still, people discovered techniques for extracting copper from ore by means of smelting, the possibility of which began with pottery firing;[910] and they also found ways to cast it into moulds. By 6000 BP copper objects were widespread in the Near East, at which time advanced horticultural society began.[911]

It is doubtful that a true Copper Age preceded a Bronze Age (bronze being an alloy of copper and tin in about an eight to one ratio). What is more probable is that, from the beginning, the pure metal and the alloy were used without distinction.[912] By 3000 BC however the advantages of an alloy of copper and tin had been recognised in India, Mesopotamia, Asia Minor and Greece.[913] In the Near East the Bronze Age ended about 1200 BC, and was followed by a dark age.[914]

Iron

Since the copper ore on Cyprus, for example, contained no more than four per cent copper but 40 per cent iron, Cypriot smiths at the time of the Bronze Age/Iron Age transition (on Cyprus, ca. 1070 BC) could obtain more usable iron than copper from such ores using the same amount of fuel. They discovered that the iron in the slag could be removed simply by hammering. Removing the iron in this way allowed them to bypass the heating process altogether, freeing them from the constraint of the availability of wood, of which there was a shortage.[915] A similar route must have been followed by the original ironsmiths of Asia.

People knew of iron at least as early as 2000 BC,[916] the first smelting of iron perhaps having occurred in China or India, after which knowledge of the process spread west.[917] According to D. W. Anthony, iron objects were being made by the central-Asian Yamnaya herders as early as 2500 BC.[918] The use of smelted iron ornaments and ceremonial weapons became common during the period extending from 1900 to 1400 BC. The process for making hardened steel from smelted iron was discovered perhaps towards the end of this period by the Chalybes, who were a subject tribe of the Hittites, living in the Caucasus Mountains. The Hittites apparently kept this process secret for some 200 years (ca. 1400–1200 BC)[919] and at the same time restricted the export of steel weapons. (It may be noted that throughout history craft lore has generally been kept secret.)[920] The oldest known article of iron shaped by hammering is a dagger found in Egypt; it was made before 1350 BC and is believed to be of Hittite origin. After the downfall of the Hittite empire (1650–1180 BC), great waves of migrants spread through the Near East and southern Europe, taking the steel-making technology with them.[921]

At first, iron was no improvement over copper and bronze. It was harder to work, required more fuel to produce, and the cutting-edge made by hammering blunted more easily. The Chalybes' discovery was that annealing (which brought the metal into contact with charcoal), followed by plunging into cold water (quenching), gave iron a hardness superior to that of bronze. The heating and hammering before quenching had transformed the surface of the wrought-iron into *steel* by the diffusion of carbon (cementation). And it is with the Chalybes' discovery and production of steel, ca. 1400 BC in northern Anatolia, that the Iron Age properly begins. Mesopotamia is considered to have entered the Iron Age around 885 BC[922] – later than Cyprus.

The properties of iron depend greatly on its carbon content. *Wrought*-iron contains less than one per cent carbon,[923] and is soft and malleable. It bends, where copper and bronze break, and it can be mended. Most of the work on it can be done by cold hammering and annealing at 700°C, and many primitive smiths work with it in this way today. The *melting* point of wrought-iron however is 1200°C, a temperature reached with the aid of charcoal. *Cast*-iron, on the other hand, contains between 1.5 and 5 per cent carbon, is hard and brittle, and has a lower melting point than wrought-iron. The melting point of *pure* iron, containing no carbon, is 1540°C, a temperature not attainable until the 1800s. *Steel* is intermediate between wrought-iron and cast-iron with respect to its carbon content, which may vary between 0.15 and 1.5 per cent, with the properties of the steel varying accordingly.[924]

The manufacture of steel did not so much lead to new tools and weapons as to *improved* tools and weapons, its first use being mainly for weapons, with the sword (invented in the Bronze Age) being the weapon that was most greatly improved. The efficiency of the sword was for the first time limited by the strength of the human arm, not that of the blade.[925]

For the technology of tools, the advent of the use of iron was extremely important. While new tools made of iron appeared early in the Iron Age, it was not until about 800 BC that they came to be used generally. Comparisons of the efficiency of stone and iron tools suggest iron contributes relatively little to the efficiency of hunting, more to the gathering and processing of wild vegetable foods, and most to farming.[926] The new tools included hinged tongs, frame saws, anvils for making nails, blocks for drawing wire, and undoubtedly improved branding instruments. Later, when it was discovered how to raise the carbon content of the iron so that tempering and hardening became possible, axes, adzes, chisels and gouges were also much improved. New kinds of drills, bits and augers were made, as well as a range of specialised hammers to suit various trades. Most of our modern hand-tools had come into existence by the end of the Iron Age[927] around 300 BC (in the Near East).

The agrarian (plough and irrigation) revolution 5000 BP

Mesopotamia and Egypt

Innovation and the development of civilisation

The *horticultural* revolution consisted in the transformation from a hunting and gathering lifestyle to the cultivation of *swidden forest plots*, which were burned and then cleared using *stone axes* by *men*, and cultivated by *women* using the *digging stick*. This later became the cultivation of *bush* plots and small vegetable *gardens*, eventually by *men* using the stone-headed *hoe*. The *agrarian* revolution can be described as the transformation from this to the growing more particularly of *cereals* in *fields* by *men* using *ploughs*. The nature of the use of the plough meant that, unlike in swiddening, the land had first to be cleared of stumps and stones. The effort involved in so doing may partly have accounted for men's taking over cultivation completely, combined with greater strength generally being required in the productive effort, plus the fact that there was a reduction in the quantity of prey that they might otherwise have spent time hunting. But already in horticultural societies we find men doing all the hard work. "Men clear the ground for the farms because women are not strong enough."[928]

Where the first significant tool of the horticultural era was the digging stick, the first significant tool in the agrarian era was the hoe used as a plough. The early plough was simply a hoe drawn through the ground, perhaps first by one man with a rope, while another walked behind, driving its point into the ground – though already by about 3500 BC cattle were being harnessed for pulling in Mesopotamia,[929] while in Egypt ploughs were being pulled by a pair of oxen at least by the time of the Old Kingdom (2686–2134 BC).[930]

The invention and use of the plough was the most important technological development after the beginning of the employment of horticulture. The second most important was artificial irrigation. Already some 8000 years ago the mastery of simple irrigation techniques allowed large quantities of people to settle in southern Mesopotamia.[931] These developments made possible the maintenance and growth of the human population of about 80 million some 5000 years ago, as well as the development of civilisation. With the plough, seed crops could be grown more widely, largely replacing that part of horticultural cultivation consisting of hand-planted tubers, and thereby reducing the variety in people's diet. The resultant harvest per unit land in terms of calories increased enormously. Worldwide, seed cultivation expanded at the expense of vegetable horticulture, while at the same time cereal crops depleted soils more quickly than did the vegetables and fruits they replaced. A characteristic shared by *all* early civilisations is their direct or indirect dependence on grains – wheat, rice, maize, barley, oats, millets, sorghums, rye, and others – as food for their growing populations.[932]

The plough was first used in Mesopotamia or Egypt somewhat before 5000 years ago. (Both Mesopotamia, with its Fertile Crescent, and Egypt, with its Nile flood-plain, were extremely fertile.) As compared with the hoe, the plough was more effective in killing weeds; and, as it was later developed, by turning them into the soil it returned nutrients to it and kept them near the surface, where they could be of use to crops. This meant that, with crop rotation, the same area could be used for longer periods, as well as that the amount of food produced per unit area was greatly increased. Also, where hoeing can only be done by humans, the use of oxen to draw ploughs made it possible to cultivate larger areas and produce more food per person. This productivity was further enhanced when irrigation was employed and manure from the oxen was used to fertilise the fields.

The first ploughs were made wholly of wood; later the ploughshares on some were covered with or made of bronze, which at that time was relatively scarce; and it was not before about 750 BC that they were generally made of steel – more than 1000 years after the first employment of steel technology. (Since then, steel has remained the primary material for making tools and weapons used for cutting.) The plough was not used in sub-Saharan Africa or the Americas until introduced there by Europeans.

Artificial irrigation was employed more in Mesopotamia than in Egypt. There it was used to retain and regulate flood waters, canal-digging being among the activities most frequently ordered by the rulers of the Sumerian city-states. According to Russell and Russell, in Sumer, as in China, overpopulation led to competition for water between city-states;[933] and in Mesopotamia irrigation disputes were a constant cause of war.[934] The first irrigation canals were constructed some 7500 years ago, and their use became widespread during the agrarian period. The most famous of all the ancient canals, the Nahrwan, which irrigated a large portion of the left side of the Tigris, was 125 m wide and over 320 km long.[935]

According to K. A. Wittfogel, the rulers of what he terms 'hydraulic civilisations' control their populations by controlling their access to water. (Cf. Boserup and Wilkinson below, re the controlling of vital resources.) Most of the first civilisations in history, such as those of Mesopotamia, Egypt, the Indus Valley, China and pre-Columbian Mexico and Peru, were hydraulic empires. In keeping with Donald Kuenen we can suggest that these civilisations began in delta areas or flood plains, where there were the most fertile soils, and then afterwards, thanks to irrigation, expanded into the more arid and even desert regions in the vicinity.[936]

An important aspect of agrarian agriculture is its production of monoculture crops, which are grown over a significantly larger area than were earlier horticultural crops. Apart from being much more demanding of the soil, they are also much more susceptible to pest invasion.[937] (Cf. the relevance of population density to the spreading of infectious diseases.) This increased susceptibility means a greater likelihood of crop failure and consequent famine.

The manufacture of more advanced *pottery* was dependent on the development of the potter's wheel.[938] The functioning of the potter's wheel is based on the principle of rotary motion, the first application of which was the bolas, and later the sling and the twisting stick, and later still the bow drill. The first potters were itinerant, a characteristic that continued at least up to the 1950s, when potters, with their 'families' and their wheel, travelled from village to village on Crete and from island to island in the Aegean, turning out in each what the local taste demanded.

V. G. Childe suggests that the development and use of such implements as the potter's wheel presupposed the existence of a *surplus* to support the potters and their families, who, like soldiers, were not themselves directly engaged in the food-acquisition process. Here we should say however that, in being engaged at least in the food-*storing* process, what the potter and his family consumed was not part of the community's surplus, but part of its *stock*, i.e. part of its vital resources needed to support the population. The existence of the potter (unlike that of the soldier) was necessary to the group's economic effort.

The potter was a full-time specialist – one whose services were much less costly than those of the coppersmith, whose profession also presupposed sufficient resources to support more than the directly-producing population. The first coppersmiths were also itinerant, the tinker in rural Europe being a survival of the same system. The use of the potter's wheel also required a sufficiently large population to maintain the potter and his family, viz., a couple of hundred households.

Potter's wheels and cart wheels, both of which employed discs revolving on a fixed axis or with an axis free to turn in a bearing, were possibly first produced in Mesopotamia, both at around the same time, wheeled vehicles being used in Sumer soon after 3500 BC, though sledges were used at least for funerals as late as 2600 BC in Mesopotamia. Though there are somewhat earlier remnants of potter's wheels than cart wheels, this may simply be due to their being more durable. The most obvious use for carts and wagons would have been to convey food from rural areas to towns and cities, and to carry manure in the opposite direction. By thus allowing a larger population to be fed at a single centre, the invention of the wheeled vehicle must have contributed to the beginnings of urbanisation.

Another invention employing rotary motion at around this time, or perhaps as early as 6000 BC, was the *spindle*, used to collect fibres so as to produce thread, first from flax, and later wool.[939] And sails, being the first harnessing of an inorganic force to provide motive power, were used in Egypt by 3000 BC.

One of the most important technological developments made by herders was the use of horses, and later camels, for transportation. This practice originated around 3500 BC, when the Proto-Indo-European Yamnaya herders of the steppes between and north of the northern parts of the Black and Caspian Seas began both riding horses and harnessing them to wagons. Thus the Yamnaya had the advantage of two kinds of mobility: wagons for slow bulk transport (water, shelter, food), and horseback riding for rapid light transport (scouting for pastures, herding, and raiding expeditions). The first chariots probably appeared on the steppes before 2000 BC, where they were used in warfare; and by around 1800 BC they were being used by the Assyrians. Chariots were designed for speed, having two spoked wheels (making them light and fast), and carried one or two standing occupants. They were pulled by bitted horses, and usually driven at a gallop.[940] The use of horses gave these herders an important military advantage over their less mobile agrarian neighbours, and enabled them to rule over much of the Near East – at least until the new chariot technology was adopted by the more numerous agrarian peoples.[941] Mounted horsemen were organised into *armies* 1000–900 BC, which led to new waves of conquest.[942]

Although there was a temporary slowing of the rate of technological innovation during the first part of the agrarian era, it accelerated again well before its end.[943] The Mesopotamian–Egyptian era included the invention of a balance for weighing grains, the drawbridge (in Egypt, by 2000 BC), the pulley (in Mesopotamia, before 1500 BC), the crank (perhaps in Syria, ca. 800 BC), the crossbow (650–500 BC), iron casting (550 BC), the lathe (perhaps in Persia, by 550 BC), and glass-blowing (probably in Syria 200–100 BC).[944]

Resource depletion

With the agrarian revolution began a further lowering of the quality of the land. The removal of trees – eventually with the aid of steel axes – to allow the creation of fields led to soil erosion, a process accelerated by the topsoil's being exposed to wind and rain due to the use of the plough. The hills of Attica were stripped bare of trees within a couple of generations, and by 590 BC in Athens the great reformer, Solon, was arguing that cultivation on steep slopes should be banned because of the amount of soil being lost.[945] And, in areas which were too hilly or deficient in nutrients for agriculture, goats were let roam as they were in the horticultural period, exterminating all but the hardiest bushes. The results of these activities remain with us today. The soils of the Mediterranean countries have been washed into the sea, with the result that, combined with the felling of timber, in Greece for example only a few per cent of the original forest cover remains.[946] In other areas, the use of irrigation to make possible or increase the supply of water to fields led to salinisation, leaving the soil unusable for cultivation for thousands of years – as can be witnessed today in the Tigris–Euphrates valley, as well as in the valley of the Indus.[947] Wastelands developed around cities, where the area was picked clean of fuel and building materials, and vegetation disappeared due to pedestrian traffic. And wild animal species were driven back from areas of human habitation, some of them becoming extinct.

The early growth of the Mesopotamian population reflects the original success of their irrigation systems in exploiting the fertile alluvial soil. The increasing soil salinity caused by this large-scale irrigation meant however that wheat had successively to be replaced by barley, a more salt-tolerant crop. Between 3500 and 2500 BC, wheat declined from contributing one-half to one-seventh of the total crop yield, and after 2000 BC it was no longer grown in Mesopotamia. Furthermore, between 2500 and 1700 BC the total crop yield in the Tigris–Euphrates valley fell by two-thirds.[948]

Population checks; religion

Population checks were practised at all times among early agrarian cultures, the most notable perhaps being infanticide. Infanticide was probably common among Egyptians at the time of Moses, was practised by the ancient Jews, and was common in early Rome.[949] Abortion was widespread: Plato and Aristotle permitted it in their ideal states,[950] and there was no law against it e.g. in Greece or during most of Roman times. Later, both pagans and Christians accepted it virtually universally. Such population checks were employed by the poor, especially in times of crop failure. On the other hand, however, a law of the Spanish Visigoths in the 600s AD punished both abortion and infanticide with death. But in medieval society both were widespread, if covert, while postponed marriages and socially condoned celibacy also functioned as checks.[951]

It should be noted at the same time however that, given the agrarian surplus of food, after the horticultural era the role of internal population checks continued to decline. More important for early agrarian societies (or their leaders) was that they have *large* populations, in order to hold their own in war against other societies. It was in this context that *religions* came into existence – what were to become moral traditions, which included sharing within the tribe and fighting outsiders, supporting the continuing existence of the relevant tribe(s) by *condoning* rather than restraining practices leading to population growth.

Population growth

With the agrarian revolution the size of the human population once again began to increase dramatically, particularly in the valleys of such large rivers as the Nile and the Tigris and Euphrates (and, elsewhere, the Indus, ca. 2500 BC, and the Yellow River, ca. 1600 BC). The first towns appeared, which, like the villages of the horticultural period, were military strongholds. Some developed into cities with over 100,000 inhabitants; and some of these became religious centres and the centres of empires. Ca. 3500 BC the Sumerian city of Uruk had a population of about 40,000,[952] and a wall around it which, with its large defensive towers, was 10 km in circumference and 4 to 5 m thick. It has been estimated that at its zenith Mesopotamia supported between 17 and 25 million inhabitants.[953]

The relatively huge populations of urban centres required an inflow of substantial quantities of food, fuel and building materials; i.e. commerce was *necessary* for the existence of cities. The complex economic network this presupposed itself required political stability and security, conditions which could be afforded (for periods of time) through the power of the rulers.[954]

Population growth was slow in the pre-horticultural period, rapid immediately after the beginning of horticulture, slower when irrigation was used to expand the productive area (5000 years later), and slower still when irrigation reached its practical limits.[955]

Social stratification: kings/pharaohs, serfs and slaves

The transition from horticulture to agrarian agriculture, with its more advanced technology and economy, meant an increase in social inequality, such societies as those whose members practise irrigation, own domesticated animals, or practise metallurgy for ornamental and ceremonial purposes generally being less egalitarian than groups without these characteristics.[956] As pointed out by Diamond, skeletons from tombs at Mycenae around 1500 BC suggest that royals enjoyed a better diet than commoners, since the royal skeletons were 5 to 8 cm taller and had better teeth (on the average, one instead of six cavities or missing teeth). Among mummies from Chilean cemeteries around 1000 AD, the elite were distinguished not only by ornaments and gold hair clips, but also by a fourfold lower rate of bone lesions stemming from infectious diseases.[957]

There were two basic classes, a small warrior nobility, including the king, and the great mass of common people. Often the king was viewed as divine or as having access to divine powers, which served to legitimate his tyrannical and exploitative practices.[958] Due to increased communication between peoples – largely thanks to trade – unlike in earlier times the powerful individuals from different nations came themselves to constitute a class, or even, in a sense, an interbreeding tribe. And in this tribe there was of course a pecking order. In the hunter-gatherer state leaders attained their position largely on the basis of their general ability. By agrarian times, leadership went more to men who evinced executive skills and low morals (cf. power corrupts), at least until a system of royal and aristocratic descendancy was established. Though *government* would already have existed in the horticultural period, it became of greater importance in the agrarian. Thus it was not only through his personal prowess but, even more, through his governing ability that the agrarian king or pharaoh was able to ensure that his subjects did his bidding.

Babylonian law as codified by Hammurabi, the first ruler of the Babylonian empire (died ca. 1750 BC), was "to secure the creditor against the debtor and consecrate the exploitation of the small producer by the possessor of money." The debtor might not only mortgage his land, but pledge his children, his wife, or his own person. Enslavement for debt was legalised; trading partnerships were regulated in the interest of the owner of capital; and fraud by the working partner was severely punished.[959] Usury, mortgages and enslaved debtors followed the use of money wherever it was introduced.

In Sumer, *writing* was originally invented to record economic transactions ca. 3000 BC; and, for the first time, land in the form of estates could be bought and sold by individuals.[960] In Mesopotamia in the days of the Empire of Akkad (ca. 2300–2100 BC) land was bought and sold like any other commodity. In Egypt under the New Kingdom (1550–1295 BC), though land was held on lease from the pharaoh and generally carried obligations to perform military service, it could nevertheless be transferred by testament or sale.

In early agrarian urban communities beggars comprised from a tenth to as much as a third of the total population.[961] And the institution of *slavery* reinforced the domestication of humans[962] through slaves' adaptation to agricultural toil and a poorer diet (though slaves in some cultures may not have reproduced as rapidly as in others).

Because politics and economics were always highly interdependent in advanced (i.e. iron-technology) agrarian societies, the people who dominated the political system also dominated the economic system.[963] Those in the governing class usually *owned* most of the peasants who worked the land.[964] Domesticated people, at least in the case of serfs or peasants, like domesticated animals, were provided only with sufficient food and shelter to allow them to work and reproduce.

The individuals who commanded the greatest military power became kings, personally owning huge tracts of land, while the vast majority of the population worked as peasants or slaves for the king's personal best interest. The commoners toiled creating the agricultural surplus that constituted his wealth; they fought in his wars; and they built megalithic structures to commemorate his greatness.[965]

In a relatively recent survey, *slavery* was found in 83 per cent of modern advanced (metal-using) horticultural societies, but in only 15 per cent of simple such societies. And class systems were reported in seven per cent of advanced horticultural societies and in only one per cent of simple.[966]

Leaders' enormous political power during the agrarian era led to the creation of what Mumford calls *megamachines* – huge machines the constituents of which are *people*, the army being a paradigmatic example. In most societies perhaps the greatest portion of the surplus was drawn into the feeding, weaponing and overall operation of the military megamachine, while for those who were drafted into it, work became a curse.[967] Among other things, the existence of such megamachines was a clear manifestation of the great degree of social stratification at this time, with proportionately very few rulers engaging thousands of their subjects in hard labour.

The physical power of such social units in the early agrarian period was stupendous. The Egyptian pyramids and all monolithic structures e.g. in Mesopotamia and the Americas were built using such megamachines, as was the Great Wall of China. Cheops' (Khufu's) Great Pyramid, completed ca. 2560 BC, measures 230 m per side and is 147 m high. It contains some 2,300,000 blocks each weighing on average 2½ tons. But the Great Wall of China is the largest alteration in the world's surface effected by man. The first Great Wall, completed in only 12 years, was built just before 200 BC. It was ca. 6400 km long and 5 to 10 m high, and claimed the lives of something like half of the one million workers who constructed it. (The second Great Wall was approximately 11,000 km long and built around the time of Christ.)

The small and often literate urban governing class lived in a strikingly different world from that of the illiterate, rural, peasant majority. Despite the fact that they were members of the same society, each group had its own distinct subculture, as intimated above. The rulers lived in walled cities that served as their fortresses, and it was they who enjoyed whatever benefits the new technology and new social system had to offer. It was the subculture of the minuscule governing class, in the context of the new phenomenon of the *city*, that incorporated the cultural refinements we identify with civilisation.[968]

Meanwhile *bronze* was originally used solely in the making of weapons and ceremonial objects for this elite class, metals playing a major role in the military, religious and artistic spheres, but, as implied earlier, not in subsistence activities.[969] And the remarkable artistic achievements of agrarian societies were a product of their harshly exploitative social system.[970] This governing-class subculture also included a contempt for physical labour of any kind, and for those who engaged in it[971] – as might be expected of hunters/warriors, and as is in keeping with increased work's being avoided in primitive societies.

Quality of life

As is clear from the above, just as the human workload grew with the move to horticulture, it grew again with the move to agrarian agriculture. With the coming of the agrarian revolution, if not earlier, work for the commoner became *labour*, i.e. work for someone else in order to obtain the resources (primarily *food*) necessary for the survival of one's family. And in early agrarian times people laboured to the point of death.

Due to a paucity of rainfall, particularly in Mesopotamia, rising population pressure meant that farming came to depend increasingly on irrigation, which in turn meant that more time constantly had to be devoted to food production. Furthermore, as populations grew, the continuing struggle to produce sufficient food to feed everyone was exacerbated not only by the salinisation of the soil, but by a reduction in the proportion of actual food-producers due to urbanisation and the division of labour, the result being that undernutrition was a chronic problem for at least part of the population.[972]

Apart from increased work, the quality of the common person's life declined further as a result of an increase in the effects of disease, whose spread was abetted by poor diets. In fact, from ca. 3000 BC until well into the 1900s, infectious disease was the major overall cause of human mortality.[973] Unlike during the horticultural period, during the agrarian era up until the medieval period the common people ate almost no meat, which was also reflected in the quality of their lives.

War

As the human population grew during the agrarian period – largely due to the increased productivity of the soil thanks to the use of the plough and irrigation – and new weapons were developed made of bronze and later iron, warfare increased while protowarfare declined. For the first time in the development of humankind, the conquest of other people became a profitable alternative to the conquest of nature. Thus, beginning in advanced horticultural societies and continuing in agrarian, almost as much energy was expended on war as on the more basic struggle to subsist. Bronze was to the conquest of people what plant and animal domestication was to the conquest of nature. But where it was demeaning to engage in subsistence activities related to agriculture, there was honour to be gained in engaging in war, just as there was honour in hunting; both activities involve the excitement of killing, and war the further excitement of getting killed. And to this may be added that urban centres emerged from just those villages that had the advantage of bronze weapons.[974]

The transition from bronze to iron weapons began around 1800 BC. In ca. 1286 BC, at the battle of Kadesh between the Hittites, who quite possibly used iron weapons, and the Egyptians, each nation had an army of about 20,000 men.[975]

As Diamond emphasises, written records testify to the frequency of *genocide* in early literate civilisations. The wars of the Greeks and Trojans, of Rome and Carthage, and of the Assyrians, Babylonians and Persians proceeded to a common end: the slaughter of the defeated irrespective of sex, or else the killing of the men and enslavement of the women. As Diamond says regarding the Old Testament (ca. 1450 BC):

> We all know the biblical account of how the walls of Jericho came tumbling down at the sound of Joshua's trumpets. Less often quoted is the sequel: Joshua obeyed the Lord's command to slaughter the inhabitants of Jericho as well as of Ai, Makkedah, Libnah, Hebron, Debir, and many other cities. This was considered so ordinary that the Book of Joshua devotes only a phrase to each slaughter, as if to say: of course he killed all the inhabitants, what else would you expect? The sole account requiring elaboration is of the slaughter at Jericho itself, where Joshua did something really unusual: he spared the lives of one family (because they had helped his messengers).

We can obtain a further idea of how war was perceived during this part of the agrarian period by considering a couple of quotations from that time. The first is that of the Assyrian king Shamshi-Adad, ca. 800 BC:

> Thirteen thousand of their warriors I cut down with the sword. Their blood like the waters of a stream I caused to run through the squares of their city. The corpses of their soldiers I piled in heaps …. His [the Babylonian king's] royal bed, his royal couch, the treasure of his palaces, his property, his gods and everything from his palace, without number, I carried away. His captive warriors were given to the soldiers of my land like grasshoppers. The city [Dur-Papsukal] I destroyed, I devastated, I burnt with fire, and I … laid claim to the whole of the country under the ancient title of 'King of Sumer and Akkad.'[976]

The second quotation is from *The Iliad*, stemming from around the same time:

> With that, just as Dolon reached up for his chin
> to cling with a frantic hand and beg for life,
> Diomedes struck him square across the neck –
> a flashing hack of the sword – both tendons snapped
> and the shrieking head went tumbling in the dust.
> They tore the weasel-cap from the head, stripped
> the wolf pelt, the reflex bow and long tough spear
> and swinging the trophies high to Pallas queen of plunder,
> exultant royal Odysseus shouted out this prayer:
> "Here Goddess, rejoice in these, they're yours!"

We see here little mercy for one's fellow, if he should happen to be of another tribe or nation. And we see the importance of myth in the moulding of people's morals. Shame for lack of bravery rather than guilt for deeds done was the ethos among men at this time, which supported the existence of the community through glorifying military prowess. Though on the whole wars did not last long, they were nevertheless an annual affair, the period following the harvest being known as 'the season when kings go forth to war.'[977]

Trade

Money as we know it was absent in the first, simple agrarian societies. At the beginning of the agrarian era, as in the horticultural era, food products constituted not only capital, but also the medium of exchange – as living animals did for herders. The horticulturists' media of exchange were numerous, including grains and perhaps other storable vegetable products such as nuts and the seeds of legumes,[978] as well as all their animals – and sometimes their dependants as well. Amongst modern horticulturists, salt and tobacco have also been used as currency.[979]

In the agrarian period, grain became the sole or central currency; and that grain was produced with the aid of ploughs and irrigation. Early on, wheat served this function in both Mesopotamia and Egypt, while soil salinisation in Mesopotamia led to barley later playing this role. Wages, rents, taxes, and various other debts were paid in specified quantities of these grains. Due to their natures, seeds store better than other vegetable products – at ordinary temperatures for about a year – and their protein and mineral content, though lower than that of many other vegetable foods, gives them sufficient nutritive value to function as staples. In any case, grain constituted a source of *energy*, and came to constitute the primary form of capital.

The bulkiness and perishableness of grains, however, led to such metals as silver and copper coming to replace them as media of exchange, being used as currency alongside grains rather early in the agrarian period. Trade, in which metal currency allowed the rise of a class of merchants or middlemen, increased throughout the era. (*Paper* currency was first used by the Chinese in the 900s AD.)[980]

The growth of monetary systems, which have always facilitated the production and exchange of goods and services of every kind, had enormous implications for societal development. The use of money greatly expands the market for various products, since they can then be sold even to people who produce nothing the producer wants in exchange, thereby increasing the opportunities for trade.

The Roman Empire
Resource depletion

As population continued to grow during the agrarian era so did the depletion of resources. For example, with the Roman civilisation came an enormous increase in metalworking,[981] which implies a commensurate increase in the extraction of metals, particularly iron. And of course increasing population means increasing consumption of food and, in the case of the Romans as in the case of earlier agrarian peoples, a reduction in the productivity of agricultural land. As regards Rome and the northern Mediterranean more generally, unlike Egypt and Mesopotamia, there was not enough fertile soil to produce the food needed for the population of a sophisticated civilisation. Thus Rome had to import grain from Sicily, Tunisia and Egypt. This need was increased when slavery diffused into northern Mediterranean societies from the herder dry belt to the east. Peoples of the northern Mediterranean could not feed slaves from their own supplies, while at the same time the proportion of slaves in their populations became extremely high.[982] By the time of Christ, Italy had about two million slaves, making up about one-third of the population.[983]

With the continuing existence of the Roman Empire, Rome's supply of grain, which was collected as *taxes* from provinces in the Empire, constantly dwindled as its grain-growing areas were over-used. At the time of Christ, Rome imported ca. 300,000 tons of grain each year from Sicily and North Africa; in 200 AD: 180,000 tons; and in 400 AD: 67,000 tons.[984] Today these regions are deserts or near-deserts.

The number of animal species also decreased under the Romans. Around 200 BC the lion and the leopard became extinct in Greece and the coastal areas of the Near East. The trapping of beavers in northern Greece led to their extinction there around this time; and by the early centuries AD, the elephant, rhinoceros and zebra were extinct in northern Africa, the hippopotamus on the lower Nile, and the tiger in northern Persia and Mesopotamia. And whales were hunted to extinction in the Mediterranean before the fall of the Roman Empire.[985]

Population checks

A number of internal checks to population growth were employed in ancient Rome. As in the case of other ancient empires, these checks were almost totally positive rather than preventive – thus at least partly accounting for Rome's high level of mortality, where average life expectancy may have been only about 20 years.[986] During the first centuries AD there were also widespread epidemics of syphilis, plague, tuberculosis, malaria, etc., which supported high mortality.[987] It was not until the latter days of the Roman Empire, when there no longer existed a surplus of imported grain, that the preventive checks of celibacy and postponement of marriage were employed.[988] The population had overshot its temporary carrying capacity.

In the Roman Empire the principal checks to population growth were war and infanticide, as well as the use of slaves.[989] War first took the form of battles to extend Roman territory at the expense of horticultural and herder groups. But the slaughter of the German horticultural barbarians by a succession of emperors did not prevent the re-growth of their populations in the next generation. Later, war was the result of Mongol herders (Huns) from Asia invading Rome.

Despite these checks, the population of Rome grew to one million by the time of Christ, a size it maintained for the next 150 years before beginning to shrink;[990] by 600 AD Rome contained no more than 50,000 people while Constantinople contained over 500,000.[991]

Social stratification: patricians and proles

Social stratification was extreme during the whole agrarian period. In the Roman imperial household, for example, the luxury was amazing, with emperors at the height of the Roman Empire having thousands of household staff.[992] The vast majority of the people belonged to the proletariat, the literal Latin meaning of the term being 'those with many offspring,' and the full ancient Roman sense of the word being 'the lowest class of a people, whose members, poor and exempt from taxes, were useful to the republic only for the procreation of children.'[993] And of course it was from this class that the Roman soldiers were originally drawn. Nevertheless, the proletariat worked only 206 days a year, and spent the remaining 159 at shows and festivals.[994]

Civil conflict

During the population crisis in Rome from 160 to 330 AD, 26 of the 37 emperors were murdered or killed in civil battles, not to speak of their relatives, high officials and dozens of pretenders.[995] And though Rome's record may have been worse than most, struggles for greater power among those who already had power were common in all advanced agrarian societies. Most of the intrasocietal conflict in agrarian societies was within one and the same class. Members of the governing elite fought with one another for power and social position, as manifest in their ruling over group territory in the form of geographical regions. At the same time, on the village level, peasants and their families competed for access to individual territory in the form of the best farmland. There was little or no conflict *between* the rich and the poor. And the replacement of one elite by another often had little effect on the lives of peasants,[996] though they died by the thousands in their leaders' wars.

Innovations

Innovations during or immediately before the Roman era include the catapult (Sicily, ca. 400 BC), the hydraulic screw (invented by Archimedes ca. 250 BC), the pump and the water clock (both in Alexandria during the 200s BC), concrete (ca. 200 BC), horseshoes (perhaps around the time of Christ), glass for windows (on a small scale in Rome, ca. the time of Christ),[997] and a work-harness for horses of the sort known as 'the dorsal yoke,'[998] the original horse collar having been invented by the Chinese between 300 and 200 BC. Another important labour-saving device first used during the time of the Roman Empire was the animal-powered Gallic reaping machine, used and probably invented in north-western Europe.[999]

Medieval Europe
Innovation

There was a constant acceleration in the development of technology during the agrarian era. Agricultural output was further increased with the invention and employment of the heavy plough and the use of crop rotation involving the planting of nitrogen-fixing legumes. The heavy quadrangular-framed wheeled plough, with a fixed mould-board for turning the furrow, was probably developed around 500 AD, and its use spread across Europe, reaching northern Europe by 700 AD at the latest.[1000] The earliest ploughs, in contrast, did not turn the soil, and left a wedge of undisturbed earth between each furrow. These simpler ploughs were not well suited to the heavier soils of northern Europe however, particularly those of the lowlands. The heavy plough that was needed required a team of eight oxen to pull it, but at the same time it enabled a much larger area to be cultivated.[1001]

The *watermill* was invented by the Greeks perhaps between 300 and 200 BC,[1002] and was slowly introduced throughout Europe beginning in Roman times, being used mainly in the north-west;[1003] it reached England about 700 AD and Scandinavia some 400 years later.[1004] The *chimney* appeared in the 900s.[1005] The *windmill* appeared in Persia (with a vertical axle) during the 600s or somewhat earlier, and in Europe during the 1100s, coming into more general use as an energy supplier around 1600, particularly for the draining of fens and marshes in England, France and the Low Countries.[1006] The European windmill, with its horizontal axle, was so different from the Persian however as to make it almost a new invention.[1007]

Vertical-axle and horizontal-axle windmills

As is perhaps most clearly evident with regard to the medieval north-western Europeans' development of windmills, E. J. Kealey claims that they were the first people to consistently combine technological expertise with natural energy to create labour-saving devices.[1008] Apart from draining, windmills were used for grinding grain, fulling (shrinking and thickening) cloth, sawing timber, extracting oil from olives, and making paper. Hand-made paper was invented by the Chinese in the 100s AD, and diffused to the Islamic world in the 700s; in the 1200s it reached north-western Europe, and production was promptly mechanised.[1009] During the 1300s block printing – which had paper money as a by-product[1010] – reached north-western Europe, coming ultimately from China. The Chinese also invented movable type during the 1000s;[1011] by 1450 it had been reinvented in Europe, and the *press* was devised for mechanised printing.[1012]

Other innovations during the medieval period include stirrups (introduced into Europe from Asia ca. 600 AD), on which the development of mounted warfare depended.[1013] Though glass was originally invented ca. 3000 BC, it first attained widespread use in windows about 1000 AD. Medieval innovations also include gunpowder (before 800), porcelain (800), the wheelbarrow (ca. 200 AD in China; around 1200 in Europe),[1014] the magnetic compass (before 1100; ca. 1400 in Europe) and the spinning wheel (1220). In 1250 Roger Bacon invented the magnifying glass; and eyeglasses were invented shortly after 1286.[1015] The mechanical clock, originally invented by the Chinese perhaps in the 1000s,[1016] was reinvented by Europeans around 1300. The gun was probably invented in Germany at the beginning of the 1300s, and the buttonhole at around the same time. The first canal lock was built ca. 1400 in Europe. Note that a large number of these innovations, including gunpowder, the compass[1017] and the spinning wheel, were first made in China, which, around 1200, had the largest, most literate and most advanced society in the world.[1018]

In Europe the techniques of both land transport and ploughing were considerably improved by the introduction of the horse collar (visible in the Bayeux Tapestry of 1066)[1019] – an improvement on the dorsal yoke – that arrived in Europe north of the Alps during the 500s AD and had made its way across the continent by the 700s. Most subsistence farmers were too poor to keep horses, however, and though oxen with their heavy yokes were less efficient, they predominated for example in England until the late 1700s, survived in Scotland into the 1800s, and may still be found in many parts of Europe.[1020]

Improvements in the construction of sailing vessels made possible the long-distance transportation not only of goods whose value was high in relation to their weight, such as spices, hides, wool and precious metals, but also the transportation of grain, wine and timber.[1021]

Resource depletion: extinctions

As regards animal species, the spread of agriculture and the human population forced many from their natural habitats and led to the extinction of some. The crane became extinct in this way sometime before 1600; other animals, such as the wild ox, were hunted to extinction (in 1627).[1022]

On Mauritius the introduction of pigs and rats together with hunting by sailors drove the ground-nesting dodo to extinction by 1681.[1023] Przewalski's wild horse (also known as the Asiatic wild horse, the Mongolian tarpan, etc.), which lived in western Europe in the Middle Ages, was exterminated in the wild in the 1700s,[1024] but has survived in domesticated form,[1025] and has recently been reintroduced to its natural habitat. And before 1800, as mentioned in Chapter 2, two species of hippopotamus from Madagascar became completely extinct. The European bison was still found in the early medieval period across a wide area of what is now Belgium and Germany. By the 1700s it was only found in eastern Europe, and the last wild animal died in 1920.[1026]

Population checks

As in the horticultural period, apart from war and disease one of the constraints on population growth in the agrarian era was infanticide in the face of the threat of severe deprivation. When crops failed and famine was imminent, families often abandoned their new-born by the roadside or left them at the door of a church or monastery in the hope that someone else might raise them. Sometimes even older children would be abandoned by their parents. The story of Hansel and Gretel is based in the reality of scarcity. In some districts in China as many as a quarter of the female infants were killed at birth – signs were put up near ponds, reading, 'Girls are not to be drowned here.'[1027] In India, as recently as at the end of the 1800s, infanticide and child neglect existed everywhere.[1028] That a population kill a quarter of its female infants was not uncommon (cf. the survival, in all populations, of only one reproducing offspring from each individual). Female infanticide at such a level was not exceptional in such cultures, but, like warfare, was rather the *norm*. The infant hadn't the status of being a member of the community, and thus its being killed was not murder. In the case of more primitive peoples, the achieving of such status required going through 'baptismal' initiation rites that admitted the individual to the band/tribe, though not as a full (adult) member. The committing of what we would call murder would not have been considered immoral; rather, as may be inferred from what has been said earlier, what would have been immoral would have been the *failure* to kill under the requisite circumstances.[1029]

Other population checks included the directing of adolescents into religious careers as priests or nuns, and setting minimum age and capital requirements for marriage. The Elizabethan Poor Law of 1601, as it was summarised in 1651, read: "[T]hey who could not maintain a wife, might not marry; for a License they could not have. [U]sually none was permitted marriage till the man were 35 at least, and the woman 30."

The population-checking effect of the spread of infectious diseases was greater than it otherwise would have been due to the constantly increasing size of the population and greater interaction between groups, coupled with worsening hygiene. The most devastating instance was the Black Death in the middle of the 1300s, which killed about a third of the population of Europe over a period of four years. (Between the mid-1200s and the mid-1400s in England the expectation of life at birth for a male dropped steadily from about 35 years to about 17 during the population crisis, and rose again to about 33 during the recovery period.)[1030] In this case, as in that of every other epidemic of bubonic plague, the disease followed the main trade routes.[1031] This increase in mortality of course implied a lowering of the quality of people's lives. It also meant a drastic reduction in the population of Europe during the 100 years after 1350 – a period during which the environment began to recover from its previous over-use.[1032]

Social stratification: kings, lords and serfs

In the beginning of the agrarian epoch the most powerful were the *chiefs*. And as the population grew these chiefs became fewer, and ruled over/controlled ever greater territories and numbers of people, becoming eventually *kings* and *pharaohs*, with those directly beneath them being *lords*.

Throughout the agrarian period, both a surplus and soldiers were provided through having a large and sufficiently healthy population which could easily be *controlled*. (In a study by Herbert Barry and others it was found that where hunter-gatherers stress assertiveness more than compliance, agriculturists stress compliance over assertiveness.)[1033] Two particular results of this requirement are the institution of *slavery*, and the creation and enforcement of *laws*, both processes having begun already in the horticultural era, and both still being with us today. As regards the laws: they are such as essentially support the maintenance of the status quo, so that those in power can remain in power. This control itself involves the using of *force*, but force *within* the community, in the form of *police* who enforce the laws, which protect the property of *individuals*.

In medieval Europe, the state was indistinguishable from the ruler and what he owned. This form of state can be traced back to horticultural societies and, in a sense, even to hunting and gathering bands, where no distinction was made between the private and public aspects of the lives and activities of leaders.[1034] For example, one Russian nobleman in the 1800s, who owned over 8000 km² of land, also *owned* nearly 300,000 serfs. Prior to the emancipation of the serfs in Russia, the Tsar owned some 27 million of them. Throughout Europe, petty thievery could be punished by death, often by cruel means;[1035] and peasants were taxed to the limit of their ability to pay.[1036]

The slavery and serfdom that were common in agrarian societies meant that large land-holdings and large numbers of slaves or serfs normally went hand in hand.[1037] It was only when death reduced the number of good workers to the point where they were scarce that the governing class was forced to bid for their services, thus raising incomes above subsistence level. Usually, however, high birth rates kept this from happening, or, when it did, high birth rates soon brought about a return to there being a surplus of labour.[1038]

Quality of life: longevity

The usual situation in Europe as elsewhere during the agrarian era was one of dietary monotony, chronic malnutrition, and periodic and sometimes massive famine. In Ireland in 1318 bodies were dug up from graves for food. An existence under the constant threat of starvation and in the face of the daily reality of an inadequate diet and malnutrition has been the common lot of humanity since the development of agriculture. Where Chinese agriculture was able to maintain a large number of people on the brink of starvation, medieval European agriculture kept a smaller number in the same condition.[1039] During the unusually cold period from about 1600 to 1800, throughout which susceptibility to infection and high death rates were the norm,[1040] severe famines occurred widely.[1041] The average lifetime of most people in the world has been estimated to have been only about 30 years in 1650,[1042] though it may have been even lower during the agrarian period as a whole. In 1700, the children of British queens and duchesses had a life expectancy of only 30 years, and nearly a third died before their fifth birthday. For the common people, conditions were even worse. Life expectancy at birth was barely 25 years in Britain at this time, and even those who reached adulthood usually died young.

In France, towards the end of the agrarian era, great masses lived in a state of chronic malnutrition, subsisting mainly on porridge made of bread and water with some occasional home-grown vegetables thrown in. They ate meat only a few times a year, on feast days or, if they did not have enough silage to feed their livestock over winter, after autumn slaughter. For most peasants village life was a struggle to keep above the line that divided the poor from the indigent.[1043]

Although in agricultural societies the year has periods of intense activity, in many of them it also has slack periods of perhaps several months, when there is little work to be done.[1044] Thus, though the vast majority of humankind was terribly poor during the agrarian era, the seasonal nature of their work nevertheless meant that they had a fair amount of leisure. During this period, even in mining communities, as late as during the 1500s more than half of the worker's days were holidays, while for Europe as a whole, the total number of holidays, including Sundays, came to around 190, a number greater than that in Imperial Rome.[1045] The French peasant in 1750, for example, worked on his farm less than 200 days a year. Until 1776, however, he was also required to do about 30 days a year labour (corvée) on roads, etc.[1046]

Conflict

War was also prevalent in medieval Europe. Until the use of gunpowder, however, there was little long-term damage due to war. Cities were sacked, but they recovered relatively quickly. But with the advent of firearms wars became constantly more devastating in their physical effects.

A survey of eleven European countries of the agrarian era found that, on the average, they were involved in some kind of conflict with neighbouring societies almost half of the time. Revolt and revolution also became more prevalent. For example, in the period from 1801 to 1861 there were some 1500 peasant uprisings in Russia.[1047]

Herder invasions

The various herder groups occupied the dry belt extending from Morocco to the centre of Asia.[1048] Most notable among them were the Mongol and Turkic horse-herders of central Asia – the horse originally being herded and domesticated by the Yamnaya on the Asian steppes.[1049] These people lived in tents, and were constantly on the move after fresh pasture for their horses, travelling north in summer, and south in winter. Their only possessions besides their animals were what could be pulled along in their wagon trains – their wagons being, like the Palaeo-Indians' sleds, 'mobile homes.' They piled their families and their wealth onto their wagons – their wealth including gold ornaments, plaques for harness, and other equipment – as well as their tents and carpets.

These Asian nomads had stocky, powerful bodies with huge arms, short, bowed legs, and flat noses, being genetically adapted to survive in a harsh climate, and, as mentioned, coming from the same stock as the Palaeo-Indian and Eskimo migrants to America. And, like certain Eskimo tribes,[1050] some Asian nomad tribes subsisted wholly on raw meat. (In comparing them with civilised people one is reminded of the distinction between the Neanderthals and the Cro-Magnons; or, as suggested by René Grousset, the proletariat and the capitalist elite.)[1051] They wore trousers and boots rather than the robes and sandals of their contemporaries, both being better suited to a life on horseback. And, around 100 BC, they invented the stirrup, which gave them a tremendous advantage over the cavalry of sedentary peoples. The stirrup was unknown to both the Greeks and the Romans, and it seems that it came into common use in the West only when the Asian nomads introduced it into Hungary. And the nomadic Sarmatians, who occupied the area north of the Black Sea, invented the saddle with raised pommel and cantle ca. 365 AD.[1052]

Herder peoples – and in particular the Asian nomads – due to constantly recurrent population crises,[1053] were always inclined to instigate mortal conflicts with their richer agrarian neighbours, attacking them *en masse*. This propensity was supported by the fact that the herders occupied inferior land, and saw sedentary people's more lush areas as a potential source of food for their herds. Furthermore their populations were constantly increasing in size. As expressed by Malthus:

> An Alaric, an Attila, or a Zingis Khan, and the chiefs around them, might fight for glory, for the fame of extensive conquests, but the true cause that set in motion the great tide of northern emigration, and that continued to propel it till it rolled at different periods against China, Persia, Italy, and even Egypt, was a scarcity of food, a population extended beyond the means of supporting it.

Here we have an instance of the poor attacking the rich – though not in civil strife.

As compared to the people they attacked they came from a completely different, and much poorer, world. But while the Asian nomads could be called poor, they were not weak. Due to the nature of their existence they had a definite military advantage over those with whom they fought. Though they were less generally technologically developed than the agrarians, they were not so as regards the quality of their weapons. This was partly the result of their relative closeness to the hunting-gathering lifestyle, which undoubtedly made them generally more proficient in making and using weapons. Their short, composite, re-curved, 'cupid' bows, invented around 1000 BC, were both powerful and at the same time short enough to be swung over the horse's rear.[1054] These bows were said to have had a range of 500 m, or twice that of the English long-bow[1055] (the latter constituting a regressive step as far as the efficiency of a projectile weapon is concerned).[1056] The bow in fact reached its highest development among the central-Asian Mongoloids, whose lateral build and strong shoulders show the greatest genetic adaptation to its use.[1057] (Remember that Eskimos, from the same stock, had sophisticated composite bows.)

Another advantage the Mongols and Turks had militarily was their mobility, due to their herd-following, hunter-gatherer lifestyle. It is an advantage for any community to be mobile when it comes to military action, and Quincy Wright tells us that horsemen and seamen tended to migrate and fight more than agriculturalists until the advent of the railroads, steamboats and organised armies with artificial means of mobility.[1058] The Vikings, for example, who were also 'barbarians' pushed by population pressure, enjoyed such an advantage thanks to their longboats. And the Asian nomads' experience in the hunt provided them with tactics they could successfully use in battle, including, for example, outflanking both wings of a retreating army as they would a fleeing herd of wild animals. Trained from childhood to drive deer at a gallop over the vast expanses of the steppe, and accustomed to patient stalking, the Asian nomad was an unbeatable foe.

The mobility of the herding nomads allowed them to move easily as a group, and was an aid both in attacking and retreating. Later on in their development this mobility was greatly increased by their ability to use their animals for transportation, pulling carts and chariots, and, in the case of e.g. horses and camels, as mounts. As pointed out by Childe,[1059] improvements in vehicular transport may well have first been developed for war-chariots, much as advances in aeronautics have often been first associated with military aircraft. This military strength was further enhanced by the fact that, given the animals at their disposal, when fighting they were never dependent on extended supply lines, but always had immediate access to a source of high-quality food.[1060] And on top of this they had the extra incentive of experiencing *vital* needs.

Herders' armed emigrations were on the scale of whole tribes, with all their bands, at once;[1061] and when they attacked, the entire male contingent of the group attacked, with their camps, including their wives and children, at their backs. When they captured a city they slaughtered everyone: men, women and children – except for the artisans, whom they kept for their skills. Their barbarity was extreme. One Tatar had the prettiest of his concubines roasted and served up for dinner.[1062]

When fighting they would rush their opponents, release their arrows, and then retreat, avoiding hand-to-hand combat and enticing their enemy into pursuing them onto the steppe, where they would completely eliminate them. These invasions took place despite the fact that, as in the case of the Chinese for example, the nomads were often outnumbered 100 to 1. And it is to be kept in mind that it is not as though the nomads' agrarian adversaries were themselves peace-loving, for they too were constantly engaged in war with one another.

Not only did the Turko-Mongol hordes continually invade sedentary peoples, but they were constantly slaughtering one another, to increase leaders' and groups' wealth by acquiring more animals, and to increase the leader's status and political power through their group's acquiring more territory and the leader's acquiring more men. The law of their life was a series of "struggles for existence" (Malthus' phrase in this context) between one nation and another,[1063] one of the results of which was to alleviate population pressure.

Their leaders' interest seemed to be in conquering *per se*, and with it the obtaining of consumables or capital in the form of loot, rather than territory for their groups. They often abandoned the lands they had defeated directly after their conquests – if they didn't turn the fields into grazing land for their animals, in effect making it an extension of the steppe. Sometimes when they retained possession of conquered cities they afterwards pillaged them even though they were their own, in order to obtain more loot to finance further military exploits. When victorious Mongols did inhabit conquered agrarian territories, however, they often became assimilated and softened, only later to be defeated by nomads still living on the steppes.

From before the time of Christ until into the 1700s Mongol and Turkic peoples from the Asian steppes repeatedly invaded the agrarian peoples of Persia, Eastern Europe and China – the Great Wall of China for example being built, rebuilt and elaborated to keep them (actually their *horses*)[1064] out. Despite its presence, the Mongols succeeded in ruling the whole of China from 1234 to 1260,[1065] and established the Yuen Dynasty (1279 to 1368), and the Ching Dynasty (1644 to 1911). Assisted by famine and disease, their armies reduced the population of China by more than 40 million between 1125 and 1290.[1066] In 1258, in Mesopotamia, they blocked the channels leading to the irrigation canals, causing the water to overflow its banks and flood the countryside, rendering the land useless.[1067]

In 1241 the Mongols had the largest steppe empire ever, stretching from Hungary to China.[1068] The power of the nomads began to wane starting in the 1500s however, when they were first met with artillery, after which they never regained their previous military might.

Colonisation and the (capitalistic) mercantile expansion *1500* AD

With the discovery of the New World, business (economics) began to take over from military force (politics) in the power structure of the world. The proceeds from the colonies of European nations went into the pockets of the businessman, not the leader of the nation's army, i.e. the king *qua* king. Of course the king himself was often engaged in business, his military clout undoubtedly giving him an advantageous bargaining position. And it was still the individual or group who had the greatest military power that had the last say. But ever-increasingly that power was being determined by economic means.

Colonisation has been a theme throughout human history, with the colonists normally taking land from native *Homo sapiens* with less advanced weaponry. The Romans, for example, sent their poorest citizens to colonies, where they might 'better their condition and multiply.'[1069]

Social stratification: economic inequality, slavery

As described by Clive Ponting, when European colonists populated the lands of less technologically sophisticated natives, the natives lost their land, livelihood, independence, culture, health and in most cases their lives. (As regards the heavy toll in lives taken by the diseases Europeans brought with them, it may be remembered that these diseases were originally contracted through domesticating animals.) Despite differences in approach, the common themes running strongly through European attitudes to the process were a disregard for the native way of life and an overwhelming urge to exploit both the land and the people. The societies of the native Indians of North and South America, the Aborigines of Australia, and the islanders of the Pacific all collapsed under European influence. The Guanches of the Canary Islands were the first people to be driven to extinction by the Europeans; they were enslaved and worked to death on the sugar plantations that the Spanish established there – the last of them dying in the 1540s.[1070] The story is one of soaring death rates brought about by disease, alcohol and exploitation together with social disruption and the decline of native cultures, especially under the influence of the missionaries. The Europeans showed little or no interest in native beliefs or customs until anthropologists in the last hundred years tried to investigate the remains of the shattered societies.[1071] And to this it may be added that, particularly as regards Africa, many of the conquered people became slaves serving European masters.

From the beginnings of horticulture, but with a particular spurt starting around 1500, the distribution of wealth (property) in the world became increasingly unequal.[1072] In medieval Europe the supply of raw materials afforded by the colonies constituted a precondition for the rise in the wealth and stature of the business class. More of the economic resources of European societies wound up in the hands of people who were interested in, and knew something about, both economics and technology. More than this, these people were oriented towards profit-making (not a typical orientation in agrarian societies) and were therefore motivated to provide financial support for technological innovations that would increase the efficiency of people and machines,[1073] with the ultimate aim of increasing their own wealth. As expressed by Mumford, the pursuit of money became a passion and an obsession, the end to which all other ends were means. In the goal-directedness of business enterprise, profit was the only goal.

Furthermore, money was the only form of power which, through its very abstraction from all other realities, knew no limits.[1074] Money is a kind of non-interest-bearing IOU accepted by the citizenry, a quantified socially-invested potential or power to acquire property.[1075] As such, it is a medium of exchange, a bill of lading. If I have a particular sum of money, then I am *owed* the consumables represented by that sum of money.[1076] The use of such a bill turns all property in connection with which it is used into *capital*. Unlike consumables, which have concrete use value, and which spoil or deteriorate when hoarded (due to entropy), abstract exchange value in the form of money can accumulate indefinitely without spoilage or storage cost.[1077]

The role played by money in its facilitating exchange must have been great at the time of the advent of global capitalism; and it has remained great ever since. It was with the introduction of money into the economic system, money for which stored commodities stood as collateral, that truly entrepreneurial efforts could be made.

One might have imagined that the influx of resources from the New World would have eased the lives of the lower strata in Europe; however, as a sign of continuing maltreatment of the weak, the height of Britons was at its lowest ever in the 1700s, when it was also the lowest of all humans during recorded history.[1078] More generally, until 1800, in every part of the world nearly everyone lived on the edge of starvation.[1079]

Between 1698 and 1775, Britain's trade with her colonies increased more than fivefold, and that was only the beginning. With the European mercantile expansion the centre of world trade shifted for the first time, as western Europe replaced the Near East in the favoured position.[1080] The growth of cities as centres of trade and handicraft production created a new class of bourgeois entrepreneurs who supplied ambitious monarchs with the funds and expertise to build strong national states, undercutting the power of the regionally-based land-owning nobility.[1081]

Geographic expansion and population growth

While the largest cities in pre-iron agrarian societies had populations of about 100,000, the more advanced agrarian societies of 1700 supported cities with as many as one million inhabitants.[1082] The discovery of America not only allowed a continuing increase in the size of the human population through emigration/geographic expansion, as would be expected given marginalisation and the pioneering principle, it also made possible an extension of the agrarian period in Europe. And not only did population increase in the New World, but it also increased in Europe thanks largely to the agricultural and other resources obtained from the New World, the result being an increase in world population from 500 to 790 million during the 250 years of the mercantile period.[1083]

Innovations

Particularly noteworthy as regards the innovations of this period is the improvement in ships, beginning with the caravel (end of the 1400s) and carrying on up to four- and five-masted schooners, all of which were driven by wind power. Contemporary innovations also include the microscope (1590) and the telescope (1608).

* * *

During the agrarian era nature became less the 'habitat' of the farmer and more a set of economic resources to be managed and manipulated by those in power. This was particularly true of cultures where the dominant class was urban-based, as in early Greece and Rome.[1084] With the continual distancing of nature from humankind came a gradual change in our conception of the world, with reality increasingly being thought of in terms of other people and their actions, as though we were not ourselves animals dependent on the biosphere for our survival.

It seems that life expectancy during the agrarian period generally varied between 20 and 25 years, longer for men than for women, and shorter in cities than in the country. As in all periods prior to the 20th century, a high rate of infant deaths markedly affected average longevity. The spread of disease was even more of a problem than during the horticultural era, due to increased population density and interaction between societies, as well as worsening sanitary conditions in cities as they grew larger.

The amassing of political and economic power in the hands of but a few, together with various technological innovations, also made possible the marked development of the arts and sciences through the agrarian period, as was manifest in painting and sculpture, music and poetry, and in literature, the theatre, history and prose more generally, and in architecture, astronomy, philosophy and science.

VCP analysis of the agrarian era

At the time of the agrarian revolution the human population's tendency to grow beyond what could be supported by existent techniques meant that, if great dieback was to be lessened or avoided, new techniques or a new technology would have to be found. A new technology was found, in the form of ploughing combined with irrigation.

With the beginning of the agrarian era, the vicious circle of population growth, technological innovation and resource depletion was about to revolve once again, on a massive scale. According to the VCP, the agrarian revolution was a result of the same forces as brought about the horticultural revolution. The need for more food on the part of a constantly growing population led to the invention and more particularly the use of new technology, in this case the plough and irrigation, the employment of both becoming new needs for humans. The result was once again a tremendous increase in the quantity of food available. And this increase continued throughout the agrarian era thanks to the continual invention of new tools and machines and the wider application of those already in use, as well as the continuous development and application of other agrarian technology, such as irrigation and crop rotation. This development allowed ever-increasing quantities of resources to be taken from the environment, with the result that by the end of the era the population had increased approximately tenfold, from 80 to 790 million. Technological development has not been the result of a seeking after a better life, but the response to experienced need resulting primarily from the presence of increasing numbers of people. The experienced need came increasingly however to be that of the powerful: during the Mesopotamian to medieval periods, from political leaders' need to defend the territory they ruled over and perhaps increase it; during the mercantile period, from capitalists' need to make a profit.

New technology

As regards the use of the plough, Wilkinson has pointed out that ploughing becomes necessary when population density has grown sufficiently to force the fallow period to be shortened until grass becomes the predominant vegetation. Grassland cannot be prepared for planting crops by burning. The turf survives fire and often grows with renewed vigour after it. The plough, like the hoe, is not introduced as a superior version of older tools to tackle familiar problems; it is introduced because grassland can only be prepared for sowing by ploughing – i.e. by the laborious procedure of cutting and turning the turf. With short fallow periods it is also likely that a considerable amount of additional work will have to be put into manuring and other techniques – such as liming and marling (the application of soil consisting of clay and carbonate of lime, a fertiliser) – to maintain soil fertility.[1085] The first use of the plough constituted an instance of economic development (change); and the increase in food that this use resulted in constituted an instance of economic growth (increase).

More intensive peasant agricultural techniques have a lower productivity per unit labour than more primitive extensive methods. As Wilkinson suggests, throughout humankind's development the problem of the increasing complexity of the tasks which the productive system has to perform has necessitated the dramatic development of new tools and productive equipment, while the increasing workload has only been absorbed by harnessing additional sources of power and adopting labour-saving techniques, including a greater division of labour. As population grows and people are forced to exploit the environment more intensively, the productive system has to develop to include trade, a greater division of labour, increasing quantities of capital equipment, and additional sources of power. Here we have a clear instance of diminishing returns to labour,[1086] i.e. a lowering of the net energy ratio in the meeting of the vital economic tasks of acquiring food and constructing dwellings.

As regards *irrigation*, the amount of work involved just in irrigating the land sometimes equalled that needed for all other tasks combined under dry cultivation, i.e. preparing the ground, sowing, weeding, harvesting and threshing.[1087] The demands of irrigation for mass labour, together with the inevitable effects of crowding, produced extreme class distinctions and rigidly hierarchical states, in which subservience was ensured by torture and appalling methods of execution.[1088]

In Mesopotamia the use of irrigation technology meant that, due to salinisation, what was first an increase in cereal production became a constant decrease. In response to this situation of diminishing returns we (those in power) turned to the handiest alternatives, extending the irrigation system and switching to a more salt-resistant crop.

The substitution of barley for wheat is typical of *Homo sapiens'* reaction to the turning of the vicious circle, and, as is suggested by the reaction principle, is the same sort of response as one might expect in other animals. We provide a short-term remedy treating the symptoms of the problem, which only exacerbates the problem in the long run, rather than provide a long-term remedy dealing with its causes. In employing irrigation we implement technology that produces a surplus at the same time as it undermines the conditions for its own functioning.

On the VCP, and as has been expressed by Boserup, more sophisticated forms of agricultural technology were not used e.g. in medieval Europe until larger urban populations required more food than could be obtained by simply farming more intensively, as was possible thanks to the extra labour afforded by the constantly growing rural population. Thus for hundreds of years it was not worthwhile for the aristocracy to make their serfs produce fodder for livestock, and so they did not do so; and fertilisation and other technological innovations were not used in agrarian Europe before a particular population density had been reached.[1089] And as technology increased in sophistication, the greater wealth (capital) of the powerful was required to produce and harness it.

Resource depletion

With this increase in production and population came an increase in resource depletion, mainly of land, as treated above, and metals. Though the mining of metals had begun in the Late Neolithic it came into its own in the agrarian epoch. What is important as regards mining is not that the resource in question is being depleted *per se*, but that society becomes *dependent* on a resource that is being depleted, with no guarantee that there will be a replacement when the amount that can be extracted begins to decline. From the point of view of the equilibrium and longevity of the human (and other) species, and from the point of view of the VCP, it would have been much better for *Homo sapiens* if non-renewable resources had never existed.

As regards resource depletion in the case of land, we have an excellent example of it with the constantly decreasing grain supplies to Rome. This must be seen as a major contributing factor to the fall of the Roman Empire.

The turning of the vicious circle also led to the *extinction* of some species of animals, and reductions in the populations of others. As expressed by William McNeill on the basis of his analysis of the agrarian period, time and again a temporary approach to stabilisation of new relationships occurred as natural limits to the ravages of humankind upon other life forms manifest themselves. Yet, sooner or later, and always within a span of time that was minuscule in comparison with the time required for biological evolution, humanity discovered new techniques allowing the fresh exploitation of hitherto inaccessible resources, thereby renewing or intensifying damage to other forms of life.[1090]

Population pressure

We humans have experienced constantly increasing population pressure and consequent stress since we first came into existence. When this pressure is particularly severe there occurs a *population crisis*. For example, each major Chinese dynasty rose and fell in a cycle of reduced population pressure, followed by population growth, and then by population overshoot and crisis. The mandarins wanted more land to be cultivated and more peasants to work it, to yield them rent (economic) as owners, and taxes (political) as officials, both paid in grain. And in the Eastern Roman Empire there was an unusually protracted population crisis which extended from the 200s to the 500s AD, ending with the pandemic of the 500s, mainly of bubonic plague, which lasted 60 years and killed between one-third and one-half of the population.[1091]

As regards the great migration to the New World, it may be mentioned that the emigrants, like the Cro-Magnons moving into Neanderthal territory, were marginalised by their own societies. As regards the most marginalised, the slaves, in total about 12 million Africans were taken to the Americas as slaves. The massive expansion of slavery in the United States in the first half of the 1800s was given impetus by the export of cotton to the growing textile industry of Britain.[1092] And upon arriving in the new territory, the European immigrants eliminated the other human inhabitants, but in this case ones of the same species, but then usually not completely. And directly afterwards they came to experience the new land as being one of plenty.

The economic effects of population pressure are clear. The periods of most acute population pressure in Britain during the later agrarian period – around 1300, 1600 and the late 1700s – were all periods of remarkable expansion and innovation as well as of unusual poverty.[1093]

Social stratification: militarism and capitalism

With the move to agriculture, society as a whole tends to become organised around the acquiring of as large a surplus as is possible from year to year; and as the agrarian epoch proceeded, the ownership of this surplus fell into fewer and fewer hands. Thus where horticultural societies were still rather egalitarian – as we see in modern horticultural societies – around the time of the agrarian revolution leaders began obtaining and employing absolute control over their subjects. According to Wilkinson, the splendour of the monuments and creations of the 'great' civilisations of the past is more a reflection of the extent of class control over people's labour than of basic wealth. Because of the greater difficulty of maintaining such control where resources are plentiful, there is a tendency for the richness of a society's creations to be inversely correlated with its basic wealth.[1094]

Here it is of interest to note that according to Adam Smith, civil government (implying *laws*), insofar as it is instituted for the security of property, is in reality instituted for the defence of the rich against the poor, or of those who have property against those who have none.[1095] This is certainly the case with regard to agrarian cultures and, we might suggest, with regard to all subsequent cultures. And we might say more generally that *government is paradigmatically nothing other than the governing of the weak by the powerful in such a way as to attain their own ends*. The sexual instincts of the minority, given the sophistication of the human species, lead to the manipulation of the social instincts of the majority.

Those on the top of the social hierarchy are always those who have the greatest power; and though that power may be manifest in terms of wealth, in the end it must consist in *arms*. This means their being the strongest militarily, which requires having good weapons and many soldiers.

Social stratification increased due not only to population growth, but also to improved communication and weapons. These improvements allowed the powerful to increase their power, both politically (militarily) and economically, by gaining greater control over the population's vital resources. In primitive societies where resources are not sufficiently scarce to control people's access to them, widespread inequality lacks an important prop. As expressed by Wilkinson, the general bulk of the population does not become exploitable until they no longer have the means (land or other productive resources) at their disposal to satisfy their own needs.[1096]

In societies of hunters and gatherers or swidden agriculturalists on the other hand, individuals and families can often break away from the main community and set up an independent group. And in such cases, where access to the requirements for subsistence can be controlled by ordinary members of the group, the power of chiefs must lie in abetting the well-being of the group as a whole, so that, for example, neither desertion nor revolution occur. Thus according to Boserup, in societies where the population is sparse and there is much fertile land that is not owned or ruled over by anyone, a social hierarchy can only be maintained through direct control of the members of the lower classes. In such societies, both conquered peoples and individual prisoners of war are therefore held in slavery.[1097]

With the beginning of the mercantile expansion some 500 years ago began a shift from power lying in a person's ruling over *land* (group territory) to power lying in the individual ownership of *capital*, or *exchangeable* property. In keeping with the view of Robert Heilbroner, we can say that the uniquely human drive to acquire this 'individual territory' in the form of wealth is the most important aspect of capitalism and the bourgeois culture in which it flourishes.[1098] Thus "need" takes on a new meaning in this culture – the need to make a *profit*. This need constituted a major spur to technological development at this time, as it still does today. Any innovation that reduces resource use, or more importantly given the size of the surplus in the present case, increases production, can increase profits in a growing economy.

In order for property to function as exchangeable capital it must be possible to *store* it, the greater its storability the greater both its use value and, consequent upon that, its exchange value. In the context of trade, i.e. on a market, the only value property is taken to have is exchange value. Thus, after the horticultural revolution, livestock and the seeds of domesticated grasses could constitute capital, while meat and fruit could not – at least to the extent that they could not be stored as long. In some areas during the agrarian period just those crops were grown whose seeds could be stored, indicating the importance the powerful attributed to acquiring capital.[1099] Here we of course have a situation where capital has use value in being edible. It may also be noted that *land*, on the other hand, while clearly being storable, was, other than in Egypt and Mesopotamia, seldom traded during the agricultural period, neither between nations nor individuals.

The growth of opportunities for *merchants* to purchase goods which they do not want for themselves but which they could sell for a profit was in part the consequence of the emergence of a monetary system. Unlike in the case of military endeavours, by effecting such transfers they were and are able to increase their power without having to engage in armed conflict. The money economy subverted many of the values of simpler societies, especially the moral values supporting cooperation within extended kin systems, including the sharing of commodities. It fostered instead a more individualistic, rationalistic and competitive approach to life, and laid a foundation for many of the attitudes and values that characterise modern industrial societies.[1100]

As the social status and political influence of the merchants in pre-industrial societies increased, so did the rate of technological and economic innovation.[1101] And governments, supporting these efforts for the wealth they brought the nation, enacted laws against theft and meted out severe punishment for breaking them, further strengthening the position of the wealthy and weakening that of the poor. In this context it is of interest to cite Thomas More, who wrote more than 200 years before the industrial revolution:

> In fact, when I consider any social system that prevails in the modern world, I can't, so help me God, see it as anything but a conspiracy of the rich to advance their own interests under the pretext of organizing society. They think up all sorts of tricks and dodges, first for keeping safe their ill-gotten gains, and then for exploiting the poor by buying their labour as cheaply as possible. Once the rich have decided that these tricks and dodges shall be officially recognized by society – which includes the poor as well as the rich – they acquire the force of law. Thus an unscrupulous minority is led by its insatiable greed to monopolize what would have been enough to supply the needs of the whole population.[1102]
>
> [T]he one essential condition for a healthy society [is] equal distribution of goods – which I suspect is impossible under capitalism. For, when everyone's entitled to get as much for himself as he can, all available property, however much there is of it, is bound to fall into the hands of a small minority, which means that everyone else is poor. ...
>
> In other words, I'm quite convinced that you'll never get a fair distribution of goods, or a satisfactory organisation of human life, until you abolish private property altogether. So long as it exists, the vast majority of the human race, and the vastly superior part of it, will inevitably go on labouring under a burden of poverty, hardship, and worry.

Apart from More's eventually being beheaded, we note here his belief that the situation he describes might be *remedied* by abolishing private property, as Marx believed. On the VCP, on the other hand, there will always be an accumulation of power in some form or other at the top of the human pecking order in *any* complex society. If one were to eliminate private property, i.e. economic power – as was done in the Soviet Union – the power of those at the top will simply take the form of political/military power employed in direct coercion (the army becomes the police). And, as suggested by our *Animal Farm* syndrome, if the weak were to replace the powerful they would act just as the powerful did. It lies in the karyotype of the human male, which, through sexual selection, differs somewhat from that of the female, that he try to attain power and/or status vis-à-vis other males, whether it be through the acquisition of individual territory or through ruling over group territory. And the same may be said of the karyotypes of the other primates. If gorillas or chimps had developed technology to the extent we have, one would expect gorilla or chimp leaders to act in the same way. The difference between complex societies and ape and hunter-gatherer societies is that in complex societies the preconditions for particular individuals' acquiring great power are better met. Most important in this regard is probably the ever-growing population. Given the relative influence of the survival and sexual instincts in economics, we should say that it is primarily *greed* that propels the capitalist, albeit greed coupled with his constantly felt need to improve his status or social position. No complex and populous society has as yet been without a hierarchical power structure, with those at the top always being concerned to maintain their position.

Surpluses and soldiers

The need for soldiers promotes the formation of a society in which a surplus is maintained and soldiers are available. Note however that men are not used as soldiers when they can be economically productive, i.e. they don't fight until after the harvest is in, and that they would have to be fed whether they were fighting or not.

Beginning near the end of the horticultural period, supported by the arrival of metal weapons, inter-tribal skirmishes came to involve armies, the bigger the better. Thus the powerful promoted there being a large population; and, together with the presence of a surplus, this reinforced the phenomenon of inter-tribal conflict supporting population growth. This growth meant an increase in experienced population pressure, which itself made the society as a whole more inclined towards engaging in armed conflict with other societies.

As regards the agrarian era, population growth was never so great as to prevent the acquisition of a surplus on the part of the powerful. *That* the powerful saw to the production of so much food was so as to increase both their own wealth as well as the numbers of the weak, thereby adding to their own economic and political power.[1103]

Disease

While the poor continued to live lives of drudgery, infectious diseases and the presence of other parasites increased among the powerful and the weak alike during the agrarian era – though better nutrition, hygiene and living conditions would have made these effects less severely felt among the powerful. The Black Death of the Middle Ages, which was preceded by economic decline, can be seen as the result of a too rapid and too lengthy growth of the European population as relative to the available resources. Slicher Van Bath attributes the high death rate of the plague largely to the prolonged malnutrition that resulted from this population growth. According to him, if there had been no plague, then the population would have been reduced by other means.[1104] Here we see the operation of the vicious circle running into the ecological wall: population growth in the context of crowding and unsanitary conditions could not continue indefinitely. Not only did population growth stop, but population overshoot was eliminated through an externally imposed positive check in the form of the Black Death.

Crime and war

According to Mumford, apart from murder and rape the most horrendous crimes punished by civilised authority stem back to that of disobedience to the sovereign. Murderous coercion was the royal formula for establishing authority, securing obedience, and collecting booty, tribute and taxes. At bottom, every royal reign was a reign of terror. With the extension of kingship, this underlying terror formed an integral part of the new technology and the new economy of abundance. In short, the hidden face of that beautiful dream was a nightmare, which civilisation has so far not been able to throw off.[1105]

In agrarian nations the lower-ranking strata (the peasant producers of agricultural commodities) are vulnerable to temptations to resist the demands of the powerful. But, as noted in the previous chapter, effective action on the part of peasantry can generally take place only when they are allied with other strata. This strategy is rarely employed, however, the usual course being recurrent upheavals on the part of the peasants by themselves.[1106]

The importance of the presence of a surplus in this context cannot be overemphasised. Without an agricultural surplus kings in the agrarian era would not have been able to build cities, or to maintain a priesthood, an army or a bureaucracy; nor could they have waged war.[1107] In periods when there was no surplus, depression ensued, and with it disappeared the money to buy arms and pay soldiers.

As in the horticultural era, in the agrarian the powerful actively worked towards maintaining or increasing their power, both inside and outside their nations. As the power of the chiefs and social stratification increased, conflict came more and more to be a matter of relations between powerful individuals, each of whom controlled his own group or population of weak individuals; and inter-tribal protowars became inter-national wars. To survive or succeed in this context requires being strong militarily, which allows one to secure the surplus (capital) for one's own uses, and to use part of it to acquire as many soldiers and produce as many high-quality weapons as possible. As regards the having of many soldiers, it is of interest to cite Aristotle:

> Again, the law which relates to the procreation of children is adverse to the correction of this inequality. For the legislator, wanting to have as many Spartans as he could, encouraged the citizens to have large families; and there is a law at Sparta that the father of three sons shall be exempt from military service, and he who has four from all the burdens of the state. Yet it is obvious that, if there were many children, the land being distributed as it is, many of them must necessarily fall into poverty.[1108]

And as suggested by Wilkinson:

> In class societies (which, from European feudalism to the Indian caste system, usually originated in territorial conquest), there is no longer a unified view of the society's subsistence problem: instead there are major inequalities between large groups of people. The rich upper classes have no need to limit their family size for fear of inadequate subsistence. This affects practices such as abortion and infanticide which are, at best, necessary evils. If the upper classes find them unnecessary, then practising them will come to be regarded as an unmitigated evil. Because the upper class has a disproportionate influence on the society's ideology and law, frequently infanticide and abortion cannot be carried on openly but become illegal, undercover activities.

Not only this, but e.g. abortion works against the desire of the powerful to have as many subjects as possible; thus the midwives who performed abortions were often believed to be, and treated as, witches.

We see the strong effect of the intratribal/inter-tribal difference continuing on into the agrarian period. Homer's Odysseus says in the *Odyssey*:

> However, let me tell you of the disastrous voyage Zeus inflicted on me
> when I started back from Troy.
> The same wind that wafted me from Ilium brought me to Ismarus,
> the city of the Cicones. I sacked this place and destroyed its menfolk.
> The women and the vast plunder that we took from the town we
> divided so that no one, as far as I could help it, should go short of his
> proper share.[1109]

Poor Odysseus and his disastrous voyage. What we see here is that strangers
don't count. Moral responsibility extends only to those in one's own group
(tribe), and is largely supported by myth, a precursor of religion.

The large number of assassinations of Roman emperors is a manifestation
of the struggle amongst the powerful for greater power. The conflict *between*
classes during the agrarian period was mainly between advanced, sedentary,
agrarian societies and nomadic *herders* from other areas. This was the case particularly when the agrarian society was already in a state of decline due to
the problem of supporting a growing population given erosion etc., i.e. due to
environmental collapse.[1110]

Given the importance of warring during the agrarian period, it is not surprising that here too the leading edge of technological development occurred in
the fabrication of weapons. It is the smith's ability to make weapons that places
his status and rewards above those of the potter.

Population checks and religion

While the powerful promoted population growth in order to have large
armies and cheap labour e.g. in the form of megamachines, war and female
infanticide worked in the opposite direction. Both infanticide and participation
in war continued from the horticultural era on into the agrarian, both being
morally obligatory, participation in war today still being morally obligatory
and often legally obligatory as well. While the killing of the members of other
communities is under certain circumstances morally quite acceptable, what is
immoral under almost *any* circumstances, and is labelled *murder*, is the killing
of members of one's *own* community.

Here we might recall the role played by religion in this context. While it
supported sharing and social cohesion generally, it did not support population-
reducing acts such as sexual abstinence, abortion or infanticide.

Under the feudal system there was a residue of landless and homeless serfs
or semi-slaves, inferior in status to the peasant, who slept around the log fire
of the manorial hall, and who had no family life and produced no children.
Likewise slavery made marriage and reproduction impossible for many of

those enslaved, and reduced women's fertility as a result of their hardship and ill-treatment.[1111] All of these factors thus functioned as population checks.

But primarily due to the magnitude of the surplus, greatly increased towards the end of the agrarian era thanks to the development and exploitation of colonies in the New World, these checks, in combination with others related e.g. to marriage customs, failed to curb the continuing growth of the human population.

The agrarian model

Scarcity and need
Innovation: **the plough, irrigation, the use of metals**
Economic development
Increase in available resources: **grain**
Prerequisite of the existence of resources
Increased total energy use: fire, domesticated animals, **labourers,
        draught animals, water and wind**
Geographic expansion made possible by technological change: **the
        New World**
Resource depletion
Renewables: **timber, soil**
Non-renewables: **metals**
Continuing species extinctions
New needs: **the plough and irrigation**
Technological development self-perpetuating
Genetic dependence on developing technology
Increased waste: **filthy cities**
Surplus: storable grains; **money (abstract; exchange value) comes
        to represent consumables (real; use value)**
Increase in total consumption
Inferior substitutes
Population checks further weakened
Morals; **religion**
Population growth
Population pressure
Migration: **centralisation; colonisation**
Increasing complexity of technology, and continuing increasing
        social complexity: division of labour
Social stratification: property; class society; **kings and emperors vs.
        vassals and slaves**

Lower quality of life **for the weak**: **labour**, poor diet, disease
Laws, crime
War
Leisure for the powerful
Cultural development: **the arts and sciences**
Economic growth
Security further decreased
Population overshoot
Diminishing returns
Undermining of technology and human existence
Economic decline
Scarcity and need

Movement from the horticultural form of the VCP to the agrarian involves a number of sophistications. The improved agricultural and mining techniques supporting the constantly growing population make possible an increase in the total consumption of both food (grain) and metals. Increased consumption means a decrease in resources, most notably the renewable resources of soil and timber and the non-renewable resources consisting of various metals. Religion takes on part of the task of maintaining social cohesion, partly countering the impersonal nature of the relations among most of the members of the population. At the same time the continual existence of a surplus tends to lead to centralisation at the points where the (edible) surplus is stored or produced, a process that began in the horticultural era but which eventually led to the existence of cities of one million inhabitants during the agrarian period. These constantly growing cities become filthy due to the accumulation of waste. Overpopulation at the points of surplus accumulation leads to migration in the form of colonisation. Money comes to represent property, thereby facilitating trade. Total energy use increases, not only through an increase in the burning of wood, but also through the use of human muscle, draught animals, and water and wind. The possession of property by only a few develops into a more clear-cut form of social stratification, where those without property work for those with property, i.e. they perform labour, already evident during the Neolithic e.g. in the building of the walls of Jericho. There are two main classes: the royalty/aristocracy and the serfs, and a third class consisting of slaves. Those possessing property have the opportunity to increase their possessions as well as enjoy increased leisure, the combination of which provides the basis for cultural development. The weak come to be governed by the strong through the latter's passing of laws, the breaking of which constitutes crime. Increased population size plus a greater surplus leads to larger wars, while larger political units lead to less killing overall. At the same time the existence of a surplus in the form of increased property leads to economic growth and an increase in total consumption.

The various technologies developed during the agrarian period supported a tremendous increase in population, at the same time as agricultural techniques drastically reduced the productivity of the land. The only way such a huge population could continue to be supported were if new resources could be exploited. Such resources existed in the form of *fossil fuels*. And the innovativeness to devise the technology necessary to mine those resources was not lacking.

The (capitalistic) industrial (fossil-fuel) revolution *1750* AD

Coal (1750)

Though increased population pressure in England contributed to encroachments on forests for pasture and firewood, the most significant cause of the reduction in their area there was the growth of early industry. The ship-building, soap, glass, iron and copper-refining industries depended on forests for their energy. Although the forest reserves of the mid-1500s had been increased by the dissolution of the monasteries in 1535, by the late 1500s shortages of wood and land had become acute.[1112]

In Britain, around 1700, there was a dearth of both land and wood, wood being used as a building material and a fuel. This 'timber famine' was manifest, for example, in the splicing of smaller pieces of wood to form masts. Even prior to this, starting in the late 1600s, mast-wood and other wood for ships was imported from Scandinavia in large quantities. As early as in 1696 warships for the Royal Navy were built in North America because of the shortage of European timber; and during the 1700s one-third of British warships were being built there.[1113] And during the 1800s great quantities of wood were imported from Canada and New England.[1114]

During the 1700s, England was thus forced to change from wood to coal as its chief source of industrial energy. The acquisition of coal was hindered however by water constantly seeping into the mines dug to extract it, a problem overcome through the invention and use of machines to pump out the water. (Coal was mined in northern China already at the time of Marco Polo's journey.)[1115] No other European country except Holland developed its coal industry to such an extent.[1116]

Coal is an inferior substitute for wood as a fuel however, a fact which required the invention of other devices and processes such as the use of coke, derived from coal, as a substitute for wood-charcoal in the iron-smelting process – the over-use of charcoal having been a major cause of the dearth of wood.

While coal eventually met the requirements of wood as a fuel, it did not meet them as a building material; and at the same time large quantities of oak were needed in the building of railroads and canals. As regards the building of canals, wood was first needed for scaffolding, after which wooden retaining walls requiring thousands of oak pilings were built on either side of the canal trench, to prevent earth from collapsing when the water was let in. Timber had also to be used for the gates of locks.[1117] The little wood that was left in Britain went primarily to such uses, and its availability further decreased. A substitute had to be found for wood as a building material. As it turned out, there was a replacement, namely *clay*, that could be fashioned into *bricks*. (Note that wood is renewable while clay is not.)

This transition from a biotically-based economy to a fossil-fuel-based economy consisted to a large extent in the development of methods for transforming the immense energy reserves in coal into sources of power and heating.

Around 1800 Britain extracted about 11 million tons of coal per year. This is equivalent to about 1.8 million km^2 of forest. By 1850 more than six times as much coal was taken from British mines as in 1800, i.e. the equivalent of about 11.2 million km^2 forest, which is around 23 per cent of the world's average total forest cover. Expressed in these terms, the coal extracted from British mines in 1850 was 0.23 *world forest equivalents* (see Glossary) or *wfe*. British coal extraction peaked in 1910 at around 0.8 wfe, 20 to 30 times the figure for 1800.[1118]

In this way, the replacement of wood with coal as a fuel represented a tremendous relief on the pressure on the land in Britain, a relief which was even greater when combined with the replacement of wood with clay in construction, and draught animals with steam-powered locomotion in transportation. Perhaps most importantly, this freeing of the land from producing fuel and animal fodder allowed an increase in agricultural production,[1119] which supported further population growth.

Innovations and economic growth

Just prior to and at the beginning of the industrial revolution a number of inventions were made in England, including the Nielsen hot blast, the Newcomen steam engine, Watt's steam engine, the water frame, and various other power-driven machines (Glossary).[1120] As regards the development of the steam engine, there were three basic steps, as manifest in the Newcomen engine of 1712, the Watt engine of 1770, and the classic Watt engine of 1796. This last engine had a technical efficiency of five per cent (a net energy ratio of 1:20), which might be compared with modern steam turbines, whose efficiency reaches 40 per cent. The commercial use of Watt's results came after 1785, and to a greater extent after 1820.

Since coal only existed in particular places, it was necessary that a means of distribution be found. This means, again involving innovation or the extended use of known technology, as well as a huge increase in capitalist economic activity, consisted in the construction of canals and, increasingly, railways, as well as the development of hard surfaces for roads. The first canal was opened in 1757, linking a Lancashire coalfield with the Mersey River.[1121] The average cost of moving coal by canal was one-quarter that of moving it by land carriage.[1122]

Some of the growth resulting from the abundance of usable energy was manifest in capitalists putting the steam engine to uses other than pumping water out of mines, particularly in the transport sector, where it was employed propelling both trains and ships. The first steamboat was built by the American Robert Fulton in 1807, and steam locomotives began to be used in England after 1815 – in coal districts, since fuel was particularly cheap there in relation to fodder for horses.[1123] All this construction activity was dependent on the use of coal, and manifest itself, among other ways, in economic growth and pollution.

Capitalists also used these and other new inventions for the manufacturing of ever more products – in England cloth and steel products in particular. In the case of cloth the material used was cotton imported from the colonies, which was woven using the flying shuttle, invented in 1733. The need for a faster method of spinning cotton into thread led to the invention of the spinning jenny in 1764. And the cotton gin, which made the removal of cotton seeds easier, was invented in 1793. Watt's steam engines also contributed significantly to the making of increased quantities of textiles, being used to power the flying shuttle and the spinning jenny. Factories all over Britain quickly picked up this method. And since the machinery was powered by coal, it was not necessary that the factories be located beside rivers, which made the production of cloth even easier. Because of all of these inventions, and the availability of cotton from the colonies, cloth production in Britain rose dramatically. By the mid-1700s, i.e. at the beginning of the industrial revolution, British cotton goods were being sold all over the world, and they constituted Britain's most valuable product.

As the cotton industry grew, it created a greatly increased demand not only for textile-making machines, but also for means of transport, dyes, fuel and building materials. The almost explosive development of the textile trade therefore led to a rapid expansion in the industries which produced these things, most notably the iron industry. Thus some 12 years after the sudden huge expansion of the British textile industry the iron industry expanded in turn.

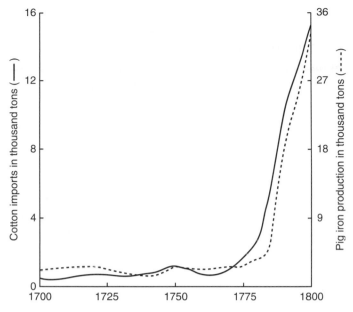

Expansion of British textile and iron industries 1700–1800[1124]

British production of wrought-iron rose by 70 per cent between 1750 and 1788, and pig iron production increased in the same period by 150 per cent.[1125] In 1784, Henry Cort's puddling and rolling process overcame all of the technical problems of using coal to refine pig iron into bar iron.[1126] With such changes and the implementation of pumps driven by various steam engines, the use of coal increased tremendously. Here it may be noted that the generation of motion by burning fuel only became possible starting in the early 1700s, with the invention of the steam engine.[1127]

Bulk steel production in Britain rose from zero in 1856 to 300,000 tons in 1870, to five million tons in 1900, at which time nearly one million tons were being exported annually, and to 7.33 million in 1930. In this way the cotton industry effectively led the whole of the British economy into the industrial revolution.[1128] Britain, the hearth of the industrial revolution, reached her economic peak in 1870, being at that time responsible for 32 per cent of the world's manufacturing production.[1129]

Agriculture and population growth

It is to be noted that the increase in manufacture incident upon the industrial revolution presupposed not only mining to extract coal and metals, but also agriculture to feed labourers.[1130] (Of course the availability of food and breeding sites are presupposed by all animal populations at all times.) The industrial revolution followed the first 'agricultural revolution,' which began in England around 1680 and doubled English yields in about 80 years. Most other European societies followed suit.[1131] Starting around 1750 there was an enormous increase in food production in Britain.[1132] During the industrial revolution domestic output of food rose almost as fast as the population, and as late as in 1868 it was estimated that no less than 80 per cent of the food consumed in the United Kingdom by a highly urbanised and industrialised population had been grown at home or in Ireland.[1133] Agriculturally, Britain drew on her Celtic periphery from 1800 to 1850, and on the whole world from 1850 to 1900.[1134]

The modern period of rapid population growth started with the industrial revolution.[1135] This was originally due, in England, to the increase in food production incumbent upon the freeing of land thanks to the use of coal; and it was abetted both by the more effective application of traditional agricultural methods – increased land use, manuring, winter feeding, crop rotation, etc., and by the cultivation of potato and maize, introduced from the New World in the 1600s. The great increase in food supplies was later extended by mechanisation, mineral fertilisers and, still later, insecticides.[1136]

But starting already in 1760 mainland English agriculture was having difficulty keeping pace with the growth of the population. Between 1800 and 1900 the population of Great Britain quadrupled, while coal consumption increased more than sixteenfold.[1137]

Population growth in north-western Europe during the first half of the 1800s led to the beginnings of a crisis, with inflation, unemployment, revolution, war, cholera epidemics, and famines, especially during the 1840s. Then, instead of crashing, the population exploded, thanks to massive increases in the food supply. These increases came from high-energy-input agriculture and from the New World, in the case of the latter, in part thanks to the use of the new steel plough there. By the 1850s Britain was importing nearly one-quarter of her wheat,[1138] and by the 1870s she was importing more than half of it, all acquired from her land-rich colonies and ex-colonies in exchange for manufactured goods.[1139]

During the industrial revolution proportionally more people married than before, marriages were earlier, and there were more illegitimate children. At the same time life expectancy in Britain and Wales increased, further supporting population growth; where in the 1670s people lived about 32 years, by the 1810s the average was almost 39.[1140] A number of innovations in medicine may have contributed marginally to this increase in longevity, such as vaccination for smallpox (first in 1796; compulsory vaccination for new-borns in 1853)

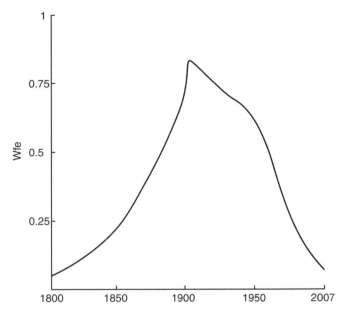

British coal extraction 1800–2007[1141]

Population growth and life expectancy at birth for England and Wales
1840–2007[1142]

and anaesthesia (first used in 1846); and antiseptic surgery was introduced in 1865.[1143]

Except for during periods of war, female life expectancy generally increased from being two years greater than male in 1841, to being six years greater in 1986, after which it tapered off to four. The year of greatest discrepancy was 1945, when female life expectancy was 13 years greater than male.

Migration/Geographic expansion

The dominant pattern of migration within Europe up through the 1700s was from less-developed to more-developed areas, and from the rural to the urban. At the same time, however, migration from densely populated Europe to the sparsely populated colonies accelerated, particularly after 1800, when populations in Europe began to press hard upon their resource base and the environment generally. In Ireland, for example, blight severely damaged the monoculture of potatoes upon which increasingly large populations had come to depend, killing about one million people between 1846 and 1849. Disaster was reduced as the less-developed world of that day – North and South America and Australia – opened to comparatively easy migration about 1850.[1144] Here we have another instance of marginalisation – migration to the colonies most often not being voluntary – where those who are marginalised end up in the better position. The least voluntary migrants were the slaves, whose position did *not* improve. Slavery was a common source of energy for developing civilisations in need of unskilled labour, and many slaves were forcibly transported from Africa to the New World. The slave trade, which is estimated to have killed a hundred million people and destroyed native societies over much of Africa, was banned internationally in 1884,[1145] but still exists in covert form today.

Social stratification: capitalists, labourers and the quality of life

With the industrial revolution came capitalism's dominant place in the world; and with it a worsened situation for labourers in Europe. In England many of these labourers were forced from the land by the 'enclosure' movement, which began in the 1400s and went on steadily and rapidly throughout the wars with France (1793–1815). That is, well-to-do landowners obtained Acts of Parliament entitling them to fence off and appropriate areas which had formerly been common pasture and woodland for local farmers.[1146]

Unlike in the case of farming, mechanised work can be done throughout the year, and with artificial lighting also at night. Not only did labour increase for both men and women with the industrial revolution, but so did it for their children, who were used, for example, in mines, where they could squeeze into spaces too small for adults.

Here we see an instance of a recurrent phenomenon: the use of children in mines. For example the town of Göltepe, in the Taurus Mountains of southern Anatolia, had been occupied from around 3290 to 1840 BC by people whose livelihood was derived from a nearby tin mine, the narrowness of whose tunnels – just 65 cm across – suggests that children had done the brutally hard work of excavating the ore, a suspicion that was borne out by the discovery in the mine of several skeletons of individuals between 12 and 15 years of age.[1147] Child-labour has a positive side, however, if one should see it as such, in that it leads to a decrease in infanticide.[1148]

Not only did working in factories mean working longer hours, but factory labour was harder and more demeaning than farm labour. The standard of living for most people in the industrial workforce up until about 1850 was far below that of a yeoman farmer. Where there was relatively little hardship in English villages in 1700,[1149] with the industrial revolution factory work was the only alternative to starving on the increasingly crowded land.[1150]

Between 1500 and 1800, in Denmark, the per person use of energy in heating and cooking dropped by 50 per cent. During the same period, national production almost doubled, while total hours per capita labour tripled, which for the labourer meant a declining hourly wage and falling standard of living. All increase in output during this period was bound to an increase in the use of labour, and it was only with the constantly growing access to non-human energy that, in modern Europe and the United States, production per man-hour labour eventually began to rise.[1151]

After the industrial revolution the quality of life of the majority of Britons worsened until the 1850s.[1152] In fact, looking over the whole of the development of humankind in Europe, one can see a relatively constant increase in workload for common people from the time at which the hunting of big game peaked about 30,000 years ago until the agrarian era, during which it plateaued and perhaps even decreased, until the industrial revolution, after which it rose again. By 1841 male life expectancy at birth in England and Wales as a whole was over 40 years.[1153] Despite an increase in longevity mortality was high, contemporary observers noting that parents were frequently indifferent to the loss of a child, and that smallpox was commonly referred to as 'the poor man's friend.'[1154] At the same time the quality of clothing also became lower, with cotton largely replacing linen, wool and leather, while the environment around factories became unspeakably filthy.[1155]

In 1850 the living conditions for most people in Britain were worse than they had been for centuries, the degree of exploitation and relative poverty being greater than ever before.[1156] The average working week in the United States at this time was 70 hours – as may be contrasted with the 20-hour or less 'working week' of modern hunter-gatherers. The first effort to limit the working day of children under 12 to ten hours per day was in 1842 in Massachusetts; and the ten-hour workday was not widespread for everybody in the United States until 1860.[1157]

Conflict

During the first hundred years after the beginning of the industrial revolution there were a number of wars between European states, and between them and their colonies, the most notable of the latter being the American Revolution, which began in 1776. This revolution has been considered the first *social* revolution, though it was hardly a revolution of the poor against the rich. We might say, more particularly, that it was a *colonial* revolution, where the strength of the colony approaches or exceeds that of the mother country when the cost of deploying armaments to the colony is taken into account.

Oil (1859)

The technology of the industrial era has been operated using huge quantities of fossil fuel, the second of major importance being oil. As regards the origins of coal and oil, we note that when the bodies of *plants* are compressed under water for long periods of time, they can undergo chemical changes so as to form *peat*, which may then become brown coal or *lignite*, then soft or *bituminous* coal, and finally hard coal or *anthracite*. Over a similarly long time, the bodies of certain *marine plants and animals* may undergo changes so as to form *petroleum*.

Where coal began to be used in response to a scarcity of wood, petroleum began to be used in response to a scarcity of whale oil.[1158] In Britain, millions of candles were made from whale oil in the mid-1700s. In London, about 5000 street lamps were lit by whale oil in the 1740s, and at the same time whale oil was the sole lubricant for industrial machinery.[1159] In 1857 the *Scientific American* reported that "The whale oils which have hitherto been much relied on in this country to furnish light, are yearly becoming more scarce, and may in time almost entirely fail, while the rapid increase in machinery demands a large proportion of the purest of these oils for lubricating." Attempts to find other sources of lubricants and illuminants, which included distilling fractions of coal, included drilling for oil, the first well in the United States being the 21-metre deep Drake well in Pennsylvania drilled in 1859 (the year of the

The Drake Well[1160]

publication of *The Origin of Species*); by the following year, some 700 oil companies had been incorporated in the state.[1161]

Energy use since the beginning of the industrial revolution

The total consumption of fossil fuels has been constantly increasing over the 250 years since the industrial revolution began. In the United States, for example, the contribution of coal, oil and gas to total energy consumption increased from being less than ten per cent in 1850 to being more than 95 per cent from 1950 to and including 1965.[1162] Today over 85 per cent of the

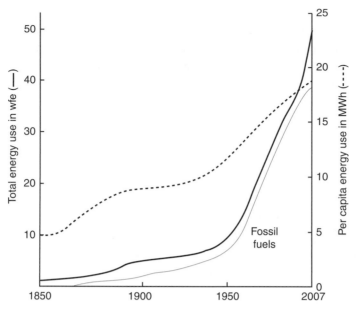

Energy use, total and per capita, 1850–2007.[1163] It would have taken at least 30,000 years of world forest growth to acquire the 300 wfe of energy used since 1850, assuming that it takes about 100 years for a forest to grow, and that *all* the wood grown is used to produce energy.

world's energy comes from fossil fuels. The world's annual consumption of coal is now about 500 times greater than it was in 1800. Oil was hardly used until 1900, but annual consumption is now about 380 times greater than it was a century ago. And the consumption of natural gas rose 175-fold in the 20th century.[1164]

Apart from fuelling engines and heating buildings, oil is used in the making of a myriad of synthetic products, including most forms of plastic and an ever-increasing proportion of textiles, both of which, in keeping with what was said in Chapter 4 regarding substitutes, being invariably of inferior quality to what they are replacing. The vast quantities of these fuels have not only meant increased production, but increased transport as well, with products of relatively low value, such as foodstuffs, being transported great distances,[1165] and by energy-inefficient means (e.g. by air).

Innovations

The technological innovations most directly responsible for supporting the huge increase in the human population in the past 150 years are related to agriculture, and almost invariably involve the use of oil. They include farm machinery, biocides and fertilisers. The mechanisation of agriculture has made it possible to increase the amount of cultivated land far beyond what could have been achieved using only human and animal labour.[1166] And the use of oil- and gas-based products has allowed much more to be reaped from the land that is cultivated. According to David Pimentel, if mineral fertilisers, irrigation (made possible in part by energy from oil) and pesticides were withdrawn, corn yields in the world, for example, would drop by about 75 per cent.[1167]

Oil is required to drive agricultural machinery and produce biocides, while natural gas is required to produce nitrogenous fertilisers (which consist of compounds of ammonia made through a process involving the extraction of hydrogen from natural gas and nitrogen from the air). As regards mineral fertilisers, mined phosphate and potassium are the primary raw materials used in the production of phosphate- and potassium-based fertilisers, while ammonia/urea is the main ingredient in nitrogenous fertilisers. The development of nitrogenous fertilisers depended on the synthesising of urea (the first human-made organic substance), which was accomplished for the first time in 1848. Further as regards nitrogenous fertilisers it may be mentioned that the production of each ton requires one ton of steel and five tons of coal,[1168] plus large amounts of electricity.[1169]

By 1870 practical types of generators were already available to produce either direct or alternating current. The first American transcontinental railroad was completed in 1869. And the supplies of foodstuffs from the continents newly settled by Europeans began to hit Europe like an avalanche in the 1880s.[1170] Other innovations after 1850 include the underground railway (in London, 1863), the Svend Foyn harpoon gun, which shoots a harpoon carrying a grenade in its tip (1868), and various machines used in manufacturing. At the Vienna Exhibition of 1873, practically all the electric appliances of modern life, such as electric stoves, pans, cushions and blankets, were demonstrated. Various forms of refrigerator were developed from 1803 on into the 1900s. The telephone was invented in 1876, the light bulb in 1879, and the electric oven in 1880 (the gas oven having been invented in 1824), the light bulb proving to be the greatest consumer of electricity of all these innovations.

Later inventions include the car (1885), the radio (ca. 1900), the aeroplane (1903), television (1920s) and the electron microscope (1933), the invention of the car and the aeroplane being dependent on the prior invention of the internal combustion engine (1876). Unlike steam engines – including nuclear reactors – which power machines indirectly (or turn electric generators) through the production of steam, internal combustion engines power machines directly through the burning of oil.

Extinctions

Extinctions of other species due to the activities of humans continued into the industrial era. Three species of ungulate have gone completely extinct over the past 200 years: the bluebuck (1800), the red gazelle (1894) and, very recently, the Arabian gazelle. The Saudi gazelle is probably also completely extinct. At least three other species have gone extinct in the wild, but continue to exist in domesticated form: Pere David's deer, the black wildebeest and the Arabian oryx (extinct in the wild in 1972 due to hunting with rifles). Roughly 20 of the 297 known mussel and clam species and 40 of about 950 fish species have disappeared in North America in the last century.[1171]

By the end of the 1800s some of the rare marsupials of Australia such as the hare-wallaby and the banded hare-wallaby had become extinct, and one of the two species of bilby (rat-sized marsupial omnivores) went extinct in the 1950s.

As regards genera of large mammals, the eight-metre long Steller's sea cow was exterminated sometime between 1742 and 1768.[1172] The Bali tiger became extinct in the 1940s; and the Caspian tiger in the 1970s.[1173]

As regards birds, the conversion of grassland to arable land, combined with extensive hunting, led to the extinction of the great bustard by 1838. The last pair of great auk, a flightless seabird, was killed in Iceland in 1844. The sea eagle was still common as late as in the 1870s, but shortly afterwards became extinct.[1174] The last wild passenger pigeon – previously the most numerous bird in the world – was shot in 1900; the last captive one died in 1914.[1175]

Many other species of bird also became extinct during the 1900s, including the Paradise Parrot (1927), the Grand Cayman Thrush (1938), the Wake Island Rail (1945), the Aldabra Warbler (1967), the Bush Wren (1972), the Colombian Grebe (1977), the Canary Islands Oystercatcher (1981), the Guam Flycatcher (1983), the Atitlán Grebe (1986), the Kaua'i 'O'o (1987)[1176] and the dusky seaside sparrow (a subspecies; 1987).[1177]

In the last 400 years some 83 mammals, 113 birds, 288 other animals and 650 plants have become extinct. But nearly all of these extinctions have occurred in the last century – of the 21 marine species known to have become extinct after 1700, 16 have done so since 1972.[1178]

Health and population growth

The increase in population from the 1700s to the present has not been due to an increase in the birth rate brought about by the withdrawal of constraints on fertility, as in the case, e.g., of sedentism at the beginning of the horticultural revolution. In fact, this is not the case in Third World countries today, nor in industrialised countries since the time that births and deaths began to be recorded by national registration; indeed, for most of this time birth rates in the world have been falling.[1179] The increase of population associated with the industrial revolution is rather the result of a decline in mortality. This decline was itself essentially due to a reduction in deaths from infectious diseases, as shown in Britain by national statistics from their first registration in 1838 until 1900. Of the non-infective causes of death, the most prominent to decline has probably been infanticide, followed by starvation, infanticide having been responsible for a high level of infant mortality until the late 1800s.

Here it may be recalled that aboriginal cultures, both horticultural and hunter-gatherer, counted on infantile death in order to keep their populations in balance with the environment (e.g. three of five children must die). While this death was often the result of infanticide, diseases also removed many of the young. When Western culture came to these areas, the missionaries prohibited infanticide. As a consequence, given the increase in food production thanks to the use of Western agricultural methods, significantly more than a replacement number of children survived. This reduction in infant mortality was later reduced still further by the use of DDT to combat malaria, a disease which affected infants particularly badly. This meant increased longevity and runaway population growth in the Third World.

Examination of the diseases mainly associated with the fall of mortality during the 1800s shows that death rates declined long before the introduction of effective immunisation or treatment. It was not until 1935, with the introduction of sulphonamides and, later, antibiotics, that medical measures became available which were sufficiently powerful to have an effect on death rates. And even since 1935 they have not been the main influence. (L. J. Henderson reckons that until about the year 1905, calling in a physician to attend a sick person *decreased* their chances of survival.)[1180]

The change probably most influential in the control of diseases is a greater abundance, quality and variety of food, malnourished populations having higher infection rates and their members being more likely to die when infected. This was abetted by better clothing, housing, and working conditions, as well as less exacting physical work and more effective protection against inclemencies of the weather. Also, in industrial countries after 1850, the introduction of hygienic measures – the purification of water, efficient sewage disposal and improved food hygiene – particularly as regards milk – substantially reduced mortality from intestinal infections. Such measures as the pasteurisation of milk (begun in Chicago in 1908) and the introduction of new techniques such as canning and refrigeration, also reduced infection rates from contaminated food, contributing to a spectacular fall in mortality from tuberculosis long before vaccination or any of the anti-tuberculosis drugs were available.[1181]

During the 1800s average life expectancy also rose in the whole of Europe, from 28 to 50 years, principally as a direct result of a dramatic fall in infant mortality towards the end of the century. The quality of labourers' lives in the industrialising countries began to improve around 1850, and on such a scale as is unique in the history of humankind. From the middle of the Upper Palaeolithic the consumption of animal protein seems to have declined for all but privileged groups,[1182] until in the middle of the 1800s labourers in industrialising countries began to eat more meat. Before 1800, Europeans consumed less than 250 grams of meat per person per year, eating mainly bread and potatoes. By 1850, however, people were consuming 45 kg per person per year![1183] By 1900, Europeans were 15 cm taller than they had been a century earlier.[1184] More generally, however, no application of technology has substantially reversed the decline in meat consumption for most of the world population.[1185]

In the late 1800s the population of the industrialising countries did not rise to the point at which food supplies again became marginal, however, as occurred e.g. at the end of the Mesolithic. Rather, food production continued to increase through the application of technology to agriculture.[1186]

Understandably, with this superior quality of life, workers' productivity increased, and, together with technical improvements in manufacturing methods, led to a rise in GNP in the industrialising nations. From around 1850, the wealth derived from Britain's predominant industrial and commercial position began to be shared to a growing extent by all sections of the British community.[1187] Ironically enough, however, labourers and their families were much better fed during the Great Depression of 1873–1887 than they were before, and with the subsequent improvement in business conditions starting in 1887, the improvement in labourers' living conditions ceased.[1188]

Since the beginning of the industrial revolution the population of the world has increased more than eightfold;[1189] and it has more than doubled in the last 50 years. On the other hand, the world's consumption of energy, mainly in the form of fossil fuels, has increased *sixty*fold in the past 250 years. In 2005 world population grew by some 74 million, 95 per cent of this growth occurring in the Third World.[1190] This population growth has been accompanied by an increase in urbanisation, with some cities now containing as many as 20 million people, as well as an increase in the problems of resource depletion and pollution.

Communication

Communication of all sorts among humans has increased over the past 50 years, including communication in the form of trade and migration. As regards migration, where during the industrial revolution it was from the less-developed rural areas to the more-developed urban, this trend was reversed when people began migrating in massive numbers from more-developed Europe to less-developed colonies. Today, however, the historic pattern of migration from less- to more-developed regions has returned, with people from the poorer

Mediterranean countries and those of Latin America, Africa and Asia migrating in ever-increasing numbers to the wealthier countries,[1191] as well as from rural to urban areas in their own countries. To this it may be added that the overall distance that humans travel each year, including back and forth to work and on business and pleasure trips, has increased tremendously since World War II.

The middle class

Colonialism, involving the departure of perhaps 50 million workers for the New World, and industrialism together eventually allowed a slightly larger segment of humanity to lead a better life, in the context of which arose the middle class. But this improvement was at least in part paid for at the cost of terrible repression, suffering and death for the vast majority,[1192] i.e. for the indigenous inhabitants of the colonies.

War

Ninety to ninety-five per cent of societies that we know of have been involved in war.[1193] It has been estimated that during the 20th century around 150 million people were killed by their fellow human beings, this century claiming more lives through war than any other hundred-year period,[1194] the great powers being the most frequent fighters.[1195] Since 1900 the involvement of civilians in wars has steadily increased, from five per cent of the violent casualties in World War I, to at least 50 per cent in World War II, and 84 per cent in the Korean War.[1196] That the world is still heavily committed to military conflict is evidenced, for example, by the fact that the United States' economy has become increasingly dependent on arms production by the military–industrial complex, which in the mid-1980s accounted for over one-third of all research and development in science and technology in the US, and two-thirds of public expenditure in these areas.[1197]

As suggested in Chapter 4, until the 20th century armies nearly always lost more soldiers to disease than they did as casualties to the enemy. In the Crimean war (1854–1856) ten times as many British soldiers died from dysentery as in fighting the Russians. The situation was similar in the American civil war; and in the Boer war at the end of the 1800s disease killed five times more British soldiers than the Boers did.[1198] Also, in the 20th century, at the end of World War I influenza, a disease we originally contracted from pigs, was responsible for one of the worst plagues in history, killing between 25 and 50 million people. And after the influenza virus came bacteria, and increased those numbers.[1199] Probably at least ten per cent of deaths in modern civilisation can be attributed directly or indirectly to war.[1200] At the same time, during the 20th century the population of the world grew from 1.65 to 6 billion, and to 6.6 billion seven years later.

Nuclear energy (1945)

The atomic bombs dropped on Hiroshima and Nagasaki in 1945 were each the equivalent of about 15,000 tons of TNT. They killed some 200,000 people within the first year after being dropped, and at least 30,000 after that. In 1952 the first *hydrogen* bomb was exploded, such bombs being easier to produce[1201] and at the same time some one thousand times more powerful than the earlier atomic bombs. At the height of the arms race, ca. 1986, the destructive capacity of the world's arsenal of nuclear warheads was equivalent to 22 billion tons of TNT. This is in turn equivalent to 1,470,000 Hiroshima and Nagasaki bombs; and, in terms of the effects of those bombs, it was sufficient to kill more than 300 billion people – about 50 times the present world population. Since the Americans' acquisition of nuclear weapons in 1945, Russia, Britain, France, China, India, Pakistan, Israel and North Korea have also acquired them, while Iran is/was attempting to do so, and Egypt, Japan, Saudi Arabia, Syria and Taiwan are prepared to begin to do so.[1202] The total number of nuclear warheads in the world in 2006 was ca. 27,000.[1203] This total may be compared with the more than 65,000 in 1986 and 20,000 in 2002. Of the 27,000 in 2006, Russia had 16,000, the United States 10,000, France 350, China 200–400, Britain 200, Israel 75–200, Pakistan 50 and India 50, while the number in North Korea is unknown.[1204]

At the end of 1998, there were over 1350 tons of plutonium in the world, or enough for more than 170,000 nuclear weapons. Most of this plutonium – about 1115 tons of it – was produced in civil nuclear reactors in 32 different countries. At the beginning of the 21st century there was an annual global increase of 65 tons.[1205] Bearing in mind that the half-life of plutonium is about 24,000 years, and that a mere half-pound of it dispersed into the atmosphere as fine particles would suffice to inflict every living mortal with lung cancer, the plutonium 'inventory' that accumulates over the years will constitute a tremendous carcinogenic hazard for one thousand human generations. Clearly a containment level of 99.99 per cent would hardly be enough to avert disaster. Yet in as much as plutonium is a necessary material for the fabrication of nuclear weapons, and is expected to be a lucrative item of illicit traffic, even the attainment of such a containment level is likely to prove unfeasible.[1206]

The use of nuclear power to produce electricity was a spin-off from programmes to manufacture nuclear weapons,[1207] the first commercial electricity-producing reactor being built at Calder Hall (Sellafield) in England in 1956. By 1964 there were 14 nuclear reactors in the world; by 1975 there were 167; and by 1985 there were 365. But after 1970 two-thirds of the nuclear plants on order were cancelled (in the US all were cancelled) due to problems related to accidents and the disposal of radioactive waste, as well as financial costs; and in many European countries it was decided to phase out nuclear power,[1208] the result of these changes being a reduction in the rate of reactor construction starting around 1985. The world's largest breeder reactor, the French Superphoenix, was completed in 1981. Around 1987 the Superphoenix, due to malfunctions, started consuming more energy than it was producing, and it was closed ten years later. As of 1991, no single large nuclear reactor had been successfully dismantled.[1209] The number of reactors peaked at 435 in 1995 and had dropped to 428 by 1999; in 2001 some 45 were planned; and as of 2005 there were 434 operating in the world. Nuclear power survived only through massive taxpayer-paid subsidies – over $100 billion in the US in the 40 years after 1950.[1210] Overall some 116 reactors totalling 34,603 megawatts have been permanently closed after average lifespans of less than 21 years. At the end of 2005, the International Atomic Energy Agency (IAEA) reported that eight reactors had been completely dismantled, with the sites released for unconditional use. A further 17 had been partly dismantled and then enclosed, 31 were being dismantled prior to eventual site release, and 30 were undergoing minimum dismantling prior to long-term enclosure, which leaves 30 closed plants that were as yet receiving no attention. One study estimates that as of 2006, 80 new nuclear power plants would have had to be ordered and built within the next ten years in order to keep the number of operating plants constant.[1211] But neither the World Bank nor the Asian Development Bank will presently fund new nuclear plants since they are not cost-effective. In the UK a government report in 2005 concluded: "Nuclear power is ... not currently competitive as a new investment option." Part of the reason for this is that there is no solution to the problem of storing the high-level nuclear waste that remains at the end of a nuclear plant's life. Nevertheless 35 plants were under construction in the world in 2000;[1212] and Britain at the beginning of 2008, despite the government report, gave the go ahead for the building of four reactors, the first of a new generation.[1213]

VCP analysis of the industrial age

New technology only used when necessary

As in the case of the horticultural revolution, the changes associated with the industrial revolution did not come about immediately upon the discovery of the new technology. Rather, as suggested by Wilkinson[1214] and as can be explained in terms of the VCP, the new technology was implemented only when there was an experience of *need*. In the case of the industrial revolution in England, rotary steam-power, for example, as in Watt's first steam engine, appeared as late as it did not because of the difficulties of invention, but because it was not needed earlier.

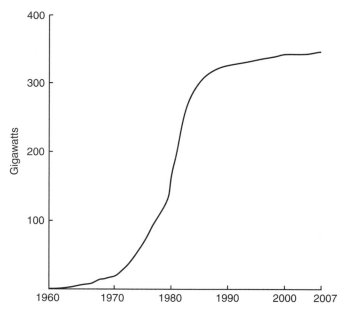

Electrical generating capacity of nuclear power plants 1960–2007[1215]

In the United States the transition to the use of coal as a fuel did not take place until local wood scarcities had made its implementation worthwhile. And steam-power was introduced into mills – as an inferior substitute – only when water-power became inadequate to meet growing needs. Similarly, agricultural innovations were not introduced in America as soon as they were discovered in Europe. Americans, despite being familiar with the alternatives, used less-sophisticated technology when they could. They clothed themselves in furs and leather rather than woven cloth; they used wood rather than coal e.g. in heating buildings and powering trains, water-power rather than steam, and herding (ranching) and primitive extensive agricultural methods rather than the more intensive European ones. In 1880 Americans were still making almost only wooden ships, while their kinsmen in Europe made them of iron.[1216]

As Wilkinson explains:

> The easiest way of clothing oneself is to make use of the skins of animals that are eaten. This is the most primitive method. When the supply of leather becomes inadequate to meet the population's growing needs, people are forced to develop textiles from natural fibres such as bark, flax, wool and cotton. The necessity of spinning and weaving these fibres greatly increases the work required to

produce clothing. As the population grows and the pressure on the land increases, artificial fibres are developed from mineral resources, leaving the land for more specialized food production. Once again the manufacturing process becomes more complex and difficult.

Of course in an even broader perspective clothing itself was first only used by humans when it was necessary.

When modern devices or methods were seen as necessary, however, they were quickly used. Thus heavy steel ploughs – 'prairie breakers,' pulled by six to twelve oxen – were developed, thereby making large-scale settlement of the American prairies possible. The use of such implements must have increased the cost of cultivation considerably, and would undoubtedly have been avoided had they not been necessary.

Coal did not become the principal source of energy in the United States until the mid-1880s, when the country's readily available wood supplies were largely exhausted; and its use grew in direct proportion to the distance at which forests had to be felled. If wood is available locally it is obviously much easier to use it rather than coal, since coal requires greater energy to acquire and, normally, due to greater distances, greater energy to transport.

The same pattern of not employing new technology until it becomes necessary can be found in the later industrialisation of Japan.[1217] Similarly, the Chinese did not develop machines of the industrial-revolution type because there was no resource crisis there that seriously threatened China's ruling class.[1218] Again in France: where the English used coke to produce iron, the French continued to use charcoal, since wood was still available there. And, in Britain, traditional agricultural methods were used until it became necessary to employ industrial-mechanical agricultural technology. In one area of production after the other, it was the shortage of resources that forced innovation and the finding of new solutions.[1219]

Here we see the peculiarities of marginalisation again. Britain, having run out of wood, can be seen as being in a relatively marginal situation vis-à-vis other countries. This meant having to find new resources and a way of acquiring them, just as it did for people living in the Hilly Flanks at the time of the horticultural revolution. And in both cases the new application of technology this required resulted in people's acquiring *more* resources than could be acquired by people using the old technology in 'non-marginal' situations.

In keeping with what was mentioned earlier, particular innovations related to vital need can perform a multiple function, only one part of which leads directly to increasing the availability of food. While the innovation of the iron axe increased the availability of firewood for turning what would otherwise be indigestible seeds into food, at the same time it increased the availability of wood for building dwellings and boats. The technological developments making petroleum and natural gas usable resources not only made possible an increase in the production of food, but also increased the possibility of transporting and clothing people, and of building, heating and cooling dwellings. And just as the use of a new technology may only make resources available that partly make up for a particular scarcity, so too might its use have the effect of increasing the accessibility of resources that were not particularly scarce in the first place.

As noted by Max Weber, people living in pre-industrial societies tend to show a 'leisure preference,' working less when they can rather than trying to increase their earnings. According to Weber, a man does not 'by nature' wish to earn more and more money, but simply to live as he is accustomed to living, and to earn as much as is necessary for that purpose. Wherever modern capitalism has increased the productivity of human labour by increasing its intensity, it has encountered the immensely stubborn resistance of this leading trait of pre-capitalist labour.[1220]

As cited in Chapter 2, Wilkinson has claimed that the innovations of the industrial era are not the fruits of a society's search for progress, but the outcome of a struggle on the part of a society with its back to the ecological wall.[1221] This is true in that without coal to replace wood there would have been a population crisis. Here, however, we point out another aspect of the situation. While it is also the case on the VCP that the ecological conditions are basic, the impetus to develop and use new technology need not always be prompted by such conditions. Thus the greater the surplus, other things being equal, the greater the likelihood that technology will produce goods and services meeting *non*-vital needs. The steam engine was not developed for the benefit of the poor, though their situation improved in the sense that more of them survived, but for the benefit of the owners of the coal mines. In neither Europe nor America was the industrial transition made in order to fill the vital needs of the weak, but to fill the non-vital needs of the powerful. And the influence of the powerful has been such that the worldview in which their needs may be met has come to be shared by the weak. Thus, while we would not say that *all* innovations are the result of ecological pressure (or that, to the extent that they are, then in some cases that pressure is indirect), we should say that the most *basic* ones are, i.e. those related to the filling of vital needs. Thus technology's role in making human life *possible* is most apparent in the hunter-gatherer model. Due to constantly increasing surpluses we therefore see through the course of human existence an ever-strengthening trend towards the development of technology in response to the non-vital and perhaps only imagined needs of the wealthy and powerful.

Non-renewable resources

From the point of view of the VCP, what is key as regards the industrial revolution is not industrialisation *per se*, but the drastic increase it meant in our use of non-renewable resources, or the non-sustainable use of resources, in the form of our use of both fossil fuels and metals. As regards fossil fuels, where the horticultural revolution constituted a shift to the domestication of plants and animals, and the agrarian era to the 'domestication' of metals, the industrial revolution is a shift to the 'domestication' of coal, oil and natural gas. Where in the Bronze and Iron Ages non-renewable *materials* were introduced, in the industrial period non-renewable *energy* was introduced. As expressed by Ponting:

> Much of the historical treatment of this period, particularly the more popular accounts, concentrates on the idea of an 'Industrial Revolution.' Attention has therefore been focused on changes in industrial technology and the adoption of new industrial processes. These were important in expanding production, utilising new materials and developing new industries, but the more fundamental change that occurred was the shift in energy sources. Although other sources of energy were important in the early stages of the process, ultimately the move to an industrialised society depended on the consumption of *non-renewable* energy resources.[1222]

As expressed by Sieferle, without coal European societies of the 1700s and 1800s would have remained agrarian.[1223] And, as noted already in 1789 by John Williams with regard to Britain: "When our coal mines are exhausted, the prosperity and glory of this flourishing and fortunate island is at an end."[1224] And in 1865, five years before Britain's economy peaked, William Stanley Jevons wrote:

> A farm, however far pushed, will under proper cultivation continue to yield for ever a constant crop. But in a mine there is no reproduction, and the produce once pushed to the utmost will soon begin to fail and sink to zero. So far, then, as our wealth and progress depend upon the superior command of coal, we must not only stop – we must go back.

What alarmed Jevons was the fact that, as he and others correctly assessed, England's current economic superiority was based on her abundant coal. He foresaw that the still more abundant coal reserves in the United States and possibly in other countries would eventually tip the balance in their favour, which they in fact did.[1225]

A report presented to the British Secretary of State for the Environment in February 1972 introduces its chapter on energy with the words:

> There is deep-seated unease revealed by the evidence sent to us about the future energy resources, both for this country and for the world as a whole. Assessments vary about the length of time that will elapse before fossil fuels are exhausted, but it is increasingly recognised that their life is limited and satisfactory alternatives must be found. The huge incipient needs of developing countries, the increases in population, the rate at which some sources of energy are being used up without much apparent thought of the consequences, the belief that future resources will be available only at ever-increasing economic cost and the hazards which nuclear power may bring in its train are all factors which contribute to the growing concern.

The importance of the existence of fossil fuels is also underlined by Abernethy, who points out that (just as the proper sorts of soil are necessary for the development of horticultural or agrarian technology) without oil, it is highly improbable that the automotive or aviation industries would ever have developed, or for that matter that we would have developed biocides and mineral fertilisers. Nor would irrigation be possible on today's scale.[1226] And, as noted by Williams and Jevons with regard to coal, of course the quantity of these non-renewable resources is constantly dwindling as we use them. As suggested by the VCP and as pointed out by Schumacher,[1227] the modern industrial system, with all its intellectual sophistication, consumes the very basis on which it has been erected. And as Sieferle notes:

> The limit set by the exhaustibility of fossil fuel deposits has a completely different character than the growth limit of the solar energy system: no stationary state is possible based on fossil energy; when this system has reached its limits, a new contraction must set in. The only escape could be a shift to novel energy carriers. Therefore, the fossil energy regime is a transitional regime and the society built upon it is a transitional society.

The passing of the peak of world oil production will probably be the greatest socially negative physical occurrence in human history to date, affecting more people in more ways than any other event. Despite this, and in keeping with the reaction principle, it wasn't until 1998 that the International Energy Agency projected a peak in world oil production; and none of the seminars at the Seventh International Oil Summit in April of 2006 addressed the problem of peak oil. A question to the Executive Director of the International Energy

Agency about the decline in existing fields received the answer that fields have always been declining and so far this has not been any problem.[1228]

With the replacement of charcoal with coke in smelting, and the implementation of Watt's steam engine, the tremendous increase in the use of previously inaccessible coal may be seen as resulting from the vicious circle moving from need – of an energy source to replace wood for heating and in construction – to innovation – in the form of the invention of the steam engine, among other devices – and on to resource depletion – through the extraction and use of non-renewable coal.

New needs

The creation of new needs incumbent on the turning of the vicious circle is manifest in at least two, overlapping, ways. One may be seen to be the result of the de facto growth of the population, which implies the constant meeting of new needs in order to feed and house increasing numbers of people. The other is a result of capitalists' constant attempt to acquire more territory in the form of capital.

Thus, as pointed out by Wilkinson, at the time of the industrial revolution – independently of whether it was due to the growing population in Britain or the avarice of the owners of coal mines (or their felt need to raise their status), or a combination of the two – the availability of coal only at particular places created the need for a national transport network capable of handling heavy bulk commodities.[1229] Similarly, modern systems of waste and sewage disposal are obviously a response to the problems of urban sanitation that became particularly pressing in the early stages of industrialisation, requiring the laying of sewers and the building of filtration plants. (Note that since the industrial revolution the idea of using human waste in agriculture, particularly from the cities, has never been entertained.) Likewise, fresh water must be supplied through pipes from pure or purified sources.[1230] And since the beginning of the 20th century in particular, the use of the car has meant the need for increasing numbers of oil tankers and roads, bridges, parking facilities and so on, which require increased labour and the depletion of resources for their construction and maintenance. As expressed by Mishan with regard to education in particular: "A very large part of modern education, both of child and adult, is a form of current expenditure necessary to replenish the stock of skilled human capital, without which the running of a highly industrialised economy would be impossible." As noted in Chapter 4, a person living in a modern society cannot do without such things as a car, refrigerator, cooker, microwave oven, hot water, telephone, cell phone, television, computer, e-mail services and so on[1231] – the provision of all of which has meant economic growth and increasing profits for capitalists, and at the same time a worsening of the state of the ecosphere and with it the preconditions for continuing human life on earth.

As expressed by Wilkinson:

> During the course of economic development man has been forced over and over again to change the resources he depended on and the methods he used to exploit them. Slowly he has had to involve himself in more and more complicated processing and production techniques as he has changed from the more easily exploitable resources to the less easily exploitable. … There are two major consequences of this pattern. The first is the need for more and increasingly sophisticated tools to undertake the processing – i.e. for more capital equipment as economic development proceeds; and the second is a tendency which one might expect to find for the productive system to demand increasing amounts of labour time and power to produce the basic per capita subsistence.

On the VCP, and in keeping with the view expressed by Ellul,[1232] technological development has increasingly been the result of *special interest*. Among other ways it is through the production and sale of technological 'improvements' that capitalists acquire competitive advantages vis-à-vis one another, and increase their relative status and power.

Thus, as is to be expected on the VCP, capitalists' non-vital needs lead them to pursue new lines of technology, and to fully exploit available resources, as long as doing so increases their profits. And this behaviour has the effect, among others, of providing paying consumers with increased material 'benefits' that overshoot their requirements. As is suggested by the pioneering and reaction principles, as long as there is a surplus there will exist an inclination to use it, and one can expect that technology will be developed to help speed up the process – whatever turns the biggest profit. As remarked by G. H. von Wright, technical innovations are marketed so as to meet desires awakened in consumers and formed by already existing technology, such desires becoming (non-vital) needs once a society is accustomed to their being met.[1233] Thousands of academically trained psychologists make their living in the advertising industry or in marketing departments by helping create such needs. So, as remarked by Ernst von Weizsäcker and his associates,

> let us not be surprised that many dissatisfactions are becoming needs in our society, and that almost all needs are becoming the targets of intensive advertising by suppliers of answers to those

needs – 'needmongers' who profit from persuading us all to try to use material means to meet nonmaterial needs.

Capitalists' influencing of the middle and working classes through advertising, together with the surplus of energy afforded by fossil fuels, has stimulated economic growth and resource consumption, in effect increasing the cost of living in terms of energy and materials required. As a result of such changes the productive task facing industrial societies has become enormous;[1234] and technological innovation at an increasing rate has become required, due to the diminishing returns of innovation itself.[1235] And consequent upon all this is of course an increase in the production of waste. An expanding population with burgeoning technological power drastically shrinks the carrying capacity of its environment.

The 'communications revolution'

As we move round with the vicious circle thanks to our innovativeness and the existence of resources amenable to that innovativeness, the situation in which we find ourselves becomes ever more complex. The latest increase in complexity has been seen by some to constitute a 'communications revolution,' which has led to a 'post-industrial' or 'post-modern' society. This change, however, was itself the result of society's acquiring new needs, and all the requirements of the agrarian (food) and industrial (artefacts) epochs are still with us. Virtually *all* of the products that were necessary before the age of computers, and certainly all of the resources, are necessary today, and the provision of virtually all products necessary after the 'communications revolution' is dependent on the use of non-renewable fossil fuels. In fact, from the beginning of the increase in the amount and sophistication of communication after World War II and up to the present, industrial production has constantly expanded, as has the consumption of fossil-fuel energy and metals.[1236] We are not living in a post-industrial age (just as, one could say, we are not living in a post-*Iron* Age) but rather, to date, in an *increasingly* industrial age.

Pollution

Pollution essentially came into existence with industrialisation, and is almost totally a side-effect of our increasing use of coal, oil, natural gas and uranium, the waste of which can be degraded into biologically neutral material only at a rate vastly slower than that at which we are emitting it into the ecosphere.

In 1980 there were 50,000 toxic waste dumps in the United States. By the late 1980s the international trade in toxic waste involved millions of tons annually, and the spectacle of rich countries paying poor ones to take their poisons aroused political resistance.[1237]

Inferior substitutes, etc.

As noted in Chapter 4, certain resources made available by technological development may provide a greater (short-term) service than did the resources they are replacing – in the present case both coal and oil, due to their greater abundance, provide a greater service than wood and whale oil. However, even when they do so, it may be the case that the service they provide is only partial,[1238] as treated earlier and as is often the case in the turning of the vicious circle. This is so as regards the replacement of wood with coal and oil, neither of which meets the requirements met by wood as a building material. Though coal could be and was used in the firing of bricks, clay for the making of the bricks was of course also required.

Not only may the newly accessible resources provide a greater service, but the *products* resultant upon the use of new technology may sometimes be superior to those they are replacing – fibre-optics comes to mind as being superior to copper wire in some applications; synthetic fibres – which don't rot – are superior to hemp and jute in fishing lines and nets. As regards whether e.g. personal computers and e-mail are superior to the typewriter, calculator, telephone and ordinary post, it is perhaps too early to say – my own experience in any case is that the frustration of dealing with recalcitrant machines certainly hasn't lessened.

What has happened in the main, however, is that scarcity has forced us to use materials which are often regarded as inferior to the traditional ones.[1239] For example, just as woollen and cotton clothing are inferior to sheepskin and linen, plastic products are inferior to those of leather and rubber; and the energy provided by coal is inferior to that provided by wood. As regards the energy provided by the splitting of the atom, considering the long-lasting effect of its waste, it is a poor substitute for the energy provided by fossil fuels, though the excessive use of either could result in the extinction of our species. More generally then, it may be suggested that the goods produced by new technology are often of *lower quality* than those they are replacing, but have to date generally been of *greater quantity*.

Population growth and checks

The only way such a huge population as existed at the end of the agrarian era could continue to be supported were if a major new resource could be made available; and, as it turned out, such a resource existed in the form of coal, later to be supplemented by other fossil fuels. The consumption of fossil fuels has indirectly sustained the eightfold increase in world population since 1750, through its role in increasing agricultural production and housing.

Though the longevity of the members of various societies cannot be determined with any degree of accuracy until around 1750, it may safely be said that throughout *sapiens'* development until recently it varied between 20 and 35 years of age.[1240] As regards population checks, the weakening of humans' *internal* checks began some 50,000 years ago, if not before. In the industrialising countries, decreasing mortality in association with the reduction of *external* checks began somewhat prior to the 1800s. Virtually all external checks being positive, their first being pushed back can perhaps be said to have been some 300,000 years ago, first with weapons, and later also with barricades, reducing the likelihood of being killed by large predators. (Note that with the pushing back of external checks the importance of internal checks rises; and in the case of humans these checks were of course constantly being eroded due to the existence of various surpluses.) Human domestication and sedentism, however, coupled with increasing population density, brought with it *small* predators in the form of parasites and infectious diseases, as well as worsening hygiene, which counteracted this development. Through a process of domestication incumbent upon technological innovation, we have constantly reduced the size of our predators; and, as in the case of prehistoric overkill, the smallest evade our technology.

Beginning with our first technological innovations, not only has the territory *actually* needed to raise a family been diminishing, but, equally important, so has the territory *perceived* as needed. Thus individual territoriality has been undermined in its capacity as a population check for humans through the vicious circle's recurrent provision of a surplus, not only in the form of food but also the materials needed to construct dwellings.

Since the advent of the industrial revolution the curve of population growth has had essentially the same form as, but not been as steep as, the curve of fossil-fuel use, each member of the human population, on the average, being dependent on increasing quantities of energy.[1241]

Here we can apply Malthus' reasoning:

> This is a rate of increase which in the nature of things cannot be permanent. It has been occasioned by the stimulus of a greatly increased demand for labour, combined with a greatly increased power of production, both in agriculture and manufactures. These are the two elements which form the most effective encouragement to a rapid increase of population. What has taken place is a striking illustration of the principle of population, and a proof that, in spite of great

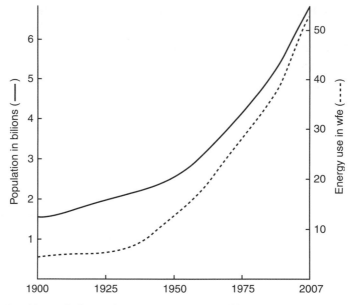

World population and energy use 1900–2007[1242]

towns, manufacturing occupations, and the gradually acquired habits
of an opulent and luxuriant people, if the resources of a country will
admit of a rapid increase, and if these resources are so advantageously
distributed as to occasion a constantly increasing demand for labour,
the population will not fail to keep pace with them.

This is a clear expression of the pioneering principle as applied to humans.
Note here capitalists' non-vital need to make a profit, and its influence on population growth through the demand for labour.

During the 1800s there was still a surfeit of people even in the short term, as
manifest in high mortality rates among the poor. The institution of foundling
hospitals at that time, which were apparently intended to care for orphans,
actually supported infant mortality, and functioned as a positive population
check. As expressed by Malthus:

> Considering the extraordinary mortality which occurs in these
> institutions, and the habits of licentiousness which they have an
> evident tendency to create, it may be said … that if a person wished
> to check population, and were not solicitous about the means, he
> could not propose a more effectual measure than the establishment
> of a sufficient number of foundling hospitals, unlimited in their
> reception of children.[1243]

In all of France some 127,500 children were abandoned in the year 1833, and anywhere from 20 to 30 per cent of all children born were left to their fate. The figures for Paris suggest that in the years 1817–1820 the 'foundlings' comprised 36 per cent of all births. In some Italian hospitals the mortality of infants under one year of age ran to 80 or 90 per cent. In Paris the *Maison de la Couche* reported that of 4779 babies admitted in 1818, 2370 died in the first three months and another 956 within the first year.[1244]

While the rapid development in fossil-fuel technology eventuated in a tremendous increase in population, as noted, technological development in medicine during the industrial era has had a relatively much smaller influence on population size. In this regard there seems to be an inverse relation between the degree of sophistication of health-related changes and their effects on mortality. As taken up above, relatively simple innovations such as the regular use of soap, the cleaning of dwellings, and improvement in diet have had a much stronger impact than the employment of such technologically sophisticated contributions as CAT-scanners and organ transplants.[1245] Here we have yet another instance of diminishing returns. And to this it may be added that it is primarily the small upper strata of human society, not the large lower, that is supported by medicine.

Note that the vicious circle principle concerns humankind *as a whole*; and in this regard the human population has been constantly growing at an accelerating rate since its inception. That population growth may have slowed or even stopped e.g. in modern hunter-gatherer societies despite the continuing presence of a surplus does not alter this fact.

In the 1800s Mill suggested, in keeping with the long quote from him in Chapter 2, that little improvement can be expected in morality until the producing of large families is regarded with the same feelings as drunkenness or any other physical excess. But, asks Mill, while the aristocracy and clergy are foremost to give the example of this kind of incontinence, what can be expected of the poor?[1246] Here we have an instance both of the Church supporting population increase – the more Catholics there are, the better for the powerful in the Church – and of the influence of the upper classes filtering down.

On the VCP, as is in keeping with Wilkinson's view, the *essence* of technological change to date has consisted in humans' development of new methods of relating to their environment such that more can be extracted from that environment in order to support a growing population. While on the whole world population growth has been accelerating since we first came into existence, with the employment of fossil fuels, thanks to the huge quantity of energy their use has provided, it began to accelerate at a truly tremendous rate.

The fact that the VCP has been operative right from our species' inception suggests that the internal homeostatic checks to human population growth have *never* been fully adequate to the maintenance of population stability – the main reason being the surplus afforded by technological development. *Homo sapiens* is a constantly pioneering species.

As Wilkinson remarks, the most primitive subsistence strategies are those which come most easily to the relatively ill-equipped hand, and each new strategy in the course of development is the easiest to adopt in the context of the tools and technology of the existing culture. The most primitive societies, with the lowest population densities, have the widest choice of resources – more resources are capable of providing them with adequate subsistence. They choose to use the things in their environment which already approximate most nearly to the consumable state. Resources and materials such as wild animals and plants for food, with skins, bones, stone-flake tools, sinews and leather thongs for other purposes, involve the minimum amount of work and processing to prepare them for use.[1247]

As Wilkinson says:

> To support a larger population man is forced to follow a course of increasing intervention into the climax community of flora and fauna to divert a larger and larger part of the natural system's productivity into forms which humans can use. Human societies must find or create a large enough ecological niche for themselves in the organic food-chains of the biosphere. They must increasingly dominate and restructure their ecosystem to support rising population numbers. Much of the history of agricultural development is the history of attempts to increase the control and manipulation of the natural environment, and to use land for the specialized and increasingly intensive production of the crops and animals which particular societies depend on.

To support our increasing numbers, we take increasing amounts of low entropy from our environment, which we convert into high entropy. Thanks to the availability of resources, each major turn of the vicious circle through the horticultural, agrarian and industrial revolutions has resulted in a surplus, thereby not only supporting the then-current population, but allowing it to continue to grow, and the vicious circle to turn. As C. G. Darwin says, "The central feature in all of these revolutions has been that they have made possible a great increase in the number of people."[1248]

Quality of life: labour, industrialisation and capitalism

As Wilkinson has suggested and as is implied by the operation of the VCP, the implementation of new technologies throughout human history has almost always meant a transition to a more onerous lifestyle for the weak,[1249] and, we might add, increased their alienation from the products of their labour. Two major steps in alienation are, first, as in horticultural and agrarian societies, not consuming your own products *immediately* (postponed gratification); and second, as in capitalist societies, not consuming them *at all*. The latter was clearly the case with the first introduction of new technology in the industrial revolution.

In the industrial revolution, an increase in the amount of labour performed by the weak before the supporting non-human infrastructure had been built up was also the rule. When new industrial techniques are first being implemented, much manpower is often required; and when it is not, the weak go unemployed. The over-efficiency of technology can mean a lack of employment and thus of income for large parts of the lower strata. (Cf. the Luddites, taken up below.) And this is still the case in the non-industrialised countries today. Thus with its requirement of greater effort to obtain the newly available resources, like earlier technological revolutions, the industrial revolution brought with it a lowering of the quality of life of the common people.

It may be countered that this has not been the case – that since the industrial revolution the hard work has been done by machines. But this is not so. Much hard work is still required on the part of those at the bottom of the social hierarchy – as is now particularly noticeable in the case of people living in Third World countries.

While each major turn of the vicious circle tends to worsen the situation for the weak, this effect becomes somewhat lessened once the new mode of production has become established – perhaps because it takes some time for the powerful to realise that it is to their advantage to have a healthy population. In any case, as noted, the standard of living of the industrial poor began to improve around 1850.

According to Wilkinson,[1250] the reason the workload was so great during the industrial revolution was because an increasingly complex productive system was being foisted by capitalists on a society at the same time as the society's original choice of resources, in particular wood, could no longer be used. It is for this reason that men, and increasing numbers of women and children, worked extraordinarily long hours. The arduous factory routine contrasted sharply with the much shorter hours and seasonal work associated with earlier periods of English history and with many Third World countries today.

As argued by Wilkinson, with the industrial revolution our lifestyle and environment had to be accommodated to an industrial mode of production as a matter of necessity. And this accommodation meant the acceptance of a new ruling class, one based on the ownership of the new resources and means of processing them, i.e. on the ownership of *capital*. Thus from occupying an agricultural niche, we moved to an industrial-agricultural niche, in which capitalists replace landowners at the top of the social hierarchy (an instance of the *Animal Farm* syndrome), while the hierarchy itself remains essentially intact.

As has been noted, social power stems inevitably from control of people's access to vital resources, including the means by which they might raise a family. Scarcity is both a precondition for exploitative relationships and the motive for development, while a state of plenty accessible to all favours stability and egalitarian community life.

The leading role of industrial capitalists in Western industrialisation would have been impossible if the mass of ordinary people had had their vital needs met. It is only in so far as people have unsatisfied needs that they will be amenable to the twin roles of labourer and consumer. Not only the methods and the technology but also the institutional organisation of capitalism in the 1800s was a response to the growth of need – a need which increasingly became that of the capitalist. Wages as the price of labour are in the end determined by the cost of food and other means of subsistence, and they tend to find an equilibrium around the minimum at which workers can maintain themselves and their offspring.[1251]

The pattern in England of labourers' preferring to work less rather than earn more changed during the second half of the 1700s. Since that time, the dominant desire has been for higher consumption, requiring greater labour, rather than for increased leisure. Whenever the basic working week is shortened people prefer to work overtime, increasing their earnings rather than their free time.

One might speculate that the reason for the change has to do with the difference in ethos between the militarily/politically powerful aristocracy on the one hand and the commercially powerful capitalists on the other, not least because the worldviews of each tend to shape those of the masses. For the agrarian warrior-lord the highest value is bravery and, in the best of cases, consequent victory in battle. It is such success that gives him power, while, as mentioned earlier, he looks down upon all kinds of manual labour. As noted by Tom Burns, the ruling groups of pre-industrial society were not themselves concerned in or with the production of goods, merely with making sure that they acquired a substantial share of them.[1252] For the businessman/capitalist, on the other hand, as already noted, what is most highly valued is to make as large a profit as possible, and the drive to do so underlies all others;[1253] and work, which in the beginning of his career might well include manual labour, is a prerequisite to the attaining of this goal. As pointed out by Weber, the classical representatives of the spirit of capitalism "were not the elegant gentlemen of Liverpool and Hamburg, with their commercial fortunes handed down for generations, but the self-made parvenus of Manchester and Westphalia, who often rose from very modest circumstances." And as pointed out by B. D. Jouvenal, the great contrast between the time at which the industrial revolution was just beginning and the present is that a ruling class exempt from material concerns has been supplanted by a ruling class entirely devoted to material concerns.[1254]

Note that on the VCP the transition to capitalism as the major social force in the world is not conceived of as being incumbent upon other social conditions – as e.g. in Weber's attempts at least partly to explain the rise of capitalism in terms of that of Protestantism.[1255] Such reasoning only leads one to ask why the earlier social condition arose. To *explain* social (or any)[1256] phenomena one must go to a deeper level. Here, assuming a modern-scientific approach, the deeper level is taken to consist of the bio-ecological preconditions of what is to be explained. Thus the explanation of the rise of capitalism on the VCP is that first the discovery of the New World with all its resources, and then the ability to acquire the huge energy reserves of fossil fuels, provided a massive surplus of such a sort as was conducive to use in production and trade, thereby legitimising not only capitalism but the Protestant ethic with it. As implied by Darwin,[1257] it is practical exigencies that mould our morals, not the other way round. Had there not existed a New World, and fossil fuels and metals, the Protestant ethic could hardly have been expected to have underlain a transition from the aristocratic system to capitalism. In any case, with the transition from the medieval agrarian system to the modern capitalist system there occurred a change in ethos felt throughout society, in which work was no longer to be looked down upon, but instead became a virtue.

For the wealthy 20 per cent of the world's population, however, there was nevertheless a decrease in hours of work from ca. 65 per week to 40 (in 1938 in the US, though not by law), though this is still more than twice as much time as modern hunter-gatherers expend in acquiring resources. Part of the reason for this reduction must include fossil fuels' not only constituting an enormous surplus of energy, but one which, in keeping with the VCP, is unequally distributed; as well as the fact that the use of machines in the productive effort is constantly increasing. Thus, as the industrial era has proceeded, the ever-increasing quantities of energy employed, at least in the industrialising countries, have constantly become less that of the muscles of the politically and financially weak, and more that of fossil fuels used to power muscle-replacing machines in agriculture, manufacture, and most other areas.

As the vicious circle turns, the proportion of a person's work that is labour, i.e. work for others, and the proportion of people labouring, have constantly increased, and the labour performed has increasingly become *paid* labour for *capitalists*. Unlike slave-owners, capitalists are completely absolved by society of any responsibility towards the people working for them *qua* people, and they act accordingly. Where slaves have to be fed even when there is no work to be done, capitalists simply lay their workers off. In fact one can see the abolition of slavery in the industrialised countries as primarily the result of the transition from agrarianism to industrialism, slavery only waning with the decline of agrarianism.[1258] For example, it is no coincidence that in the American Civil War the powerful in the Southern slave-states obtained their power from agriculture, while the powerful in the Northern free-states obtained it from industry. Comparing their slaves with the 'wage-slave' factory workers in the North, wealthy Southerners claimed them to be better cared for.

As K. W. Kapp says with respect to social costs,

> the entrepreneur, in his desire to reduce costs of production as far as possible, will generally be reluctant to consider the impairment of the physical and mental capacities of his laborers as part of the costs of his enterprise. [T]his is in contrast with the conditions prevailing under slavery systems and feudalism, where the producer is likely to consider any impairment of the human factor by the productive process as a depreciation of the capital value of his property and thus as part of the costs of production. His unwillingness to do so will be the greater the easier he finds it to replace 'worn-out' workers by new laborers.

According to J. M. Clark, the individual worker "is, under our social system, a free being, responsible for his own continuous support and that of his family; hence his maintenance is his own burden and not an obligation of industry, except so far as he can exact wages that will cover it."[1259] This is in marked contrast with the costs of machines and the fixed charges of the borrowed capital which have to be met by industry regardless of business conditions.[1260]

> It is at this point that the social costs of unemployment become apparent. For, once the fixed overhead costs of labor have been converted into variable costs, the entrepreneur is able to disregard the fixed costs of labor completely. A decline of business will be met first by a reduction of the 'variable' costs of labor and thus tends to give rise to a wave of unemployment. This procedure is not only the most convenient for the entrepreneur, but in view of the fixity

of most capital outlays it is the only method of reducing costs of production. Periods of depression will thus give rise to a general shift of the fixed burden of labor to the individual worker, his family or the community.

The same disregard of the fixed overhead costs of labour marks all entrepreneurial decisions concerned with the introduction of technological improvements.

One major reason for slavery's not being associated with industrialism may be that many industrial jobs require skilled labour, while jobs in agriculture (and mining) do not, and can even be performed by children. Thus by not having slaves the industrialist avoids having to pay both for training his labourers and for rearing and educating their children. And he can select the ones he wants after the taxpayer-funded education is complete.

In the industrial era the existence of labour has come to presuppose the presence of a capitalist economy – consider e.g. the unions, or the Labour Party in Britain, whose existence presupposes that of a capitalist system. And where in the agrarian era labourers were paid in real *kind*, in the industrial era they are paid in abstract *coin*.

Here we see a second major reason industrialists would rather have paid labourers than slaves. In order to rise in the hierarchy the capitalist needs a *market*. Given the huge surplus, and its nature, selling only to other powerful people does not allow him to make as large a profit as possible. A larger market is necessary. A balance is thus developed between the capitalist paying a salary on the one hand, and his receiving an income through labourers buying his goods on the other, in such a way that he can make the largest possible profit. This experienced need to have a market for one's products is a situation that did not exist when wealth consisted in the ownership of land. Thus the capitalist, through trying to make as great a profit as possible, provides the labourer, whom he needs, with material goods. And the more goods the labourers buy from a particular capitalist, the better for that capitalist. All of the money eventually goes in a circle from capitalists to labourers and back to the capitalists again, with the capitalists constantly hanging on to a larger and larger portion. As suggested by Roger Terry: "In order to make a profit, businesses must pay employees a diminishing portion of the system's total wealth, must prevent them from purchasing as much as they produce."[1261]

In this way then the increasing surplus of energy available after the industrial revolution, together with the profit motive of capitalists, led to a general improvement in the quality of life of labourers in industrialising countries after 1850, and at the same time gave rise to a middle class of white-collar workers.

Here we have virtually the only instance in which the trickle-down effect has actually been operative.

As suggested by Wilkinson, it seems likely that the division of labour, carried to the extent that it is in industrial societies, represents a lowering of the quality of people's lives. The vast majority of jobs resultant upon the division of labour of complex, capitalist society are quite unfulfilling, being either monotonous and/or involving the production of entities from which the worker is alienated. Interconnected with such issues, but perhaps most closely related to urbanisation, is the break-up of small communities and 'significant other' societies, this probably being, according to Wilkinson, one of the most serious cultural losses associated with industrialisation.[1262] As expressed by Ted Trainer, the more we commercialise our lives, the more we whittle away those non-economic social relationships which ensure a warm and supportive community spirit.[1263] The breakdown of the local community – in cities, the breakdown of the *neighbourhood* as a social reality – created a need both for new forms of entertainment to replace the disappearing social activities and traditions, and for transport and communications systems to allow contact between friends and relatives who lived in different places. The passive reception of manufactured images, fiction and fantasy, increasingly become substitutes for a disappearing social reality.[1264]

With regard to the purely economic well-being of ordinary people living in the industrialised world and elsewhere, it is of interest to cite Adam Smith and his notion of an *invisible hand*, which is also the origin of the notion of trickle down:

> The rich only select from the heap what is most precious and
> agreeable. [A]nd in spite of their natural selfishness and rapacity,
> though they mean only their own convenience, [and] the sole end
> which they propose from the labours of all the thousands whom they
> employ [is] the gratification of their own vain and insatiable desires,
> they divide with the poor the produce of all their improvements. They
> are led by an invisible hand to make nearly the same distribution of
> the necessaries of life, which would have been made, had the earth
> been divided into equal portions among all its inhabitants, and thus
> without intending it, without knowing it, advance the interest of the
> society, and afford means to the multiplication of the species.

We note here that Smith restricts his application of his invisible hand to the distribution of *vital* resources ("the necessaries of life"), realising that virtually all of the non-vital resources go to the rich.[1265] It is true that the employees of the rich have their vital needs of food and shelter met and can thereby multiply.

This must be seen to, or they would not be able to continue providing the rich with cheap labour. But the incomes of the rich can be hundreds of times greater than those of their labourers. And in thereby being able to select for themselves "what is precious and agreeable," they not only see to the meeting of their non-vital needs, but improve the filling of their vital needs as well, e.g. through access to medicine and better quality housing and food. Thus neither are the vital needs of the poor met to the same extent as those of the rich, and Smith is quite wrong in suggesting that there is nearly the same distribution of vital resources in society as would be the case if the earth had been divided into equal portions among all its inhabitants. And, particularly with regard to the non-vital needs only of the rich being filled, he is equally wrong in suggesting that the rich, through gratifying their own vain and insatiable desires, advance the interests of society.

Furthermore, while all who are engaged in this system may receive the necessities of life, those who are not engaged in it are left to starve, beg, steal, or be taken care of from outside the system, with the result that everyone scrambles to find a place within it. At the time of the industrial revolution there was no trickle-down effect via which the increase in profits which accrued to the capitalist made their way to the jobless – just as there is no such effect today benefiting the starving in the Third World.

The social effects of mechanisation

The lower echelons in society did not benefit from the introduction of labour-saving devices such as the flying shuttle and the spinning jenny in industry. The original effect of the introduction of these machines, apart from affording an increase in production, was to reduce the capitalist's costs for labour. This can be done by reducing the number of jobs or by increasing the size of the labour market through making it possible for what was once skilled work to be performed by the unskilled. As Kapp says:

> the introduction of machinery during the initial stages of the
> Industrial Revolution, by opening the way to woman and child
> labor in the manufacturing process, led to a continuous increase of
> competition among workers, depressing wages often to the level of
> starvation. ... Virtually every new technological advance tends to
> render particular skills and occupations obsolete and increases the
> competition among workers.

As emphasised by Douthwaite, mechanisation *does* destroy jobs.[1266] And not only that, it lowers the value of labour and thereby the wages of those who still have work.

Capitalists' pursuit of profits has meant a constantly increasing use of automation, i.e. as mentioned above, the employment of devices that use fossil-fuel energy to do work that otherwise would have had to be done by more-expensive muscle power,[1267] the introduction of each new such device bringing with it a wave of unemployment. The machine-breaking campaigns of the Luddites (Glossary) in the early 1800s were the expression of a real problem faced by the workers. In the words of a contemporary: "The steam engine and spinning machines, with the endless mechanical inventions to which they have given rise, have, however inflicted evils on society, which now greatly overbalance the benefits which are derived from them." Of course capitalism requires some paid labourers to constitute a market, but such labourers might just as well be paid by some capitalist other than oneself.

Here we might also consider the requirement of modern society of constantly dealing with strangers. Note that in novels and television programmes people often know and call by their first name others they meet e.g. in their tenement building; they shop in corner stores where they know the name of the proprietor, who has owned the place since they were children; they know the police officer in the patrol car, typically having gone to high school with him. In residential districts consisting of houses, the protagonist knows the names and habits of his or her neighbours – all of this seldom being the case in reality. The having of personal relations with the ones one meets adds intimacy and interest – it provides a context in which social relations are meaningful.

The middle class

The middle class, an essentially new cultural phenomenon, came into existence after the industrial revolution. It arose as the result of perhaps two main factors related to the VCP. One is the physical necessity that there exist a sufficiently great surplus to support such a class, which in the present case has existed in the form of the huge amount of usable energy contained in fossil fuels, plus metals, particularly iron, to which that energy could be applied. And the second is the increasing complexity of society and the need to keep it organised in order for capitalists to reap as large profits as possible. The existence of a middle class is an outgrowth of the upswing for labourers in the middle of the 1800s combined with the increasing social/economic complexity requiring a bureaucracy. Paid labour, which became widespread first with the industrial revolution, was the first step towards the coming to be of the middle class.

Given globalisation and the tremendous surplus of energy afforded by fossil fuels, the position of the middle class in industrialised nations has risen above that of only having sufficient means to reproduce. Thus its members also have access to leisure and luxury goods – their experienced needs being non-vital – though not of course to the same extent as the members of the upper class of capitalists. But while there was a clear improvement in the standard of living of the worker in industrialising societies from about 1850 to 1890, it may be pointed out that there has not been any fundamental improvement in the standard of living of the industrial middle class since the 1920s, including as regards their hours of work.[1268] Supported by the work ethic, as mentioned, the dominant desire in the middle class has been and is to work more in order to be able to increase consumption. It seems likely that the total productive workload required per head of the population has not decreased in the last 80 years, during which it may actually have risen – despite automation. As regards society as a whole, as suggested by Wilkinson, there can be little question but that its productive task is growing. And while the amount of per capita labour is not decreasing, the amount of energy used in production *is* increasing, with automated and semi-automated labour-saving machinery constantly becoming more common. As the workload increased, the productive system became unmanageable without the introduction of labour-saving equipment.

With respect to all the people to whom the members of the middle class are related – the producers of their food, clothing, etc. – they are on a stratum above the vast majority. The middle class is not located in the *middle* of the social ladder, but quite close to the top. Where before the labouring class lived in the same region as the middle class, with globalisation virtually the whole of the Third World has become part of the labouring class.

Territoriality

Territory in the form of capital came into being after the Mesolithic period, first in the form of herded animals. Afterwards, with human sedentism, territory took the form of arable land, not itself capital, but the harvests from which were. With the mercantile expansion that began around 1500, while ruling over land/territory still constituted the basis of *political* power, the increasing sale of commodities and manufactured products led to ever greater *economic* power shifting to those who controlled such activities, such that today power obtained through business transactions has virtually come to eclipse political power. (The agricultural states and empires were all at roughly the same level of economic development, and the differences in wealth between them were marginal – cf. today's difference between the industrialised nations and the Third World.) As Reiman points out, Marx wrote that the bourgeoisie – the class of owners of businesses and factories, the class of capitalists – has "conquered for itself, in the modern representative State, exclusive political sway. The executive of the modern State is but a committee for managing the common affairs of the whole bourgeoisie."[1269] And as Reiman himself continues: "Anyone who thinks this is a ridiculous idea ought to look at the backgrounds of our political leaders. The vast majority of the president's cabinet, the administrators of the federal regulatory agencies, and the members of the two houses of Congress come from the ranks of business or are lawyers who serve business."

Historically, the turning of the vicious circle has taken us through a period in which the *property* of kings and emperors acquired on the basis of their territorial instinct consisted in *real* territory in the form of land ('real estate'),[1270] to a period (the industrial) in which capitalists' property consists in *nominal* territory in the form of capital. Beginning already in the horticultural era and continuing ever since, given our surplus of food, major successes and failures have consisted in the results of exchanges on the market or on the battlefield, rather than in the results of the hunt. And just as kings benefit from large populations in order to have soldiers for their armies – as noted with regard to the period after 1500 by Carr-Saunders[1271] – capitalists benefit from large populations in order to have labourers for their factories. In the case of the military, the greater the number of soldiers, the stronger the army, and thus the greater the likelihood of the nation's acquiring more land and its leader's acquiring greater power and status. In the case of capitalism, the greater the number of labourers, the lower the wage they have to be paid, and the greater the capital that may be accumulated. And just as a surplus is required for a king to be able to supply his army, so is a surplus required for a capitalist to make a profit. Note here that there can be a surplus for the powerful at the same time as the weak experience dire necessity, the powerful e.g. taking from the weak in situations where the general surplus is being/has been significantly eroded or is gone. Another parallel is that just as it is to the soldier's benefit that his king's army be victorious – the better the army does the greater the likelihood that the soldier will not only survive but himself obtain booty, while the worse it does, the greater the likelihood of his getting killed – it is to the labourer's benefit that the capitalist's business do well: the better the business does, the greater the likelihood that the labourer will keep his job and perhaps even get a pay rise, while the worse it does, the greater his chance of losing his job or having to take a pay cut. It is in this way, i.e. via military disputes amongst the powerful over land, or economic competition amongst them over capital, that kings and capitalists increase (or lose) their property (territory).

Though the economic strength of the capitalist is ultimately not on a par with the military strength of the king or politician, often economic and military powers reside in the same persons, or the ruler is dependent on the capitalist for e.g. the development and provision of weapons. At the same time, however, for both kings or other political leaders as well as for capitalists, in keeping with what was suggested in Chapter 4, technological development in the industrialised nations in the form of superior weapons and automated machinery has meant a decreased reliance on the existence of a large able-bodied population, and thus a reduction in the push to population growth in such nations.

War

Civilised battles have been becoming *less* deadly over the last four centuries;[1272] and the proportion of the population killed in war and causes related to war has diminished since horticultural times. It would seem that, à la Hobbes, this is due to the fact that larger societies, through setting laws, keep order within themselves (allowing capitalists to do business), order that does not exist when there are many independent tribes. Thus there are broad areas in which peace reigns. (*Pax Romana* again.)

On the other hand, increasing surpluses have meant that since the industrial revolution the wars that have occurred have become progressively larger and more destructive, the last two major ones being *world* wars. This is partly due to the invention and use of more powerful weapons, the most powerful on the long list being nuclear devices. As expressed by Quincy Wright, "If the world's population is divided into many small groups, these oppositions are likely to be moderate, whereas if there are few large groups they are likely to be intense. In the latter case, while conflicts will be less frequent, they will be more violent." Though, we should add, the total amount of killing will be greater if the world's population were divided into many small groups.

The constantly increasing proportion of civilians that are being killed means an increase in the effectiveness of war as a population check through its removal of females.[1273] Nevertheless, despite this, the role of wars as checks to population growth has as yet been minimal in post-horticultural societies. The reason for this in industrial societies on the VCP is mainly that the tremendous amount of energy available in the form of fossil fuels, together with the need of the powerful for cheap labour and soldiers, have supported population growth. These factors have, primarily through their application to agricultural production, made it possible to support a *growing* population, so that the species' need for population control has not been strongly experienced.

Economic growth

As expressed by Sieferle,

> it becomes obvious at a glance why the Industrial Revolution is associated with a colossal acceleration of material consumption in the sense of 'economic growth.' Suddenly and very quickly much

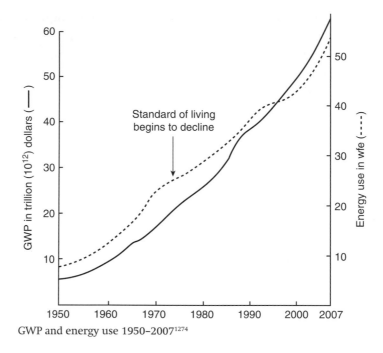

GWP and energy use 1950–2007[1274]

more energy became available than could have been provided by the agrarian solar energy system.[1275]

While the over-consumption of a renewable resource in the form of wood made it necessary to expend more energy in order to procure a non-renewable substitute in the form of coal, the usable energy in the coal itself more than repaid this expenditure. More energy, again repaid by coal, was also needed in *distributing* the coal, energy which would not have to have been spent if wood were the energy source. And it may be said that this increase in resource consumption and energy transformation, with its effects in the form of increasing pollution and species extinctions, would not have been necessary if economic forces and population pressure had not in the first place led to a failure to husband wood resources.

Just as coal proved to be the means to support an otherwise impracticable expansion of industry in the 19th century, the availability of oil as an energy source has been the principal sustainer of continued economic growth in the 20th century[1276] and on into the 21st.

While this economic growth did not benefit the poor, it did benefit the rich (including the middle class, at least in the 1950s–1960s), providing them with large *profits*. And those who benefited most were those who *owned* the surplus,[1277] particularly in the form of reserves and stocks (not resources). It was through the use of this energy that economic growth took place. In fact right from the beginning of the industrial revolution economic output has varied directly with fossil-fuel use in both capitalist and other social systems.[1278] And in the context of the VCP we can see that the additional power afforded by fossil fuels was not a fortuitous discovery which allowed the culture to blossom into new richness and splendour. Rather, the acquisition of fossil fuels was a response to various situations of need,[1279] the meeting of which involved the entrepreneurial spirit of capitalists.

In practice, during this era, which continues today, economic growth has taken the form of a per capita increase in the quantities of goods and services provided, an increase resulting from the use of increasing quantities of fossil fuel to power mechanical devices. This constantly increasing production and construction of material artefacts has not only meant a reduction in resources, but, when applied to agriculture and housing, has prompted tremendous population growth.

As pointed out by John Gever and his associates, technological improvements have only rarely reduced the amount of energy needed to exploit a resource.[1280] And as noted in Chapter 4, as resources become less accessible – the most accessible having been taken first – more energy is required to obtain them. Part of this energy is invested in manufacturing the more sophisticated technology that obtaining the resources requires, and in the construction and powering of the machines needed to acquire it.

* * *

In sum, during the industrial era the vicious circle has moved from the experienced need of a particular resource, namely energy, to technological innovation, making that resource available, to the exploitation and consequent depletion of the resource, which provides a surplus, which leads to population growth and war, and economic growth and increased consumption, as well as increased waste, giving rise to an impending period of need.

At the time of the horticultural revolution there were some five million people living on earth; at the time of the industrial revolution there were 160 times as many; and today there are 1130 times as many. Due to the accessibility of a huge though finite quantity of fossil fuels, the use of which supports a tremendous increase in the world's population at the same time as it gives rise to environmental damage on a global scale, the vicious circle of technological development is making its greatest and possibly its last loop.

The industrial model

Scarcity and need
Innovation: **the steam engine and the internal combustion
engine; mineral fertilisers and biocides**
Economic development
Increase in available resources
Prerequisite of the existence of resources
Increased total energy use: **fossil fuels**
Resource depletion
Renewables: timber, soil, **aquifers**
Non-renewables: metals, **fossil fuels**
Continuing species extinctions
New needs: **fuel-driven machines, etc.**
Technological development self-perpetuating
Genetic dependence on developing technology
Increased waste: **pollution**
Surplus:

> of resources: **energy**
> of consumables: **marketable products**

Accelerating energy use
Increase in total consumption
Inferior substitutes
Population checks further weakened
Morals
Accelerating population growth
Population pressure
Migration; **further** centralisation
Continuing increasing technological and social complexity: division
of labour
Social stratification: property; class society; **capitalists/business-
men, the middle class, labourers, and the poor**
Lower quality of life for the weak
Laws, crime, **police**
Revolt, terrorism and revolution
War
Cultural development
Medicine
Accelerating economic growth
Security further decreased

Population overshoot
Diminishing returns
Undermining of technology and human existence
Economic decline
Scarcity and need
Drastic population reduction

Steam engines and internal combustion engines are used to transform a huge surplus of fossil fuels into usable energy, resulting in the accelerating production of enormous quantities of goods as well as pollution and the depletion of aquifers. Food production is increased thanks to the use of fossil fuels in providing irrigation, biocides and fertilisers, fossil fuels and fuel-driven machines becoming new needs. Centralisation continues as businessmen obtain power by acquiring wealth through trade. Four strata of society develop: the wealthy capitalists, the middle class, the labouring class, and the unemployed poor. Police come into being to protect the wealth of the two upper classes, the unequal distribution of which leads to revolt, terrorism and revolution against the bourgeoisie. Medicine is developed, but has little effect on health or longevity. The economic growth made possible by the huge fossil-fuel surplus accelerates, as does population growth, further overshooting the carrying capacity of the land, the result of which can be expected to be a drastic reduction in the population occasioned by the decreasing availability of fossil fuels.

Looking at the development of humankind as a whole, we see the vicious circle beginning to turn already with the protohominids' use of tools, and continuing to do so at an accelerating rate right up to the present.

6

The vicious circle today

In the previous chapter the vicious circle principle was applied to the whole of our species' development up to the present. In this chapter a closer look will be taken at how it has manifest itself today in particular. Of key importance in this regard is how, from the beginning of the industrial revolution up until the present, the increasing availability of energy and the non-renewable materials that it has made available have meant that the effects of the turning of the vicious circle have also constantly been increasing.

The situation in which we find ourselves today is unique as regards the magnitude and rate of growth of the human enterprise. The turn now being taken by the vicious circle is tremendous. As expressed by Ellul, "In spite of all the worthy persons who reassure themselves by saying that all historical epochs are alike, that the crises of the fourth century resembled those of the ninth, and so on, the fact is that no one ever before saw world economies or world wars, or world and national populations which, on the average, doubled every forty-five years."[1281]

The mastery of fire, one of humans' first instances of technological development, continues to be of paramount importance today with the burning of fossil fuels. And so the vicious circle of the development of humankind churns on, and does so with ever greater momentum due to the constantly increasing consumption of fossil fuels and metals, with only the tiniest sign of resistance in the form of the efforts of environmental organisations and green political parties.

Our use of minerals

Where the surplus of the agrarian era was *grain*, that of the industrial era, up to and including the present, has been *energy*. And where the lifestyle of the agrarian era was not sustainable at least due to its dependence on metals, the lifestyle of the present epoch is unsustainable at least due to its dependence on both metals *and* fossil fuels.

Fossil fuels

Though the surplus of fossil fuels has been tremendous compared with that of any earlier source of energy, we humans have managed to use up a large portion of it in 250 years, and have been doing so at a constantly increasing rate. In 2005 at least 87 per cent of the energy used in the world came from non-renewable sources: 35 per cent from oil, 24 per cent from coal, 21 per cent from natural gas (making a total of 80 per cent from fossil fuels), and 7 per cent from nuclear sources. The remaining 13 per cent came from: burnable renewables (10.5 per cent), hydro (2 per cent) and solar and wind sources (0.5 per cent).[1282] In industrialised countries, transport, taken to include the making of vehicles, consumes half of all energy used.[1283]

Of the various non-renewable sources of energy we presently use, oil is of particular importance. Not only is it our largest source at present, but it also has the advantage, unlike electricity, of being easily stored, and, unlike coal and gas as well as electricity, of being in a liquid form. Furthermore, it is easier to remove the sulphur content of oil than that of coal, and thus to reduce its direct polluting effects. Its being in a liquid form not only means that it can easily be transported over land by pipeline, but has led to its being almost the only energy source used for fuelling aircraft, ships and road and farm vehicles. Thus, while present coal resources constitute a source of energy that is more than 16 times greater than that of oil,[1284] the advantages oil has over coal make it a generally much more useful fuel.

Of the 10 wfe of fossil fuels consumed in 1955, 2.4 consisted of oil, while during that year 18 wfe of oil were discovered. An enormous reserve was accumulated, while at the same time the highest quality oil (lowest in sulphur content, etc.) was used first. The peak global oil-finding year was 1962. Since then, the discovery rate has dropped sharply, with no new major oil provinces having been found since 1980.[1285] By 2005, 7.6 times as much oil (18.4 wfe) was being extracted each year as in 1955, while the discovery rate (6 wfe per year) was less than one-third of what it had been then.[1286]

Looking at the situation from a slightly different angle, in the 1940s, 100 times as much usable energy was extracted from the average oil well as had been expended in finding and developing it, i.e. it had a net energy ratio close to 100. By the 1970s the ratio had fallen to 23,[1287] due to increasing quantities of energy having to be used – in the form of labour-saving machinery[1288] – to acquire it.

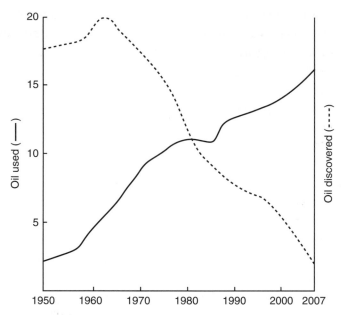

Estimated oil used vs. oil discovered in wfe 1950–2007[1289]

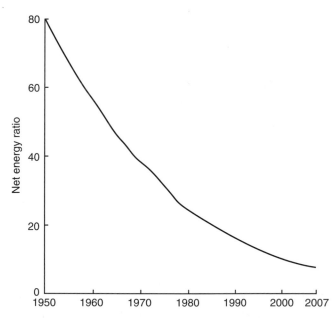

Estimated net energy ratio for oil 1950–2007.[1290] Net energy ratio: the amount of energy obtained divided by the amount expended obtaining it.

Metals

From 1750 to 1900, the world's use of inorganic minerals increased tenfold; and since 1900 it has risen at least 13-fold. In 1776 the amount of pig iron produced in the world was about 360,000 tons. In 1976 it was 560 *million* tons.[1291] Today about 30 times as much iron is produced as the next most common metal, aluminium, the largest use of which in the United States is for beverage cans.[1292]

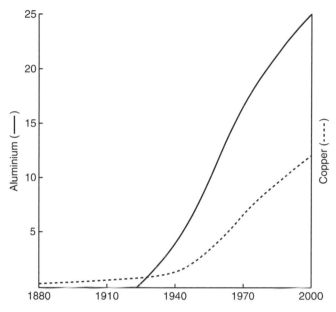

Aluminium and copper extraction in million tons 1880–2000[1293]

Three-quarters of all the gold ever mined was extracted in the 20th century, while the rate of extraction of nickel rose over 80-fold.[1294] From 1970 to 2000 the world extracted three to four times the total volume of inorganic minerals mined earlier.[1295]

Biotic consumption

Human biotic consumption consists in our use of potentially renewable resources, including both plants and animals. Our two basic requirements, inherited from Palaeolithic times, are food and fire. As regards the acquisition of food, or more particularly grains, we can see a development from the use of the digging stick, to the wooden hoe, and on to the stone hoe, wooden plough, irrigation, manuring, steel plough, crop rotation, tractor, and the use of mineral fertiliser and poison. Seventy per cent of all crops in industrialised countries are presently being fed to livestock,[1296] and since 1970 animal feedlots in the United States have produced more organic waste than the total sewage from all American municipalities.[1297]

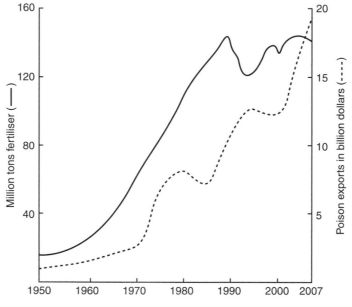

Mineral fertiliser[1298] and poison[1299] use 1950–2007

Energy use

When the agricultural system is understood to include the processing, packaging and distribution of food, it is tremendously inefficient in terms of edible food energy produced per unit fuel input. In the case of the United States, about three times as much energy is consumed in these off-farm activities as is used on farms.[1300] (Similarly, and in keeping with the great amount of energy used by vehicles, it costs capitalists on average more to distribute goods than to make them, in about a 60/40 ratio.)[1301] As succinctly expressed by A. A. Bartlett, modern agriculture may be seen as the use of land to convert petroleum into food.[1302] And this it does, we might add, in such a way as provides capitalists with the largest possible profits.

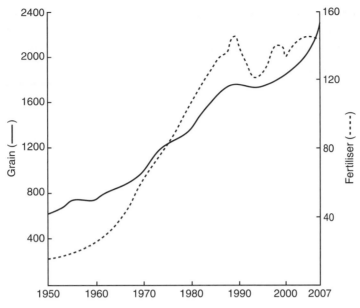

Grain production[1303] and mineral fertiliser use[1304] in million tons 1950–2007

Livestock population, million head 1890–1990[1305]

Year	Cattle	Sheep	Goats	Pigs	Horses	Poultry
1890	319	356	52	90	51	706
1910	391	418	83	115	73	828
1930	513	567	153	187	88	1203
1950	644	631	187	300	69	1372
1970	1016	1001	325	634	81	2734
1990	1294	1216	587	856	61	10,770

Mechanisation

Mechanisation was introduced into agriculture when it became profitable to do so. Tractors could do much more work than draught animals, and there was plenty of steel to make them and oil to fuel them. There were certain problems, however. Tractors were uneconomical on small farms, and so required the ownership of larger areas of land, a fact which e.g. in England reinforced the system of enclosures. This of course may be seen on the VCP as also being the result of the striving on the part of the powerful constantly to obtain more power, or rise in the pecking order. Another problem was that without draught animals there would be no manure to fertilise the fields, and such fertilisation was necessary for the production of food in the traditional system. Further following the VCP we see that a *substitute* was found, namely mineral fertilisers. This substitute, from the point of view of the long-term use of the land, while being greater in quantity, was of poorer quality, mineral fertilisers being susceptible to leaching. This is undoubtedly related to the fact that while some mineral fertilisers are organic, unlike manure, they are not biotic. Somewhat more than half of such fertilisers end up in nearby waters, from which they spread, contributing to the eutrophication of rivers, lakes and seas. Those fertilisers that stay in the soil often lead to problems in micronutrient supply, handicapping rather than helping farming.[1306]

Mineral fertilisers

In 1950, on average worldwide, one additional ton of mineral fertiliser used on grain crops produced 46 more tons of grain; in 1965, one ton produced only 23 more tons; and by the early 1980s, it produced only about 13 more tons.[1307] Here we have a clear instance of diminishing returns in agriculture.

During the latter half of the 20th century fertiliser use grew more than 900 per cent, due primarily to agricultural researchers promoting the 'Green Revolution' during the 1950s and 1960s. Sixty-four per cent of world fertiliser use in 2002 was on the part of Third World countries.[1308] As underlined by L. R. Brown, the growth in the world mineral fertiliser industry after World War II was spectacular, its use climbing from 1950 to 1989. From 1989 it declined until 1995, at which time it began to rise again.

From 1950 to 1990 the world's grain farmers raised the productivity of their land at a rate slightly higher than that of world population growth, the agricultural yield *per area land* climbing by 2.1 per cent a year. But from 1990 to 2000 this rate dropped to 1.1 per cent, due, in some countries, to the loss of irrigation water.[1309]

Land degradation

Worldwide, close to 4000 km² of land, much of it cropland, are presently being paved over each year for roads, highways and parking lots.[1310] Between 1945 and 1975 farmland of the area of the United Kingdom was paved over.[1311] By 1990, in North America, Europe and Japan, the space allotted to cars constituted from five to ten per cent of the land surface.[1312] By 2007, worldwide, traffic accidents were killing about 1.2 million people each year and injuring as many as 50 million.[1313] Not only has fertile agricultural land been lost forever, but the rapid water runoff from these non-vegetated surfaces has caused flooding, especially in low-lying areas.[1314]

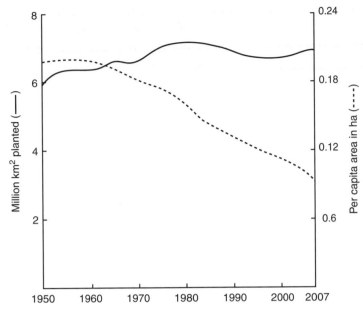

Grain area vs. area per capita 1950–2007[1315]

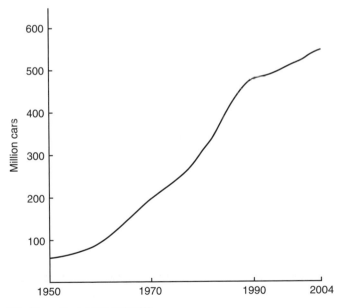

World car fleet 1950–2004[1316]

Roughly one-tenth of the earth's land surface is presently being used to produce crops. One-fifth is grassland of varying degrees of productivity, and another fifth is forest. The remaining half of the land is either desert, mountains, or covered with ice. The desert area is expanding, largely at the expense of grassland and cropland.[1317]

Between 1945 and 1990 we lost 17 per cent of the arable land in the world;[1318] the per capita area of available cropland shrank between 1950 and 1998 by almost half, dropping from 0.23 hectares to 0.12 hectares per person. In the past ten years, the amount of farmland per person has declined 20 per cent.[1319] Since the 1980s more land has been going out of production because of degradation (erosion, soil compaction, salinisation from irrigation, etc.) than has been opened up.[1320] In the late 1980s, global soil losses in excess of new soil formation were about 25 billion tons per year.[1321] Each year, some 50,000–120,000 km² of farmland – about 0.3 to 1 per cent of the world's arable land – go out of production as a result of degradation.[1322] In the early 1990s the sediment carried in the rivers of the world due to human activity (largely farming and forestry) was ten times the natural burden.[1323]

Keeping in mind that in fact *all* soil has been degraded to varying degrees, an area of about 4.3 million km², half the size of China, has been 'irreversibly destroyed' by accelerated erosion.[1324] African erosion rates are nine times as high as those in Europe. And by around 1980, quarrying moved more earth than did natural erosion.[1325]

Deforestation

World forest cover was some 70 million km² 10,000 years ago, while it is somewhat more than 35 million today.[1326] The world has lost about half of its trees since the horticultural revolution, and of those it has lost, more than half have been cut down since 1970.[1327] As of the mid-1990s over 200,000 km² of tropical forests were vanishing each year, at which rate the forests would disappear by 2040.[1328]

The two main reasons people remove forests are to use the land or to use the timber. As regards using the land, the continuing erosion of cropland constantly increases the pressure to clear forested land. The maximum sustainable yield of a forest is the amount which can be taken without exposing the uncovered forest soil itself to erosion from sun and rain,[1329] and this yield was clearly surpassed. Over the course of the 1900s, logging motivated perhaps one-sixth of the deforestation, and land conversion the rest.[1330] Agricultural expansion is the single most important cause of forest loss, involved in 96 per cent of a sample of 152 cases of deforestation between 1880 and 2000. However, agricultural expansion was most often connected with timber harvesting and road building.[1331]

Food production

As pointed out by Brown, *world grain production* is a basic indicator of dietary adequacy at the individual level and of overall food security at the global level. Worldwide, the area in grain expanded from 5.9 million km² in 1950 to its historical peak of 7.3 million km² in 1981. By 2004, it had fallen to 6.7 million km². As the world's population continues to grow, the area available for producing grain is constantly shrinking.[1332]

The absolute world grain harvest, after nearly tripling from 1950 to 1996, stayed flat for seven years in a row, through 2003; and from 1999 to 2003 production fell short of consumption. The shortfalls of nearly 100 million tons in 2002 and again in 2003 were the largest on record. But in 2004 there was a record harvest, in which the two-billion-ton mark was passed. The 2005 harvest was 2.5 per cent lower however, and the 2006 harvest was roughly one per cent lower than 2005.[1333] In 2007 the harvest was a record 2.3 billion tons; despite this however cereal stocks at that time were at their lowest level in 30 years.[1334]

World grain production expanded faster than population from just after World War II until the mid-1980s, raising the amount of grain produced per capita from 250 kg in 1950 to the historical peak of 343 kg in 1984, an increase of 37 per cent. Since 1984, grain harvest growth has fallen behind that of population, however, and the amount of grain produced per person dropped to 303 kg in 2006, down 11 per cent from 1984.

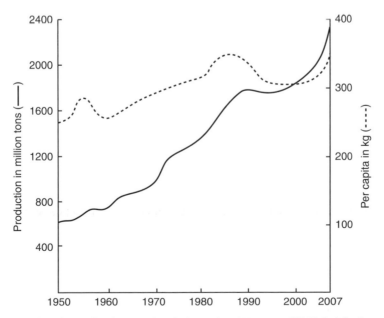

World grain production, total and per capita, 1950–2007.[1335] Global food production per person peaked in 1986,[1336] and has more or less plateaued since then.

Modern wheat varieties, which are bred to convert more than 50 per cent of the plant's photosynthetic product into seed, have little room left to increase yields before they reach the scientifically estimated absolute limit of 62 per cent. Anything beyond that amount deprives the rest of the plant of energy needed to sustain leaves, stems and roots.[1337]

Grains dominate the world's diet and agricultural landscapes. They account for 47 per cent of our calories and 42 per cent of our protein. Farmers plant grains on half of the world's cropland and on nearly two-thirds of the irrigated land.[1338]

The total global production of *root crops* peaked in 1984; and *meat* and *milk* plateaued during 1986–1991.[1339] Since 1950, world meat consumption has climbed from 44 million to 253 million tons, more than a fivefold jump, while the population more than doubled. Worldwide, the average person consumed 41 kg of meat in 2003, more than twice the figure of 50 years earlier.[1340]

Total *fish* consumption plateaued during 1986–1991.[1341] Numerous important fisheries collapsed in the 1900s, generally the most valuable ones. The catch of the 1980s and 1990s included great portions of previously uneconomic fish ('trash fish') sought out because cod, herring, haddock and tuna, among others, became harder to find. While the Peruvian anchoveta collapse in the early 1970s is the most spectacular in the history of fisheries, after the 1970s a general crisis afflicted almost all the world's fishing grounds. The collapses brought (taxpayer-paid) subsidies, which deepened the problem, and regulation, which did not solve it.[1342]

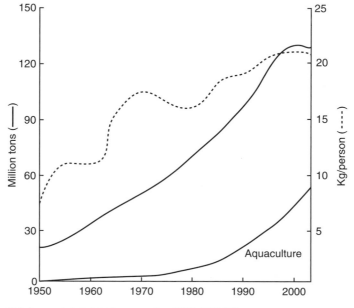

Fish harvest, total and per capita, 1950–2003[1343]

Despite knowledge of the anchoveta and other fishery collapses, in order to exploit its newly won dominion over the Grand Banks, and to boost the economy of Newfoundland, Canada in the 1980s subsidised a great expansion and technological upgrade of its fishing fleet.[1344] Within ten years of this the 500-year-old Canadian cod fishery failed, putting some 40,000 fishermen and fish processors out of work. Fisheries off the coast of New England were not far behind. And in Europe, cod fisheries are presently in decline, approaching a free fall. Like the Canadian cod fishery, those in Europe may have been depleted to the point of no return.[1345] Since 1950, with the onset of industrialised fisheries, we have rapidly reduced the resource base to less than 10 per cent – not just in some areas and not just for some stocks, but for entire communities of large-fish species from the tropics to the poles.[1346] Fisheries are collapsing throughout the world.

A paradigmatic example of the influence of mechanisation on biotic production that is positive in the short run and negative in the long is American west-coast sardine fishing. Here we have a clear example, like the moose on Isle Royale, of the effect of the pioneering principle. In 1916, 28,000 tons of sardines were fished up. The fishing fleet was quickly enlarged and catches climbed rapidly until 1934, when 600,000 tons were caught. Between 1934 and 1945 catches were relatively constant around this number, despite the construction and use of more fishing boats. After 1945 the size of catches rapidly decreased. In 1962 only 20,000 tons were caught, while at the same time increasingly modern equipment and improved fishing methods were constantly being introduced.[1347]

Global per capita production of all the *basic renewable resource systems* (forests, fisheries, croplands and grasslands) has peaked and begun to decline. In the modern world, the production of potential renewables has become dependent on the extraction of non-renewables. Though the decline in food production worldwide is not due to reduced energy availability, when that availability begins to drop there will be a deceleration not only in the rate of extraction of non-renewables, but in the production of renewables, including food.[1348] As expressed generally in entropy/systems terms by Daly, in increasing the entropy of the non-human part of the biosphere we interfere with its ability to function, since it also runs on low entropy. The fact that such interferences are now much more noticeable than in the past indicates that low entropy is becoming increasingly scarce.[1349]

Irrigation and hydroelectricity

Another example of the negative biotic impact of the employment of machinery is its use in providing irrigation. Irrigation made possible thanks to the use of fossil fuels has constantly increased from the beginning of the industrial revolution up to the present.[1350] A total of some 80,000 km² were being irrigated in 1800;[1351] by 2000, thanks to the building of dams, some 2.75 million km² were being irrigated – an area about the size of Argentina – thus growing at approximately the same rate as world population. From 1950 to 1980, the amount of irrigated land in the world, for which 70 per cent of all diverted freshwater is used,[1352] expanded more than two times. After that, expansion slowed; and it presently appears to have plateaued – mainly due to salinisation and increasing water scarcity: another instance of diminishing returns.[1353]

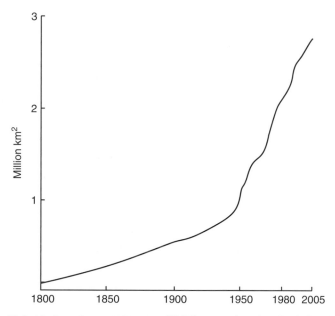

Global irrigated area 1800–2005.[1354] When produced under irrigation, 1 kg of corn requires 1400 litres of water; 1 kg of sugar beets 1900 litres; 1 kg of rice 4600 litres; and 1 kg of cotton 17,000 litres.[1355]

Irrigated fields accounted for 16 per cent of the world's cultivated area in 1990, and for some 30 per cent of total food production. By 1996 salinisation due to irrigation ruined land as fast as engineers could irrigate new areas.[1356]

The expansion of irrigation proved a bonanza not only for engineers, contractors and capitalists, but also for certain disease vectors, notably snails and mosquitoes.[1357] In the meantime, water tables are falling on every continent – in the southern Great Plains of the United States, the south-western United States, much of North Africa and the Near East, most of India, and almost everywhere in China that the land is flat.[1358]

Farmers are collectively overpumping groundwater sources by at least 160 km³ a year – the amount of water used to produce nearly one-tenth of the world's current grain supply. In the United States, several decades of heavy pumping have depleted the Ogallala aquifer stretching from Texas to South Dakota by 325 km³, a volume equal to the annual flow of 18 Colorado rivers.[1359]

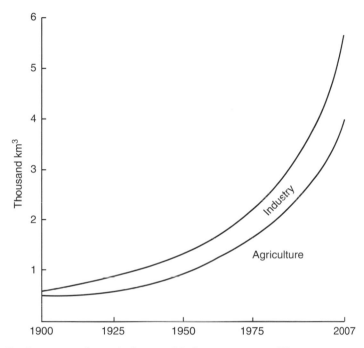

Freshwater use for agriculture and industry 1900–2007[1360]

Irrigation is one side of a coin the other side of which is *hydroelectricity*, the material of the coin itself being *the building of dams*. (The first hydroelectric power station, however, built in 1886 at Niagara Falls, made use of naturally falling water.) By 1950 there existed some 5000 large dams in the world; in the year 2000 there were 40,000. In 2006 hydroelectricity constituted two to three per cent of the world's total commercial energy and 20 per cent of its electricity, with fossil-fuel-generated electricity accounting for more than 50 per cent. The overwhelming majority of electricity has always been generated by fossil-fuel-fired power stations – at first coal and then oil and natural gas. In 1950 electricity took up 10 per cent of the world's fossil-fuel production, by 2000 it was 40 per cent.[1361]

Pollution

From World War II to 1990, and presumably since then, the quantity of pollution emitted (like energy used, and, of course, *because* of energy used) has increased faster than both population and GNP,[1362] an effect of which is that almost every aspect of the environment is worse today than it was 50 years ago.[1363] This of course means that constantly increasing quantities of non-biodegradable material have been accumulating in the air, water and soil. Particular pollution-related phenomena of a global nature include global warming, the thinning of the ozone layer, agricultural pollution and the spreading of radioactive waste.

Global warming

The 'greenhouse effect' was hypothetically postulated already in 1896,[1364] and was recognised actually to be occurring, and was called by that name, already in the 1960s.[1365] The most common 'greenhouse gas' is carbon dioxide, the concentration of which in the atmosphere has been rising since 1850, being 295 parts per million (ppm) in 1870 and 380 ppm in 2005. The rise in 2005 was the largest annual increase ever recorded to that time.[1366] At present the atmospheric CO_2 concentration is 385 ppm, which is substantially more than at any other time in the last 800,000 years,[1367] this being the first time in 650,000 years that it has risen above 300 ppm.[1368] This increase in atmospheric CO_2 may be seen as being mainly the result of our burning of fossil fuels. R. P. Brennan suggests that a rise in global temperature of two degrees Celsius could make our planet warmer than it has been at any time during the past 100 million years, and perhaps warmer than it has ever been since life moved to the land some 400 million years ago.[1369]

The average global temperature in 2005 was 14.6 degrees Celsius, making it the warmest year ever recorded. The four warmest years recorded before 2005 were 1998, 2002, 2003 and 2004.[1370] The average global temperature has risen nearly one degree Celsius in the past century. More than half of that warming – a rise of 0.6 degrees – has occurred in the past 30 years.[1371]

From the polar regions to high mountain glaciers, the earth's ice cover is melting at a rapid rate. By one estimate, the mountain glaciers of the world lose at least 90 km³ of ice annually – as much water as all US homes, factories and farms use every four months.[1372] Greenland's polar glaciers alone lost nearly 230 km³ of ice in 2005,[1373] while there has been an estimated sea level rise of 10–25 cm over the past century.[1374] In September 2005 sea ice in the northern hemisphere was at its lowest levels in recorded history,[1375] and it is predicted that the Arctic ice cap may have disappeared already in the next few years.

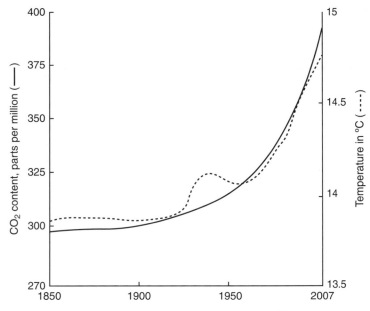

Atmospheric CO_2 content and temperature 1850–2007[1376]

The thinning of the ozone layer

The thinning of the layer of ozone (O_3) in the stratosphere, which filters out ultraviolet radiation from the sun, is mainly due to humans' use of chlorofluorocarbons (CFCs), a type of molecule. Since 1979, stratospheric ozone levels have fallen about five per cent per decade.[1377] In 1999, the ozone layer in the northern hemisphere was the thinnest recorded since measurements began in the 1970s.[1378] The hole in the ozone layer in the Antarctic grew from about one million km² in 1979 to over 20 million km² by the early 1990s. In 2006 it reached the record size of 26 million km² – larger than North America.[1379]

Agricultural pollution

In the mid-1970s the use of pesticides resulted in about 45,000 human poisonings each year in the United States, the killing of beneficial insects, earthworms and other invertebrates, as well as many million fish, birds and mammals, the reduction in activity of essential bacteria, fungi and algae, and the destruction of crops and trees. A conservative estimate is that the use of biocides was causing at least one billion dollars in environmental and social costs each year in the early 1980s.[1380]

In 2004, world exports of pesticides reached $15.9 billion, a new high. Pesticide use has risen dramatically worldwide since 1961, from 0.49 kg per hectare to 2 kg in 2004. According to WHO, some three million people a year suffer from severe pesticide poisoning. The chemicals in biocides are now found at unacceptable levels in the bodies of people worldwide, and can cause cancer, birth defects, and damage to the nervous system.[1381]

Nuclear pollution

The major US nuclear weapons site at Hanford in Washington State leaked almost two million litres of radioactive material from storage tanks between 1945 and 1973. Over half a ton of plutonium has been buried around the site. A full clean-up is impossible but the latest estimates of the cost of even a partial clean-up amount to somewhere between $100 billion and $1000 billion (one trillion) over a 75-year period. The USSR nuclear programme produced a much greater ecological disaster around the Mayak nuclear fuel reprocessing plant in the Ob river basin in Siberia. Liquid nuclear waste (including 26 tons of plutonium – 50 times the amount dumped at Hanford) was released into the Techna River, a tributary of the Ob. By the 1950s the highly radioactive material had reached Lake Karachay where it slowly accumulated. The heat generated by the waste, accompanied by a drought in the region, evaporated the lake and exposed the highly radioactive lake bed. It released radioactivity equivalent to 3000 Hiroshima-sized bombs, which was blown over 500,000 people. The lake bed was covered in concrete to stop the wind blowing more away. Standing on the lake shore for half an hour today is enough to give a lethal dose of radiation.[1382]

In 1957 the core of one of the reactors at *Windscale* (Sellafield) caught fire and there was a major release of radioactivity across much of Britain.[1383] In 1979 hydrogen explosions and meltdown severely damaged a large fraction of the core of the reactor at *Three Mile Island* in the US, and caused major plant contamination. Releases from the plant were not measurable as most of its off-site monitors were not working on the night of the accident;[1384] and none of the its six safety systems were operational.[1385] The total release of radiation from the *Chernobyl* accident in 1986, officially put at 90 million curies, was hundreds of times greater than that given off by the bombs at Hiroshima and Nagasaki, which continued to cause health problems for decades after detonation.[1386] By 2000, 32,000 people had died as a direct result of the Chernobyl explosion, and increasing numbers of people have been contracting fatal cancers. And there have been more major nuclear accidents after Chernobyl, e.g. in Japan.[1387] At Sellafield, since 1994 when used fuel from Europe and Japan started to be reprocessed, increasing quantities of radioactive waste have been discharged into the Irish Sea.[1388] And nuclear reactors, by their very nature, leak radiation, as does every uranium mine, mill tailing pile, and nuclear waste dump.[1389]

Extinctions

As the use of energy accelerates with the turning of the vicious circle, so does the rate at which species of plants and animals become extinct, every ecosystem in the world presently being in decline.[1390] In fact extinction rates are increasing more or less in step with energy use, due to its polluting effects. Thus where human-induced species extinctions were originally the direct or indirect result of hunting, they are now the result of habitat destruction through pollution and such things as soil erosion, deforestation, and replacement by introduced species.[1391]

Extinction 'background rates' suggest that on average over aeons one to three species disappear every year. Every 200 years or so one of the currently ca. 5000 mammal species ought to go extinct. Human exploration, settlement of new lands, forest clearance, and hunting quickened this pace. Since 1600, at least 484 animal species and 654 plant species have become extinct. Most extinctions have occurred on islands and in freshwater lakes and rivers, i.e. in isolated habitats. The 1900s extinction rate for mammals was about 40 times the background rate; for birds about 1000 times. Roughly one per cent of the birds and mammals extant in 1900 were extinct by 1995.[1392] It has been estimated that by the year 2000 some 30,000 species per year were becoming extinct, and that the earth had lost close to one million species due to human activity – about ten per cent of the world total.[1393] The present rate of extinctions for land and freshwater vertebrates is about 1000 times greater (give or take an order of magnitude) than in the fossil record.[1394]

This present wave of extinctions – which I would suggest began with *Homo sapiens'* first migration northward some 100,000 years ago – is one of the six greatest to have occurred during the 570-million-year history of complex life on earth,[1395] and the only one to be caused by a species of organism. The last of the previous five took place about 65 million years ago, at the end of Cretaceous period, wiping out three-quarters of all extant species, including all dinosaurs and most mammals.[1396] The apparent cause of all five[1397] of these earlier mass extinctions was in each case an asteroid hitting the earth, whereas the cause of the present wave is the activity of humans.

Population growth and checks; morals

In the period since 1950, population has increased at roughly 10,000 times the pace that prevailed before the invention of agriculture, and 50 to 100 times the pace that followed. The number of humans now exceeds the total population of the other 232 primate species combined.

From a biological/ecological point of view, the growth of the human population can be likened to that of lemming populations prior to their periodic migrations. Such a rate of growth is unprecedented in any large animal species. The global aggregate weight of humans is today ca. 350 million tons – well ahead of any other category except cattle. The human population is 'swarming;'[1398] we are behaving like an r-selected species.

The greatest proportion of today's population growth occurs in the Third World – that part of the world which undoubtedly had the lowest population growth prior to its contact with Europeans, and whose change to having the highest can be seen as a result of the influence of European culture.[1399]

Just as the money economy subverted many of the values of simpler societies, so did population growth, as mentioned in Chapter 4. As expressed by Baschetti and discussed in that chapter, the excessive increase in the size of human groups, by diluting the socially harmful consequences of immoral actions among millions of untouched and distant individuals, has allowed immorality to become less easily recognisable than it was in small primitive groups, in which all the members knew each other and promptly recognised the actions that could menace the survival of their communities.[1400] It is not surprising that one's incentive to act morally, moulded as it is to ensure the survival of the small-scale societies in which humans lived for most of our prehistory, began to fade when the size of human societies increased enormously after the horticultural revolution.

Apart from anonymity, the increasing availability of the direct benefits of the huge energy surplus to the middle class, combined with other factors, has led to a change in morals in the world, particularly in the industrialised countries. Where in earlier cultures population growth was curtailed by such customs as the requirement of dowries, and marriages only after a certain age, today extramarital relations and children resulting from such relations are hardly frowned upon. As expressed by Mumford, this can be seen as due to the lack of any effective taboos[1401] – taboos which, were it not for the great surplus, might still be in place.

Migration

Today, on average, each person has 0.02 km² in which to live,[1402] the density of humans having increased a thousandfold since hunter-gatherer times. With increasing population and use of technology there is a push towards centralisation. In the agrarian era centralisation favoured both defence (castles) and attack (armies), though the agrarian mode of production nevertheless required a large rural population. This trend increased with the rise of capitalism. Centralisation favours production; it is when centralised that capital finds its highest returns.[1403] The centralisation of production requires a centralisation of labour, so there is mass migration to the cities where there are jobs. When sufficiently cheaper employment can be obtained in the Third World, capital finds higher returns by centralising there. All this is part of the turning of the vicious circle.

The number of people who fled from their homelands due to armed conflict or human-rights violations doubled between 1985 and 1993. As of mid-2007 there were more than two million Iraqi refugees as a result of the American invasion of 2003. And the forced movement of people in Third World countries has been further increased by the large-scale projects financed by the industrialised countries. For example, the total number of people displaced by large dams in India and China alone has been estimated to have been between 26 and 58 million during the years 1950 to 1990.[1404]

Power begets more power: capitalism

Those who have power, of whatever kind, are those in the best position to increase that power; this they constantly do until such time as this behaviour leads to the next social and/or technological revolution – starting another turn round the vicious circle – after which the process should repeat itself, so long as a surplus can be produced. And so we have a situation where, in the modern world, not only are the people with the most capital the most powerful, but they are also constantly increasing that power to the detriment of those who are weaker. As noted earlier, this process was already evident to More in the 1500s, though since the mid-1800s it has been hidden to some extent due to the huge energy surplus and the subsequent formation of a middle class.

That the having of capital is of great advantage in the accruing of more capital is attested to by Herman Daly and John B. Cobb. As they point out, competition in business involves winning and losing. Last year's winners find it easier to be this year's winners; winners tend to continue winning and losers disappear. Over time many firms become few firms, competition is eroded, and monopoly power increases. And since competition is self-eliminating in this way it must constantly be re-established by the political move of trustbusting. Without trustbusting there would in the end be but one giant conglomerate, indistinguishable from a centrally planned economy. Thus the maintenance of competitive markets requires the abandonment of laissez-faire at least to the extent that government must assume the extra-systemic role of limiting monopoly and the excessive size of businesses.[1405]

From 1979 to 2000, the richest one per cent of Americans had real salary increases of 157 per cent; those in the bottom 20 per cent were making $100 *less* a year.[1406] In Sweden, during the 1990s the directors of companies on the stock market had nominal increases in salary that were more than ten times greater than those of industrial workers.[1407] Eighty per cent of the world lives in countries where economic disparity increased between 1990 and 2003.[1408] Income differences have been widening in almost every country since the energy crises of the 1970s.[1409] Other examples of essentially the same phenomenon: the profits of the world's richest 200 companies almost quadrupled between 1983 and 2004; their combined sales are now higher than the combined GNP of all but ten nations.[1410] In the 1950s, taxes from corporations constituted 27 per cent of the revenues of the United States government; in 2004 they made up less than 10 per cent.[1411]

The Third World

Technological development – including increasingly extensive communication – together with human territoriality, has resulted in the world's societies increasingly intermeshing with each other, so that we now live in a *globally* stratified society, where few people have much power and many have little or none, and where the powerful constantly become fewer while the weak multiply.

Where Malthus' considerations concerned Europe or individual countries in Europe, and the lowest classes he considered were European, given the nature of agrarianism, all such cultures had about the same standard of living, and so he could generalise his findings. But today this is not the case. In the global community it is the members of the Third World who are the lowest class. Thus the powerful societies of the north dominate the weak societies of the south thanks to their having control over technology – technology, by the way, on which they are dependent to a much greater extent than are the southerners.

As noted by Jack Parsons more than 30 years ago, population pressure is throughout the world forcing hundreds of millions of people off the land and out of a traditional occupation and way of life into a wretched existence of unemployment, poverty and desolation in urban shanty towns. The necessities of life simply do not exist for these people, and probably never will while the flood of numbers continues unabated.[1412]

On the vicious circle principle one can expect that with constantly increasing population there will always be people whose incomes and wealth are merely sufficient for them to work and produce children, and this only to the extent that they serve a function for the upper echelons. This view is similar to that of Malthus, who also assumed there always to be a lower class living on the verge of starvation. But it differs from Malthus' view in that it provides a reason for this state of affairs, namely territoriality coupled with the social stratification incumbent on the turning of the vicious circle; and it also differs in that it doesn't take this state of affairs as having to be the case.

The wealth and income gap constantly increasing

As the bourgeoisie become richer and relatively less numerous, those in the proletariat remain equally poor and become both absolutely and relatively more numerous.[1413] According to Noam Chomsky, this growing economic inequality is the result of the industrialised nations feeding off the wealth of the non- or less industrialised nations,[1414] and is expressed in terms of a growing difference in both income and wealth. Much of this is the result of corporate globalisation and the role it plays through such organisations as the World Trade Organisation, the International Monetary Fund (IMF) and the World Bank.[1415]

In 1990 the 20 per cent in the world with the lowest incomes were twice as poor as they were in 1960 vis-à-vis the 20 per cent with the highest.[1416] The per capita GNP gap between the richest and poorest nations more than doubled between 1960 and 1995 – to 37 times.[1417] The income of the richest ten per cent of the world's population is today roughly 117 times higher than that of the poorest ten per cent. This is a huge jump from the ratio in 1980, when it was about 79 times higher.[1418] By the end of the 20th century, the world's three richest individuals had a combined wealth greater than that of the world's poorest 48 countries.[1419]

Apart from China and India, most of the 'developing' world has been stagnant or shrinking economically, while most of the rich world has been growing.[1420] If the gap between the rich and the poor had stayed proportionally the same as it was in 1900, all human beings would be ten times 'wealthier' today. Yet the number in abject poverty today is as great as that of all humankind in 1900.[1421]

The United Nations estimates that a child born today in the United States, Britain or France will in its lifetime consume and pollute more than would 50 children in the poor nations. It also estimates that, in 1998, only $40 billion was needed to bring basic health, education, clean water and sanitation to the world's poorest citizens.[1422]

Third World aid, debt and trade

In 1970 the United Nations set the target of a mere 0.7 per cent of each industrialised country's GNP to be given as aid to Third World countries.

This target was to be met by 1975. Instead of reaching or surpassing this figure however, since 1970 the amount of aid has always been around 0.3 per cent. Recently 0.7 per cent was again set as the goal, this time to be reached by 2015 – 40 years later.

In 1997 total aid from the rich countries was $47.5 billion. Seventy-two per cent of United States bilateral aid commitments during 1999 were tied to the purchase of American goods and services.[1423] And more than half is today being spent in middle income countries in the Near East.[1424] As regards Africa, in the mid-1980s more than half of the seven to eight billion dollars spent yearly by development donors went to finance 80,000 expatriates working for public agencies under official aid programmes.[1425]

In the period from 1966 to 1978, American transnational corporations invested $11 billion in 'underdeveloped' countries, while the return flow to the United States on this investment was $56 billion. In 1976 the Third World exported to the rich countries 20 per cent of the food traded in the world, while importing 12 per cent from them. Furthermore, the food exported to the rich countries was and is of higher quality than that imported. Third World imports are mostly of grain, but their exports include peanuts, vegetables, fruit and meat, and many luxury goods such as coffee, strawberries and flowers.[1426] As summarised by Ponting: "One part of the world can now be dubbed 'affluent,' while the great majority of the world's population still live, as they always have done in the past, in conditions of absolute poverty."

The taxpayers of the United States, Europe and Japan together spend $350 billion each year on domestic agricultural subsidies (seven times as much as *global* aid to poor countries), creating gluts that lower commodity prices and erode the standard of living of the world's poorest people.[1427] In 1998 the average African's monthly expenditure was only $14, and, in the early 2000s, 52 per cent of the people in sub-Saharan Africa lived on less than one dollar a day.[1428] The number of people in the world who have to live on less than one dollar a day rose from 1.2 billion in 1987 to 1.5 billion in 2000.[1429] Today half of the people in the world live on two dollars a day or less, and spend up to 70 per cent of their income on food.[1430]

Biplab Dasgupta describes how banks pushed loans at the Third World at the end of the 1960s: "[T]he international banks were now flush with funds and were willing to lend to anybody. Between 1970 and 1980 the amount borrowed multiplied by nine." The banks were competing with each other to lend, so "they threw the usual conservative banking norms to the wind."[1431] But since then, according to Peter Self, "[t]he weapon of debt obligations has been wielded skilfully to keep [Third World debtors] tied to the global economic system."

During the colonial period, Third World countries' main source of income in the market economy consisted in payment received for resources exported to the industrialised countries, by far the largest proportion of this income going to Western companies operating in the Third World. Today, the situation is the same. Until the 1930s, the acquisition of minerals accounted for two-thirds of all European investment in Africa; and African mineral exports rose sevenfold between 1897 and 1935, at which time they made up half of all African exports. Today minerals still provide over 90 per cent of the exports of both Zambia and Mauritania, such countries becoming increasingly important in world mineral provision.[1432] Each major industrial region looks to a corresponding part of the Third World for most of its minerals. Many of the best reserves now lie in developing countries, since industrial nations have a much longer history of mining.[1433] Between 1913 and 1970 the share of the world's iron mined in the developing world rose from three per cent to 39 per cent and the amount of bauxite from less than 0.5 per cent to over 60 per cent.[1434] The achievement of political independence did not bring economic independence: the structure of the world economy had already been established.[1435]

With the removal of European political sovereignty the people in the Third World weren't left alone to return to their tribal ways, but were boxed into artificial states, in which different tribes were to cooperate. One cultural effect of this was to prevent the strengthening of tribal customs, including population checks. The only change with the 'liberation' of these states was that particular natives – most often dictators – were in the position to acquire some of this income, while the majority still went to Western corporations. With liberation, the idea essentially forced on Third World governments by Western capitalist interests was that through borrowing Western money their countries would all be able to *increase* their export of resources and goods (note: *not* be able to create manufacturing industries) – which was to provide natives with paid employment – the profits from which would allow the repayment of the loans with ease. But this did not happen, as probably few expected it would; and no one asks why it did not happen, or how to remedy the situation. What did happen was that at least as many natives went unemployed as before, while Western capitalists made greater profits than otherwise. One form these profits took was through the World Bank and the IMF – capitalist organisations dealing only in debt (money) – driving in the interest on their loans. This is a source from which they are still reaping profits today, while at the same time helping keep Third World people poor. Another form is through the industrial cartel's seeing to it that the prices of resources have always been inordinately low. This combination means Third World countries always having to catch up, providing the West with their resources as quickly as possible to increase their incomes so as to pay their debts. And after more than 50 years of failure as a means of improving the lives of those living in the Third World, given the power-seeking nature of the primate male it is not surprising that the borrowing of Western money is still being pushed at them as a need they have.

The total debt of the 60 poorest countries in the world in 1970 was ca. $15 billion; in 2002 it was ca. $510 billion.[1436] In 1990 African governments were $272 billion in debt, which approximately equalled the total wealth of the continent, and was about three times the value of its exports for one year.[1437] These governments have to spend two-fifths of their revenues to service that essentially unpayable debt. As a result, they are forced to divert scarce resources away from spending on health, education, environmental protection and other vital social services. For the 46 countries of sub-Saharan Africa, foreign debt service in 1996 was, at $15 billion, four times their combined spending on health and education. Between 1982 and 1986 the countries of Latin America paid $26 billion in debt interest annually which, after taking account of aid, meant a net transfer to the industrialised world of $21 billion each year,[1438] in keeping with Chomsky's assessment. In each of 1995, 1997 and 1998 the IMF received from severely indebted low-income countries as a whole one billion dollars more in debt and interest payments than it loaned them.[1439] The 'developing' countries as a whole repaid (in capital and interest) a total of $350 billion in 1999, i.e. seven times more than the $50 billion in Public Development Aid they received that year. (Here it may be kept in mind that the little aid the Third World countries receive comes from taxpayers, while the interest they repay on their debts goes to capitalists.) In 2005 they repaid $245 billion. And today many poor countries pay more to the World Bank and IMF each year than they borrow.

The UN Human Development Report estimated that the payment of this amount of debt each year caused seven million deaths annually in Africa alone.[1440] According to Carol Welch, by the end of the 1990s the World Bank and the IMF had been responsible for tremendous economic and social damage wrought on Third World economies for over two decades.[1441] The United Nations Children's Emergency Fund (UNICEF) estimates that 500,000 children die in the Third World each year because of the debt crisis and the cruel and counter-productive policies imposed by the IMF.

The reforms that rich countries forced on Africa had the result that imports increased massively while exports went up only slightly. The growth in exports only partially compensated African producers for the loss of local markets, and they were left worse off.[1442]

In the late 1980s, the per capita resource consumption of the 25 per cent of the world's people who lived in the industrialised countries was about 17 times that of the world's poorest 50 per cent.[1443] In 1990 about seven times as much energy per capita was used by the one billion or so people in the north as compared to the more than four billion in the south.

It is the powerful nations that formulate trade and aid policies. Their representatives' concern is not with poverty, but with maintaining sources of cheap labour and cheaper goods to sell to populations back home, increasing their own wealth, and maintaining power over others in various ways.[1444]

Malnutrition and mortality

At the beginning of the 1990s, about 13 million people died each year of malnutrition or starvation. Deaths due to malnutrition decreased from some 15 million annually in 1972 to about nine million in 2002.[1445] The number of malnourished people dropped from 900 million in 1972 to 800 million in 2000,[1446] but rose to 1.2 billion in 2006, malnutrition being chronic for some 850 million people. Africa is the only continent where food production per capita declined after 1960. At the same time 1.2 billion people in the world were over-nourished and overweight in 2006, most of them suffering from excessive caloric intake and exercise deprivation.[1447] (Cf. our discussion of hunter-gatherers and there being an *optimal* level of energy expenditure, as determined by our species' karyotype.)

The average lifespan at birth in the world in 1900 was around 40 years; in 1950 it was 48; in 1973, 58, and in 2007, 67. The average life expectancy in Swaziland in 2005 was 33.[1448] In the early 1990s, each year half a million children died of malaria in sub-Saharan Africa.[1449]

In 2003, UN demographers announced that the HIV/AIDS epidemic, the existence of the disease being discovered in 1981, had reduced life expectancy for the 700 million people of sub-Saharan Africa from 62 to 46 years over that time. For the first time in the modern era, the rise in life expectancy has been reversed for a large segment of humanity.[1450] This precipitous drop was primarily the result of governments' failure to check the spread of the HIV virus. While industrial countries held HIV infection rates among adults to under one per cent, in some African countries they climbed to over 30 per cent.[1451] By 1990 an estimated ten million people were infected. By the end of 2004 the number had climbed to 78 million. Of these, 38 million had died and 39 million were living with the disease.[1452] In 2005 there were approximately five million new HIV infections, 95 per cent of those infected living in the Third World.[1453] Of the ca. three million people who died of AIDS-related illnesses in 2005 nearly 600,000 were children under the age of 15.

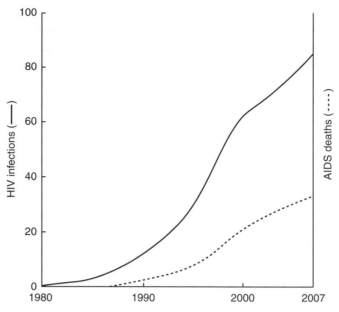

Estimated cumulative HIV infections and AIDS deaths in millions 1980–2007[1454]

Today the variation in life expectancy between different countries is wider than at any time in history, ranging from a low of 33 in Swaziland, 34 in Zimbabwe, and 37 in Botswana to a high of 81 in Iceland and 82 in Japan. Not surprisingly, life expectancy usually correlates with income levels except where the distribution of income is heavily skewed.[1455]

Third World military spending, etc.

Apart from debt repayment, the per capita incomes of people living in the Third World are being dragged down by the tremendous and increasing expenditure on armaments in these areas. As pointed out by Al Gore, fully half of all Third World debt has been accumulated not to start new enterprises, but to purchase weapons with which to wage war among themselves.[1456] These armaments are produced in industrialised countries and their sale of course benefits the capitalists of the global military-industrial complex who produce them.

For example, between 1972 and 1982 Latin American arms imports rose by 13 per cent a year, and African by 18 per cent a year.[1457] And from 1960 to 1992, military spending in the Third World was *20 times* greater than the total given as development assistance.[1458] In 1986 the Third World spent more than four times as much on weaponry and upkeep of military forces as it did on health care – $150 billion compared to $38 billion.[1459] Between 1973 and 1978 the United States gave to the ten nations with the worst repression and human-rights records one billion dollars in military aid and sold them an additional $18 billion worth of military equipment.[1460] Since World War II the United States has consistently provided military support to *capitalist* countries, irrespective of whether they are democracies, and has in fact supported the overthrow of democracies (e.g. in Chile, Guatemala and other Third World countries) when they showed signs of moving away from capitalism. As P. A. Baran says,

> social revolutions, regardless of where and how they unfold, meet with the implacable hostility of the ruling class – the most powerful citadel of reaction in the world today. No regime is too corrupt, no government too criminally negligent of the vital interests of its people, no dictatorship too retrograde and cruel to be denied the economic, military, and moral support of the leading power of the 'free world' – as long as it proves its allegiance to the anti-socialist Holy Alliance.[1461]

Third World countries spend an estimated $22 billion each year on arms, an annual sum the Control Arms Campaign says would be enough to end extreme poverty by 2015.[1462] According to a report from the UN Office of the Special Representative of the Secretary General for Children and Armed Conflict, the permanent members of the UN Security Council – France, China, Russia, Great Britain and the United States – together account for 88 per cent of the total export of conventional weapons in the world.[1463]

Civil wars and sporadic violence in failing states produced enormous refugee populations throughout Africa during the mid-1990s.[1464] And, in the early 2000s, armed conflict, mass displacement of people, torture, ill-treatment and exemption from punishment continue to be rife in the African region.[1465] As of 2005, approximately 30 million illegal weapons were in circulation in Africa south of the Sahara. Nasibu Bilali, of the Action Network on Small Arms and Light Weapons, believes that the conflict in his country (Democratic Republic of the Congo) would not have been able to continue had it not been for the spread of weapons, illegal and legal. According to his organisation there are about 500,000 illegal weapons in the country.[1466]

Global military spending and war

With the increase in the importance of commerce in the world one can expect an increase in violence, for there is a profit to be made in the selling of weapons, and the bought weapons will most often be used. Thus, for example, the powerful capitalists in the military-industrial complex favour war, which is in keeping with the fact that there have been more than 160 wars and armed conflicts in the world since 1945.[1467] And of course the destructive nature of war means a constant demand for weapons – ammunition disappears as soon as it is used, as does whatever it hits.

In 1913, US defence expenditure was about 0.7 per cent of its total expenditure. By 1940 it was two per cent. By 1950 it had grown to 7.5 per cent, and by 1970 it was over ten per cent.[1468] Even without *using* their weapons, the United States and the Soviet Union spent over $10 trillion ($10,000 billion) on the Cold War, enough money (other things being equal) to replace the entire infrastructure of the world, every school, hospital, roadway, building and farm.[1469] In 1986 global military expenditure was one trillion dollars,[1470] and the cost this meant for the Soviet Union was probably the main reason for its dissolution. By 1998, due to the end of the Cold War, global military spending had dropped to about $750 billion per year. But after the 11 September attack in the United States, expenditures began to rise markedly, reaching $1.12 trillion in 2005, more than at the end of the Cold War, and beginning to approach the levels at the end of World War II. From 1996 to 2005 there was a 34 per cent increase in arms sales.[1471] In 2004 high-income countries, containing only 16 per cent of the world's people, accounted for 75 per cent of these expenditures.[1472] In a world where billions of people struggle to survive on one to two dollars a day, governments spend on average $166 per person on weapons and soldiers. At a time when endemic poverty, health epidemics, climate change and mass unemployment cry out for attention, the continued growth in military budgets reflects a troubling set of priorities and a failure to address the underlying reasons for much of the world's instability.[1473]

As noted by Quincy Wright, the particularly rapid population growth of the 20th century has meant increasing population pressure, particularly in the less-developed two-thirds of the world. With this comes a rise in tensions, while at the same time the increase in population augments contacts between people of different cultural and political allegiances. This in turn increases the number of opportunities for friction between nations, and the result is a less stable world. At the same time armies have also grown in size during the modern period both absolutely and in proportion to the population.[1474]

We see similar genocidal behaviour to that during the horticultural period, though on a larger scale, in modern times. Major genocides in the first half of the 1900s include Stalin's killing 13 million of his countrymen during the 'purges' (1934–1939), Hitler's killing 12 million civilians in concentration camps

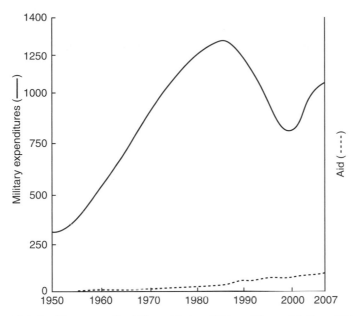

Global military spending[1475] vs. aid given[1476] in millions of dollars 1950–2007

and elsewhere during World War II, and the killing of five million civilians by Hideki Tojo of Japan between 1941 and 1944. Since 1950 there have been many episodes of genocide, including Mao Zedong's killing of some 49,000,000 during the 'great leap forward' (1958–1961) and the 'cultural revolution' (1966–1976), and the killing of 1,700,000 by Pol Pot of Cambodia (1975–1979).[1477] To these may be added the most recent genocides in Rwanda in 1994 and in Darfur, Sudan, which is taking place at present, each of which has claimed hundreds of thousands of lives.

As suggested by Diamond, one might expect on first thought that no horror could grip public attention as much as the intentional, collective and savage killing of many people. But in reality, genocides rarely grip the public's attention in countries where they are not happening, and even more rarely are they interrupted by foreign intervention. "Who paid much attention to the slaughter of Zanzibar's Arabs in 1964 or of Paraguay's Aché Indians in the 1970s?"

Compare our non-response to these and all the other genocides of recent decades with our strong reaction to that perpetrated by the Nazis on the Jews. In this case the victims were whites, with whom other whites identify; the perpetrators were our war enemies, whom we were encouraged to hate as evil;

and there are articulate survivors in the United States who expend much effort forcing us to remember. Thus, it takes a rather special constellation of circumstances to get third parties to focus on genocide.[1478] Here we have an expression of the social instincts, which in primitive people are manifest as seeing your tribal mates as friends and people from other tribes as potential enemies.

As regards population pressure, Paul Leyhausen suggests that in situations of crowding the privacies of territory and the courtesies of relative hierarchy disappear, and absolute hierarchy prevails in its most unrelieved form, as a natural consequence of shortage of space (cf. the dominance behaviour of mice and moles in situations of crowding, taken up in Chapter 1). As the difference between ranks becomes steeper, subordinates eventually become desperate, and violence becomes endemic, with massive social reorientation and the persecution of outcasts. Among the results of crowding, the emergence of dictatorship is one of the more striking.[1479]

Since agrarian times, pushed by the VCP, the destructive capacity of weapons has constantly increased, and channels of communication have constantly been extended. This has had the effect of steadily increasing the power of the politically strong and lessening that of the weak, at the same time as it increases the complexity and reduces the security of society as a whole. And the war drain on the natural wealth of the world has been terrific, taking a heavy toll on forests, oil and coal, as well as on iron and other metals, and adding to the already heavy depletion of soil resources.[1480] This development, involving weapon and communication improvement, globalisation, centralisation and population growth, is a manifestation of the VCP, according to which a surplus of the vital resources of food and shelter should lead to population growth, which in turn should give rise to increased complexity and violence.

Economic growth

On the VCP, increased energy use goes hand in hand with technological development. This interaction produces economic growth,[1481] which supports population growth and results in increasing pollution.

What we see in the world today is that government is constantly losing ground to business, particularly after the fall of the Iron Curtain, with, e.g. the privatisation of public utilities in such 'socialistic' Western countries as Great Britain and Sweden. Capitalism has not only become virtually worldwide, but is strengthening its position all the time. In 1991, the ten largest businesses in the world had collective revenues of $801 billion, a greater turnover than the smallest one hundred countries. The five hundred largest companies in the

Increases (and decreases) in various items during the 20th century[1482]

Item	Increased times
World population	4
Urban proportion of world population	3
World urban population	13
World economy	14
Industrial output	40
Energy use	16
Coal production	7
Air pollution	ca. 5
Carbon dioxide emissions	17
Sulphur dioxide emissions	13
Lead emissions to the atmosphere	ca. 8
Water use	9
Marine fish catch	4
Pig population	9
Horse population	1.1
Blue whale population in the southern oceans	0.0025 (99.75% decrease)
Bird and mammal species	0.99 (1% decrease)
Irrigated area	0.8 (20% decrease)
Cropland	2

world in 1992 controlled 25 per cent of the world's gross output (GWP) while employing 0.05 of one per cent of the world's population.[1483] And in all financial sectors a tendency to monopolisation is the norm. The economic power leading to monopolisation has been growing constantly stronger vis-à-vis the political power required to stop it. Just to take one example, in 1908, 23 years after the invention of the car, there were over 250 American car manufacturers;[1484] in 1919 there were 108. Ten years later the number was 44. By the end of the 1950s it was eight, and by 2004 it was two and a half.[1485] Furthermore, the taxpayer has footed much of the bill for this monopolisation process; as Kapp points out: "the development of all privately operated land, water and air transportation would have been impossible without active intervention and public aid by government authorities."

Disease

Just as economic activity has been constantly increasing, so has the movement of people over large distances, which has been accompanied by an increase in the spread of communicable diseases such as tuberculosis, polio, the Ebola virus, HIV-AIDS, transmissible spongiform encephalopathy (TSE), foot-and-mouth disease, SARS, avian influenza, scabies, haemorrhagic dengue fever, lassa fever and the Marburg virus. Disruptions of tropical ecology have loosed various of these infections on humankind. Haemorrhagic dengue fever first appeared in South-east Asia in the 1940s; the extremely lethal Marburg virus turned up in Central Africa in 1967; lassa fever was first identified in Nigeria in 1969; and Ebola was recorded first in Zaire (Democratic Republic of the Congo) in 1976. In 1998 AIDS ranked fifth among causes of death in the world.[1486]

However, from a social point of view, of all the infectious diseases, malaria has to date caused the greatest harm to the greatest number. From an ecological point of view, on the other hand, malaria may be seen more clinically as having been the greatest external check to human population growth until the end of World War II.[1487] Keep in mind that there has also always been infant mortality due to *internal* checks, such as infanticide. During the 1960s the number of malaria victims in the world was drastically reduced thanks to the use of DDT, which, however, proved to be poisonous when ingested by humans and other complex life forms, and was banned in the United States in 1972 and in Britain in 1984. But even where DDT was still being used malaria resurged. India in 1977 had roughly 60 times more malaria cases than in 1960. The disease is now staging a strong global comeback, with mortality rates increasing by 13 per cent between 1970 and 1999.[1488] By 1990, there were 24 types of mosquito resistant to DDT.[1489] Today more than 500 million cases of malaria claim more than one million lives each year, 89 per cent of them in Africa.[1490]

In many ways, the global programme of the 1960s has made the modern malaria problem far worse. It introduced the dynamics of insecticide and drug resistance, it encouraged some vectors to change their behaviour, it virtually eliminated malariology as a speciality, it created a void in interest and funding for malaria control that is only now turning round, and it engendered the idea of DDT as a first resort against mosquitoes.[1491]

According to WHO in a report published in 1996, the world is on the brink of an infectious disease crisis.[1492] Others who have investigated the situation would concur.[1493] For example, a totally resistant form of tuberculosis has recently been discovered.[1494] In northern Nigeria there were about 225 polio cases in 2005; but already in the first half of 2006, 467 new cases were noted, placing the global plan to completely extirpate the disease in jeopardy.[1495]

The 1950s–1960s peak and the subsequent lowering of the quality of life of the middle class

The presence of a surplus is to the material benefit of those who have access to it. In the case of the great energy surplus provided by fossil fuels, and the surplus of metals this energy has made it possible to extract, this benefit has accrued to the populations of the industrialised nations – particularly to the most wealthy among them. The benefits to individuals and their families in the West that this surplus underlay peaked during the 1950s and 1960s. As Terry suggests, there was a trickle down in absolute though not relative terms until the early 1970s.[1496] Due to diminishing returns generally, subsequent high levels of growth have instead been accompanied by a worsening in the quality of people's lives even in industrialised countries.

As suggested by C. G. Darwin, people in the industrialised countries were living in a golden age at the beginning of the 1950s, which could show itself to be the greatest golden age of all time.[1497] In the United States, from the end of World War II to the early 1970s every year brought higher wages, more leisure and a greater stock of material possessions.[1498] Thus, as Trainer says, it is important not to regard the achievements of the 1950–1970 period as normal. In many ways this was the most remarkable and atypical period in the entire history of the world capitalist system. These achievements are more sensibly taken as marking the upper limits to realistic expectations regarding the potential of conventional development under extraordinarily favourable conditions.[1499]

But all was not well during this period all the same. Rising standards of living and education in all the richer countries were accompanied by a rising incidence of crimes of violence, notably among the young. Crime in the West has grown without interruption since the end of World War II.[1500] In north-western Europe and the United States, the central problem during the 1950s and 1960s was not perceived as increasing violence however, but of supplying the growing population with adequate housing (rather than stopping it from growing so that such housing would not be required and violence would decrease – there's no money to be made in that). It is above all through the density and quality of housing (breeding sites) that *crowding*, and the increase in conflict it gives rise to, express themselves in civilised societies.[1501] Perhaps in part because of this, in 1970 Americans considered themselves to be less happy than they did in 1957.[1502]

Most if not all social problems have been getting worse since 1950: crime, vandalism, homelessness, alcoholism, stress diseases, depression, analgesic consumption, the hard drug problem and child abuse.[1503] According to WHO, the worldwide annual suicide rate among men rose between 1950 and 2000 from approximately 17 persons per 100,000 to about 30.[1504] From 1973 to the early 1990s, despite a 74 per cent rise in GWP, our standard of living did not improve, real wages did not rise, and, for the first time since the industrial revolution, our work week began getting longer, not shorter.[1505] And people who previously did not have to earn a wage – most notably housewives – were increasingly pushed into the labour market. Since the 1970s, the actual social effect of economic growth has been to diminish welfare.

Turning to a particular country: during the 1980s Australia averaged 3.2 per cent economic growth per annum, over which time virtually all economic problems in the country became more serious. Though Australia's GNP per capita in real terms more than doubled between 1950 and the late 1990s, and increased by one-third in the 1980s, various indices show that in that decade social inequality increased by 25 per cent, poverty increased by 70 per cent, the unemployment rate almost doubled, rural debt multiplied by nine, and foreign debt multiplied by ten! The picture would be worse if it were possible to add numerical indices for environmental quality and quality of life. For instance, the Australian rate of youth suicide in 1977 had doubled in a generation, and problems of stress-anxiety and depression were reported to be ten times as prevalent as they had been a generation earlier.[1506]

Another example: in Britain between 1955 and 1988, during which the per capita national income doubled, almost every social indicator deteriorated. Six times as many marriages broke up in 1988 as in 1955, and 25 per cent of all children were born out of wedlock, compared with five per cent in 1955. Advertising expenditures in real terms rose more than fivefold between 1950 and 1986. Between 1955 and 1988 the per capita incidence of recorded crime increased eightfold, while chronic disease increased and unemployment soared. Between 1960 and 1990 the per capita suicide rate in Britain doubled. And children stopped getting taller after the 1970s. As regards transportation, there were more than five times as many road vehicles in 1988 as in 1955; and between 1952 and 1988 car travel increased tenfold while bicycle use dropped by 80 per cent and bus travel by 50 per cent. Almost all of the extra resources the growth had created were being used to keep the system functioning in an increasingly inefficient way. During this period economic growth in fact made life considerably worse for people in Britain.[1507] Thus we see that even in situations where there is high economic growth, its existence does not imply social improvement.

As regards the United States, a congressional study from the early 1990s showed that at that time adults in 80 per cent of two-parent American families with children worked more hours in 1989 than in 1979, while at the same time their incomes didn't rise commensurately. Hourly wages were lower in 1988 than in any other year after 1966, after adjusting for inflation. In 1950, home ownership in the United States took only 14 per cent of a typical wage-earner's income, compared with 44 per cent in 1990, and a smaller proportion of Americans owned their own homes in 1990 than in 1950.[1508] Between 1979 and 1989 the median weekly wage of workers dropped from $409 to $399; and real wages have declined over the past 25 years.[1509] Where in our parents' generation a household could be supported by only one working person, now two are required; and not only is the length of the per capita working day increasing, so is the stress involved in work.

Furthermore, the disparity between the rich and the poor in the United States in 1990, for example, was increasing,[1510] and it is undoubtedly still doing so today. Between 1977 and 1989 the average after-tax incomes of American families on the bottom fifth of the income ladder fell some nine per cent, the next fifth fell 6.5 per cent, and the middle fifth about 4.5 per cent. Only the top fifth was spared. In fact the incomes of the higher reaches of the top fifth soared, with those of the top one per cent doubling.[1511] Between 1978 and 1995 the mean income of the richest five per cent in the United States rose to 24 times that of the poorest 20 per cent, a ratio unparalleled in the industrial world.[1512] The ratio between the salary of a CEO and that of a shop-floor worker went from 39:1 in the late 1970s to about 1000:1 today.[1513]

While productivity in the United States rose about one per cent per year between 1980 and 1995, the national debt increased at an annual rate of about 12.5 per cent – a rate far higher than inflation, population increase or real economic output. Americans' frantic push to increase productivity is not improving their economic lot.[1514]

Americans could afford much less per hour worked in 1995 than they could 20 years earlier – in spite of their producing more per unit time. Housing, food, health care and transportation were relatively much more expensive in 1995 than they were in 1975. In fact, the middle class in the United States is disappearing, and a two-tiered society is taking its place. In the future, if there is a future, the common man in Europe and America will be living on the verge of existence, just as he did before the industrial revolution.

Thus, despite a constantly accelerating use of energy and a constant increase in the size of the world economy, the period since the 1960s has been one of gradual but accelerating decline. What had been a steady increase in average American real wages up to 1973 afterwards became a steady decrease for everyone without a postgraduate education.[1515] As expressed by Daly and Cobb, and in keeping with the view of Wilkinson, this decline in wages is itself a sign of rising real resource costs, particularly that of energy.[1516] The reason for wages falling despite the growing economy is, according to Douthwaite, that the production system has simply been using more energy to achieve less.[1517] The industrial world is meeting diminishing returns, *before* its energy surplus has peaked.

7

... and too dumb to change

The fundamental problem as regards the continuing existence of the human species is that, while we are 'smarter' than other species in our ability to develop technology, we, like them, follow the reaction, pioneering and overshoot principles when it comes to dealing with situations of sudden, continuous or great surplus. In keeping with this, and also like other animals, we are not karyotypically built so as to care about coming generations, other than those with which we have direct contact. As Georgescu-Roegen says, the (rat) race of economic development that is the hallmark of modern civilisation leaves no doubt about humans' lack of foresight. Even if made aware of the entropic problem of the human species, humankind would not be willing to give up its present luxuries in order to ease the life of future generations.[1518] When problems arise we turn to the nearest solution to hand, and do not take into account the long-term consequences of our actions. In this regard we act irrationally. We humans, in whatever situation, will gladly use irreplaceable resources to produce a technological fix if it fills an immediate need. The longest we are prepared to put off gratification is perhaps a year, where in certain societies, though people may be dying of starvation, seeds are saved for the next year's planting.

From the point of view of evolution, to react spontaneously to one's immediate environment has been the best policy for all species up to now. But now, in our case, in acting spontaneously we are not only worsening the situation for our own species, but for all other complex species as well.

To react directly to our surroundings is how we *instinctively* react; it is built into our karyotype, just as it is built into the karyotypes of other species. And if it were at all possible to overcome this predilection, it would seem that we, *as a species*, would have to *act* on the basis of that very intelligence that has landed us in this situation in the first place. Overcoming our instincts with our intelligence would be a difficult task to say the least however, as is evident from the fact that we haven't made the least effort to do so despite being well aware of the problem for many years.

To use our intelligence in this way would require our manifesting, as a species, *social instincts* which could override our survival and sexual instincts. But for such instincts to work, first they must be appropriately manifest, and with tremendous force due to the extent of our species' disequilibrium; second, in order to be effective they must be manifest globally; and third, their being manifest globally presupposes world stability and the creation of population-checking *traditions*. As regards this last point, for modern hunter-gatherers, the constancy of the life-situation, including the lack of technological development, made it possible to implement adequate population checks through the creation of traditions so as to keep the size of their populations oscillating about a mean. But the present human ecological situation is unstable due to constant technological development, and will continue to be so as long as technological development continues. (Cf. Wynne-Edwards' comments regarding animals in unstable or transitory environments, cited in Chapter 1.) Change is occurring at a faster rate than ever; and this change not only prevents the creation of new traditions, but means the disappearance of those that are already established. Among other things, this change has meant that the various environmental triggers (epideictic phenomena) for the appropriate social instincts are now lacking, while at the same time our own genetic domestication has made us disinclined to manifest such instincts, or inclined to manifest them in counter-productive ways.

That our survival as a species is in jeopardy, and that we must act with an eye to the long-term future, has been realised by educated people at least since the beginning of the 1970s. At that time we already knew of the greenhouse effect, as mentioned, and of acid rain.[1519] And people generally became aware of our dire situation with the publication and wide distribution of such works as Commoner's *The Closing Circle* in 1971, the Meadows team's *Limits to Growth*, and Edward Goldsmith and others' *Blueprint for Survival* – both in 1972, and Schumacher's *Small is Beautiful* in 1973, as well as by the holding of the United Nations Conference on the Human Environment in Stockholm in 1972. And the 1972 Peruvian anchoveta collapse coupled with the 1973 oil crisis should have driven the point home.

In *A Blueprint for Survival*, for example, Goldsmith and his co-authors claim that at that time, i.e. more than 35 years ago, humankind was faced with a total ecological crisis, and that with the chaos to come there would be social disruption and a great likelihood of war. They also claimed that governments at that time either refused to face the ecological facts or played down their seriousness.

"A measure of political reality is that government has yet to acknowledge the impending crisis." This is a situation that has not changed in the ensuing 35 years. And they draw the conclusion that we are on our way to extinction as a species.[1520]

Similar views were expressed around the same time and earlier by such authors as Boulding, Schumacher, Hans Palmstierna, Hardin, Commoner, Georgescu-Roegen, Forrester, Donella and Dennis Meadows, and Wilkinson. These are the wise people of what may be the last age of humanity, none of their ever so important warnings being refuted (or acted upon). Rather, they have since been corroborated in the works of Daly, Mishan, Orio Giarini and Henri Loubergé, William Catton, Hazel Henderson, Michael Redclift, Trainer, Lester Milbrath, Mary Clark, Ponting, Douthwaite, Diamond, Abernethy, Anthony McMichael, James O'Connor, J. W. Smith and Sieferle. As Dennis Meadows has recently said, "The message that current growth trends cannot be sustained is now reconfirmed every year by thousands of headlines, hundreds of conferences, and dozens of new scientific studies."

Furthermore, all computer simulations of humankind's development into the future since that time, including the original ones of *Limits to Growth*, show not only that the present system will decline, but that it will *crash*, and that the longer it continues the greater the crash will be. In terms of Schumacher's metaphor: *we're stampeding over a cliff*. So the fact that our situation is terribly threatening has been known to decision makers for more than 30 years, and this quite independently of an awareness of the operation of the vicious circle principle. What an understanding of the VCP adds is a realisation both of *how* we have come to this pass, as well as *why* we in fact have made no serious attempt to remedy the situation despite our being aware of it.

As P. R. and A. H. Ehrlich also noted even before the first Gulf War, the world might well come to be engaged in *nuclear* war over the oil resources in the Gulf area.[1521] The inclination to acquire (further) power, or the inclination of the powerful to act offensively rather than defensively when possible, inclines leaders to attempt to secure sources of energy rather than make their societies independent of such sources. No laws are enacted to make non-practical use of oil illegal – such use as one sees in motor sports, for example. Again, this is because the powers that be in today's world are economic, and for power-hungry or greedy capitalists increasing consumption means increasing profits.

According to the VCP the individual territorial instincts of the powerful override whatever other instincts they may have as support the well-being of the species, and it is they who determine the course taken. And, it seems to me, there's not much we can do about it. The revealing of the nature of the situation, such as is attempted in this book, is not going to make any noticeable difference.

Perspectives and worldviews

I believe that, from a socio-psychological point of view, there are three major factors that account for the failure on the part of the vast majority of educated people to admit what is happening, and none of them is that they are unaware of it. The first, and perhaps most important, is that it does not accord with the way they see the world (which is largely determined by short-term contingencies, in particular capitalists' desire to make as large a profit as possible); the second is that the facts themselves are unsavoury; and the third is that the effort that would be required to change them is gargantuan. Thus, though most educated people are aware that something is terribly amiss in the human situation, and that it bodes ill for our children, the negative view of the human condition as is implied by environmental research of the past 50 years or so is both psychologically and practically repressed.

That the present state of the world is not in keeping with today's dominant worldview can be considered more closely via an analysis in terms of the notion of *perspective*,[1522] a notion introduced already when I spoke of the new views in anthropology, archaeology and economics. A perspective, as I mean the term, is a way of *conceiving* of reality. Moreover, it is a way that, in many cases, can be compared with other ways (perspectives) so as to determine which of them is in fact more in keeping with how things really are (and/or how they should be, given that a particular end is desired).[1523] At the core of each perspective lie a person's presuppositions or beliefs concerning the nature of the reality the perspective is on – its *intended domain*. And when that domain is all-encompassing, the perspective constitutes a *worldview*.[1524]

Everyone has a worldview, at the core of which are his or her most fundamental beliefs concerning the nature of reality. Some of these beliefs are shared by everyone – e.g. that nature follows laws, that nothing happens without a cause, and that nothing comes from nothing – and some beliefs differ from culture to culture – e.g. that there exists a god or gods with powers enabling it or them to transgress the limits set by the other fundamental beliefs.

The increasingly dominant worldview of today is that of Western society, with its belief in the validity and importance of modern science, its emphasis on economic growth and technological development, and its view of humans as primarily socio-psychological entities, and only secondarily if at all as biological entities. As regards our relation to our ecological/biological situation, as expressed by Schumacher: "Modern man does not experience himself as a part of nature but as an outside force destined to dominate and conquer it."[1525] And as Commoner says: "It becomes clear, then, that we are concerned not with some fault in technology which is only coincident to its value, but with a failure that derives from its basic *success* in industrial and agricultural production."

Among notable worldviews differing from the Western, we must mention the non-materialistic worldview originating in the Far East. But the worldview of greatest interest here as a contrast to that of the Western is the *ecological*, of which, as mentioned in Chapter 2, the new views in anthropology, archaeology and economics may be seen as each being a part, as may the worldviews of people living in hunter-gatherer societies to a certain extent. In this book, the vicious circle principle is being advanced as the conceptual core of the ecological perspective or worldview. The fact that the ecological worldview can be given a scientific basis of course speaks in its favour as far as its adaptability to Western thinking is concerned.

From a psychological point of view, since worldviews emanate from one's fundamental beliefs they are very difficult to break out of, and people may even defend them as they defend their own lives.[1526] Thus, for example, for the vast majority of people educated along Western lines, be they ever so intelligent in a conventional sense, to take the theory developed in this book seriously, despite its scientific nature, would require their making such a great break with the socio-economic aspect of their worldview that it would be virtually impossible for them to do.

From a biological point of view a further difficulty is that the Western worldview is supported by human territoriality and the reaction and pioneering principles. I am reminded of our addiction to technological fixes by the end of Kurt Vonnegut's book, *Player Piano*, where, after the crash of technology, some idle hands find an orange-drink machine which doesn't work, and begin fixing it. It's in our blood to tinker, and we're exceedingly good at it.

And as regards the limits of the application of technology, as Daly points out, it is dangerous to assume that the content of new knowledge will abolish old limits faster than it discovers new ones.[1527] Examples mentioned by Daly and Cobb include nuclear energy, the applicability of which was reduced when we learned of the dangers of radioactivity, and asbestos, the use of which became illegal when its carcinogenic properties became known.[1528] We might also mention such 'backward' innovations as those of freons, PVC plastics, DDT, thalidomide, creosote and so on.

As Schumacher says, we are here concerned not only with *social* malaise but also, and most urgently, with a malaise of the *ecosystem* or *biosphere*, a malaise that threatens the very survival of the human race. To 'leave it to the experts' means to side with the people of the forward stampede. Today, the main content of politics is economics, and the main content of economics is technology. If politics cannot be left to the experts, neither can economics or technology.[1529]

In what follows I shall present some paradigmatic examples involving our inability to turn reason into action, often abetted by the predominance of the Western worldview, in the case of certain aspects of the vicious circle that threaten our existence as a species.

Planning

As Schumacher stressed already in 1968, if we are at all concerned about the continuing existence of humankind we must ask and answer such questions as what the appropriate size of a city is, and what the appropriate size of a country might be.[1530] And Palmstierna suggested already in 1967 that we must demand an economic system where the richer don't simply get richer and the poor die of starvation. The time for liberal laissez-faire, laissez-aller is over. There is no way to surmount the crisis if we continue to let everything follow its own path. Only a true internationalism with a strict planned conservation of the earth's resources can let us get through.

Development through the centuries has, if as yet incompletely, created systems that give citizens the opportunity to influence and control their governments. Large international concerns on the other hand, which operate only according to their own short-term goals, lack inbuilt homeostatic checks. We have no insight into their operations, and thus cannot connect them to sensible paths that are of use to *everybody*.[1531]

To give a striking example: In 1936 General Motors, Standard Oil of California and Firestone formed a company called National City Lines, whose sole purpose was to buy up alternative transport systems and then close them. Within 20 years, over 100 urban electric surface-rail systems in 45 cities had been bought and closed. The company's biggest operation was the purchase in 1940 of part of the Pacific Electric system, which carried over 110 million passengers a year and served 56 communities. Over 1760 km of track were ripped up, and by 1961 the whole network was closed.[1532]

Failure to act in anything like the right manner is evident on every level of decision making. As regards internal affairs, for example, as O'Connor points out, there is no state agency or corporatist-type planning mechanism in any developed capitalist country that engages in overall ecological, urban and social planning. The idea of an ecological capitalism, or a sustainable capitalism, has not even been coherently conceptualised, not to speak of becoming embodied in an institutional infrastructure. Where is the state that has a rational environmental plan? Intra- and inter-urban planning? Health and education planning organically linked to environmental and urban planning? Nowhere. Instead, there are piecemeal approaches, fragments of regional planning at best, and irrational political spoils allotment systems at worst. Every day, therefore, new headlines announce another health care crisis, another environmental crisis, and another urban crisis.[1533] And, we might add, the casualties from yet another war. In the case of the United States, as Leonard Duhl says, "all you have to do is grope your way through the legislation of the United States government and then look at the way the laws are carried out, and you see how disconnected all the programmes are. They are not connected in any way, either in concept or in specification or in the groups of people who are given the responsibility to do something about the problems."

Our development as a species never has been, and certainly is not at present, under any sort of rational control. While the same may also be said of wild species, it is not they but we who, through the use of our intelligence, are undermining the preconditions of our own existence. Such things as constantly growing world population, accelerating pollution, increasing arms manufacture, nuclear proliferation, the increasing gap between the rich and the poor, the emphasis on quantity over quality – all of these things not only suggest our having virtually no regard for our species' survival, but at the same time speak against our ever coming to have such regard – or to act on it if we did. As pointed out by Ehrlich and Ehrlich, few people protested against the Reagan administration's pro-natalist population policies, its relaxation of fuel-efficiency requirements for new cars, or its dismantling of research and development programmes for energy efficiency and alternative energy sources. All of Reagan's policies were senseless when viewed in a long-term perspective,[1534] but provided American capitalists with greater profits.

As Baran says, no planning worth the name is possible in a society in which the means of production remain under the control of private interests which administer them with a view to their owners' maximum profits or other private advantage. It lies in the essence of comprehensive planning for economic development that the pattern of allocation and utilisation of resources that results from it *differ* from the pattern prevailing under the status quo. Since, however, the prevailing pattern of resource allocation and utilisation corresponds, at least in the short run, to the interests of the dominant class, it is inevitable that any serious planning endeavour should come into sharp conflict with that dominant class and its allies at home and abroad.[1535]

The pursuit of economic growth

What constitutes economic growth?

Economic growth is growth in the GNP of a country or the GWP of the world, each of which, as mentioned in Chapter 4, is taken to be a measure of *welfare*. But Boulding, beginning already in 1949, repeatedly suggested that gross national *product* is essentially the same thing as gross national *cost*, a question taken up in Chapter 2. And in this vein Kapp, in 1950, said that, "in their present form, national income indices not only fail to subtract ... social costs, but include money spent to repair the damages caused by productive activities of the past and present."[1536] And Daly, while he admits that part of the GWP may be an economic gain, sees part of it as definitely a cost – one due to its own production.[1537] This view, which has as yet to be refuted,[1538] may be easily understood on the VCP in terms of renewable and non-renewable resources. As Boulding suggests, it should be possible to distinguish that part of the GNP which is derived from exhaustible and that which is derived from reproducible resources, as well as that part of consumption which represents effluvia and that which represents reintroduction into the productive system,[1539] and from there be able to create a better measure of welfare.[1540]

The preconditions of economic growth

Another distinction to be made in considering economic growth is that between resources and what is produced using the resources, the existence of the latter presupposing that of the former. As pointed out by Schumacher, an expansion of our ability to produce goods and services is useless unless preceded by an expansion of our ability to obtain resources, for humans are not really producers (due to the conservation principles of physics and chemistry) but converters, and for every instance of conversion we need resources (low entropy) to convert. And without a surplus of resources there can be no growth. In particular, our ability to convert depends on the accessibility of *energy*, which immediately points to the need to distinguish between renewable and non-renewable sources of energy. The market knows nothing of these distinctions however. It provides a price tag in terms of money for *all* goods and services indiscriminately, and thereby enables us to pretend that their exchange value is their true value.[1541] Take various alternative fuels, such as coal, oil, wood and waterpower: the only difference between them recognised by modern economics is relative cost per unit power produced. The cheapest is automatically the one to be chosen, as to choose otherwise would be irrational and 'uneconomic.' As Schumacher also says, the economic growth of the industrial era could just as well be seen as a measure of the rate at which we are consuming geological capital, while counting it as income.[1542]

The continuing existence of economic growth implies a number of processes, some of which are physical/biological and some of which are social.

The physical effects of economic growth

Regarding the physical implications, as succinctly expressed by Daly, after we deplete one resource, we redesign our machines and set about depleting another. The assumption is that in the aggregate resources are infinite, that when one flow dries up there will always be another, and that technology will always find cheap ways to exploit the next resource. When the whales are gone, we will hunt dolphins, and so on until we are farming plankton.[1543] And, we might add, when that's no longer possible we'll find a way of getting nutrients from bottom sludge.

Economic growth, so long as it is based on a non-renewable surplus, erodes that surplus at a faster rate than would occur otherwise, shortening the time to its eventual disappearance – as would be expected given Georgescu-Roegen's considerations regarding entropy.[1544] It also produces increasing quantities of non-biodegradable waste that increasingly pollute the environment. And if the surplus is of *renewables*, economic growth will tend to convert them into non-renewables: their being drawn into the economic system to a constantly increasing degree will mean their being used at a successively higher rate until that rate exceeds their ability to reproduce themselves. As Schumacher says, it is clear that the rich are in the process of stripping the world of its once-and-for-all endowment of relatively cheap and simple fuels, their continuing pursuit of economic growth producing ever more exorbitant demands.[1545]

Thus, as Schumacher noted already in 1970 with regard to economic growth and sources and sinks, we find that the idea of unlimited economic growth – more and more until everybody is saturated with wealth – needs to be seriously questioned on at least two counts: the availability of basic resources and, alternatively or additionally, the capacity of the environment to cope with the degree of interference implied.[1546] In this latter regard, as expressed by Commoner, all this 'progress' has greatly increased humans' impact on the environment. The postwar technological transformation of the United States economy produced not only the much-heralded 126 per cent rise in GNP (from ca. 1946 to 1970), but also, at a rate about ten times faster than the growth of the GNP, rising levels of pollution.[1547] "The very system of enhancing profit in this industry is precisely the cause of its intense, detrimental impact on the environment." According to Milbrath and in keeping with *The Limits to Growth* view, the economy will sooner or later begin an irreversible period of contraction that will probably be seen at first as another recession. Maintaining the expansive economy means eventually starving from a dearth of resources or choking on a superabundance of garbage.[1548]

As Goldsmith says,

> If development means the continued substitution of machinery and
> other inputs for human labour, so as to increase *per capita* production
> and hence consumption and thereby foster economic growth – then
> surely the answer is that there is no possibility of development for
> very much further without ecodisasters. I think we have reached a
> stage where it is extremely difficult for serious people to continue to
> assume that somehow, with the aid of science and technology, we can
> go on indefinitely enjoying economic growth without changing the
> climate in a disastrous way, without large-scale famines, annihilation
> of wildlife, etc. In fact we are heading towards a whole series of crises
> which can only be exacerbated by further economic growth.

In keeping with Wilkinson, and more particularly with Schumacher, we
could also see economic growth, through its mirroring physical 'production,'
as a measure of the failure of humans to live in ecological equilibrium with
their environment. A rising GNP may just as well be taken as evidence of the
increasing needs and problems which make higher consumption necessary.[1549]
As put by Mishan, continued technological and economic growth will act to
worsen the human condition.[1550]

Were it not for the pursuit of economic growth, we would not have population growth nor technological development nor pollution; nor need our society be highly stratified. And population pressure would be less, which could
decrease the likelihood and/or severity of wars.

The social effects of economic growth

As pointed out by Kapp, a system of investment for profit cannot be
expected to proceed in any way other than by trying to reduce its costs whenever possible and by ignoring those losses that can be shifted to third persons
or to society at large. Capitalists' costs do not include the social (or ecological)
costs of production and, therefore, are no adequate measure of total costs. The
competitive cost-price calculus is not merely meaningless, but nothing more
than an institutionalised cover under which it is possible for private enterprise
to shift part of its costs to the shoulders of others, and to practise a form of
large-scale spoliation.[1551]

As Kapp further suggests, the disruption of the environment and the resultant social costs have a tendency to increase both absolutely and relatively
as production and consumption and hence factor input and residual wastes
increase, and as these wastes are emitted or dumped into the environment

without adequate prior treatment and prior assessment of the consequences. In both capitalist and socialist countries, pressure groups use their political influence to perpetuate existing methods of doing business, which is to the detriment of society as a whole.[1552] "[C]apitalism must be regarded as an economy of unpaid costs, 'unpaid' in so far as a substantial proportion of the actual costs of production remain unaccounted for in entrepreneurial outlays; instead they are shifted to, and ultimately borne by, third persons, or by the community as a whole."

While the intended social effect of economic growth is ostensibly to improve welfare, as Commoner points out, the growth of the American economy between 1946 and the early 1970s had a surprisingly small effect on the degree to which individual needs for basic economic goods (vital resources) were met. In each case of 20th-century technological development there was a drastic change in the technology of production. A detergent is no better an economic good than soap – but it is, far more than soap, an ecological 'bad'.[1553] But its use is pushed by advertising, as detergents increase capitalists' profits more than does soap. As Goldsmith and his associates see it, economic growth constitutes the principal defect in the industrial way of life.[1554]

One social effect of constant economic growth is increasing psychological stress, the result of being caught up in the rat race, and constantly having to run faster and faster. Lorenz traces this stress back to its biological preconditions, suggesting that: "The rushed existence into which industrialized man has precipitated himself is actually a good example of an inexpedient development caused entirely by competition between members of the same species." Economic growth and the rat race are the result of intraspecific male competition over territory in a context where technology allows more and more to be taken from the environment.

Here we might consider excerpts from Forrester's analysis of economic growth. As he sees it, such growth is bringing pressure on every facet of human existence. The American federal policy in the mid-1970s (and the policy of all nations today) was to attempt to relieve the pressure that results from growth while at the same time trying to accelerate that growth. But continuing growth, far from solving problems, is the primary generator of our growing social distress. The ultimate pressures resulting from growth are of a social nature – crime, civil disorder, declining mental health, war, drug addiction, and the collapse of goals and values. For each technical goal that is met, some social or economic goal must be forgone. Though population growth and migration have been controlled at all times, that control has often been guided by short-term considerations, with unexpected and undesirable long-term results. (Cf. Mishan: "Rapid economic development over the last century has been responsible not only for an unprecedented expansion of populations the world over but also, and especially in the richer countries, for the growing mobility of their populations.")[1555] The issue is not one of control or no control. The issue is the kind of control and towards what end. No group can be expected to exert the self-discipline now necessary to limit population size and the environmental demands of industrialisation unless there is a way to keep the future advantages of such self-discipline from being swallowed up by inward migration.[1556]

The view that at least a good part of GNP represents a cost may be seen as being *society's* point of view, or, ultimately, the point of view of the *species*. But it is not the point of view of individual capitalists. As suggested earlier, war, for example, which represents a cost for society, is a source of profit to capitalists. In this way we can partly understand e.g. the American military expenditures in the Persian Gulf area. Already before the first Gulf War, i.e. in 1985, the United States spent $47 billion projecting power into the region. If seen as being spent to obtain Gulf oil, it amounted to $468 per barrel, or 18 times the $27 or so that at that time was paid for the oil itself. In fact, if Americans had spent as much to make buildings heat-tight as they spent in *one year* at the end of the 1980s on the military forces meant to protect the Middle Eastern oil fields, they could have eliminated the need to import oil from the Middle East.[1557] So why have they not done so? Because, while the $468 per barrel may be seen as being a *cost* the American *taxpayers* had to bear, and a negative social effect those living in the Gulf area had to bear, it meant only *profits* for American capitalists.

But there are large economic discrepancies even within wealthy nations. As Schumacher[1558] asks regarding the United States, how could there be public squalor in the richest country in the world, and in fact much more of it than in many other countries whose per capita GNP is markedly smaller? Schumacher wrote at the beginning of the 1970s, but the subsequent 35 years of continuous growth haven't helped the situation. As Reiman remarks, the richest nation in the world continues to produce massive poverty,[1559] 37 million people in the US – over 12 per cent of the population – today being classed as poor.[1560] The life expectancy of the average African-American is lower than that of the average Chinese, and the infant mortality rate in cities such as Washington, Baltimore and St. Louis is higher than in cities such as Bangkok and Cairo.[1561] As Schumacher says, if economic growth to the present American level has been unable to get rid of public squalor – or has even been accompanied by its increase – how could one reasonably expect that further such growth would mitigate or remove it? It is strange indeed that the conventional wisdom of present-day economics can do nothing to help the poor. Invariably it proves that only such policies are viable as have in fact the result of making those already rich and powerful, richer and more powerful. The conventional wisdom of what is now taught as economics bypasses the poor, the very people for whom development is really needed. The economics of giantism and automation is totally incapable of solving any of the real problems of today.[1562]

As regards free trade, Douthwaite points out that international free trade inescapably leads to a levelling down. It means that salaries and wages will tend to converge at Third World levels, and social security provisions in industrial countries will continue to be cut, since these are an overhead that economies cannot bear if they are to compete successfully with countries without them. Only the owners of the surviving transnational companies and of natural resources will escape the general impoverishment. Already the islands of prosperity are growing steadily smaller in an otherwise sick, dilapidated and hungry world.[1563]

As expressed by Schumacher, one does not have to be a believer in total equality, whatever that may mean, to be able to see that the existence of inordinately rich people in any society today is a very great evil. As is implied by what has been said earlier, excessive wealth, like power (and as a form of it), tends to corrupt. Even if the rich are not 'idle rich,' even if they work harder than anyone else, they corrupt themselves by practising greed, and they corrupt the rest of society by provoking envy.[1564]

> In the excitement over the unfolding of his scientific and technical powers, modern man has built a system of production that ravishes nature and a type of society that mutilates man. If only there were more and more wealth, everything else, it is thought, would fall into place. Money is considered to be all-powerful; if it could not actually buy non-material values, such as justice, harmony, beauty or even health, it could circumvent the need for them or compensate for their loss. The development of production and the acquisition of wealth have thus become the highest goals of the modern world in relation to which all other goals, no matter how much lip-service may still be paid to them, have come to take second place. The highest goals require no justification; all secondary goals have finally to justify themselves in terms of the service their attainment renders to the attainment of the highest.
>
> This is the philosophy of [social] materialism, and it is this philosophy – or metaphysic – which is now being challenged by events.

What prompts economic growth?

As intimated above, the primary force behind economic growth is capitalists' drive to make a profit. Thus when it comes to *need*, the increasing complexity and stratification of society have led to the experienced non-vital needs of the powerful, wealthy part of the population playing a role much greater than their relative numbers would otherwise suggest. In vulgar terms, it means that the greed or hunger for power (the filling of non-vital needs) on the part of the powerful far outweighs the suffering (vital needs) of the weak.

But the existence of economic growth is prompted not simply by the powerful's need to make a profit, but also by the worldview of which this is an integral part, and the acceptance of that worldview on the part of the masses. In this regard, as remarked by Marx, "the ideas of the ruling class are in every epoch the ruling ideas: i.e. the class which is the ruling material force of society is at the same time its ruling intellectual force. The class which has the means of material production at its disposal has control at the same time over the means of mental production."[1565]

As Daly points out, following Schumacher, economic growth is the most universally accepted goal in the world. Capitalists, communists, fascists and socialists all want economic growth and strive to maximise it. The system that grows fastest is considered best. According to Daly, the appeals of growth are that it is the basis of national power and that it is an excuse for not sharing as a means of combating poverty.[1566] It offers – in conflict with the entropy principle – the prospect of more for all with sacrifice by none.[1567]

The conceptual core of the worldview that supports economic growth as a means to welfare is depicted in the academic discipline of neoclassical economics, where such growth is taken as a *sine qua non* for the functioning of the economic system. As Commoner says:

> In a private enterprise system, the no-growth condition means
> no further accumulation of capital. If, as seems to be the case,
> accumulation of capital, through profit, is the basic driving force of
> this system, it is difficult to see how it can continue to operate under
> conditions of no growth. At this point, it can be argued that some
> new form of growth can be introduced, such as increases in services.
> However, nearly all services represent the result of human labor
> expended through the agency of some form of capital goods. Any
> increase of services designed to achieve economic growth would have
> to be accomplished without increasing the amount of these
> service-oriented capital goods, if the ecological requirements are
> to be met.

Neoclassical economics is one of the most 'scientific' of the disciplines in the social sciences, due to the fact that capital can be measured in terms of money (exchange value).[1568] What this measurement requirement means however is that goods and services that cannot be bought or sold are valueless from the point of view of neoclassical economics.[1569] According to J. M. Clark, this conception of value has "failed to keep pace with the facts of production;" and according to Kapp the fact that capitalists are able to shift part of the total costs of production to other persons, or to the community as a whole, points to one of the most important limitations of the present scope of neoclassical value theory. Because it is confined to exchange value, neoclassical theory is incapable of including in its reasoning many of the costs (and returns) which cannot easily be expressed in dollars and cents.[1570] Through academic economics depicting a situation that favours the economically powerful, the subject of the discipline has had a strong influence not only on economic but also on political thinking.[1571]

As noted by Douthwaite, the idea that economic growth itself constitutes progress or a precondition for progress is a dangerous illusion which the better-off have been only too keen to propagate.[1572] Through their control of the media, the powerful have had no difficulty in getting the populace to accept their worldview, which includes such beliefs as that growth is necessary for full employment. In the 1960 American presidential elections, for example, economic growth was taken on both sides to promote employment and security. And at the time of the Cold War economic growth was put forward as being necessary for winning the arms race.[1573]

If we look at the economic-growth aspect of the traditional or Western perspective from the point of view of supply and demand, we find that on the demand side, as Daly points out, growth is stimulated by greed and acquisitiveness, while on the supply side it is reinforced by technocratic scientism, which proclaims the possibility of limitless expansion and preaches a reductionistic, mechanistic philosophy. This philosophy, as also pointed out by Daly, in spite of its success as a research programme, has serious shortcomings as a worldview.[1574] As a research programme, it furthers the acquisition of power and the control of nature, but as a worldview it leaves no room for purpose, much less for any distinction between good and bad purposes.[1575] Furthermore, as pointed out by Schumacher: "An attitude to life which seeks fulfilment in the single-minded pursuit of wealth – in short, materialism – does not fit into this world, because it contains within itself no limiting principle, while the environment in which it is placed is strictly limited."

By being able to meet the needs foisted upon the populace by advertising, the capitalist increases his power and improves his social position. This takes the form of his receiving *capital* for his efforts in providing the populace with goods and services, which gives rise to a competition to see who can accumulate as much capital as possible. This in turn has stimulated economic growth and resource use, and the production of waste, all this in the context of a surplus of resources; and it implies consumers' access to at least part of that surplus. This is the main reason – coupled with genetically/karyotypically inbuilt short-sightedness as implied by the reaction principle – that the leaders of nation-states, who are financially supported by profits from commerce, themselves politically support the growth ideology of capitalists.

According to Mishan, the most powerful economic group in the modern nation-state, i.e. that whose members have 'entrenched interests,' is unwaveringly committed to continued economic growth. The bulk of the working class – using the term to cover both blue- and white-collar workers, skilled and unskilled – have not even thought of questioning the growth gospel. The Western world will have to suffer some pretty terrifying experiences before the unions will be ready to give their official blessing to the idea of a steady-state economy. In an affluent society (in which all vital needs are met) people's satisfaction, as Thorstein Veblen observed, depends not only on the innate or perceived utility of the goods they come to possess, but also on the status associated with their possession. In Mishan's terms, "in high income societies an individual's satisfaction comes to depend less on his absolute income and more on his income relative to the incomes of others."[1576] Relative income (keeping up with the Joneses) matters more as affluence increases[1577] (and non-vital needs replace vital needs).

On the problem of leaving economic growth behind us, Daly says that internal class conflicts, as well as international enmity, will make agreement difficult and predispose the relevant parties to accept the wishful thinking of technological optimists, who advocate that we have faith in the Great Breakthrough that will invalidate the theorem that a US-style growth-dominated economy is impossible for a world of what is now more than six-and-a-half billion people.

> All that is needed, they say, are larger research and development budgets, greater offerings to the Technological Priesthood who gave us the Green Revolution, Nuclear Power, and Space Travel. That these technological saviors have created more problems than they have solved is conveniently overlooked. The mythology of technological omnipotence is by itself very strong, but when backed by class interests in avoiding the radical policies required by the steady state, it becomes a full-fledged idolatry.[1578]

Here we have growthmania.

> As long as we remain trapped by the ideology of competitive growth, there is no solution. … The value of growth is rigidly held in first place, and we are trapped into a system of increasing environmental disruption and gross injustices by our inability to reorder values.

The role of technological development

Having economic growth as a goal leads to technological development on the part of capitalists, either to provide customers with products that appear superior to those of their competitors, or to produce their own products more cost-effectively.[1579] Terry notes that business develops technologies for the express purpose of becoming more competitive, capturing market share, and increasing profits.[1580] As Douthwaite says: "If I do develop scruples, the growth imperative forbids me to delay introducing my technique, because if I do not innovate, someone else will, reaping the pioneer's profits, and unless I follow suit I will be driven out of business." (Cf. the benefits accrued to England during the industrial revolution.) And if someone else should introduce a new technique, our capitalist, if he does not want to be left behind, has no option but to adapt to it or devise a better one.[1581] As Kapp notes, capitalists who delay the introduction of new techniques until they have fully amortised the original cost of their old equipment will be forced out of business by new firms not burdened with obsolete equipment.[1582] As expressed by Baran, in addition to offering the carrot of extra-profits, the competitive system brandishes the stick of bankruptcy.[1583] When it comes to the introduction of technological innovations, the capitalist has no choice. In this way a wave of new needs is created throughout society. And we see through the industrial era a constantly increasing trend towards the development of technology in response to the perceived needs of the wealthy and powerful.

As Douthwaite clarifies, the growth imperative works on Darwinian principles: it ensures that only the fastest-growing businesses and nations survive. And just as evolutionary survival is due to the possession of more appropriate genes, the key to commercial growth is the adoption of more appropriate technology.[1584]

Commoner cites the Manufacturing Chemists' Association in this regard, which points out that: "The maintenance of above average profit margins requires the continuous discovery of new products and specialties on which high profit margins may be earned while the former products in that category evolve into commodity chemicals with lower margins." In other words, in the context of a surplus in a capitalist economy, higher prices can be obtained for newly introduced products than for those that have been on the market for some time – particularly when one has a monopoly (thanks to patents) on the product in question.

Regarding patents, according to Kapp the patent system is a necessary and essential outgrowth of the institution of private property, which, we might add, was strengthened by the institution of money. And an exclusive privilege of making use of an innovation or improvement puts all other producers at a competitive disadvantage. Several patents combined provide a convenient and effective means of closing the industry to all newcomers.[1585]

As Commoner points out with regard to the period between 1946 and 1970 in the United States, while production for most vital needs – food, clothing, housing – just about kept up with the 40 to 50 per cent increase in population, the *kinds* of goods produced to meet these needs changed drastically. New production technologies displaced old ones. Soap powder was displaced by synthetic detergents; natural fibres (cotton and wool) were displaced by synthetics; steel and lumber were displaced by aluminium, plastics and concrete; railroad freight was displaced by truck freight; returnable bottles were displaced by non-returnable ones. On the road, the low-powered car engines of the 1920s and 1930s were displaced by high-powered ones. As Kapp notes, instead of developing a new engine and a new type of car designed to minimise atmospheric pollution and increase the safety of the driver, car manufacturers have devoted the major part of their research and development expenditures to purposes they consider more profitable in terms of their micro-economic calculations of net returns.[1586] On the farm, while per capita production of food remained about constant, the number of harvested hectares decreased; in effect, fertiliser displaced land. Older methods of insect control were displaced by synthetic insecticides such as DDT, and for controlling weeds the cultivator was displaced by the herbicide spray. And the range-feeding of livestock was displaced by feedlots.[1587] In all of these cases we see replacement by substitutes that increase ecological disequilibrium.

As Terry explains, since real wealth is perishable, if we do not consume it, use it up, or trade it to someone else who can use it, it becomes valueless. It is expedient, therefore, that we purchase all the goods we produce with our expanding productivity. Consumption must keep up with production. For example, in the United States the 35 hours it took auto workers to assemble a car in 1977 has now been reduced to eight – we must buy more cars. If the time it takes to produce a given quantity of goods steadily decreases, then, if the capitalist is to make a profit, we must also consume those goods more rapidly.[1588] Obsolescence is therefore built into them.

As Kapp says, one form the plundering of consumers has taken is the systematic degradation of the quality of consumers' goods in the interest of a quicker and repeated turnover. This interest of the monopolist in selling products of a shorter life for the sole purpose of increasing sales and profits is well exemplified by the following extract from a proposal made to the General Electric Company to reduce "the life of flashlight lamps from the old basis on which one lamp was supposed to outlast three batteries to a point where the life of the lamp and the life of the battery under service conditions would be approximately equal …. If this were done we estimate that it would result in increasing our flashlight business approximately 60 per cent. We can see no logical reason, either from our viewpoint or that of the battery manufacturer, why such a change should not be made at this time."[1589] It is safe to assume that other manufacturers, too, fail to see why they should not reduce the average life and quality of their products if by so doing they are able to increase the volume of their sales. (In the struggle for commercial survival it is often the most unscrupulous who set the standard for moral restraint.)

Everything speeds up (the rat race), and average Joe has to run faster and faster just to say in one place.[1590] ('Now *here*, you see,' said the Red Queen to Alice, 'it takes all the running you can do, to keep in the same place.') But increased productivity does not make *societies* wealthier. They simply produce and consume at a faster rate. There is no more wealth in the system. It merely appears so, because money and products change hands faster,[1591] which means an increase in the GNP.

So in the event that there is no real need for a product, an imagined need must be created among consumers via advertising. People must come to need the constantly changing stream of articles capitalists produce, not because each successive one is better, but because it is *different*. Such products are advertised as *New!* Fashions develop.[1592] Due to the monotony of the working life of the average consumer, and his or her desire to keep up with the Joneses, these products are bought, the profit going to the capitalist.

With this type of 'growth,' once you reach a high level of productivity, each additional increase becomes more difficult. A consumer society, driven by the demands of an ever-expanding technology, is inherently self-destructive. As new technologies replace old at an accelerating pace, few companies can keep up. Only the biggest and wealthiest survive, which results in the concentration of both power and capital in society.[1593]

An end in itself

The central unquestioned belief of the Western worldview, which has spread from the world's rich to the world's poor mainly during the past 50 years, is the belief in the inherent goodness of economic growth. As Mishan says, and as intimated above, the economic growth rate is the one indicator of progress to which politicians of every party pay homage. As a criterion by which the nation's overall performance is judged, and by which the current values of nations are compared, the index of economic growth has achieved international recognition as the common standard of virtue.[1594] Practically everybody in the world who has thought about the situation believes that economic growth is to be striven for, despite the fact that common sense indicates that such growth cannot continue indefinitely. An increase in production, or constantly increasing trade, will, through raising our standard of living, improve the quality of our lives.

In 1960 President Kennedy promised a five per cent annual growth rate,[1595] and in 1961 the United States and 19 other countries agreed to aim at a 50 per cent growth in GNP during the 1960s. Everyone in the United States and Britain supported economic growth. In Britain, since the end of World War II up to the present, both political parties have considered economic growth to be great, and have seen to it that the only acceptable solutions to any social problems are consonant with it.[1596] The Organisation for European Economic Co-operation (OEEC) council recommended in 1958 that member countries "should now pursue policies encouraging sustainable [*sic*] economic growth." In 1960 the OEEC became the Organisation for Economic Co-operation *and Development* (OECD).

Our Common Future, the report of the UN World Commission on Environment and Development of 1987, in which the notion of sustainable development was brought to the fore, suggested strong action should be taken to make development sustainable – while at the same time urging a stepping up of economic growth![1597] One gets the impression that the thought had not even crossed the minds of the members of the Commission that, as pointed out by Trainer, it is the present economic system with its growth imperative that *generates* the problems for which sustainable development is to be the answer.[1598] Similarly Principle 12 of the Rio Declaration on Environment and Development, which came out of the huge United Nations Conference on Environment and Development held in Rio in 1992, says that: "States should cooperate to promote a supportive and open international economic system that would lead to economic growth and sustainable development in all countries, to better address the problems of environmental degradation." But seeing as economic growth increases environmental degradation, this line is quite absurd.

As expressed by Schumacher, the economist-turned-econometrician, having established by his purely quantitative methods that the GNP of a country has risen by, say, five per cent, is unwilling, and generally unable, to face the question of whether this is to be taken as a good thing or a bad thing. He would lose all his certainties if he even entertained such a question: growth of GNP must be a good thing, irrespective of what has grown and who, if anyone, has benefited. The idea that there could be pathological growth, unhealthy growth, disruptive or destructive growth is to him a perverse idea which must not be allowed to surface.[1599] As Schumacher says regarding those who have the Western worldview: "Anything that is found to be an impediment to economic growth is a shameful thing, and if people cling to it, they are thought of as either saboteurs or fools." On the Western, neoclassical worldview, the concept of finite resources is meaningless.[1600] With regard to who is a fool, we note Boulding's suggestion that: "Anyone who believes exponential growth can go on forever in a finite world is either a madman or an economist."

Throughout the world, economic growth has become an end in itself.[1601] This is what development *means*. We have lost sight of what economic growth was to accomplish for us, namely to improve the welfare of society as a whole. But it does not do this; what it does is quite the reverse. *Economic growth has come to be seen as the greatest social good, when in fact it is the greatest social evil.* Its presence constitutes *progress*, never mind that the quality of life or standard of living of the average person may be declining through the closing of hospital wards, increasing criminality and mental illness, or rising unemployment. As long as the economy continues to grow, it is thought that we are moving in the right direction, when in fact we are moving in the worst possible direction.

As Mishan suggests:

> How, in fact, we are likely to respond if and when 'the awful truth' dawns upon us, even though we might well understand what we ought to be doing, is a matter or rueful conjecture. For our economic institutions are all polarised along the growth axis; they all pull in the growth direction. Our vast educational establishments, from infant schools to postgraduate research departments, have acquired an orientation which produces the research mentality. The spirit of innovation, in matters great and small, pervades our society. Habitually, unthinkingly almost, we seek more efficient ways of doing things. To use the current slang, we are 'hooked on' efficiency. We are corrupted by an unquestioned imperative to work 'improvements' into everything; into ever article, every substance, every material, every machine, every process, every idea. Indeed, what keeps our civilisation going are successive 'shots' of innovation. Nothing compels our attention so much as the oft-claimed 'scientific breakthrough.'

The limits to economic growth

In 1958 John Kenneth Galbraith recognised the problem of resource depletion, and complained of the Western attitude of consumption. He noted that if the economy were not to stagnate at that time, there would be an absolute contraction later. And in anticipation of the Meadows team, he stated that "there must be some limit to 'growth'."[1602] Sooner or later, with or without a major war, economic growth will stop. No activity that is at least partly physical can grow indefinitely if the system of which it is a part is finite, as the planet Earth is.[1603] Everyone should realise this if they thought about it – which, unsurprisingly given the dominance of the Western worldview, very few have.

Other things being equal, one might have expected research to have been done regarding the conditions under which such growth can no longer continue, and how best to prepare for those conditions. I doubt that there is one economist in the world who is engaged in working on this obviously extremely important question. We will be completely unprepared when economic growth cannot continue.

Largely as a result of our ability in the past to find technological solutions to subsistence problems, we have come to believe that new technological solutions will save the day both now and in the future, while giving little thought to either which resources are to be exploited or the potential environmental effects of exploiting them. As expressed by Hardin, it's un-American to say *no* to technology.[1604] And as this American/Western worldview encompasses the globe, it becomes *un-human* to say no to technology.

But growth can even come to a standstill given a surplus (in our case, of fossil-fuel energy). If the surplus is finite, continuing growth will eventually run into problems of decreasing returns that make it counterproductive (cf. e.g. salinisation due to irrigation). This is manifest, among other ways, in the lowering standard of living of the populations of industrial countries since the 1960s.

The existence of a surplus is of course not the only requirement for continuing economic growth. Even if we were to consider an imaginary situation in which the surplus was practically infinite – if, as regards energy for example, we were able to devise an effective way to convert sea water into fuel – this would only increase the devastation of the turning of the vicious circle. As has been pointed out by Heilbroner, it would lead to an increase in the production of heat, if nothing else, which itself would constitute an environmental problem.[1605] More generally, it would increase the rate at which the circle leads to the elimination of species, the erosion of land, global warming, and so on, any one of which can alone bring economic growth to a halt. And this is precisely the effect that the current worldwide attempt on the part of political and business leaders to increase economic growth is having. What is generally missed in this context, again thanks to the Western worldview, is that the economic system is part of the biological system, and presupposes it for its functioning.

What it seems is required to attain such a steady state would mean in some way suppressing or redirecting human male territoriality, and with it the ingenuity humans employ in order to rise in the social hierarchy, a hierarchy which will undoubtedly continue to be global for some time to come. If not by other means, perhaps this steady state will come about naturally, when the only available resources are once again renewable – which of course presupposes that our species still exist at such a time.

Innovation

On the Western worldview, technological innovation is conceived of in an exceedingly positive light. Children, for example, are encouraged to be innovative and find new solutions to various technical problems; the natural sciences provide their practitioners with higher status than do the social sciences; educations obtained at institutes of technology lead to better-paying jobs than those obtained at universities; medical research concentrates on technical responses to illness rather than on its prevention; the prestigious Nobel Prize is often given for technological advances related to science, and so on. (In 1948 the Nobel Prize for chemistry was awarded to the discoverer of DDT, which was hailed as a miracle chemical.) The orientation is one of providing technological solutions to today's problems, rather than avoiding the arising of tomorrow's. As is in keeping with the VCP, this orientation is one in which humans, following the reaction principle, reach for the nearest 'solution.' And when today's solutions themselves give rise to tomorrow's problems, the new problems are to be dealt with as they arise – again, by technological means. And though in some cases these means have yet to be devised, when devised they will provide some capitalist with a profit.

The attitude of reacting with technology rather than acting with prudence goes back to our very beginnings 200,000 years ago, and even earlier. The central position the provision of technological solutions has for our species suggests that we shall continue to try to provide such solutions to the problems that our earlier technological solutions gave rise to until we just cannot do so any longer. The reason we will be unable to do so will be because we no longer have sufficient resources to employ, or no technological solution can be found (as in the case of AIDS), or we no longer exist due to our eradicating our resources or choking ourselves with waste. As pointed out by Georgescu-Roegen, population pressure and technological progress bring the career of the human species nearer to its end.[1606] Though our intelligence tells us that prevention is to be preferred to cure, the reaction principle leads us to devote our efforts to curing rather than preventing. As long as we behave in this way, technological development will continue, as will its undermining the preconditions of our existence as well as its own.

As Schumacher says, we should instead be implementing the noiseless, low-energy, elegant and economical solutions normally applied in nature, rather than the noisy, high-energy, brutal, wasteful and clumsy solutions of our present-day sciences.[1607] But this is not the way it is. Human intelligence combined with primate territoriality stand in the way.

Nuclear energy

As suggested by Frans Berkhout, no nuclear power plant anywhere makes commercial sense: they all survive or have survived on an 'insane' economics of massive taxpayer subsidy. In Britain, which privatised the electricity industry in the late 1980s, there were no takers for nuclear power plants.[1608]

Apart from the effects of nuclear war, of all the changes introduced by man, as remarked by Schumacher, large-scale nuclear fission presently constitutes the greatest threat to our survival.[1609] The idea of replacing thousands of millions of tons of fossil fuels, every year, by nuclear energy, means 'solving' the fuel problem by creating an environmental and ecological problem of a monstrous magnitude. The original problem is just shifted to another sphere – there to become an infinitely bigger problem.[1610] As Daly says, "Fission energy is probably the biggest mistake we could make, and we seem determined to make it." Regarding the bomb, as Schumacher suggested in 1973, and as has since shown itself to be the case, with the spread of plutonium we cannot hope to prevent its proliferation.[1611] And all the weighty opinions of leading geneticists, doctors, physicists and students of politics

> play no part in the debate on whether we should go immediately for a large 'second nuclear programme' or stick a bit longer to the conventional fuels which, whatever may be said against them, do not involve us in entirely novel and admittedly incalculable risks. None of them are even mentioned [in the nuclear programme debate]: the whole argument, which may vitally affect the very future of the human race, is conducted exclusively in terms of immediate advantage, as if two rag and bone merchants were trying to agree on a quantity discount.

Regarding uranium, the official Working Party's report on the Control of Pollution,[1612] presented to the British Secretary of State for the Environment, stated already in 1972:

In effect, we are consciously and deliberately accumulating a toxic substance on the off-chance that it may be possible to get rid of it at a later date. We are committing future generations to tackle a problem which we do not know how to handle. ...

Since planned demand for electricity cannot be satisfied without nuclear power, many responsible people believe that mankind must develop societies which are less extravagant in their use of electricity and other forms of energy. Moreover, they see the need for this change of direction as immediate and urgent.

As Douthwaite reasons, if a zero discount rate had been used to compare the costs and benefits of nuclear power stations with those of other energy sources, the ridiculous course of building nuclear plants that consume more energy than they produce would have been avoided. If an energy analysis of any particular nuclear plant had been undertaken, the energy involved from century to century in disposing of the waste would not have been discounted but added up. If this had been done, it would immediately have become apparent that less usable energy was going to come out of the system than went in.[1613] This implies that, to completely clean up after a nuclear reactor's lifetime has ended – *if this were possible* – would on average require more energy than the reactor produced. Here it should be pointed out that the discount rate's being as it is is the result of the assumed continuing presence of a surplus such that economic growth can continue.

In the United States, as of 2006, the cost of decommissioning nuclear facilities was estimated at an average of $325 million to $500 million and up per reactor.[1614] As regards the practical problems involved, since the essential mechanism is highly radioactive by the time of decommissioning, it must be disassembled by remotely controlled robots, a tricky operation at best. And if disassembly proves impossible, the only measure left is to encase the 'dead' but dangerous reactor in concrete.[1615] To encapsulate a plant is *not* to clean up after its use, but to relinquish the idea of cleaning up. Not only does the area taken up by the plant remain highly toxic, being removed from human use forever, but the plant itself constitutes a time bomb for future generations. Encapsulation prevents the solution of the problem by making it permanent, with the assurance (thanks to the entropy principle) of its getting worse.

As mentioned, atomic fission presently supplies only about seven per cent of the energy used in the world. According to the IAEA there are about 3.5 million tons of uranium reserves at present, sufficient for 50 more years at the present rate of use.[1616] Present global uranium reserves with mining costs up to $80 per kg amount to about two million tons. These reserves, used in light water reactors, correspond to about 90 wfe,[1617] i.e. slightly more than two years' use of fossil fuels at 2006 levels. There have been no significant discoveries of uranium in the United States since 1965, despite intensified search.[1618] As with everything else being mined, the costs – both economic and environmental – of mining uranium are constantly rising.

Agriculture

From the point of view of the survival of the human species, our ability to produce food takes precedence over all other forms of production. As expressed by Schumacher, human life can continue without industry, but it cannot continue without agriculture.[1619] Of course modern agriculture is not at all sustainable, due primarily to its reliance on fossil fuels, in particular oil, as well as metals. Also, the use of heavy machinery results in soil compaction (after 1930, and particularly after 1950, when tractor sizes grew rapidly), while the worldwide non-sustainable use of aquifers in irrigation not only leads to their diminution, but in many cases to their experiencing salt-water intrusion. As expressed by Georgescu-Roegen, the advantages of mechanisation can be obtained only by eating more quickly into the 'capital' of low entropy with which our planet is endowed. That, indeed, is the price we have paid and still pay not only for the mechanisation of agriculture but for every instance of technological progress.[1620]

As suggested by Gever and his associates, as well as by Bartlett as cited in the previous chapter, technological ingenuity has been implemented in agriculture by funnelling ever-increasing flows of fossil fuels into production. And to this it may be added that the present use of fossil fuels in agriculture is highly *subsidised*. If we were to remove the fossil-fuel subsidies paid by taxpayers – in the form of financial support for tractors, fertilisers, poisons and the other paraphernalia of modern agriculture – the agricultural system would lack virtually everything that makes it more productive than the system was in agrarian times.[1621]

Rather than making preparations for the fast-approaching time when the availability of oil begins to diminish, we are making ourselves ever more dependent on it. Not only this, but we should be looking at the nature of the effects of fossil fuels on the production effort. Particularly notable in this latter regard are the effects of poisons and mineral fertilisers. As regards poisons, already back in 1962 Rachel Carson pointed out the devastating effect of biocides on bird life in her widely-read book *Silent Spring* – birds themselves constituting a natural form of insect control. Have we reacted wisely to her warning? No. Since the publication of her book until ca. 1990, the American production

of synthetic organic chemicals increased fivefold; and thirty times more biocides were used in the United States in 1990 than in the 1950s, while the use
of mineral fertilisers, which pollute surface and subterranean aquifers, had
increased ninefold between 1960 and 1990.[1622] Taking a broader look, roughly
150,000 different synthetic chemical compounds have been used in the world
since 1900.[1623]

As regards modern agriculture's reliance on *poisons*, there are perhaps three
main problems. First, pests will constantly develop resistance to whatever new
poisons we introduce; second, the continuing use of poisons means increasing
sterility of the soil, so that harvests will constantly diminish; and third, some
of the poisons are invariably ingested by humans (and other animals), resulting in such states as acute poisoning and cancer. In this last regard, in a recent
World Wildlife Fund study of sets of 27 ordinary grocery products from several
EU countries, poisons were found in *all* of the products.[1624] When it comes to
considering how dumb we are, the fact that we allow what we are going to
eat to be treated with carcinogenic poisons is perhaps particularly noteworthy. Furthermore, the high-yielding cultivars that are dependent on the energy-
intensive approach are extinguishing significant future options through the
extermination of diverse traditional strains.[1625]

The constantly increasing use of poisons and mineral fertilisers has not
reduced crop losses which, in the United States for example, rose from 32 to 37
per cent between the 1940s and the 1980s;[1626] and from the 1950s to the 1990s
there was a tenfold increase in the number of biocide-resistant species.[1627]

The Green Revolution

On the VCP, population growth generally is the result of there being a
surplus of vital resources, which leads to or is combined with a weakening of
internal population checks. This growth then eats away at the surplus until the
population arrives at a state where vital resources are scarce. The higher the
level of fertility and/or the lower the level of mortality, the faster this state of
affairs will come about, and the more pronounced it will be.

Not only has world population since the 1950s grown fastest in the Third
World, but the vast majority of the people living there are already at the bottom
of the global power-hierarchy, making the effects of population growth even
worse for them. The result has been a high mortality rate and much suffering.

Given the VCP, the reasonable attempt at an antidote to this state of affairs
would be to try to establish or re-establish internal population checks so as to
reduce the size of the population and bring it into equilibrium with its source
of food. The path actually followed, however, was one that simply took the
Third World further round with the vicious circle. With the ostensible ultimate

aim of reducing Third World hunger by producing more food (cereals, starting with rice), in the late 1950s the Ford and Rockefeller Foundations set up the International Rice Research Institute (IRRI) in the (US-controlled) Philippines, which has since grown to be the world's largest rice research agency. In this regard both common sense and the VCP tell us that, without the reinstatement of internal population checks, given sufficient breeding sites an increase in the amount of food in the Third World would only be pouring oil on the fire, and lead to population growth together with a *further* weakening of whatever checks as might still exist, with the result that the same problem should simply recur, only on a more intractable scale. As C. G. Darwin suggested already before the Green Revolution, if a larger quantity of food should at some time be accessible thanks to some discovery, for example in agriculture, then the size of the population will quickly rise to the new level, and afterwards development will continue as before, with the difference that the marginal starving group will constitute a larger proportion of the greater population.[1628] What Darwin describes is of course an expression of the pioneering principle, manifest through the vicious circle's moving from the having of a surplus of vital resources on to population growth.

This seems so obvious that one can wonder whether the ostensible reason for the IRRI project was the real reason. And it becomes clear that it was not. The real reason for the project was not to help the poor, but to increase the power of the capitalist political bloc centred on the United States, and the personal wealth of the capitalists involved. Thus with these ultimate ends in view, the direct aim of the IRRI was, using extant Third World varieties of rice, to breed more highly productive strains. Control of these strains was to fall into the hands of American capitalists;[1629] and control of the countries producing them into the hands of the capitalist bloc. From their point of view population growth in the relevant countries was good, not bad. It ensured a market for the capitalists' products, and provided manpower if a large military force were needed in conflicts with socialist states. What they forgot was that whatever patent they may have had on these strains didn't hold in the communist bloc, so the communists could and did produce them themselves. We shall see a change in this regard in the next agricultural revolution – to genetically modified organisms (GMOs).

What were produced were rice varieties that required copious quantities of mineral fertilisers and poisons, large amounts of which American companies were manufacturing in the postwar years, at the same time as they were scouting for markets overseas. So started the Green Revolution.[1630]

Outside of Mexico, the Green Revolution received its greatest support on the frontiers of the communist world, from Turkey to Korea, where it recommended itself as a way of blunting the appeal of socialist revolution, at its height in the 1960s. The rice programme in particular largely stemmed from American anxieties about the possible spread of Chinese communism after 1949. Meanwhile, socialist societies – China, Vietnam, and Cuba – embraced the idea of scientifically improved crops with equal vigour. High-yield rice strengthened communist China as much as it did Asia's island fringe, which America relied upon to contain China.[1631] In several of its manifestations, then, the Green Revolution was a child of the Cold War,[1632] and may be said to have achieved its economic but not its political goal.

Where the Green Revolution was implemented, farmers came to use heavier and heavier doses of biocides. This efficiently selected for resistant pests – as antibiotics did for bacteria. And as of 1985, roughly one million people had suffered acute poisoning from pesticides, two-thirds of them agricultural workers.[1633] The vast fertiliser requirements of the Green Revolution led to the eutrophication of lakes and rivers. Meanwhile the necessary irrigation helped drive the huge dam-building programmes of China, India, Mexico and elsewhere. Before the Green Revolution, farmers raised thousands of strains of wheat around the world. After it, they increasingly used only a few, and became fettered to a system based on a necessarily diminishing source of energy which required constantly increasing quantities of water.[1634]

The Green Revolution did not engineer an income redistribution towards Third World farmers; nor did it achieve food independence except for a few countries. Until 1981 the Third World had long been a net exporter of food, after 1981 it was a net importer.[1635]

Of course the people on whom the Western capitalists foisted the Green Revolution were themselves much better attuned to their long-term needs than the capitalists were, not that the capitalists really cared. Western power simply usurped the ecologically more benevolent lifestyle.[1636]

With its new strains, and the fertilisers, biocides, mechanisation and increased irrigation they required, world grain production doubled between 1960 and the late 1980s. Most of the world had been 'saved' by becoming more energy-intensive, complex and polluting. And for this, the scientist who led the teams responsible won the 1970 Nobel Peace Prize. This is highly ironic, for the increase in food, leading to an increase in population and thereby population pressure, worked rather towards *decreasing* the likelihood of peace.

Not unaware of the problematic nature of the results of his efforts, the winner of the Prize himself said: "Perhaps through this development we can buy 25 to 30 years of time. [But u]nless there is a breakthrough in slowing population growth on a world-wide basis, the world will disintegrate."[1637] Yes, the world will disintegrate; but you should have thought about slowing population growth *before* introducing your more productive seeds and their poisons on the market.

As clearly expressed by Forrester (and as implied by Malthus[1638]):

> Many programs – for example the development of more productive grains and agricultural methods – are spoken of as 'buying time' until population control becomes effective. But the process of buying time reduces the pressures that force population control.
>
> … Trying to raise quality of life without intentionally creating compensating pressures to prevent a rise in population density will be self-defeating. Efforts to improve quality of life will fail until effective means have been implemented for limiting both population *and* industrialization.
>
> [I]f we persist in treating only the symptoms and not the causes, the result will be to increase the magnitude of the ultimate threat and reduce our capability to respond when we no longer have more space and resources to invade.

Another negative aspect of this 'saving' was that its use of poisons required monoculture cultivation, opening crops to potential destruction by e.g. weather, at the same time as it reduced biodiversity. Also, the ploughing that was often required raised the temperature of the soil in the spring. In temperate regions this would have increased the activity of beneficial soil organisms; but in the tropics and subtropics it had the opposite effect,[1639] and is largely responsible for the nine times greater soil erosion there. Tropical soils are not amenable to sustainable agrarian agriculture, only to horticulture, just as were the non-riverine soils in Mesopotamia; once again, the over-exploitation of soil resources using agrarian agriculture results in soil degradation. Capitalists, spurred by the profit motive, nevertheless support the implementation of agrarian technology in the tropics.

Nevertheless, as noted, world grain production doubled in the short term thanks to these efforts. And population growth followed suit. As Catton puts it, the Green Revolution burdened the 20th century with almost another doubling of world population.[1640]

In the cradle of the Green Revolution in India there are today vast stretches of land where grass will no longer grow, the water is no longer drinkable due to contamination from mineral fertilisers, aquifers have dried up, soils are degraded, and biodiversity is fast vanishing, the agricultural result being declining rice yields. In 2001, in Wayanad, millions of fish died because of the presence in the water of the copper-based fungicide Furadan, sprayed on pepper gardens to control the wilt disease. And at the same time pests developed resistance to the poisons, leading to the development and use of new ones.

The Green Revolution not only increased the profits of the capitalists who owned the more productive seeds, but it also increased the profits of the large-scale landowners in the Third World, for whom the major financial investments required in e.g. tractors were both possible and paid off at least in the short term. In India, the poorest farmers, each of whom tilled perhaps half a hectare of land, could not afford these extras, and were forced to sell their farms and migrate to the cities, while the richer farmers increased the size of their holdings at the expense of the poor, and became even richer.[1641] (This brings to mind the definition of foreign aid as the money poor people in rich countries give to rich people in poor countries.)[1642] Thus another 'achievement' of the Green Revolution was to enrich two or three per cent of the wheat and rice farmers enormously, leaving the vast majority of subsistence farmers in the lurch. The increasing incidences of suicide among farmers in India lend testimony to this failure of high-tech agriculture.[1643] Thus, as in the horticultural and agrarian eras, while the poor continue to live barely above subsistence level – and some of them under it – the increase in the amount of food led to population growth.

Genetically modified organisms

Despite more than two centuries of scientific crop-breeding, as well as the so-called Green Revolution of the 1960s and the genetic engineering of the 1990s, it may be recalled that we have not domesticated any significant new animal species in the past 5000 years, nor have we added one new staple to our repertoire of crops.[1644]

As regards genetic engineering and the development of GMOs, perhaps people think that the reason for their being developed is primarily to increase the quantity harvested from each unit area of land. But this is not so. Similarly to the crops of the Green Revolution, their primary *raison d'être* is to secure and enhance the power positions of their capitalist producers. Just as the World Bank and the IMF continue to push loans at the countries of the Third World, knowing that their being accepted will ultimately lead to a worsened situation for the people living there, the IRRI has now shifted from pushing Green Revolution seeds at the people of the Third World to pushing the seeds of GMOs at them, also knowing that this too will worsen the situation for them. In 2000 the IRRI planned to field-test the controversial genetically modified bacterial blight rice, BB-rice. Their information officer at that time maintained that BB-rice is needed because of the world food crisis.[1645] "To ensure food security and to continue the advancement against poverty in rice consuming countries of the world, farmers will have to produce 40 to 50 percent more rice to meet the consumer demand in 2025." What can one say?

GMOs, or transgenics, are organisms whose genetic structure has been altered in order to maximise one or more of its beneficial aspects. Some alterations in crop genetics include splicing in genes which produce plants resistant to heavy weather, optimise the size and shape of the crop, or increase the number of seeds.

The most common type of GMO that has been developed and commercialised consists of genetically modified crop-plant species, such as maize, soybean, oil-seed rape and cotton varieties. Worldwide, nearly 30 per cent of transgenic cropland (often grains) is planted in varieties designed to secrete an insecticide, and almost all of the rest is in crops genetically engineered to resist total herbicides. The poison-secreting transgenics basically replace an insecticide that is sprayed on the crop with one that is packaged inside it.[1646] The plant's relevant genetic structure and the chemical formula of the poison it produces or has sprayed on it are patented, so that only the manufacturer can produce and market the seeds and the poison. Seeds of the plant can then in principle be sold, in the case of herbicide-resistant crops, together with the poison.

But the story doesn't end there. The organism is further genetically modified using 'terminator technology,' which prevents harvested seeds from germinating. (Cf. the communist bloc's use of the seeds of the Green Revolution.) The suicidal terminator genes are activated by dousing the seed with the poison tetracycline.[1647]

If one should want to use different, fertile, seeds in the same ground the GMOs have been grown in, the plants from those seeds would first have to be resistant to tetracycline. Even if they were, they would then be attacked either by the insect pests the GMO secretes poison to reject, or, if the other sort of GMO has been grown, by the herbicide sprayed on it that remains in the soil.[1648] In areas where Monsanto's genetically modified soya, which tolerates large quantities of herbicide, is grown, it is the only thing that can grow. Thus the farmer cannot use his own seeds, should he have any, but is forced to *buy* his seeds for next year's crop from the same capitalist. And after the plant has been harvested, as intimated above, whatever of the insecticide or herbicide that has not been sucked up by the plant, some of which we eat, remains in the soil. While there, it creates allergic reactions amongst humans and reduces the soil's biodiversity before it is eventually leached into the nearest aquifer, where it kills aquatic biota and makes the water unfit for drinking. A very clever arrangement.

Also, the effects of the poisons associated with genetically modified organisms will lead to resistance in pests just as the use of other poisons in agriculture do, and/or cause similar problems. Genetically modified cotton, which contains a 'poisonous' insecticide gene repelling the bollworm, led to an immediate increase in cotton production in China, while at the same time the previous poisons used against the bollworm could be dispensed with. However, over the years, the effects of other pests, in particular mirids, successively increased. Mirids were previously held in check by the spraying of insecticides the main target of which was the bollworm but which killed them as well. At present, the quantity of genetically modified cotton produced in China is no greater than that of the cotton it replaced, and cotton growers have to spray insecticides as much as they did before.[1649]

Furthermore, as Douthwaite points out, all commentators accept that pollen from genetically modified plants such as sugar beets will fertilise ordinary plants and so spread the herbicide-resistant gene, causing, for example, 'super weeds' to develop.[1650] Thus the growing of GMOs places the germplasms of *all* plants and thus all *crops* at risk to genetic contamination, jeopardising food security on a huge scale.[1651]

Fossil fuels and poison resistance

As regards the use of fossil fuels in agriculture, Pimentel states that approximately 90 per cent of the energy in modern crop production comes from oil and natural gas. About one-third of the energy employed has the effect of reducing labour input in grain production from 1235 to ten hours per hectare. About two-thirds of the energy is for production, of which about one-third is for fertilisers.[1652]

The number of poison-resistant insect species increased from 25 in the mid-1950s to over 450 by the 1980s.[1653] And between the 1940s and the early 1990s world crop losses to insects doubled, while farmers increased their use of poisons tenfold.[1654] By 1990, 25 of the 36 pests that attack cotton were resistant to the poisons used on them.[1655] As expressed by Gever and his associates, we are caught in a vicious circle: intense cultivation techniques are used to make up for the effects of agricultural degradation, but at the same time they worsen that degradation, requiring even more intense cultivation to keep food production up.[1656] And while some of the biocides most harmful to wildlife, such as DDT, have been banned in the industrialised countries, they are still produced there and exported to the Third World. (In September 2006, WHO announced that the UN will once again use DDT against malaria, despite its carcinogenic effects and the development of resistance in mosquitoes.) For example, in 1990 one-third of all US biocide exports were of types whose use was banned in the United States.[1657] Once again we see the drive to increase power on the part of wealthy individuals working against what is best for the human species.

As has been pointed out by Hardin, *every* biocide selects for its own failure through the development of resistance in the types of organisms it is intended to kill.[1658] In using such poisons, we humans are thus in fact increasing the rate of development and in certain cases perhaps even evolution of simpler forms of organism,[1659] thereby *domesticating* them. The shorter a species' breeding time and the greater the number of organisms produced, the greater the chance of its genetically or karyotypically mutating in such a way as to produce an organism that is resistant to any particular poison. And while we are increasing the rate of development and/or evolution of certain simple species, we are increasing the rate of *extinction* of more complex species (mainly, K-selected species), who reproduce too slowly to be able to adapt genetically to the poisons we introduce into the biosphere.

Medicine

Medicine constitutes an interesting and important example of increasing technological dependence. As noted earlier, the human domestication of plants and animals includes the domestication of humans themselves, and domestication consists essentially in genetic adaptation to technology. Part of this domestication in the case of humans and animals takes the form of gene-pool changes resultant upon the employment of medicine. Such intervention affects organismic survival and ability to reproduce, thereby increasing the dependency of succeeding generations on medical technology.

As mentioned earlier, the main external check to population growth throughout history has been infectious disease. The combination of environmental disease control, and antibiotics and vaccination, effected a partial epidemiological transition in which infectious disease declined as a cause of death, to be replaced by non-communicable diseases such as cancer and heart disease; and chronic diseases have outpaced infectious diseases in most parts of the world.[1660] These typically affect older people, so the transition helped add about 20 years to worldwide human life expectancy between 1920 and 1990.[1661]

As Mishan points out, however, and as is in keeping with the treatment of longevity in Chapter 2, "the figures given for longevity tend to mislead. True, infantile mortality has declined over the last hundred years, and it is certainly lower today than it was in pre-industrial times. To that extent the overall expectation of life has increased. But if we exclude infant mortality and, instead, compare the expected longevity of those surviving their fifth or tenth year, this marked superiority of modern times disappears."

Though the incidence of communicable diseases has been lowered in industrialised countries, even there it is once again on the rise. In this regard, humans, as compared to other animal species, have suffered a disproportionately large share of mortality from infectious and parasitic diseases. Population sizes and densities of most animals (excluding herding animals and insects) are probably too low to maintain micro-organisms in the absence of an intermediate host.[1662] Had we not domesticated wild (herding) animals, nor allowed our total population to become so dense, we would not have had to experience anywhere near as high a level of mortality incident upon infectious diseases. And though the worldwide increase in average longevity since the industrial revolution may be seen as being a result of the reduction in these diseases, as discussed earlier, little of this reduction is thanks to medicine.

The value of medical technology, even in the short term, may be questioned however. For example, since 1900 medical intervention in the United States has probably contributed as little as 3.5 per cent of the total to the overall decline in mortality.[1663] Thus we are inclined to ask, as have Joan Davis and Samuel Mauch:

> Why is medicine considered so important? Today, it represents another prime example of an artificial scientific-technological sub-system designed to correct the side-effects of a problem existing somewhere else: to treat some 30,000 surviving victims of car accidents per year [in Switzerland], to cure cancers due to smoking, to get heart attack victims back to their managerial jobs as quickly as possible, to take care of alcoholics and drug addicts, or to undo psychological damage.[1664]

Davis and Mauch's point here, as has already been expressed above, is that through technology – in the present case medical technology – we tend to treat the manifestations of problems rather than the problems themselves: we follow the reaction principle. And, due to the turning of the vicious circle, part of this reaction involves doctors becoming increasingly dependent on sophisticated machines for diagnosis and treatment.[1665]

Genetic load

While medicine has little positive effect in the short term, it has a large negative effect in the long term, due to (among other things) its effect on our genes. Patients suffering from genetic diseases now survive – and transmit their genetic defects to their descendants.[1666] As already noted by Darwin in 1874:

> There is reason to believe that vaccination has preserved thousands, who from a weak constitution would formerly have succumbed to smallpox. Thus the weak members of civilized societies propagate

their kind. No one who has attended to the breeding of domestic animals will doubt that this must be highly injurious to the race of man.

As expressed more recently and more generally by Bengt Hubendick, humanitarian measures in the form of child and hospital care, to say nothing of technically advanced medical efforts, can lead to the retention of gene-carriers who otherwise would have been eliminated before reproduction. Humankind has to a certain extent put nature's normal genetically-cleansing mechanism out of operation, so that our gene pool is being supplied with an increasing proportion of burdening genes.[1667]

Darwin continues shortly after the above quotation: "[E]xcepting in the case of man himself, hardly anyone is so ignorant as to allow his worst animals to breed. ... We must therefore bear the undoubtedly bad effects of the weak surviving and propagating their kind." The incessant push of medical technology has now brought us to the state where today we are just as ignorant in our treatment of our domesticated animals as we are in our treatment of ourselves, breeding lines whose survival is dependent on various forms of techno-medical support (cf. e.g. Belgian Blue cattle, that often require caesarean section to give birth). Thus not only are humans today treated with antibiotics, but pets, farm animals and race horses are as well. For example, the American livestock industry after the early 1950s fed huge quantities of antibiotics to American cattle and pigs to keep the animals healthy and help them grow faster in the epidemiologically hazardous feedlots of postwar America.[1668] Even wild animals are today being vaccinated, e.g. against rabies. And surgery is regularly being performed on pets and animals in zoos, with e.g. pet poodles being given pacemakers.

The negative influence of medicine on the gene pool of the human race has been expressed in greater detail by H. J. Muller in terms of *genetic load* (or *burden*, referred to by Hubendick above), i.e. the proportion of destructive genes in a genome (Glossary), which is rising in the case of the human genome. For example, assume that medical help enables people who are afflicted with a particular genetically-based disease, and who on the average each have had 0.5 children, to raise that number to 0.9. The frequency of the detrimental gene will then be five times what it was previously – i.e. the genetic load of this particular type of gene will be five times as great – and, without the medical intervention, the disease would be five times as common.[1669] The more hereditary diseases that are 'cured' in one generation, the more of them there will be to be cured in the next. Since modern medicine and protected living increase the likelihood of the carriers of some of these diseases and malformations procreating, in future generations their influence will be increasing, particularly in technologically advanced countries.[1670]

C. G. Darwin gives another example of essentially the same phenomenon, but involving non-medical technological development. Some 17 generations ago eyeglasses were invented, and suddenly certain eye defects such as nearsightedness ceased to be a serious impediment in social relations. As a consequence these defects were no longer selected against to the same degree, and it is rather certain that in future generations they will be even more common than at present.[1671] The reasoning involved is essentially the same as that regarding human genetic adaptation to fire, and genes for weak eyesight would spread through the whole population just as have genes presupposing the availability of fire. All this is part of the domestication of humans, which is part of the turning of the vicious circle.

People with genetic impairments, thanks to technological support, have more children than they otherwise would, children who themselves have the same genetic impairments. Thus the number of cases requiring technological support will constantly increase. Technology, in domesticating us at an accelerating pace, not only weakens our genetic constitutions, but makes us increasingly dependent on the development of technology itself. At the same time there is no guarantee that the requisite technological solutions will be available in the future, or, if available, to the extent necessary to treat everyone – as they are not today.

Resistance

But the story in the case of medical intervention does not end there. Not only has the human race become increasingly dependent on various forms of medical technology, but we have become dependent on *developing* medical technology. The use of vaccines (e.g. Edward Jenner's vaccine for smallpox) and antibiotics in medicine leads to the creation of *resistance* to those very drugs on the part of the micro-organisms responsible for the disease in question. (Antibiotics offer no protection against viruses, while vaccines do; both are essentially *poisons* from the point of view of what they're used against.) The situation is essentially the same as that involving the resistance of agricultural pests to biocides. When sulfa drugs first appeared in the 1940s it was possible in the beginning to cure all cases of gonorrhoea. Ten years later the sulfa drugs were useless, since virtually all gonorrhoea bacteria had become resistant to them; and in the 1950s we changed over to penicillin.[1672] Gonorrhoea then became resistant to penicillin in the 1970s, and is constantly becoming more difficult to treat.[1673] Similarly, staphylococcal bacteria were all vulnerable to penicillin in 1941, while as of the mid-1990s 95 per cent of these strains showed at least some resistance to the drug.[1674] By the 1980s there were more than 100 poison-resistant bacteria and viruses.[1675] Throughout the world the over-use or misuse of antibiotics, particularly in treating small children and hospitalised patients, has prompted the emergence of resistant mutant organisms,[1676] which has benefited drug companies and their research staffs, for whom it has served the same purpose as planned obsolescence in cars.[1677]

As Mishan points out, the evolution of more resistant pests and microbes in response to the widespread use of more powerful drugs constitutes an insidious threat to human health and survival. Through the mechanism of natural selection, nature responds to man's 'miracle drugs' by producing 'miracle microbes' as well as increasingly pest-resistant insects, and swifter and hardier rodents.[1678] Furthermore, even in the case of curing an infectious disease, as suggested by Georgescu-Roegen, any such cure "vacates an ecological niche for other micro-organisms, which may turn out to be much more dangerous than the dislocated ones. Incredible though it may seem to the uninitiated, a famous microbiologist gave this counsel to his equally distinguished colleagues at a symposium: 'If a universal antibiotic is found, immediately organize societies to prevent its use'."

Genetic load and microbe and virus resistance are both genetic phenomena, and they in a sense mirror one another. Where modern medical technology on the one hand genetically weakens the human species, on the other it genetically strengthens the various species of micro-organism that prey on humans. More generally, the technological domestication of pests, whether they be rodents, insects, weeds, microbes or viruses, results from making life harder for them; the toughest (fittest) of them survive, and the species as a whole becomes hardier. The domestication of plants, animals and humans, on the other hand, results from making life easier for them; the *weaker* (less fit) survive, and the species in question becomes weaker.

Epidemics

In order to replicate, all micro-organisms require a certain minimum population size in their host. For example the size of the human population requisite for the outbreak of measles is 250,000 to 400,000, and a population of 500,000 is necessary to ensure a large enough input of new susceptibles for the disease to remain in the population. The denser the population, the more quickly an epidemic will spread, while the less dense, the longer it will last. At the same time however it is virtually impossible for infectious disease to which there is some resistance to cause extinction, for the disease normally declines with the population, until only scattered, resistant individuals remain.[1679] Due to the heavier genetic reliance on medical technology in the north, in the event of a future pandemic of a disease that does not have one hundred per cent mortality, despite the fact that infectious diseases are more prevalent the warmer the climate, it is highly likely that northern populations will be more severely hit than those of the Third World.

Just as we didn't harvest the large mammals of the Pleistocene in a sustainable way, but killed as many as we could for as long as we could in order to meet our immediate needs, in the case of medicine we do not attempt to avoid the effects of disease in an ecologically sustainable way, but react to their occurrence with short-term technical solutions such as the development of antibiotics – which worsen the situation in the long run – and reward those who come up with such species-undermining innovations with awards such as the Nobel Prize.

Resource depletion

As pointed out by Catton, it was thanks to two non-repeatable events that the wealthy countries have experienced five centuries of 'magnificent progress.' First, the discovery of a second hemisphere, and second, the development of technology that could unearth the planet's fossil-fuel deposits.[1680] In both cases we have to do with huge quantities of resources the exploitation of which has provided the human race with tremendous quantities of food, material and energy. But neither in the case of the discovery of the New World nor the tapping of fossil-fuel reserves was any effort made to husband the newly accessible resources. In both cases the attempt was and is being made to use them as quickly as possible, in the case of fossil fuels their price a few years back being little more than the cost of getting them out of the ground. This, says Douthwaite, is about as sensible as valuing the money one withdraws from the bank at the cost of the bus fare to go and get it.[1681] And the result of this new resource-availability has been a tremendous increase in consumption, pollution and the size of the human population. Calling ourselves *Homo sapiens* – the wise human – is quite unwarranted; it would have been much better to call ourselves *Homo artifex* – the clever human.

In 2006, the peak in world oil extraction was estimated to occur in 2010.[1682] Of course the passing of this peak is bound to have severe repercussions on the world's economy[1683] and thus on its political stability, and will constitute the first real and not just political worldwide energy crisis. In this regard a search has been made of the Internet archives of 11 sources of news for articles published during the three years 2001–2003 treating of this subject. The sources searched were: *The New York Times*, *The Los Angeles Times*, CBS, ABC, NBC, CNN, 60 Minutes, *The Washington Post*, *Time Magazine*, *Newsweek* and the BBC. Only *two* articles were uncovered published on the topic by these sources during the three-year period.[1684] (This lack of action with regard to states of affairs vitally relevant to the survival of the human species is of course just another manifestation of the reaction principle.)

To cite Schumacher in this regard:

> There are still people who say that if oil prices rose too much (whatever that may mean) oil would price itself out of the market; but it is perfectly obvious that there is no ready substitute for oil to take its place on a quantitatively significant scale, so that oil, in fact, cannot price itself out of the market.
>
> [I]t is in the real longer-term interest of both the oil exporting and the oil importing countries that the 'life-span' of oil should be prolonged as much as possible. The former need time to develop alternative sources of livelihood and the latter need time to adjust their oil-dependent economies to a situation – which is absolutely certain to arise within the lifetime of most people living today – when oil will be scarce and very dear. The greatest danger to both is a continuation of rapid growth in oil production and consumption throughout the world. Catastrophic developments on the oil front could be avoided only if the basic harmony of the long-term interests of both groups of countries came to be fully realised and concerted action were taken to stabilise and gradually reduce the annual flow of oil into consumption.

(Note that what may or may not be best for countries is, in today's world, overshadowed by what is best for corporations.)

As Schumacher says, noted earlier, if we took proper account of the use value of geological or natural capital, in the form of fossil fuels for example, we would be concerned with conservation; we would do everything in our power to try to minimise their rate of use (throughput). We might decide, for instance, that the money obtained from the realisation of these irreplaceable assets must be placed into a special fund to be devoted exclusively to the evolution of production methods and patterns of living which do *not* depend on non-renewables at all, or depend on them only to a very slight extent. But we do not do this. What we do is the exact opposite, and maximise rather than minimise our rates of use.[1685]

Considering that our use of non-renewables is actually *accelerating*, the situation is of course even worse than it would have been if our use remained at a steady rate. In the case of *renewables*, what constantly increasing use means is that they quite simply haven't the time to regenerate, so that they too are diminishing. Considering both renewable and non-renewable resources it is thus clear from even a superficial glance that, quite independently of the VCP, throughout our existence we humans have been using and are continuing to use both kinds of resource at a rate that sets our existence as a species in jeopardy. As pointed out by Daly, as depletion forces us to exploit progressively poorer-grade resources, the gross throughput of matter and energy will have to increase in order to yield the same net throughput of the minerals required to maintain stocks constant.[1686] And as Mishan points out, in keeping with Ellul, we have had no previous experience of running out of a large number of currently important materials on a world scale.[1687]

However, few contemporary writers stop to consider the fact that, at least in the case of non-renewables, the ultimate eradication of the resource in question is only to be expected, and that for the sake of the long-term survival of the human species it would have been better had we never begun using such resources in the first place. Moreover, now that we realise this, we should also realise that we ought with all haste to stop using them. From the point of view of the survival of our species we were too smart to leave well enough alone; but we're apparently not smart enough to stop doing what we are doing in order to avoid disaster.

The viability of the entropy principle means that a sustainable society can exist only on what it can reap from the constant stream of solar energy on a *year-to-year* basis. There can be no dipping into stores of energy or material amassed over long periods in the past. More generally, in a sustainable society there can be no *mining*.

As expressed by Ronny Pettersson, during the last 3000 years growth of production (thanks to technological innovation) has tended over and over again to surpass increase in population, there having been a constant race between these two magnitudes.[1688] As Wynne-Edwards says, growing skills in resource development have, except momentarily, always outstripped the demands of a progressive increase in the size of the human population, which also manifests itself in the form of economic growth.[1689] But this see-saw effect has only been possible because the resources amenable to technological innovation have been there to be exploited. That the vicious circle has continually given rise to surpluses is completely dependent on conditions external to the circle itself – it is not self-feeding, but depends on the existence of natural resources. Human ingenuity in a resource vacuum would have got nowhere. And the possible continuation of the turning of the vicious circle in the future is dependent on what nature has left to offer. When fossil-fuel extraction as a whole peaks, if not before, a world population dependent on its use will be facing a situation in which billions of people will experience vital need, a need which there will be no way of satisfying. Our non-sustainable use of resources to date suggests that a time will come when there are no more resources left to exploit, and there we will be with our useless ingenuity.

There must come a day when humans cannot use fossil fuels, uranium or metals, the ultimate limit for fuels being where the amount of energy required to extract them exceeds the amount of energy they provide, i.e. when the net energy ratio falls below one. At that time, if the human species lasts long enough to bring it about, we shall have to make do without them; and to the extent that technology still exists, it will have to be created out of and work upon renewables such as wood (fire, and rubber and cellulose-based plastics) and other vegetable matter (food and drugs), the skins, bones and sinews of animals, and such non-renewables as exist in a virtually limitless quantity (such as salt, and sand to make glass). In this sense, then, we will have to revert to the early Stone Age, i.e. to the human situation before we began mining metals some 6000 years ago. These are the preconditions for a sustainable human society and a steady-state economy, and when one speaks of sustainable development, such development must either be towards this state, or presuppose its existence.[1690]

All this should be self-evident and is quite independent of the VCP. However, no decision maker in the world has given it any consideration. Rather, the vicious circle churns on, making us ever more dependent[1691] on a technology based on limited non-renewables.

Taken as providing insight into how society should develop in the future for humankind's best, the VCP places supreme value on our species' attaining ecological equilibrium. This is taken to imply our imposing on ourselves population checks first reducing the size of the human population, and then maintaining it at a constant level – which is close to being the case in what Sahlins has referred to as the original affluent society. Unlike in hunter-gather societies, however, we should be able to maintain that constant level without having to resort to abortion or infanticide, using instead our greater knowledge to help us attain this end by preventive means.

The use of non-renewable resources constitutes a key problem as regards humankind's ever breaking out of the vicious circle, for it is doubtful that we would ever stop using them (or any others) as long as it is immediately beneficial to us to do so. When the development of new weapons-technology such as the javelin made it possible to hunt animals which were earlier beyond our reach, we did not stop hunting them when they approached the point of becoming extinct, any more than we ceased hunting wild oxen before, in modern times, we killed the last one. Nor did we stop using irrigation systems that increased soil salinity until we could no longer eke enough from the soil to make our continuing to irrigate it worthwhile in the short run. We have never stopped taking any mineral from a mine until it was no longer economically advantageous in the short term to do so, nor have we ever stopped pumping oil from a well until we couldn't get any more out. Nor did we stop felling trees on Easter Island till there were no more trees to fell – despite its being quite obvious just when the last tree *was* the last tree. In recent times, we did not stop fishing cod off the coast of Newfoundland until there were no more cod to fish; and we haven't stopped hunting whales, despite all the international treaties in which we agree to do so, as well as the fact that a prime minister of one of the whaling countries was responsible for the above-mentioned 1987 UN environmental report. And we have not stopped pursuing economic growth, and probably shall not until it is no longer possible to do so. In fact our attitude has always been and will probably continue to be one of trying to deplete the world of its resources as quickly as we can – the pioneering principle applied through technology on a global scale. We humans have, via our technological innovations, been eradicating at an increasing rate not only each of the niches we've occupied, but those of virtually all types of other species as well.

Pollution

A major physical and social disadvantage with this system that is focused on by Commoner is that in practice it means the introduction of ever more polluting technology, part of the reason being that new technology normally requires more energy for its construction and operation. In these cases, pollution is an unintended side-effect of the natural drive of the capitalist economic system to introduce new technologies that increase profits. Aluminium, for example – the production of which rose almost a hundred-fold in the 1900s,[1692] increasingly displacing steel and lumber as a construction material – requires for its production about 15 times more fuel energy than steel and about 150 times more than lumber. Even taking into account that less aluminium by weight is needed for a given purpose than steel, the power discrepancy remains. For example, the energy required to produce metal for an aluminium beer can is 6.3 times that needed to produce a steel beer can. According to Commoner, most of the sharp increase in pollution is due not so much to population growth or increasing affluence as to this constant introduction of unnecessary new technology. Such changes have benefited the capitalist, but not society.[1693]

But, it may be thought, technology is often introduced that *reduces* resource use and pollution. First we should here point out that this is only done when it increases the profits accrued by the capitalist(s) in question, and second that it can easily be and normally is erased by increases in scale. A car that requires less energy to produce and operate will not benefit the environment nor society if in the end more energy is used in producing and operating a greater number of cars. As suggested by Goldsmith and his associates: "If we develop relatively clean technologies but do not end economic growth, then sooner or later we will find ourselves with as great a pollution problem as before but without the means of tackling it."

The principle of the conservation of matter makes it clear that to reduce pollution we must reduce the transformation of raw materials. Even if it were the case that there existed ten times as much oil as there actually does, there are limits to our ability to use this oil to 'improve' the human condition. Only so many cars can be fuelled by petrol before the negative effects of so doing are greater than the positive (as has long been the case); only so many fertilisers and poisons can be applied to agricultural areas until their effects do not improve harvests even in the short run, but worsen them; and only so many synthetic products can be produced until the environmental costs of producing them outweigh the benefits their existence provides.

However, largely due to the economic-growth ethos, the fundamental fact that a reduction in pollution implies a reduction in consumption is ignored by decision makers, who suggest technological ways of reducing pollution – for example with the use of filters and scrubbers – apparently without realising that this only means a slowing down of the inevitable effects of the entropy principle. Pollution and resource use are two sides of the same coin, and the idea of reducing pollution without stopping economic growth, which implies increasing resource use, is chimerical. As Kuenen says, economic growth will mean further increases in the rate of exploitation of non-renewable resources: metals, oil, coal, etc.[1694] As it appears that we will not consciously stop economic growth, the rate of pollution will lessen only after that growth has stopped of its own accord – that is, when it is impossible to continue to increase the rate at which we use resources.

Energy conservation

Jevons applied the same reasoning regarding more efficient technologies using more fuel to the case of steam engines: the better the degree of efficiency, the more steam engines would be built, and the more coal consumption would rise. And Sieferle suggests that if energy were conserved in factories, production costs would decline, leading to a price reduction of energy costs per unit produced, which would cause more to be produced and on balance more energy to be used. These conclusions have been corroborated in the real world. Thus conserving energy through increasing efficiency has not in practice slowed down resource use, but, on the contrary, speeded it up.[1695]

Alternative sources of energy

We might begin by noting with Jay Hanson that no so-called renewable energy system has the potential to generate more than a tiny fraction of the power now being generated by fossil fuels.[1696] But as we have seen, *any* energy system that could generate anywhere near the same amount of energy would only be of detriment to the species in any case.

When it comes to alternative sources of energy, a weakness exists in our thinking which is similar to that of conserving energy by manufacturing less-energy-demanding units. Here the entropy principle tells us that in the broader context all energy transformation is from higher to lower forms, each transition involving the giving off of heat. So the only way to reduce this production of low-quality energy or heat, as in the case of reducing pollution more generally, is to reduce energy transformation. Thus, given the problem of global warming, the idea of converting the same quantity of high-quality energy as we do today using, say, biological sources, apart from being practically impossible, is thoroughly wrong-headed. The challenge is not to find new sources of energy, but to reduce energy use so as eventually to come into ecological equilibrium with our environment. Thus we should recognise that e.g. if the new fusion-energy project in Europe were to succeed (though it is already known that it will not), this would not solve the problem of heat production, while at the same time it would give rise to many other problems.

Population growth

As Digby McLaren remarks,

> If an unseen intelligent being from somewhere else in our galaxy were to visit the Earth, perhaps the most incomprehensible phenomenon it could observe would be that the planet's apparently wise and competent dominant beings are totally ignorant of the life-support system they are destined to live within. They are, furthermore, unaware that their uncontrolled reproductive capacity has grown to the extent that it is rapidly destroying this system, while they fight among themselves to preserve their freedom to do so.[1697]

And as another commentator puts it:

> One would have thought that it was even more necessary to limit population than property; and that the limit should be fixed by calculating the chances of mortality in the children, and of sterility in married persons. The neglect of this subject, which in existing states

is so common, is a never-failing cause of poverty among the citizens; and poverty is the parent of revolution and crime.

This commentator is Aristotle, who wrote these words more than 2300 years ago, when the population of the world was less than 300 million, or one-twentieth of what it is today. They are more important today than ever before. Aristotle's thinking is echoed in the report of the Club of Rome from 1972 and in the findings of the Food and Agriculture Organisation (FAO) of the United Nations at that time, when the population of the world was only somewhat more than half of today's:

> The greatest possible impediment to more equal distribution of the world's resources is population growth. It seems to be a universal observation ... that, as the number of people over whom a fixed resource must be distributed increases, the equality of distribution decreases. 'Analysis of distribution curves shows that when the food supplies of a group diminish, inequalities in intake are accentuated, while the number of undernourished families increases more than in proportion to the deviation from the mean. Moreover, the food intake deficit grows with the size of households so that large families, and their children in particular, are statistically the most likely to be underfed.'[1698]

But this represents only a small part of the current problem of population growth, and only one sector swept out by the vicious circle.

Already in 1972 it was recommended by the President's Commission on Population Growth and the American Future that the United States "welcome and plan for a stabilised population."[1699] But the interests of the powerful have worked against this, to the extent of even taking it off the discussion agenda.

This in itself bodes ill for the future of the species. But to this it must be added that it is not as though more than six and a half billion people on the planet are consuming energy at the same rate as modern humans did when we first came into existence. The average modern person's use of energy is of a much greater order of magnitude. In fact *per capita* energy use is presently constantly increasing. And, as mentioned earlier, almost 90 per cent of that energy is being obtained from non-renewable resources, which we started using when the human population was about one per cent of the size it is today. Just using this figure as a guideline, it suggests that a sustainable human population, if it should ever be attained, will have to be something like 100 times smaller than today. And this too is independent of the vicious circle principle. As Kuenen also says, "If the human population continues to increase, we have but few options before us: starvation, large-scale wars, or catastrophic epidemics."[1700]

What is the United Nations doing about this situation? In 1967 some 30 nations belonging to the UN agreed to the following:

> The Universal Declaration of Human Rights describes the family as the natural and fundamental unit of society. It follows that any choice and decision with regard to the size of the family must irrevocably rest with the family itself, and cannot be made by someone else.[1701]

In this vein, the UN Rio conference of 1992, apart from supporting economic growth, made no mention of population in any of the conference goals or related documentation; nor does the term appear in any of the titles of the 22 related meetings that were held all over the world between April 1991 and June 1992;[1702] nor does it appear in the principles drawn up after the conference. And the report coming out of the United Nations World Population Conference held in Cairo in 1994, which similarly emphasises the need for *sustained* economic growth, advocates attaining population control through family planning based on free individual choice of family size.[1703] The idea here is that people will realise that for their own sakes it is better to have small families, and that they will begin doing so quickly enough to sufficiently control the increased starvation and other effects of overpopulation that are to be expected. This is hardly a recipe for success, considering that for many people in the Third World the advantages of having large families outweigh the disadvantages, and, as noted in Chapter 4, purely voluntary population control selects for its own failure in any case. And after 1994 the idea that individuals should be free to decide on the size of their own families became the United Nations' official stance. But it is not the poor people who benefit from this liberty, it is the rich, who are supplied with more labour while the poor increase in numbers. If this is the best the world can do, it is not good enough.

As pointed out by Hardin, the question of the size of the human population has been a taboo topic for at least the last 200 years. While thousands of articles are written every year about the pathological effects of overpopulation – traffic congestion, deforestation, military conflict, crime, homelessness, species loss, soil erosion and air pollution – population growth as an important contributing cause is never mentioned.[1704] In response to traffic congestion, rather than reducing the number of vehicles we build more roads; in response to homelessness, rather than reducing the size of the population we build more dwellings. The reaction principle rules supreme. As McLaren notes, in virtually all high-level and inter-governmental discussion the question of population is ignored or receives minimum attention, it being unclear to him whether this seeming conspiracy of silence is due to ignorance or design.[1705]

Here we should say that it is due more to design than ignorance, though the design is probably not conscious. It simply does not lie in the immediate interest of the powers that be to reduce the size of the population. The pressures they receive to act are not from the poor, but, in the case of politicians, from lobbyists representing capitalists, and in the case of capitalists themselves, for whom a growing population means ready access to cheap labour, from the profit motive. And population growth means a growing market e.g. for constructors and engineers, who are constantly engaged in increasing the size of the infrastructure, including the amount of housing; and it means increasing property values for real-estate speculators.[1706] We see evidence of this design, for example, in the United States, with the only half-hearted attempt on the part of the government to stop illegal immigration and the failure to systematically check the population for working permits; in the constantly increasing use of cheap Third World labour on the part of transnational companies; and in the increase in the population of the European Union through its inclusion of countries with low-wage workers. As expressed by Daly and Cobb, "It is our laboring class that pays the bill for the generosity of the capitalists who want to let poor immigrants in."

According to the VCP, population growth is only part of a problem intimately related to technological development. While it both pushes such development, as suggested by Boserup, and is promoted by it, had it not been for technological development the population problem would never have arisen. The size of the human population, like those of other complex species, would have been naturally curtailed by internal and external checks.

Conflict

A key aspect of the VCP is its seeing increases and decreases in violence within and between societies as being directly related to population pressure. Eric Wolf takes population pressure to be the cause of revolutions occurring in Mexico in 1910, Russia in 1917, China in the 1930s, Cuba in 1953, Vietnam in 1962, and Algeria in 1963.[1707] And, in keeping with Malthus as cited in Chapter 5, according to Parsons the piracy of the Vikings, the Barbarian invasions of Europe, and the Crusades were all caused at least in part by overpopulation. So too was the militarism of the Japanese in the 1930s, and the territorial ambitions of the Germans in the two world wars.[1708]

Crime

As regards civil violence in contemporary society, studies done by the US Defense Department show that the poorer a country is the more it is involved in major outbreaks of civil violence. Here we see the distinction between individual and group territoriality: the weak, through crime, fight over individual territories (access to capital), while the powerful, through war, fight over group territories. Only four per cent of the countries with a per capita income of $750 or more a year experienced civil violence, as compared with 84 per cent of those whose population had an average annual income of less than $100. As expressed by Parsons, if we accept that population pressure causes poverty and that poverty (above starvation level) causes violence, then population pressure causes violence in man as in many other creatures.[1709]

Apart from population pressure, increased violence is supported in contemporary society by the high mobility of the members of its population, particularly through the use of the car. This mobility reduces the chance of good community relations being established, which in turn increases the likelihood of crime.[1710] And to this we may add that with the advent of police we needn't be moral anymore, we just have to avoid getting caught. Similarly, as pointed out by Paul Hawken with regard to corporations, the legal costs of fines and settlements to resolve litigation can be written off against profits – essentially, as a cost of doing business. The leaders of corporations figure out whether it is cheaper to break the law and pay the fine or to follow the law, and do whatever is cheaper.[1711] And of course part of their calculation includes considering the chances of their actually being prosecuted.

William Bonger believes that there is a causal link between crime and economic and social conditions. He maintains that crime is social in origin and a normal response to prevailing cultural conditions, and contends that in more primitive societies survival requires more selfless altruism within the community. But once agricultural technology improved and a surplus of food was generated, systems of exchange and barter began which offered the opportunity for selfishness. As capitalism emerged, there were social forces of competition and wealth resulting in avarice, individualism and an unequal distribution of resources. And once self-interest and more egoistic impulses assert themselves, crime emerges.[1712] And, in the same regard, David Gordon, like Bonger, believes that *capitalism* tends to provoke crime in all economic strata.[1713]

As regards the United States, and as is in keeping with the view of Adam Smith, Reiman points out that there is considerable evidence that the American criminal justice system has been used throughout its history in rather unsubtle ways to protect the interests of the powerful against the lower classes and political dissenters.[1714] For example:

> We know that, by and large, privately retained counsel will have
> more incentive to put in the time and effort to get their clients off the

> hook, and we know that this results in a situation in which *for equal crimes* those who can retain their own counsel are more likely to be acquitted than those who cannot. The present system of allocating assigned counsel or public defenders to the poor and privately retained lawyers to the affluent is little more than a parody of the [American] constitutional guarantee of equal protection under the law.

What this means is that the justice system is itself unjust.

And as the vicious circle turns, its momentum picks up, and the distance between the powerful and the weak increases. Still looking at the American justice system, this is evident from the fact that the prison population in the United States has risen rapidly since the 1970s, when state and federal governments began to require mandatory and increasingly lengthy prison sentences for drug possession, a victimless crime. The population of state and federal prisons grew from fewer than 200,000 in 1970 to 1.2 million in 1998, with another 600,000 in local jails. At the beginning of the 1990s there were about one million people in prison in the United States. Ten years later there were *two* million.[1715] This means a tenfold jump in 30 years! Some 36 per cent of prisoners entering state prisons and 71 per cent of those in federal prisons were convicted of drug offences. And almost all of them came from society's poorest class, with minority groups being over-represented in this class. Seventy per cent of prison inmates in the US are reported to be illiterate. And the drug-driven rapid increase in prison populations has led to widespread overcrowding; California's prison system, for example, is running at twice its intended capacity – despite the construction of 21 new prisons in the past 20 years.[1716] Here it may be pointed out that, if intended to reduce crime, the prison system is ludicrous since it itself creates a society of criminals, thereby reinforcing criminal activity.

As Reiman suggests, if our intention were truly to deal with crime, we should end poverty, criminalise the really dangerous acts of the well off, and decriminalise victimless crimes. This would reduce crime and protect society, freeing up our police and prisons for the fight against the criminals who really threaten our lives and limbs.

Reiman points out that the criminal law enshrines the *established* institutions as equivalent to the minimum requirements for *any* decent social existence – and it brands the individual who attacks those institutions as someone who has declared war on *all* organised society, and who must therefore be met with the weapons of war. As Reiman says, to look only at individual responsibility is to look away from social responsibility.

As regards crime and the lower classes, Reiman notes that by holding the poor crook legally and morally responsible, the rest of society effectively denies that it is the beneficiary of an economic system that exacts such a high toll in frustration and suffering.[1717] The poor commit crime out of need or a sense of injustice. But it is those with power who exercise control and impose punishment, equating the definition of crime with harm or threat of harm to the property and business interests of their own class. Although the inherent activities comprising, say, a theft, may be identical, theft by the poor will be given greater emphasis than theft by the rich. This will increase the pressure for survival in an unequal society, as well as the sense of alienation among the poor. Crime in the streets is a result of the miserable conditions in which workers live in competition with one another.[1718]

> That Americans continue to tolerate the comparatively gentle treatment meted out to white-collar criminals, corporate price fixers, industrial polluters, and political-influence peddlers, while voting in droves to lock up more poor people faster and longer indicates the degree to which they harbor illusions as to who most threatens them.[1719]
>
> For some poor ghetto youth who robs a liquor store, five years in a penitentiary is our idea of tempering justice with mercy. When a handful of public officials try to walk off with the US Constitution [in the Watergate affair], a few months in a minimum security prison will suffice. If the public official is high enough, say president of the United States, resignation from office and public disgrace tempered with a $148,000-a-year pension is punishment enough.

War

Why are world leaders not uniting in an effort to put an end to war in this period of globalisation? They cannot even agree to end the arms race.[1720] The reason for both is that war is a manifestation of male social *instincts*. And besides, there's money to be made in war (individual territoriality; the sexual instincts), and capitalism is constantly becoming more entwined in politics.

As Lorenz says:

> Unreasoning and unreasonable human nature causes two nations to compete, though no economic necessity compels them to do so; it induces two political parties or religions with amazingly similar programmes of salvation to fight each other bitterly, and it impels an Alexander or a Napoleon to sacrifice millions of lives in his

attempt to unite the world under his sceptre. We have been taught to regard some of the persons who have committed these and similar absurdities with respect, even as 'great' men, we are wont to yield to the political wisdom of those in charge, and we are all so accustomed to these phenomena that most of us fail to realize how abjectly stupid and undesirable the historical mass behaviour of humanity actually is.

On the VCP, however, the "absurdities" committed by the 'great' men through their military campaigns increased group territory and reduced population pressure, benefiting the survival of the group, and that is why their exploits are extolled in legends and myth. And, in keeping with the line of Russell and Russell, the extent to which the "unreasoning and unreasonable human nature" Lorenz refers to functions as a population check, and the extent to which it is manifest, depend on the degree of population pressure.

Of course, as Lorenz realises, in our hunter-gatherer, ecologically sounder, state we killed each other off at a tremendous rate, so a return to that state would not reduce conflict-induced mortality. But we will have to return to some primitive state, with or without much killing. As intimated in Chapter 4, you could of course have a situation where intraspecific violence, including e.g. feuding, war, abortion and infanticide, helps *keep* a population in ecological equilibrium. But then polluting armaments could not be used; and the amount of violence would still increase with increasing population pressure.

But it must be admitted with Lorenz that our engaging in war with one another, with its incident suffering and loss of life as well as material destruction, is one of the 'stupidest' aspects of human behaviour. The existence of war is easily explained on our theory, it being the result of group territoriality, the invention of weapons, the existence of a surplus, population growth, social stratification and the ability to communicate (give orders) via language. And its constantly increasing potential for physical destruction is also explicable as the result of continual technological innovation in the form of improved weaponry. This situation is further exacerbated by capitalism, for e.g. as regards nuclear weapons, the owners of nuclear know-how are eager to make a profit from that know-how by selling it to others, thus facilitating their proliferation. But this applies to all weapons. As Roland von Malmborg says, and as is reminiscent of the relative status of the smith and potter: "If one had also been able to use windpower as a weapon, the world would be lousy with windpower!" And, as regards nuclear energy in particular: "That nations invested these fantasy sums in nuclear energy depended completely on their attempts to acquire nuclear weapons – even in Sweden, just as in present-day oil-rich Iraq and Iran."[1721]

And as Schumacher realised already in 1970, fossil-fuel resources being limited and very unevenly distributed, any shortage of supplies, no matter how slight, would immediately divide the world into 'haves' and 'have-nots' along entirely novel lines. The specially favoured areas, such as the Middle East, would attract envious attention on a scale scarcely imaginable, while some high consumption areas, such as Western Europe and Japan, would move into the unenviable position of being residual legatees. Here, claimed Schumacher, is a source of conflict if ever there was one. The exploitation of fossil fuels at an ever-increasing rate is an act of violence against nature which must almost inevitably lead to violence between men.[1722]

Terrorism

The strength of the territorial aggressive instinct over reason in situations of population pressure is apparent in today's world in e.g. the reaction of American and Israeli leaders to acts of terrorism. As everyone realises, reacting to violence with violence only leads to more violence, unless the original source of violence is completely eradicated – which in the present case could mean the elimination of one-and-a-half billion Muslims – and steps are taken to ensure that new sources do not take its place. The elimination of particular terrorist leaders only results in a rush for others to fill the martyrs' shoes. As has been said about the latest American invasion of Iraq as part of its 'war on terrorism,' the primary effect of removing Iraq's dictator from power has been to make the country an ideal incubator for terrorism. (At the same time however it gives the United States and her allies greater control over Iraqi oil, and the military expenditure provides military-industrialist capitalists with a profit ultimately paid for by taxpayers.) But this is the sort of activity that political leaders concentrate on, following the bidding of their capitalist sponsors. They devote absolutely no attention to determining what the underlying causes of the terrorism in question are, and dealing with those causes in a non-violent fashion.

Not only has terrorism increased with advances in weapon technology and increasing population, but so has social revolt more generally – whether or not this is due to some of the weak's having access to superior weapons or for other reasons, such as increased stratification.

The Third World

As Schumacher says, problems grow faster than their solutions – in the rich countries as much as the poor. Following him, we should say that there is nothing in the experience of the last 50 years to suggest that modern technology can really help us alleviate such problems as that of world poverty, not to mention the problem of unemployment.[1723] As Hawken has pointed out, literally thousands of native cultures around the world have been destroyed by economic development. Lost with those cultures have been languages, art and crafts, family structures, land claims, traditional rites and oral histories, and traditional methods of healing, obtaining food, and population control.[1724] And, as pointed out above, the improvement of the situation of the poor in the Third World is not even the intention of those that stand behind decisions to implement large capital-intensive projects there. The intention, rather, is to make as much money as possible. This applies both to the wealthy capitalists in industrialised states who invest in Third World projects, and to those who have power in the Third World.

As suggested by Baran (in 1957), the 'backward' world has always represented the indispensable hinterland of the highly developed capitalist West, supplying it with many important raw materials, thereby providing their corporations with vast profits and investment outlets. Thus the ruling class in the United States (and elsewhere) is bitterly opposed to the industrialisation of the so-called 'source countries,' and to the emergence of integrated processing economies in colonial and semi-colonial areas. This opposition appears regardless of the nature of the regime in the underdeveloped country that seeks to reduce the foreign grip on its economy and provide for a measure of independent development.[1725]

As regards the phenomenon of increasing social inequality, it is important to appreciate that it should continue so long as the vicious circle is able to continue turning without hindrance. However, such events as an inordinate increase in a society's surplus, as in ancient Athens and modern industrialised nations, or a social revolution, as in France and Russia, can lead to an increase in social equality. But such increases are invariably directly followed by constant decreases, unless and until such an event should occur again.

That the non-vital needs of the powerful living in Third World countries also strongly influence those countries' domestic economics is emphasised by Georgescu-Roegen. He noted, already in 1971, that Third World countries' economic plans, claimed to bring economic progress through industrialisation, are, more often than not, rationalisations of the ulterior motives of the elite in the country in question. The inflation in Latin America at that time, for example, did not answer 'the aspiration of the masses to improve their standard of consumption,' as one economic expert claimed, but the aspirations of the upper classes for a still more luxurious lifestyle. Similarly, the leaders of underdeveloped countries are not anxious to limit the populations of their own lower-class majorities, because cheap and abundant labour is a benefit to the ruling class.[1726] According to Georgescu-Roegen, and in keeping with the VCP, the same lip service to the welfare of the masses concealed the aspirations of the powerful classes in many a planned economy at the beginning of the 1970s,[1727] and, we might add, the phenomenon has continued to the present day.

The majority of today's underdeveloped nations are destined never to become developed,[1728] and the Third World would have been better off without international investment and aid.[1729] As Goldsmith says: "The fact is that trade with the Third World is negative aid – it involves selling the indispensable in exchange for the totally superfluous. If I were running a Third World country, the first thing I would do would be to cut myself off from the industrial world and foster self-sufficiency at every level down to that of the village. In fact, one should not be developing the Third World but de-developing it."[1730] And as noted by Carr-Saunders, "there is a considerable amount of evidence to the effect that upon the whole before the advent of the white man the African races were healthy and long-lived."

There has been no appreciable improvement in the economies of Third World countries after World War II. As Schumacher noted already in 1965: "In many places in the world today the poor are getting poorer while the rich are getting richer, and the established processes of foreign aid and development planning appear to be unable to overcome this tendency;"[1731] and again in 1973: "For two-thirds of mankind, the aim of a 'full and happy life' with steady improvements of their lot, if not actually receding, seems to be as far away as ever." As aptly put by Boulding in 1972: "The interesting thing about developing countries is that they are not developing."[1732] And, more than 35 years later, they are still not developing.

Some 50 years ago these countries were politely and optimistically named the 'developing countries,' and the 1960s were to be known as the 'Development Decade.'[1733] But 'development' here meant growth in GNP, which was to be accomplished through *increasing* resource exportation – as taken up in the previous chapter. This growth was to be supported by growth in the GNP of the industrialised countries[1734] – the more the industrialised countries grew, the more resources they would import from the Third World, thus benefiting Third World economies. Thus, for example, the Report of the 1970 Commission on International Development (the 'Pearson Report') submitted to the World Bank considered the expansion of exports – mainly non-renewable minerals, including oil – the main criterion of success for 'developing' countries. African and other Third World countries were to develop economically through the wealthy people in each country making increasing profits by exporting ever greater quantities of their respective country's resources, and creating jobs for labourers in the process. But as Malthus said already in 1798:

> Foreign commerce adds to the wealth of a state, according to Dr
> Adam Smith's definition, though not according to the definition of
> the [French] economists. Its principal use, and the reason, probably,
> that it has in general been held in such high estimation is that it
> adds greatly to the external power of a nation or to its power of
> commanding the labour of other countries; but it will be found, upon
> a near examination, to contribute but little to the increase of the
> internal funds for the maintenance of labour, and consequently but
> little to the happiness of the greatest part of society.[1735]

Malthus' reasoning here is that it is only a growth in the quantity of vital
resources available to the poor that can improve their lot (and then, of course,
only in the short term). What we have is the making of each Third World country into a banana republic, which may here be understood to be a poor country
economically dependent on exporting unprocessed goods/resources to industrialised countries.

As Daly says:

> [T]he vision of globalization requires the rich to grow rapidly in
> order to provide markets in which the poor can sell their exports. It
> is thought that the only option poor countries have is to export to
> the rich, and to do that they have to accept foreign investment from
> corporations who know how to produce the high-quality stuff that
> the rich want.[1736]

And as Trainer says, if most money can be made producing carnations to airfreight to European supermarkets, or fattening cattle to airfreight to American
hamburger chains, then in a market system that is what will be done.[1737] And
Kuenen: "At present the technologically underequipped nations are selling
their natural wealth for short-term gains."

It is ironic however that governments call for economic growth to reduce
poverty while, as noted earlier, there has been massive poverty in the richest
nation in the world throughout a 65-year period of tremendous and unrepeat-
able economic growth. (It may be noted that the wealthiest man in America
owns more than the poorest 100 million Americans combined.)[1738] How then is
economic growth, in particular such growth as is based on exports, to reduce
poverty? For the people living in these countries, what they produce for themselves and for each other is of infinitely greater importance to them than what
they produce for foreigners.[1739] The promotion of export-oriented development
has been one of the most disastrous Third World policies in the past two decades, in fact increasing poverty.[1740]

The whole thing is a scam – part of the larger scam of the world's need for economic growth – that allows powerful capitalists to make profits stripping the Third World of what it has to offer. Thus the status quo from colonial times is maintained, with the economically most powerful making the largest possible profits. Only now it is transnational corporations that are sucking as much as they can out of these (and all other) countries, rather than such nationally-bound companies as the East India and Hudson's Bay Companies.

The fundamental 'mistake' which neoclassical economic theory makes with regard to the Third World is the assumption that simply encouraging as much economic growth as possible will result in satisfactory development. In fact the indiscriminate, sheer-growth conception of development causes immense havoc among the poor. In the form of increasing exports, it has stripped them from the land and moved them to urban slums, it has made large numbers poorer and hungrier, and it has destroyed their forests through the building of dams.[1741] According to a report of the international Institute for Agriculture and Trade Policy, below-cost imports drive Third World farmers out of their local markets, and if they do not have access to a safety net, they have to abandon their land in search of other employment.[1742]

When it comes to the provision of aid, the West has given with one hand, and taken more with the other.[1743] In effect, *more aid is going to the rich than to the poor*.[1744] It is the normal functioning of the global market-economy which delivers the available resources to a few and deprives the majority. The drive to maximise output, sales and returns on investment inevitably leads to the focusing of productive capacity on the already rich.

The conventional growth and trickle-down view accelerates the operation of the very mechanism that is responsible for the problem of poverty. As expressed by John Browett, in keeping with Boulding, while transnational corporations may be developing, the people living in the newly industrialising countries are not.[1745] While trickle down occurred in the industrialised countries from 1850, it has never extended to the Third World. And as suggested by Trainer in 1989, the lack of trickle down in the Third World may well be the most clearly established proposition to have emerged from three decades of development research. In fact, as intimated above, conventional growth strategies often result in the very opposite of trickle down, an effect most tragically evident when the 'modernisation' of agriculture enriches planters, who then increase export crops by terminating the leases of peasant farmers.[1746]

The conception of development as growth through increasing exports does not best serve the interests of classes other than the elite. What is required here is not that the rich world charitably redistribute some of its wealth to the poor; it is that it should stop taking such a disproportionate share of what the world has to offer.

Conventional development theory and practice is *capitalist* (*bourgeois*) development theory and practice. To conceive of development as indiscriminate economic growth is to opt for the view which most suits the capitalist class, since it is in their interest to maximise the amount of capital being exchanged, and not have to bother about whether capital really ought to go into things that are appropriate but not very profitable, and not into things that are inappropriate but profitable. Foreign investors *never* go into the Third World to invest in clean drinking water, mobile health clinics, or cheap staple foods for impoverished people – because there is little profit to be made from these sorts of ventures.[1747]

It should also be pointed out here that the projects funded in the name of aiding the Third World, apart from economically supporting those engaged in carrying them out, are large-scale, unsustainable and in fact ecologically destructive. These projects, such as the building of large dams, are drafted in an atmosphere in which economic growth is to be striven for as the ultimate goal, and ecological consequences are either ignored or dismissed.

C. G. Darwin provides an example: the Sukkur dam (completed by the British in 1932) spread the water of the Indus over a great area and transformed a large part of the desert into a garden. According to generally accepted values, this was a great blessing for humankind, since people who earlier were on the verge of starvation could now be fed. But this was not what happened; after a few years the only effect was, as in the case of the Green Revolution, that there was a large rather than a small number of people on the verge of starvation.[1748]

Similarly, the Aswan High Dam, designed by Soviet engineers in the late 1950s, stops 98 per cent of the silt that had formerly coated the inhabited part of Egypt. Without this top dressing of fertile silt, Egyptian agriculture had to turn to mineral fertilisers, of which Egypt became one of the world's top users, with much of the Aswan's electric power going to fertiliser factories.[1749] The Nile Delta began to shrink. The lack of silt nutrients destroyed sardine and shrimp fisheries in the Mediterranean that had employed 30,000 Egyptians. Without the flushing of the flood, the irrigation canals of Egypt became an ideal habitat for the water hyacinth, a beautiful but pernicious weed. The snails that carry schistosomiasis – a debilitating disease that attacks the liver, urinary tract, or intestines – love water hyacinth, need stagnant water, and consequently flourished in the new Egypt. Schistosomiasis infection rates increased five- to tenfold among rural Egyptians with the transition to perennial irrigation, and after 1975 approached 100 per cent in many communities.[1750] The dam also swamped and corroded the cultural heritage of the Nile Valley. However it at the same time eliminated the costly consequences of irregular Nile floods, and supported a doubling of the Egyptian population. Thus was destroyed the only large, ecologically sustainable irrigation system that ever existed – one which had maintained millions for five millennia and made Egypt the richest land in the Mediterranean from the Pharaohs to the industrial revolution.[1751]

The modernisation of Third World agriculture also means the increasing commercialisation of food production, and can consist in little more than converting land from production by the poor for use by the poor, to production by rich farmers for use by the rich in the Third World and by consumers in the rich world.[1752]

Overshoot and the ecological revolution

The next major revolution in the development of humankind – which might be termed, like the perspective being presented in this book, the *ecological* – will be the first to be connected to a general *decrease* in available resources. That is to say, it will be the first and largest revolution involving the slowing down and eventual stopping of the vicious circle. The coming overshoot will not be that of just one turn of the vicious circle, but of the circle itself. The repercussions will be tremendous.

In systems terms, with each new species in our development there was a change not only of structure but of organisation, i.e. there was a constant introduction of new systems. And each of these systems has been out of dynamic equilibrium with its environment (otherwise it wouldn't have disappeared) – and each to an increasing degree. As regards our own system and the point of no return after which dynamic equilibrium cannot be regained, it was probably passed with the beginning of the mining of metals, if not before.

With the reduction in resource use, technology will no longer be able to provide a surplus that supports continued population growth, but will rather be fighting off a deficit that inflicts constantly increasing mortality. World peaks in population, standard of living, energy use, food production, trade, and rate of innovation will all be behind us. The social effects of reduced resource consumption will include an increase in experienced population pressure and consequent violence, including war.

As Dennis Meadows says when asked about peak oil in an interview:

> In practice what will happen, as it becomes clear that peak oil is a
> reality, is that the rich and powerful will grab as much as they can,
> and not worry much about the poor and the weak. ... It isn't going
> to be the ushering in of a golden age for humanity. We also have five

or six other peaks to contend with, leave aside peak oil: peak water, peak food, and so forth, and collectively they are going to cause a lot of problems. ... If you think about the degree of change you've seen in the last 100 years, social, technical, cultural, political, environmental, all those changes, they're less than what you'll be seeing in the next *twenty* years. What will they be? I would venture that there'll be a lot less people twenty years from now, or forty years from now, it's hard to tell .... We are moving into a conflict-filled time, where people are going to do terrible things to each other.[1753]

As suggested, the slowing down of the vicious circle will involve such things as decreasing resource use and the end of population growth. But, thanks to our instincts and the reaction principle, as long as technological innovation is to the advantage of capable people, the circle will continue to turn.

Conclusion

Much speaks for the viability of the theory of the development of humankind based on the vicious circle principle. It is in keeping with all of the principles of modern science, including Darwin's principle of evolution, and in fact may be said to constitute a particular application of that principle, providing a coherent view of the whole of our development right up to today. More than this, it explains such important *unique* phenomena as how we came into existence in the first place, the prehistoric overkill phenomenon, the food crisis in prehistory, the horticultural and agrarian revolutions, and the industrial revolution. And it explains such *constantly present* phenomena as human population growth, war, the striving for economic growth, and accelerating energy use, as well as the phenomena of increasing resource depletion and environmental destruction. As regards the views of others, the theory also explains and vindicates the new views in anthropology, archaeology and economics, as well as the views of Malthus, Boserup and Wilkinson with regard to population growth and economic development.

Humankind's following the VCP works counter to our survival as a species, involving as it does constantly increasing consumption, population and quantities of waste, all of which tend to move us further out of equilibrium with our surroundings, thereby increasing the likelihood of our becoming extinct. Key to this whole process is our constantly meeting vital needs through technological development, combined with the fact that there have to date always existed resources amenable to that development.

As I see it, our situation is much like that depicted by Joseph Wayne Smith and Gary Sauer-Thompson where they suggest that due to the power/interest structures of global capitalism and the juggernaut-like momentum of the global economy, it is most unlikely that any of the radical changes to society and the economy proposed by environmentalists – especially changes in philosophies and worldviews – will be adopted in time. Consequently human civilisation – primarily Western techno-industrial urban society – will self-destruct, producing massive environmental damage, social chaos and megadeath.[1754] We are entering a new dark age, with great dieback.[1755]

The only question that remains is whether we will survive this dark age, and if so, for how much longer.

Glossary

Acheulean: Humans' second tool-making tradition, preceded by the **Oldowan** and followed by the **Mousterian**. Characterised by hand-axes and cleavers. Stones were shaped by striking off flakes with another stone, or with bone, antler or wood, until the final form was achieved. Widespread in Africa, Europe and parts of Asia from around 1.65 million to 120,000 BP, and associated with *Homo erectus*, **archaic** *Homo* and the hunting of big game.

archaic *Homo*: A number of **species** with features intermediate between *Homo erectus* and *Homo sapiens* that lived at various times from about 500,000 or more to 28,000 BP. Archaic *Homo* has recently come to be considered to constitute a genealogical line independent of that leading to *Homo sapiens*. There is disagreement about which fossils belong in this category. The archaic *Homo* line extends from directly after *Homo erectus* to and including the **Neanderthals**. *Homo heidelbergensis* is considered to be the earliest archaic *Homo*, and e.g. *Homo rhodesiensis* and *Homo soloensis* are also considered members of the order.

Archaic period: See **Mesolithic**.

Australopithecus: 'Southern ape.' The genus to which the first **hominids** belong. Existed at least between 4 and 2 million years ago – perhaps as early as 4.4 million years ago and as late as 1.25 million years ago. Australopithecines were bipedal (hominid bipedalism having existed from around 6 million BP), with prognathic faces and chimp-sized bodies and brains. The brain size of the gracile *Australopithecus* was about 450 cm^3 or somewhat larger, but so small as is believed to have made them incapable of speech. Lived in southern and eastern Africa.

biome: A biome is the synthesis of all living organisms in a geographically defined and biotically homogeneous area or region, such as a mountain valley, lake, steppe, forest or desert. A biome can be thought of as consisting of many different **ecosystems** – minus their purely physical aspects – taken as a whole.

biosphere: The term has been given various meanings. In this book it means the synthesis consisting of all living organisms (*biota*), as may be compared with the *lithosphere* (consisting of all rocks), the *hydrosphere* (water) and the *atmosphere* (air). The synthesis of these four spheres constitutes the *ecosphere*, as may be contrasted with the *technosphere*.

biosystem: See **ecosystem**.

biotic: Living. Cf. **organic**.

blade technology: Introduced by ***Homo sapiens*** ca. 40,000 BP, i.e. at the beginning of the **Upper Palaeolithic**. A blade, produced by flaking, is twice as long as it is wide, and is made from a *core* prepared in such a way as to produce many similar blades. Blades are often referred to as *projectile points*, being appropriate for weapons that are thrown (javelins) or projected (darts and arrows). Cf. **flake technology**.

brucellosis: An infectious disease that affects sheep, goats, cattle, deer, elk, pigs, dogs, and several other animals, including humans; symptoms are like those of influenza, and can become chronic.

Bushmen: G/wi; Dobe-area (including the !Kung); Auen; Heikum: Modern hunter-gatherers in Namibia, Botswana, Angola and South Africa. At one time spread over almost the whole of South Africa, they are today confined principally to the Kalahari Desert. The total number of Bushmen still in existence in 1930 was about eight to ten thousand, their numbers constantly diminishing both before and after this time. Most Bushmen are in a state of servitude to the Bantu peoples among whom they live, and have not been fully dependent on hunting and gathering for their subsistence in historic times if not for much longer.

capital: Storable, exchangeable, owned property, including property that can be used in trade (paradigmatically the *means of production* of consumables). Thus e.g. **stocks**, **reserves** and *money* all constitute capital. Capital in the form of stocks or reserves has a physical dimension, while in the form of money, i.e. debt, it does not. Physical capital normally has a **use value** as well as an exchange value, while debt has only an exchange value. *Commodities* are capital with a use value.

capitalism: Capitalism is the pursuit of profit by means of continuous, rational, economic enterprise. The acquisition of a profit presupposes the existence of a *surplus*, both of **resources** (and/or **reserves** or **stocks**) and of labour.

capitalist: A person engaged in exchanging **capital** for profit.

carrying capacity: The ability of a particular region to support a certain number of individuals of a particular **species** *for an indefinite period of time.*

> **short-term carrying capacity:** The ability of a region to support a certain number of individuals for a particular period.

Cenozoic: The era from 65 million BP, when the dinosaurs became extinct, to the present; also called the Age of Mammals.

Clovis toolkit: The toolkit of humans who lived in North America ca. 11,500–10,800 years ago. Clovis points are made by a **Mousterian flake technology**.

They were used to tip lances and possibly javelins, are on average 7 cm long, and have been found across the whole of the coterminous states of the US and down into Central America. Clovis (and **Folsom**) points are *fluted* (see **flake technology**) – unlike the points of the same era in Siberia. No Clovis points have been found in the Old World, Alaska, Canada or South America. The workmanship on the Clovis toolkit was superb, reflecting a high level of expertise. See also **Folsom toolkit**.

cotton gin: Invented by Eli Whitney in 1793, the cotton gin afforded an easy way of removing cotton seeds from cotton by means of a combing process.

Cro-Magnon: *Homo sapiens* living in Europe during the **Palaeolithic**. Cro-Magnon's brain size was on the average 1350 cm^3, while that of *Homo sapiens* living today is 1330 cm^3. One Cro-Magnon cranium enclosed a brain of 1600 cm^3. See also **modern *Homo***.

economic development: A change in the means by which goods or services are acquired or distributed. Economic development paradigmatically involves the use of more sophisticated technology capable of taking more from the environment than the technology it is replacing at the time it is replacing it.

economic growth: Increase in the quantity of goods being supplied and/or services being employed in a human **population**. Presently measured in terms of increase in the amount of money changing hands.

ecosphere: See **biosphere**.

ecosystem: Ecosystems are much smaller than **biomes**, and include purely physical **systems**. An ecosystem is constituted by the interaction of its subsystems in such a way that the exchange of materials between living and non-living parts follows a circular path. A *biosystem* is an ecosystem minus its non-living aspects. Cf. **biosphere**.

energy, free: As versus *bound* energy. The capacity to do work, thwarting the natural tendency of **systems** towards disorder. Also termed *exergy*.

exchange value: See **use value**.

Fertile Crescent: Stretched through the Levant and Iraq down to the mouths of the Tigris and Euphrates rivers, and included the *Hilly Flanks* of the Zagros Mountains to the east. The Fertile Crescent is sometimes taken to include the Nile Valley. The Hilly Flanks were a zone of woodlands and grasses (wheat, barley) where the first instances of horticultural sedentism occurred. The ancestors of many domesticated plants and animals came from this area.

fertility: The number of viable offspring produced. May be contrasted with *fecundity*, which is the number of viable offspring that would be produced under ideal conditions. Cf. **mortality**.

feuding: Intratribal lethal conflict typically between bands; may last over generations.

fitness, Darwinian: An organism's, group's, or **species'** ability to adapt to its environment, as manifest in its relative number of reproducing offspring and near relatives, or, on the micro level, in the dissemination of its particular **gene types** or instantiations of its **karyotype**.

flake technology: The most primitive technique of working stone, used from the first appearance of humans 2.5 million BP right up until the Bronze Age. Includes **Acheulean** and **Mousterian** technologies. The **Clovis** and **Folsom** toolkits included (Mousterian) points made using flake technology. *Flakes* are removed from a piece of stone termed a ridge or *amorphous core*, fashioning it into a point. Such cores are not like those used in making **blades**, where a number of blades are made from the same core. Flaked points are distinguished by flaking over the whole surface, while hollows may be made at the base by the removal of a single flake from each side – called *fluting* (also used in blade technology) – presumably to aid hafting. Cf. **blade technology**, **Levallois**.

flying shuttle: Or fly shuttle. An innovation in looms patented by John Kay in 1733. Using the flying-shuttle system, where the shuttle is operated by cords rather than being passed from one hand to the other, a weaver could produce wider cloth at almost twice the speed per quantity produced. (In 1753 Kay's house was ransacked by a mob of textile workers who feared that his machines would destroy their livelihood.) See also **Luddites**.

Folsom toolkit: Later (11,000–10,000 BP) than the **Clovis** toolkit; both Clovis and Folsom points were **flaked** and fluted, but the Folsom points were smaller, and were fluted almost to the tip. The Folsom peoples perfected the art of *pressure-flaking* fluted stone projectile points, extending the beauty and workmanship of the Clovis point to a degree never equalled. (The Mexican Indians – in 1615 – used as a flaking tool the shaft of a lance, about 130 cm in length. It was held in both hands in such a manner that it rested at one end against a block of obsidian – grasped between the feet as in a vice – and at the other against the chest of the worker. By bending forwards great pressure is applied until a flake flies off.) Some of the points were crescent shaped and were possibly used to stun birds. The toolkit also included scrapers, thin knives, crude choppers and **spearthrowers**. Around 10,800 BP those using the Folsom toolkit did not hunt e.g. mammoth (which by this time were extinct) but rather bison, picking stragglers out from the rest of the herd, or driving groups into cul-de-sacs or marshes and then killing them *en masse*.

gene type or allele type: Each gene is the manifestation of a particular gene type, such as the type for blue eyes. What are passed on from parents to children are not genes, but gene types. Cf. **genotype**, **karyotype**.

genetic constitution: See **genome**.

genome: A *species'* genome consists of those *kinds* of genes that can occupy particular places on the species' chromosomes. It is part of the species *karyome*, which is the instantiation of the species' **karyotype**. For an *individual*, the genome consists of those *types* of genes that can occupy particular places on the chromosomes. Thus a change in the species' genome implies a change in the species, while a change in individuals' genomes does not.

genotype: There are individual genotypes and **species** genotypes. The individual genotype consists in the combination of **gene *types*** (e.g. for *blue* eyes) that go into the making of an individual organism. It is the organism's genotype that makes it distinct from other organisms of the same species. The *species* genotype consists in the combination of gene *kinds* (e.g. for eye *colour*) that fill the **karyotype** of a species.

glaciation, glacial period: An ice age (though sometimes the term *ice age* is taken to cover all glaciations over the past 1.8 million years); a manifestation of climatic changes caused by changes in the earth's orbit. Such changes have intensified during the past million years. During the various glaciations glaciers and ice caps were much more extensive than they are today, the world's climate was colder and drier, and sea levels were sometimes up to 200 m lower. Since 600,000 BP there have been four major glaciations, though at least eight significant glacial–interglacial cycles have occurred during the past 800,000 years, with many less-intense cycles before this. The warmer interglacials occupy only about 10 per cent of this time. As regards the last ice age (the Würm-Weichsel-Wisconsin), ice sheets began to grow around 115,000 BP, while the glaciation itself lasted from 80,000 to 15,000–13,000 BP. From 50,000 to 40,000 years ago there was a temperate period. Glaciation reached its greatest extent about 18,000 BP, when the ice was 3.5 to 4 km thick.

The land bridges that formed when sea levels fell during the last glaciation played a large part in the spread of ***Homo sapiens*** around the world. Such land bridges existed e.g. between New Guinea and Australia and between Siberia and Alaska. The former allowed people from south-east Asia to enter Australia (though they first had to cross some 100 km of sea between mainland Asia and New Guinea) around 55,000–50,000 years ago; and the latter allowed a group of Mongoloid people from north-east Asia to enter the Americas about 11,500 years ago.

gross national product (GNP): The total monetary (i.e. exchange) value of all goods and services produced by a nation and sold on the open market during a given period. **Economic growth** or shrinking in a country is measured as an increase or decrease in its GNP. The more self-sufficient people are, the lower their GNP will be at a given level of consumption. One can similarly speak of *gross world product* (GWP).

habitat: See **niche**.

Hadza: A small group of modern hunter-gatherers living in the vicinity of Lake Eyasi, a salt, Rift Valley lake just south of the equator in Tanzania.

Hilly Flanks: See **Fertile Crescent**.

Holocene: Period after the last **glaciation**, from about 10,000 BP to the present.

homeostasis: The tendency of a **system** to maintain dynamic equilibrium (to counter entropy) thanks to mechanisms operating within it.

hominid: Primate tool-user. Hominids include humans and all earlier regular tool-using primates. They thus include all **species** of ***Australopithecus***

and *Homo*. *Early* hominids are hominids before ***Homo habilis***, and consist essentially of members of species of *Australopithecus*. Exactly what is to count as a hominid species is much debated; a total of some 16 **protohominid** and hominid species lived in Africa between 8 and 3 million years ago.

Homo: The genus to which humans belong. Came into existence 2.5 million years BP, the first named form being ***Homo habilis***. The number of extinct *Homo* **species** is uncertain, but at least five forms are recognised, including: *Homo habilis* small (= gracile) and large (= robust) forms, which were probably different species, ***Homo erectus***, the various **archaic *Homo*** species, and ***Homo sapiens***.

Homo erectus: Lived from 1.8 million to ca. 300,000 BP. First **hominid** to move out of Africa. Used fire 1 million years ago and perhaps even earlier. Its brain size was about 1000 cm^3. Includes Java Man (*Pithecanthropus erectus*; brain size ca. 850 cm^3) and Peking Man (*Homo erectus pekinensis*; 915 to 1225 cm^3). Cf. **archaic *Homo***.

Homo habilis: The first *Homo* = human. 2–1.5 million years BP; lived exclusively in Africa. Considered the first **hominid** to make stone tools. *Habilis'* brain size was 600–750 cm^3. See also **Oldowan**.

Homo neanderthalensis: The Neanderthals. A **species** of ***Homo*** that existed between 230,000 and 28,000 years ago. The Neanderthals inhabited Europe, including southern England but not Scandinavia, eastern Asia to the Caspian Sea, and the northern part of Africa, though not the coast. Associated with **Mousterian** and **Levallois** tools – intermediate between **Acheulean** and **blade technologies**. The Neanderthal's body proportions are similar to those of modern **populations**, such as Eskimo, who endure cold – stocky and compact to minimise loss of body heat. (Cf. *Allen's rule* and *Bergmann's rule*, both concerning warm-blooded animals living in cold climates, the former stating that they usually have shorter limbs than similar animals from warmer climates, and the latter that they tend to have larger bodies. Both rules have the effect that animals living farther north have less skin exposed per unit body weight.) Neanderthals' average brain capacity was 1500 cm^3, whereas the modern average, calculated for all populations, is around 1350 cm^3. The Neanderthals belong to the **archaic *Homo*** line, which is independent of the **modern *Homo*** line.

Homo sapiens: Our own **species**. Originated in Africa some 200,000 years ago. Since mitochondrial evidence suggests that the line leading to *sapiens*, viz. the **modern *Homo*** line, broke off from the **archaic *Homo*** line 500,000 or more years ago, the only species to which the term '*sapiens*' is applied in this work is our own.

Early *Homo sapiens*: *Sapiens* from our origin to the beginnings of horticulture. See also **Cro-Magnon**.

hydatid cysts: Result from parasites for which dogs are the definitive host; sheep, cattle, goats and humans can be intermediate hosts. They are a symptom of hydatid disease or hydatidosis. Cysts form in the liver and elsewhere, e.g. the brain. The disease is acquired through eating a cyst.

ice age: See **glaciation**.

ice sheet: A continental glacier covering more than 50,000 km². The *Laurentide* ice
sheet covered most of Canada and a large portion of the northern US between
ca. 95,000 and 15,000 years ago. Its southern margin included the sites of the
cities of New York and Chicago and the present course of the Missouri River
up to the foothills of the Rocky Mountains, beyond which it merged with the
Cordilleran ice sheet, which covered the north-western part of North America.

karyotype: The chromosomal structure of a **species** that determines, in
conjunction with particular genes and the environment, that species'
phenotype. The karyotype is determined by the number of chromosomes in
the species' DNA, the length of each chromosome, and the kinds (not types)
of gene that can be incorporated at various places on the chromosomes.
Karyotypic mutation, i.e. the potential creation of a new species, need not
involve the introduction of new genetic material.

K-selected species: Species selected for 'Krowding tolerance,' whence the "K."
Their members are characterised by large size, slow growth and reproduction,
long lives (more than 1 year), few offspring with low **mortality**, parental
care, relatively constant **population** size, and an existence which is easily
jeopardised by a new predation threat. The existence of K-selected species
is maintained via the production of individual offspring each of which has
a relatively high likelihood of survival. The more complex of the K-selected
species may evince sharing behaviour and territoriality, and in relation to the
latter employ internal population checks, such behaviour supporting a form
of population stability not found in **r-selected species**. K-selected species
are more flexible than r-selected species in that their members can live in a
variety of habitats. Most but not all mammals are K-selected, as are trees. In
the evolution of the vertebrates from fish to mammal, the trend has been
away from r-selection and towards K-selection, implying an increase in the
complexity of vertebrate species as a whole, and making humans the most
highly developed of K-selected species.

Levallois: A flint-knapping technique at the end of or following the **Mousterian**.
Marks the beginning of the **Middle Palaeolithic** (ca. 250,000 BP) Originated
with **archaic *Homo* species** but was also employed by *sapiens*. Involves
shaping a large piece of flint (the *core*) by flaking off pieces in a centripetal
manner, beginning at the edge and then spiralling towards the centre, until
the flint has the turtle-shell shape required for the removal of one or two
usable flakes. These flakes, which are lighter and have a sharper cutting edge
than the results of the Mousterian method, are rather round as compared
to the blades produced by **blade technology**. This technique is thought to
have been used by Neanderthals in Europe, beginning when they first came
into existence about 230,000 years ago, and to have been perfected by them
100,000 years ago.

Luddites: The Luddites constituted a social movement of English workers who
protested – often by destroying textile machines – against the mechanisation
of production that threatened or took their jobs. The movement began in

1811 and was named after a probably mythical leader, Ned Ludd. For a short time the Luddites were so strong that they fought battles with the British Army. Measures against them taken by the government included a mass trial at York in 1813 that resulted in many death penalties and transportations. See also **flying shuttle**.

marginal benefit: The benefit gained from consuming one more unit of a good or service.

Mesolithic: The Stone Age epoch that followed the **Palaeolithic** and preceded the **Neolithic** in Europe; referred to as the *Archaic* in the New World. It lasted from about 12,000 to 10,000 BP in Europe and until 3000 BP in some places in the Americas. The Mesolithic is characterised by the use of small stone tools (**microliths**) and a broad-based hunting and gathering economy.

microlith: A stone tool, max. 5 cm long, usually made of flint or chert, consisting of a segment of **blade** hafted into an arrow-shaft or handle as a projectile or other tool. The first half of the **Mesolithic** is characterised by larger and more robust microliths. These become smaller and more symmetrical (for use on projectiles) later in the period. The very late Mesolithic has exceptionally small microliths known as *microlithic rods*. The constantly decreasing size of microliths can be associated with the hunting of ever smaller prey.

minerals: Obtained by *mining*. May or may not be **organic**. Metals, salt and phosphorus are inorganic minerals; coal, oil and natural gas are organic.

modern *Homo*: A term I use to cover *Homo sapiens* and their ancestors after *Homo erectus*. The last common ancestor of Neanderthals and modern humans existed between 800,000 and 550,000 years ago.

mortality: Death rate. Varies inversely to *longevity*. Cf. **fertility**.

Mousterian: A **flake tradition** more primitive than the **Levallois**, in which the points produced were smaller than in the preceding **Acheulean**. Associated typically with the **Neanderthals**, but the modern humans of 100,000 BP in the Near East also had a Mousterian technology, as did all African *sapiens*. The tradition continued in Asia, such that the **Clovis** and **Folsom** toolkits of the Palaeo-Indians were Mousterian. The Mousterian tradition is based chiefly on the use of flint flakes about 4–7 cm long, and is characterised by small hand-axes, side-scrapers and triangular points. In Europe, archaeologists apply the name Mousterian to the remains from almost all sites dating from about 120,000 (end of the last interglacial) to about 35,000 years ago (the arrival of *Homo sapiens* in Europe), thus counting Levallois sites as Mousterian. Apart from in Europe, Mousterian sites (and Neanderthal remains) have been found in a few areas of North Africa and as far east as Afghanistan. The Mousterian is sometimes considered to be the last phase of the Acheulean, both being flake technologies.

Neanderthals: See *Homo neanderthalensis*.

neoclassical economics: The conventional view of the economic world, which sees the production of goods and services as a closed, self-regulating cycle of money and goods between households and firms. Neoclassical economics has its intellectual roots in the writings of Adam Smith and David Ricardo.

Neolithic era: New **Stone Age** = horticultural period, ca. 10,000–5000 BP.

Late Neolithic: Horticultural period after the beginning of mining.

neo-Malthusianism: The view that **population** size can be controlled by the voluntary use of contraceptives, i.e. the employment of particular preventive checks.

net energy principle: The energy gained from an energy **resource** must be more than the energy expended in making the resource usable. In the case of food, the amount of bodily energy used to obtain food cannot be greater than the bodily energy used from the food obtained.

net energy ratio: Also termed 'energy/profit ratio,' and 'energy return on investment.' The ratio of usable energy contained in a fuel/food divided by the energy required to extract it from the environment, refine it to a useful state, and deliver it to consumers. The notion also applies to engines, which (due to the entropy principle) always have a net energy ratio less than 1, i.e. an energy efficiency of less than 100 per cent.

Newcomen's steam engine: In Newcomen's engine steam was condensed in an open cylinder in which there was a piston. The suction created pulled down the piston. Its **net energy ratio** was below 0.01, i.e. its energy efficiency was below 1 per cent.

niche: Where **populations** and their members occupy *habitats*, **species** occupy *niches*, and compete over them with certain other species. Habitats are physical locations, whereas niches are abstract, indicating the ecological position of a species vis-à-vis other species in terms of the unique preconditions for its existence, including what it *consumes*. Any existing species occupies at least one particular niche at a time. The niches of most species are constant, but humans have continually been occupying new and/ or broader niches. The fact that a species occupies a niche does not imply that it is in ecological equilibrium.

Nielsen hot blast: James Nielsen's hot blast furnace from 1829 made the smelting of second-grade iron ore possible.

Oldowan: The first human technology. The tools made by humans in the guise of *Homo habilis* belong to this tradition, appearing first in the Gona and Omo Basins in Ethiopia about 2.5 million years ago. The key innovation was the technique of chipping stones to create a chopping or cutting edge. Most Oldowan tools were made by a single blow of one rock against another to create a sharp-edged **flake**. The best flakes were made from crystalline stones such as basalt, quartz or chert, and the prevalence of these tools indicates that early humans had learned the differences between types of rock. Fossils of crushed animal bones indicate that one use of these tools was to break open marrow cavities. Oldowan deposits include pieces of bone or horn showing scratch marks that indicate they were used as diggers to unearth tubers or insects.

organic: Stemming from a **biotic** source. Coal and oil are organic, but not biotic. Phosphorus, for example, while not being a metal and being necessary to life, is an *element*, and is thus *inorganic*.

overshoot: Caused by delays or faults in feedback that prevent a **system** regulated by internal checks from employing those checks so as to maintain dynamic equilibrium. Overshoot is affected by the speed of change of the system – a feedback delay that can be accommodated at low speed may cause overshoot at high speed.

Palaeolithic: The Old **Stone Age**, the first and longest part of the Stone Age. It began some 2.5 million years BP in Africa with the **Oldowan** tradition and ended around 12,000 BP with the beginning of the **Mesolithic**. It is usually divided in Europe into the Lower, Middle and Upper Palaeolithic. Agriculture was not practised during the Palaeolithic.

> **Lower Palaeolithic:** from 2.5 million to ca. 250,000 BP; **Middle Palaeolithic** (in Europe and the Near East): from 250,000 to ca. 40,000 BP, characterised first by **Mousterian** and then **Levallois flake technology**. **Upper Palaeolithic**: Part of the Old Stone Age in Europe, North Africa and parts of Asia from about 40,000 to 12,000 BP, characterised by **blade technology**.

phenotype: The macro physical and behavioural aspects of an organism or **population**. The phenotype is determined partly by the organism's or population's **karyotype** and partly by its **genotype**, and is influenced by its environment.

Pleistocene: A geological period from 1.8 million to 10,000 years ago. The Pleistocene is characterised by a series of **glaciations** and interglacials, beginning with the first ice age.

> **Late (Upper) Pleistocene:** From beginning of the last **glaciation** ca. 80,000 BP to end of the Pleistocene.

population: Living members of a **species** that interbreed. A species may have more that one population at any one time.

> **total population:** All the living members of a species, i.e. all of a species' populations at a particular time.

population pressure: Arises when a **population's** consumption exceeds its **carrying capacity**, due either to population growth or a reduction in available **resources** or a combination of the two. Population pressure is *experienced* when the population's consumption exceeds its **short-term carrying capacity**. Experienced population pressure due to a diminution in the amount of land relative to the size of the population is coincident with *crowding*.

pre-humans: Protohominids and **early hominids**. Includes *Australopithecus*. Cf. **hominid** and **protohominids**.

primate: Highest order of mammals, including humans and other apes, as well as monkeys and lemurs.

protohominids: **Primates** in those stages of evolution between apes and *Australopithecus*. Lived between about 7 or 8 and 4 million years ago. Cf. **hominid** and **pre-humans**.

protowar: A lethal conflict between *tribes*, not employing soldiers in armies but *warriors* in *protoarmies*. Cf. **feuding** and **war**.

puddling and rolling process: Involves the decarbonisation of crude pig iron in a reverberatory furnace to produce malleable wrought-iron – as opposed to cast iron – that can be rolled out in a rolling mill.

Pygmies: There are at least 10 ethnolinguistically distinct populations of Pygmies in Central Africa. These hunter-gatherers employ either bows, spears or nets in hunting. The Efe and Mbuti Pygmies, who are closely related, are from the Ituri forest in the eastern Congo basin. The Efe live in small camps of fewer than 50 people. Mbuti camps are usually two to three times as large due to the Mbuti's net hunting, which requires communal participation. Pygmies inhabiting the Ituri forest are collectively called the Bambuti, of which there remain no more than 20,000 who are pure-blooded. All Pygmies depend rather heavily on the domestic crops (e.g. manioc, corn, etc.) they obtain from nearby villages.

quality of life: How people experience their lives, positively and/or negatively. Cf. **standard of living**.

Quaternary period: From the beginning of the **Pleistocene** to the present.

reserves: Accumulable **resources** which are available given current technology. Reserves are distinct both from *flows*, such as the flow of energy from the sun, which are not accumulable, and from *resources per se*, which may be flows, and which need not be available given current technology. See also **stocks**.

resources: That part of nature which is potentially of use to a **population** of a **species**.

> **renewable vs. reusable vs. recyclable vs. non-renewable resources:** Renewable resources are typically **biotic** while reusable, recyclable and non-renewable resources are typically non-biotic. Food and wood are renewable resources; and the potentially permanent resources of air and water are reusable – while wood is recyclable. **Minerals** vary: such inorganic minerals as metals are non-renewable but recyclable; others, such as phosphorus, cannot be recycled; nor can **organic** minerals, which include fossil fuels (though some fossil-fuel-based plastics may be recyclable). Animal and plant **species** are also non-renewable, despite their being biotic. Resources requisite to survival are termed *vital* resources. See also **stocks**.

r-selected species: Characterised by small size, rapid growth and reproduction, short lives (less than 1 year), numerous offspring with high **mortality**, little or no parental care, a lack of territoriality, and **populations** that experience exponential growth followed by crashes. Insects and annual plants are typical r-selected species, but rodents other than beavers are also r-selected. r-selected species may be seen as ensuring their continued existence through their members' having a high rate of reproduction (whence the "r"). See also **K-selected species**.

spear: In this work the term *spear* is used to refer to both the *lance*, which is used in a thrusting motion, and the *javelin*, which is thrown.

spearthrower or atlatl: Used to throw javelins or darts, darts being lighter than javelins and heavier than arrows. In use ca. 20,000 BP but before the bow, which is thought to have evolved from it. Examples are the Eskimo harpoon thrower, and the looped rope attached to javelins by the ancient Greeks. Stone points were used on the projectiles (as on other missiles) due to their having a notably greater mass than the shaft, which gave the missile greater momentum at the same speed. Concerning dimensions, a dart of 140 cm, for example, would be thrown using a 45–50 cm atlatl, and would have a range of 20–25 m. Australian Aborigines use darts up to 3.5 m long, with atlatls of more than 1 m. Atlatls could have weights attached near their midsections, which put spring into the dart in diving-board fashion, improving accuracy and increasing velocity somewhat. A dart can easily be thrown over 100 m. The current world-record throw is around 250 m.

species: The characteristics of a species are primarily determined by its **karyotype**, which is unique to that species. Variations in individuals of a species are determined by differences in the genes (alleles) that occupy particular places in the **genome**. **Populations** of sexually reproducing species interbreed and have the same ancestors. Species are the basic units of Darwinian natural selection that have evolved since the first appearance of life on earth.

spinning jenny: A wooden device for producing yarn. Invented by James Hargreaves in 1764. It allowed its operator to spin several threads at once, incorporating from 8 to 80 spinning-wheel mechanisms and spindles. Cf. **water frame**.

standard of living: The quality and quantity of goods and services available to people, generally measured in terms of real income per person. Cf. **quality of life**.

stocks: Reserves removed from nature which are in a form such that they can be directly *used*. They include articles manufactured from reserves before those articles become waste. Thus, for example, a future forest constitutes a **resource**, a forest constitutes a *reserve*, as does a stack of logs, while a set of chairs made from such logs constitutes a *stock*. If the logs are being used e.g. in the building of a log cabin, then they too are a stock. All oil in the ground that will someday have been amenable to extraction by human technology constitutes a *resource*; all oil in the ground that we know of that can be pumped up using today's technology or that has already been pumped up and stored but is not usable as is, is a *reserve*; and all refined oil constitutes a *stock*, as do e.g. all usable commodities made of oil-based plastics.

Stone Age: The period in the development of *Homo* during which his essential tools were made of stone, wood and/or bone. This extends from when *Homo* first appeared (2.5 million BP) until about 6000 years ago, by which time there was a widespread use of copper in the Near East for the making of *sapiens* tools (weapons) and ornaments. The end of the Stone Age marks the last time humans lived in a potentially sustainable way. Cf. **Palaeolithic**.
New Stone Age: See **Neolithic era**.

sustainable development: Either social change that can continue indefinitely, or social change leading to a (possibly changing) state that can exist indefinitely. I.e. sustainable development is human development towards or presupposing ecological equilibrium.

system: A causally interrelated group of entities whose interaction persists or tends to persist.

use value: The value of a commodity or service in its being used, as compared to its value in being traded, i.e. its *exchange* value.

war: Conflict between *kingdoms, empires, states* or *nations* involving armies and the killing of humans. Cf. **protowar**.

water frame: Or Arkwright's machine, or spinning frame. A machine producing yarn using a water-wheel. The water frame, first constructed in 1771, was superior to the earlier **spinning jenny**, among other ways in producing a stronger thread. By 1782 Richard Arkwright had the water frame in mass operation and in the following years Samuel Crompton united the best parts of it and the spinning jenny to produce his hybrid *mule*.

world forest equivalent (wfe): The amount of free **energy** as would be released by burning all of the world's forests at one time, taking the world's forests to cover 50 million km^2 (roughly averaging between maximum **Holocene** forest cover and that of today) and their burning to produce energy equal to 2.5 billion MWh.

Notes

1 On the principles of modern science as a whole, see Dilworth (2007).

2 Cf. Georgescu-Roegen (1971a), p. 80.

3 Cf. Georgescu-Roegen (1971b), p. 294.

4 Holdren (1990), pp. 158–159.

5 Cf. Georgescu-Roegen (1971b), p. 279.

6 Georgescu-Roegen (1971a), p. 79.

7 Georgescu-Roegen (1971b), p. 279; see also his (1971a), p. 80.

8 Daly (1977), pp. 24 and 42.

9 Cf. Georgescu-Roegen (1971b), p. 280; quote following, (1972), p. 16.

10 Georgescu-Roegen (1971a), p. 80; (1971b), pp. 277–278, 281–282.

11 Cf. Georgescu-Roegen (1972), p. 15.

12 Cf. e.g. Childe (1954), p. 32; Stanford (1999), p. 163.

13 *Homeostasis* is Greek for 'to remain the same.' While the concept was created by Claude Bernard in 1865, the term was coined – ca. 1929 – by W. B. Cannon, with reference to living beings' maintenance of internal body states, such as temperature. The word *cybernetics*, like the term *government*, derives from the Greek word for helmsman (*kubernaô*).

14 A conflation of these two notions can be seen e.g. in Hardin (1993), p. 239. Discussions of this point, with references to other instances where the distinction has been missed, may be found in Dilworth (2005), p. 127&n., and Dilworth (2007), pp. 163&n.–164&n.

15 In this regard see Dilworth (2007), Ch. 7, Sect. 3.

16 In this regard, see Dilworth (2005); see also e.g. Richerson & Boyd (2001), ca. pp. 453–454 (pp. 39–40 in Internet version).

17 Cf. Mayr (1984), p. 80 (last two sentences).

18 Cipolla (1978), p. 41 (last clause).

19 Cf. Bartholomew & Birdsell (1953), p. 25.

20 Cf. ibid., p. 26.

21 Deevey (1960), p. 202.

22 Goldsmith et al. (1972), p. 99.

23 In this regard, cf. Georgescu-Roegen (1971b), p. 304.

24 Darwin (1859), pp. 163–165; next quote, p. 157.

25 In this latter regard cf. Foley (1987), p. 120.

26 In this regard, cf. Commoner (1971), p. 41; Daly (1990a), p. 269; and Wilkinson (1973), p. 91, quoted below.

27 In this regard, cf. Goldsmith et al. (1972), pp. 19–20.

28 Wynne-Edwards (1963), p. 44.

29 Burnet & White (1972), p. 10.

30 Commoner (1971), p. 34; next quote, Bartholomew & Birdsell (1953), p. 27.

31 Commoner (1971), pp. 36–37.

32 In the same vein, cf. D. H. Chitty, cited in Stott (1962), p. 356.

33 Donella Meadows et al. (1992), p. 25 (this sentence to here).

34 D. H. Chitty, as cited in Stott (1962), p. 356.

35 Cf. J. D. Clark, cited in Malmberg (1980), p. 69.

36 Dawkins (1989), p. 113.

37 Stott (1962), p. 358.

38 Wynne-Edwards (1965), p. 1543.

39 Cf. Stott (1962), p. 364.

40 Cf. ibid.; cf. also Silberbauer (1981a), p. 457.

41 Though the contraceptive function of long lactation is generally accepted, its efficacy has been questioned. According to C. A. Reed (1977a, p. 895), in sedentary societies lactation has not in fact been protection against another pregnancy, the relevant factor being rather an absence of stored fat.

42 Lack (1954), pp. 206 and 214.

43 Deevey (1960), p. 197.

44 Stott (1962), pp. 356 and 359 (this paragraph to here).

45 Lack (1954), p. 215.

46 Cf. Dilworth (2005).

47 Wynne-Edwards (1965), p. 1543.

48 Ibid., pp. 1543–1544.

49 Cf. Wynne-Edwards, e.g. ibid., p. 1546. In this regard see also Stott (1962), p. 372.

50 Wynne-Edwards (1963), p. 45.

51 Stott (1962), p. 360.

52 J. B. Calhoun, as quoted in ibid., p. 361.

53 J. B. Calhoun, as reported in Malmberg (1980), p. 43.

54 J. B. Calhoun, as reported in ibid.), pp. 360–361 (whole paragraph to here).

55 P. L. Errington, as cited in ibid., p. 357; cf. also p. 358.

56 Wynne-Edwards (1983), p. 389.

57 P. L. Errington, as cited in Stott (1962), p. 357.

58 Ibid., p. 356; following quote, Lack (1966), p. 301.

59 Diamond (1992), p. 312.

60 P. L. Errington, as reported in Stott (1962), p. 360; quote following, D. H. Chitty, in ibid.

61 Ibid., p. 364.

62 Wynne-Edwards (1965), p. 1546.

63 Cf. Stott (1962), p. 362.

64 Cf. D. M. Vowles, as quoted in Malmberg (1980), p. 43.

65 Darwin (1874), p. 123; next quote, p. 128.

66 Hägerström (1953), p. 350.

67 Lorenz (1963), p. 231; quote following, Wynne-Edwards (1964), my emphasis.

68 Hägerström (1953), p. 352.

69 D. Lack, as reported in Stott (1962), p. 362.

70 E. Mayr, as quoted in Malmberg (1980), p. 40.

71 Wynne-Edwards (1986), p. 177.

72 Carpenter (1958), pp. 225–226, 229 and 230, with many references (most of the previous two paragraphs).

73 M. A. Bennett, as cited in ibid., p. 234.

74 Wynne-Edwards (1983), p. 384; following quote, Darwin (1874), p. 518.

75 W. H. Burt, as cited in Bartholomew & Birdsell (1953), p. 24.

76 Wynne-Edwards (1983), p. 383.

77 Cf. M. W. C. Fox and P. H. Kopfler, each quoted in Malmberg (1980), pp. 43–44.

78 Wynne-Edwards (1983), pp. 382–389.

79 Cf. S. A. Barnett, cited in Malmberg (1980), p. 44.

80 Darwin (1874), p. 217; quote following, p. 227.

81 Ibid., p. 520; next quote, p. 249.

82 G. K. Noble, as cited in Bartholomew & Birdsell (1953), p. 23.

83 Stanford (1999), p. 94, with reference to K. Stewart and A. Harcourt.

84 Ibid., p. 58.

85 Cf. Russell & Russell (1968), pp. 75–77, 90.

86 Ibid., p. 278.

87 Cf. Carpenter (1958), p. 245.

88 Cf. E. J. Soja, cited in Malmberg (1980), p. 45.

89 Wynne-Edwards (1986), p. 180.

90 Wynne-Edwards (1965), p. 1544.

91 Russell & Russell (1968), pp. 57–59, 90 and 119.

92 Wynne-Edwards (1965), p. 1545.

93 Harding & Teleki (1981), p. 5.

94 Stanford (1999), p. 28, with reference to S. C. Strum.

95 Ibid., p. 29, with reference to S. C. Strum.

96 Strum (1981), p. 258.

97 Ibid., p. 276.

98 Teleki (1981), p. 328.

99 Isaac & Crader (1981), p. 85.

100 Stanford (1999), p. 95.

101 Heinrich (2001), p. 169–170.

102 Teleki (1981), p. 331.

103 Stanford (1999), p. 68.

104 Ibid., p. 70.

105 Ibid., pp. 66 and 121.

106 Teleki (1981), p. 329.

107 Stanford (1999), p. 123.

108 Teleki (1981), p. 327, with reference to R. B. Lee.

109 Stanford (1999), p. 66.

110 Teleki (1981), p. 336.

111 Stanford (1999), p. 52.

112 Baschetti (2005), p. 241.

113 Stanford (1999), p. 29, with reference to S. C. Strum.

114 Ibid., pp. 8–11, 30 and 70.

115 Cf. F. de Waal, as cited in Baschetti (2005), pp. 241–243.

116 Russell & Russell (1968), p. 84.

117 This paragraph and the next two are from Diamond (1992), pp. 291–294.

118 Russell & Russell (1968), p. 252; and (1999), pp. 1 and 2. Parenthetical quote following, Lorenz (1963), p. 115.

119 Russell & Russell (1968), p. 259.

120 Ibid., p. 286.

121 J. B. Calhoun, as reported in Malmberg (1980), p. 43.

122 Russell & Russell (1968), pp. 149–150 (with reference to J. R. Clarke and to J. B. Calhoun), and 162 and 265 (this paragraph).

123 D. Lack, as cited in Stott (1962), pp. 362–363.

124 Cf. Wynne-Edwards (1983), pp. 382–389; and (1963), p. 44.

125 Russell & Russell (1968), p. 281.

126 Lack (1954), pp. 13 and 170, with numerous references.

127 Cf. Wynne-Edwards (1965), p. 1544.

128 Malthus (1798), p. 71.

129 Bartholomew & Birdsell (1953), p. 27.

130 In this regard, cf. Dilworth (2007), Ch. 7.

131 Commoner (1971), p. 43.

132 Foley (1987), pp. 58–59.

133 Hobbes (1651), pp. 186 and 189 (Part I, Ch. 13).

134 Childe (1952), pp. 1–2; next quote, p. 110.

135 Crowley (2001); see also, e.g., Vidal-Naquet ed. (1987), p. 38.

136 Johnson & Earle (1987), p. 4.

137 Bartholomew & Birdsell (1953), p. 35.

138 Carr-Saunders (1922), pp. 158, 214 and 230–231 (with reference to H. Spencer).

139 Classic here is Lee & DeVore eds. (1968).

140 Cf. e.g. Hamburg & McCown eds. (1979).

141 The *locus classicus* here is Martin & Wright eds. (1967).

142 Environmental economics is an outgrowth of the traditional perspective; for more that is relevant to the distinction between ecological

and environmental economics, see Dilworth (2008), App. VI.

143 Lee (1968), p. 37.

144 Wilkinson (1973), p. 5.

145 Cf. Lee & DeVore (1968), p. 6.

146 Lee (1968), p. 35n. and Schapera (1930), p. 93.

147 Schapera (1930), p. 93.

148 Hayden (1981), p. 358.

149 Sollas (1915), p. 226.

150 Stanford (1999), p. 119.

151 Heinrich (2001), p. 180.

152 Sollas (1915), pp. 243–244, 411.

153 Warner (1969), p. 5.

154 Sollas (1915), p. 229–230.

155 Stanford (1999), p. 40, with reference to G. P. Murdock.

156 Estioko-Griffin & Bion Griffin (1981), p. 129.

157 Watanabe (1968), p. 74; includes references and further examples.

158 Heinrich (2001), p. 179.

159 Warner (1969), p. 129.

160 Silberbauer (1981a), p. 494.

161 Lee (1968), p. 36.

162 Draper (1976), p. 210.

163 Cf. e.g. Woodburn (1968), p. 51.

164 Lee & DeVore (1968), p. 12.

165 Wilkinson (1973), p. 47; quote following, Woodburn (1968), p. 54.

166 © Chris Knight/Survival

167 In this regard see Schumacher (1966), p. 40; and (1970), p. 23.

168 Turnbull (1968), p. 132.

169 Harako (1981), p. 515.

170 Woodburn (1968), p. 51.

171 Photograph reproduced by courtesy of the Powell-Cotton Museum.

172 Lee (1968), p. 30.

173 Woodburn (1968), p. 49.

174 Russell & Russell (1968), p. 207.

175 Acsádi & Nemeskéri (1970), p. 187.

176 A. M. Carr-Saunders, as quoted in Wilkinson (1973), p. 32.

177 Washburn & Lancaster (1968), p. 301.

178 Sollas (1915), p. 238.

179 Warner (1969), p. 85.

180 Schapera (1930), pp. 114, 116.

181 Carr-Saunders (1922), pp. 201, 214 and 476.

182 Stott (1962), p. 365.

183 Binford (1968), p. 326; quote following, Carr-Saunders (1922), p. 213.

184 Binford (1968), p. 326, with many references; next quote, Moran (1982), p. 277.

185 Cf. Acsádi & Nemeskéri (1970), p. 200.

186 G. Bara Kolata, as cited in Lenski et al. (1995), p. 111.

187 Marshall Thomas (1989), p. 159.

188 Howell (1976), p. 147.

189 Marshall Thomas (1989), p. 160.

190 Silberbauer (1981b), p. 250.

191 Coleman (1986), p. 30.

192 Rose (1968), p. 204 (much of this paragraph).

193 Sollas (1915), p. 233.

194 Wilkinson (1973), p. 39.

195 W. Immenroth, as quoted in Malmberg (1980), p. 75, where reference is also made to M. J. Herskovits and H. J. Heinz.

196 Schapera (1930), pp. 81–83.

197 C. S. Coon, as quoted in Malmberg (1980), p. 72.

198 J. B. Birdsell, as reported in Lee & DeVore (1968), p. 11; quote following, Mumford (1967), p. 84.

199 Malmberg (1980), pp. 79–83, with numerous references (this paragraph).

200 N. Peterson, as cited in ibid., pp. 225–226.

201 By M. Mauss, referred to in Lee & DeVore (1968), p. 12.

202 Schapera (1930), p. 101; quote following, Carr-Saunders (1922), p. 212.

203 F. Eggan, in Lee & DeVore eds. (1968), p. 85.

204 Schapera (1930), p. 148.

205 Ibid., p. 127.
206 Silberbauer (1981b), p. 177.
207 Endicott (1988), pp. 112 and 116.
208 As argued for in Dilworth (2005).
209 Woodburn (1968), p. 52.
210 Shostak (1976), p. 262.
211 Schapera (1930), p. 149–151, with reference to D. F. Bleek and others; cf. also p. 158.
212 Silberbauer (1981a), p. 462.
213 Lee & DeVore (1968), p. 12.
214 Warner (1969), p. 136.
215 Cf. Suttles (1968), p. 68.
216 Hayden (1981), p. 394.
217 Sollas (1915), p. 495.
218 Ibid., p. 497.
219 Silberbauer (1981a), p. 469.
220 Reed (1977a), p. 895.
221 Cf. e.g. Woodburn (1968), p. 54.
222 Truswell & Hansen (1976), pp. 171, 182, 192.
223 Konner (1976), p. 231.
224 Woodburn (1968), p. 54.
225 A. Alland, p. 244 in Discussions, Part v, pp. 241–249 of Lee & DeVore eds. (1968).
226 Lee (1968), p. 43; next quote, Carr-Saunders (1922), pp. 158–159, citing various sources.
227 Photo courtesy of Bush Ways Safaris, Botswana.
228 Lee (1976), p. 5.
229 Silberbauer (1981a), p. 456; quote following, Lee (1968), pp. 35–36.
230 Lenski et al. (1995), p. 152.
231 Warner (1969), pp. 17, 144, 151.
232 Tremearne (1912), p. 160.
233 Ibid.
234 Schapera (1930), p. 154.
235 Warner (1969), pp. 5–151.
236 Keeley (1996), pp. 29, 133 and 135; and Sollas (1915), p. 370.
237 This paragraph and the following four are due to Schapera (1930), pp. 77 and 152–158, with further references.
238 Diamond (1992), pp. 228–233 and 296; quote following, p. 298.
239 Walker (2001), p. 587.
240 Tremearne (1912), p. 183.
241 Sollas (1915), pp. 179 and 229; following quote, Sumner (1907), p. 331.
242 Tremearne (1912), p. 183.
243 Divale (1972), pp. 226–229.
244 Ibid., pp. 224–225.
245 Turnbull (1961), p. 188; following quote, Lee & DeVore (1968), p. 12.
246 As reported in Lee & DeVore (1968), p. 9.
247 Ibid.
248 Ibid.
249 Harako (1981), p. 543; regarding this question, see also Hewlett (undated).
250 R. C. Bailey and N. R. Peacock, as cited in Stanford (1999), p. 139.
251 Keeley (1996), p. 132, with reference to R. C. Bailey and associates.
252 Silberbauer (1981b), p. 250; next quote, Wilkinson (1973), p. 85.
253 Sahlins (1968), pp. 85 and 89.
254 Lee (1968), pp. 40–41, with reference to W. S. Laughlin.
255 Sahlins (1972), p. 14; next quote, ibid., p. 27. The quote in the quote is from a 1966 manuscript by J. Woodburn.
256 Wilkinson (1973), p. 85; next quote, Sahlins (1972), pp. 37–38.
257 Lee & DeVore (1968), p. 6.
258 Lee (1968), p. 33.
259 Ibid., p. 39; following quote, Lee (1979), p. 455.
260 Woodburn (1968), p. 52; next quote Wilkinson (1973), pp. 32–33.
261 Martin & Steadman (1999), p. 17.
262 Martin (1966), pp. 339, 340, 342; (1967), p. 79 (this sentence).
263 Martin (1967), p. 111.
264 Media Release (2001), with reference to R. G. Roberts, T. F. Flannery and others.
265 After Martin (1967), p. 115.

266 Ponting (1991), p. 34; cf. also Martin (1967), p. 111.

267 Martin & Steadman (1999), p. 18.

268 Martin (1984), p. 161.

269 Martin & Steadman (1999), p. 38.

270 Martin (1967), p. 111.

271 Edmeades (2006).

272 Cf. Flannery & Roberts (1999), p. 242; and Australia in the News (2007).

273 Ponting (1991), p. 34; Stuart (1999), p. 258.

274 Baerreis (1980), p. 356.

275 Martin (1984), p. 170.

276 Photo courtesy of the Royal BC Museum.

277 Wikipedia; on the Internet.

278 © Michael Long/Natural History Museum, London.

279 © Michael Long/Natural History Museum, London.

280 Holdaway (1999), p. 202.

281 © Natural History Museum, London.

282 Cf. Ponting (1991), p. 34 (the latter half of this sentence).

283 Martin (1966), p. 341.

284 Cf. Ponting (1991), p. 34; and Martin & Steadman (1999), p. 44.

285 Holdaway (1999), pp. 216, 218 (this sentence).

286 © Michael Long/Natural History Museum, London.

287 Martin (1973), pp. 969 and 972; next paragraph, cf. Martin (1967), p. 97.

288 After Martin (1973), p. 972.

289 Cf. Cohen (1977a), pp. 184 and 196.

290 Ibid., p. 187.

291 Clutton-Brock (1987), p. 174. Note however modern hunter-gatherers' apparent awareness of this necessity.

292 The archaeological record in Africa is extremely sparse: Lewin (1993), p. 125.

293 Cf. Martin (1966); and Martin & Leakey (1967).

294 T. F. Flannery, as reported in Australia in the News (2007).

295 After Martin (1966), p. 341.

296 Ponting (1991), p. 34.

297 Ibid.

298 Cf. McMichael (1993), p. 32.

299 Martin (1984), p. 183.

300 Alroy (1999), p. 125.

301 Ponting (1991), p. 35.

302 Cohen (1977a), p. 185, with reference to W. E. Edwards and P. J. Mehringer.

303 Alroy (1999), p. 127.

304 P. J. Mehringer, as cited in Cohen (1977a), p. 186; see also Alroy (1999), p. 135.

305 V. L. Smith (1992), p. 5. In this regard, see also e.g. Haynes (1966), p. 112; Barnowsky (1989), p. 247; and Holling (1994), p. 95.

306 Martin (1967), p. 79.

307 Cf. Edwards (1967), p. 148; and Martin (1967), p. 79.

308 Martin (1966), p. 342.

309 T. F. Flannery, in Williams (2001).

310 Cohen (1977a), p. 186.

311 Alroy (1999), pp. 106 and 139.

312 W. E. Edwards, as cited in Cohen (1977a), pp. 186–187.

313 Martin (1967), p. 110.

314 Washburn & Lancaster (1968), p. 299.

315 In this regard cf. e.g. Martin & Steadman (1999), p. 45; and Eldredge (1999), p. 4.

316 Cf. Martin (1966), p. 341.

317 P. S. Martin, as cited in MacPhee ed. (1999).

318 A. Gibbons and others, as cited in Martin & Steadman (1999), p. 28.

319 P. S. Martin, as cited in Cohen (1977a), pp. 182–183.

320 Cf. van der Merwe (1992), p. 370; following quote, Cohen (1975), p. 471.

321 van der Merwe (1992), p. 370; following quote, Cohen (1977a), p. 280.

322 Cohen (1977a), p. 194; next quote, Dersin & Mitchell eds. (1992), p. 154.

323 Cohen (1977a), p. 146; next quote, pp. 279–280.

324 Colinvaux (1973), p. 579.

325 Much of this subsection to here is from Cohen (1977a).

326 Wilkinson (1973), p. 185; next quote, p. 126.

327 Ibid., p. 185.

328 Daly (1977), pp. 30 and 102.

329 Boulding (1966), p. 9; quote following, Schumacher (1973), pp. 127–128.

330 Daly (1977), p. 11.

331 Boulding (1966), p. 7.

332 Georgescu-Roegen (1971b), pp. 281, 283; next quote, ibid., p. 81.

333 Ibid., pp. 277–278.

334 Cf. Georgescu-Roegen (1975), pp. 95–102, and p. 106, n. 16.

335 Ibid., p. 95; next quote, p. 97.

336 Cf. e.g. Daly (1980), pp. 378–380.

337 K. E. Boulding, as quoted in Daly (1985), p. 203.

338 Schumacher (1973), p. 129; quote following, (1966), p. 42.

339 Daly (1990b), p. 259.

340 Boulding (1966), pp. 9–10; Boulding first spelled out the idea of throughput, though under a different name, in technical papers in 1945 and 1949.

341 N. Georgescu-Roegen, as cited in Daly (1990b), pp. 243–244; next quote, Daly (1971), p. 41.

342 Daly (1977), p. 35.

343 J. S. Mill, Of the Stationary State, Bk. IV, Ch. VI of his *Principles of Political Economy*, Vol. 2 (London: John W. Parker and Son, 1857), pp. 320–326; as quoted in Daly (1971), pp. 27–28.

344 Boulding (1966), p. 9. Quote following, Daly (1977), p. 17; see also p. 114; and Boulding (1970), p. 41.

345 Georgescu-Roegen (1971b), p. 229.

346 Malthus (1798), p. 199.

347 Malthus (1830), p. 225.

348 Darwin (1892), pp. 42–43; see also Darwin's *The Variations of Animals and Plants under Domestication*, London: John

Murray, 1868, Vol. 1, p. 10. Next quote, Darwin (1859), p. 117.

349 Darwin (1874), p. 48. Next quote, ibid., Ch. 5, p. 219 in 1901 reprint; quoted in Montagu (1962), pp. xii–xiii.

350 A. R. Wallace, as quoted in Flew (1970), p. 50.

351 In this regard cf. A. J. Lotka, as cited in Georgescu-Roegen (1971b), p. 307.

352 Malthus (1830), p. 247. See also e.g. Malthus (1798), pp. 120 and 134.

353 In this regard cf. Wynne-Edwards (1983), p. 379; Carr-Saunders (1922), p. 476; and Hardin (1993), p. 107.

354 E.g. in Darwin (1874), pp. 189–190.

355 Cf. Catton (1980), p. 128; cf. also Foley (1987), p. 52.

356 Malthus (1798), pp. 89–91, 103 and 107.

357 Cf. e.g. Malthus (1830), pp. 249–250; and pp. 23 and 25 of Flew (1970).

358 Malthus (1830), p. 247.

359 Ibid., p. 248; next quote, p. 249.

360 Malthus (1798), p. 77.

361 Ibid., p. 71; next quote, Malthus (1830), p. 250.

362 Ellul (1964), p. 48.

363 Malthus (1798), p. 203.

364 Boserup (1965), p. 14; quote following, p. 29.

365 Wilkinson (1973), p. 5.

366 Ibid., pp. 5–6; next quote, p. 57.

367 Ibid., pp. 63, 99, 105.

368 The vicious circle principle was first presented in Dilworth (1994).

369 It might be mentioned here that the vicious circle of the development of humankind bears a certain affinity to the Wheel of Samsara of Buddhist teaching.

370 Schumacher (1973), p. 99; Wilkinson (1973), p. 54. See also Binford (1968), p. 328.

371 Cf. Dilworth (2007), pp. 83–92.

372 In this regard cf. Johnson & Earle (1987), p. 5.

373 Cf. Wilkinson (1973), pp. 171–172.

374 Cf. Mumford (1967), pp. 139–141, 143, 250.

375 Cf. Hole (1992), p. 377.

376 Cf. Bronson (1977), p. 26.

377 Darwin (1874), p. 198.

378 A. Huxley, as quoted in Daly (1977), p. 47.

379 Commoner (1971), p. 80.

380 Hardin (1972), p. 38.

381 Schumacher (1970), p. 17; E. Sevareid as cited in Bartlett (1994), p. 23; Forrester (1975), p. 279.

382 Ellul (1964), p. 105.

383 B. Commoner, as cited in Schumacher (1970), p. 17.

384 Boulding (1980), p. 311.

385 Forrester (1971), p. 94.

386 Wilkinson (1973), p. 91.

387 Ibid., pp. 4–91 (previous two paragraphs).

388 Ibid., p. 99.

389 Cf. ibid., pp. 7, 8.

390 Schumacher (1973), p. 120; next quote, pp. 120–121.

391 Cf. Giarini & Loubergé (1978), p. 85.

392 In this regard cf. Goldsmith et al. (1972), p. 95.

393 Wilkinson (1973), p. 47.

394 Ibid., pp. 4, 128–129.

395 Ibid., p. 4.

396 Daly (1977), p. 28.

397 See Dilworth (2005).

398 C. G. Darwin (1952), Ch. VIII, p. 133; cf. also Ch. IX, p. 164; quote following, G. Clark, as quoted in Mumford (1967), p. 217.

399 Dubos (1966), p. 28 (this sentence).

400 Russell & Russell (1968), pp. 231, 233.

401 Divale (1972), p. 222.

402 Wilkinson (1973), pp. 27–28.

403 For a detailed presentation and defence of the conception of morals being presented here, see Dilworth (2005).

404 Darwin (1874), p. 121; next quote, p. 102.

405 Baschetti (2005), pp. 241–243, with many references.

406 C. Darwin; quotation reproduced in ibid., p. 242; next quote, Mayr (1988), pp. 79 and 81.

407 Mishan (1977), pp. 201 and 208.

408 C. G. Darwin, as cited in Hardin (1993), p. 255.

409 In this regard, see also Darwin (1874), pp. 641–642.

410 Russell & Russell (1968), p. 6.

411 Johnson & Earle (1987), p. 5.

412 Cohen (1977b), p. 142.

413 Abernethy (1993), pp. 17–20, 45.

414 Cf. Carr-Saunders (1922), p. 201; Binford (1968), pp. 329ff.

415 Ellul (1964), p. 194.

416 Wilkinson (1973), p. 109.

417 Cf. Boulding (1964), p. 139.

418 J. Ruskin, as cited in Terry (1995), p. 105.

419 Daly (1977), p. 159.

420 Bartholomew & Birdsell (1953), p. 24.

421 Cf. Darwin (1874), p. 84.

422 In this regard see Wynne-Edwards (1983), p. 389; and P. Leyhausen, cited in Malmberg (1980), p. 252.

423 The 'Joneses effect' was named by J. S. Dusenberry in 1949.

424 Ellul (1964), p. 203.

425 Wynne-Edwards (1965), p. 1545.

426 Lorenz (1963), pp. 237–238.

427 Wilkinson (1973), p. 4.

428 In this regard, cf. Dobkowski & Wallimann eds. (1998).

429 Abernethy (1998), pp. 108–109.

430 In this regard cf. Johnson & Earle (2000), p. 23.

431 Lack (1954), pp. 169, 178.

432 Wilkinson (1973), pp. 29–30.

433 Cf. Russell & Russell (1999), pp. 1 and 18; and (1968), p. 3.

434 Cf. J. L. Brown, as cited in Malmberg (1980), p. 45.

435 Abernethy (1998), p. 102; cf. also Morales (1998), p. 247.

436 J.-J. Rousseau, *Discours sur l'inégalité*, Pléiade edition, Vol. 3, p. 178, 1755, as quoted in Cranston (1968), p. 21.

437 J.-J. Rousseau, as presented in ibid.

438 Plato, *Republic*, 1.338c.

439 Reiman (1996), p. 176.

440 Tainter (1988), p. 121.

441 What follows of this subsection is from Georgescu-Roegen (1971b), pp. 311–359.

442 M. Djilas, quoted in ibid., p. 311n.

443 Ibid.; next two quotes, pp. 349 and 359.

444 Cf. M. Bates, as quoted in Malmberg (1980), p. 106.

445 Keeley (1996), p. 161.

446 Diamond (1992), p. 299.

447 Cf. A. Strachey, as quoted in Malmberg (1980), p. 106.

448 Cf. Plato's *Republic*, 2.372c and 2.373c–374a.

449 Rousseau (1762), p. 56.

450 Q. Wright (1964), p. 206.

451 Schumacher (1970), p. 19.

452 Cf. Plato, *Symposium*, 178e3–179b3.

453 In this regard, cf. Divale (1972), pp. 224, 225.

454 Russell & Russell (1968), pp. 172–173.

455 In this regard cf. M. Meggitt, as cited in Keeley (1996), p. 90.

456 Cf. Burnet & White (1972), p. 12.

457 Moran (1982), p. 277; following quote, H. Cox, as quoted in Q. Wright (1964), p. 280.

458 H. Cox, as cited in Q. Wright (1964), p. 282.

459 Russell & Russell (1968), p. 205.

460 Malmberg (1980), p. 254.

461 Cf. Lorenz (1963), p. 229.

462 Cf. E. T. Hall, as quoted in Malmberg (1980), p. 226; and Wilkinson (1973), p. 12.

463 Lorenz (1963), p. 252.

464 In this regard, see Georgescu-Roegen (1971b), p. 359; and Wilkinson (1973), pp. 9–10. Two following quotes, Wilkinson (1973), p. 11.

465 In this regard, cf. T. Dobzhansky, as cited in Malmberg (1980), pp. 48–49.

466 Georgescu-Roegen (1971b), p. 361.

467 Wilkinson (1973), p. 10.

468 For a contrary view, see Mumford (1967); in this regard see also Schumacher (1973). Parenthetical quote following, Ellul (1964), p. 312.

469 In this regard cf. Terry (1995), p. 53.

470 Schumacher (1970), pp. 16–17.

471 Cf. Dilworth (2007), pp. 200–202.

472 Which is *not* in keeping with Greek atomism, which sees quantity of motion as constant. As regards Hobbes and motion, cf. Macpherson (1968), p. 19. Following quote, A. Smith (1776), quoted without page reference on p. 176 of Hardin (1972).

473 In this regard, see Dilworth (2008), App. VI.

474 Bonar (1885), p. 225.

475 Keeley (1996), p. 121; see p. 122 for many modern examples.

476 In this regard cf. Russell & Russell (1999), p. 37.

477 Ibid., pp. 51 and 83.

478 Carr-Saunders (1922), pp. 199–200.

479 www.bartleby.com/67/19.html.

480 N. Pattersson, as cited in Kornfeldt (2006).

481 In this regard, cf. Whybrow (1992); Strimmer & Haeseler (1997); and Mayr (1984), p. 82.

482 Cf. King & Wilson (1975), p. 113.

483 Mason (1979), p. 289.

484 Oakley (1951), p. 6, with reference to W. Köhler.

485 Teleki (1981), p. 339.

486 Cf. Bartholomew & Birdsell (1953), pp. 23, 25 and 27–28.

487 Cf. ibid., p. 22.

488 V. L. Smith (1992), p. 1.

489 J. D. Clark, as cited in Malmberg (1980), p. 69.

490 Washburn (1960), p. 16.

491 Oakley (1951), p. 8.

492 Stanford (1999), pp. 106–107 and 145.

493 Ibid., p. 165.

494 Cf. e.g. Bartholomew & Birdsell (1953), p. 29.

495 Stanford (1999), pp. 116 and 149.

496 Isaac & Crader (1981), p. 85.

497 Stanford (1999), pp. 132 and 149.

498 Harding & Teleki (1981), p. 6.

499 Teleki (1981), p. 329.

500 Cf. J. D. Clark, as cited in Malmberg (1980), p. 69.

501 In this regard, cf. S. L. Washburn and I. DeVore, quoted in ibid., p. 70.

502 Cf. S. L. Washburn and V. Avis, cited in ibid., p. 69.

503 Stanford (1999), p. 128.

504 R. A. Dart, as cited in Brues (1960), p. 207.

505 Heinrich (2001), pp. 165–166 (this sentence to here).

506 Isaac (1987), p. 3.

507 Silberbauer (1981a), p. 479.

508 Sollas (1915), p. 89.

509 Bartholomew & Birdsell (1953), p. 21.

510 Cf. Laughlin (1968), p. 319.

511 Cf. Wolpoff (1999), p. 142.

512 Heinrich (2001), p. 169.

513 Photo courtesy of Kenneth Garrett/ NationalGeographicStock.com.

514 Brues (1960), p. 206; see also C. B. Ruff, as cited in Bramble & Lieberman (2004), p. 349.

515 R. McNeill, as cited in Wolpoff (1999), p. 722.

516 Cf. V. L. Smith (1992), p. 1.

517 Cf. P. Wheeler, as cited in Arsuaga & Martínez (1998), p. 78; and Heinrich (2001), p. 173.

518 Heinrich (2001), p. 166.

519 Cf. V. L. Smith (1992), p. 1.

520 Heinrich (2001), p. 166.

521 Darwin (1874), p. 582; next quote, ibid., p. 54.

522 Cf. Bartholomew & Birdsell (1953), p. 29.

523 Washburn (1960), p. 17.

524 Bartholomew & Birdsell (1953), p. 29.

525 After Oakley (1975), p. 21.

526 Heinrich (2001), pp. 157, 158.

527 Cf. Bartholomew & Birdsell (1953), p. 21; and Oakley (1951), p. 8.

528 In this regard, cf. Washburn (1960), p. 16.

529 Cf. Bartholomew & Birdsell (1953), p. 22 (whole paragraph).

530 M. W. Marzke, as cited in Wolpoff (1999), p. 275.

531 Gowlett (1992a), p. 14.

532 Washburn (1960), p. 18.

533 Cf. Gowlett (1992a), p. 14.

534 Wolpoff (1999), p. 209, with reference to K. Gibson.

535 Washburn (1960), p. 19.

536 Washburn & Lancaster (1968), p. 300.

537 Bartholomew & Birdsell (1953), p. 22.

538 Cf. Isaac & Crader (1981), p. 91.

539 Teleki (1981), p. 337.

540 S. L. Washburn, as cited in Wolpoff (1999), p. 195.

541 Bramble & Lieberman (2004), p. 349.

542 Wolpoff (1999), p. 422 (this sentence).

543 Bramble & Lieberman (2004), p. 350.

544 V. Niinimaa and others, as cited in ibid., p. 349.

545 Heinrich (2001), pp. 96, 165–175 (this sentence and much of the previous two paragraphs). Re the particular animals Bushmen could run down, see also Silberbauer (1981a), p. 479.

546 Laughlin (1968), p. 312. As regards endurance running and hominid evolution, see also Carrier (1984).

547 Cf. Nabokov (1981).

548 Heinrich (2001), pp. 175, 177; re running down kangaroos, see also Sollas (1915), p. 225.

549 Foley (1987), p. 269.

550 Cf. Washburn (1960), p. 15.

551 Cf. Bartholomew & Birdsell (1953), pp. 29, 31.

552 Teleki (1981), p. 330.

553 Bartholomew & Birdsell (1953), p. 30.

554 Ibid., pp. 29–30 (this paragraph to here); cf. also Oakley (1951), p. 6; and Hole (1968), p. 249.

555 Tanaka (1976), p. 101.

556 Oakley (1951), p. 9.

557 Ibid., p. 10.

558 Stanford (1999), p. 105, with reference to P. Shipman and A. Walker.

559 Heinrich (2001), p. 209 (previous two sentences).

560 Cf. Oakley (1951), pp. 11–12.

561 Bartholomew & Birdsell (1953), p. 31.

562 Cf. Lee & DeVore (1968), p. 7.

563 Cf. Oakley (1975), p. 21.

564 Washburn & Lancaster (1968), p. 301.

565 Ibid.

566 Stanford (1999), p. 203.

567 Ibid., p. 202.

568 Teleki (1981), p. 336.

569 Heinrich (2001), pp. 167, 180, 182 (previous three sentences); following quote, pp. 181–182.

570 Cf. Stanford (1999), p. 47.

571 Cf. M. Bates, cited in Malmberg (1980), p. 69.

572 Cf. M. Bates, cited in ibid.

573 In this regard, cf. Divale (1972), p. 225, where reference is made to Q. Wright and others.

574 Cf. Dilworth (2005).

575 Lorenz (1963), p. 238.

576 Eldredge (1999), p. 7.

577 Montagu (1962), p. x.

578 Cf. V. L. Smith (1992), p. 3.

579 Ibid., p. 2.

580 Lewin (1993), p. 31.

581 Cohen (1977a), pp. 88, 89.

582 Cf. C. G. Darwin (1952), Ch. III, p. 43.

583 Cf. Gowlett (1992b), p. 344.

584 *Science Daily* (1999).

585 Cf. Walker (2001), p. 584.

586 Sieferle (2001), p. 7.

587 © John Sibbick/Natural History Museum, London.

588 Cf. Oakley (1951), pp. 11 and 21 (this sentence).

589 Mumford (1967), p. 124.

590 Bartholomew & Birdsell (1953), p. 33.

591 A. H. Schultz, as quoted in ibid., p. 35.

592 Walker (2001), p. 585, with reference to F. Weidenreich.

593 Heinrich (2001), p. 173.

594 This is the view expressed in Stanford (1999), p. 131.

595 V. L. Smith (1992), p. 2.

596 Heinrich (2001), p. 173 (latter half of this sentence).

597 Foley (1987), p. 263.

598 As suggested by R. B. Goldschmidt, cited in Washburn & Lancaster (1968), p. 300.

599 Ibid.

600 Rasmuson (2000).

601 Oakley (1951), p. 7, with reference to W. Köhler; quote following, Darwin (1874), pp. 89–90.

602 Mumford (1967), p. 92; concerning the nature of magic, see Dilworth (2007), App. II.

603 Oakley (1951), p. 7.

604 Mumford (1967), p. 80.

605 Bartholomew & Birdsell (1953), p. 30.

606 Wolpoff (1999), p. 659.

607 Ibid., with reference to M. C. Stiner.

608 M. C. Stiner and S. L. Kuhn, as cited in ibid., p. 693.

609 Ibid., p. 659.

610 Lewin (1993), p. 3.

611 Wolpoff (1999), p. 662.

612 © American Museum of Natural History.

613 Clutton-Brock (1987), p. 12.

614 Heinrich (2001), p. 130.

615 Ibid., p. 187.

616 Clutton-Brock (1987), p. 13 (much of the previous paragraph and the present one to here).

617 Cf. Brues (1960), p. 211.

618 Oakley (1975), p. 31.

619 F. Weidenreich and L. Krzywicki, as cited in Cipolla (1978), p. 86.

620 Keeley (1996), p. 37.

621 Walker (2001), pp. 585, 586.

622 Acsádi & Nemeskéri (1970), pp. 146, 152, 153.

623 Ibid., p. 197.

624 Ibid., p. 204.

625 Sollas (1915), p. 188.

626 © skullsunlimited.com

627 Ibid., p. 3.

628 Picture used with permission of www.CartoonStock.com.

629 Ibid., p. 122; and Wolpoff (1999), p. 616.

630 Harpending et al. (1998), p. 1961.

631 Cf. Rasmuson (2000).

632 In this regard, cf. Divale (1972), p. 232.

633 Cf. Bartholomew & Birdsell (1953), p. 34.

634 Cf. Brues (1960), pp. 212–213.

635 Lewin (1993), p. 112.

636 Ibid.

637 M. Stoneking and R. L. Cann, as quoted in Wolpoff (1989), p. 110.

638 Cf. Binford (1992), p. 366.

639 Last fact, Rasmuson (2000); for the time of our entry into Europe, cf. Graham (2003).

640 Ponting (2007), p. 23.

641 J. B. Birdsell, as quoted in Binford (1968), p. 330 (this sentence).

642 Coleman (1986), p. 31.

643 Walker (2001), p. 586.

644 Cf. Birdsell, as cited in Stott (1962), p. 366 (with reference to modern Australian Aboriginals). Cf. also M. Bates, as cited in Malmberg (1980), p. 69.

645 Darwin (1874), pp. 583–586.

646 Cf. Washburn & Lancaster (1968), p. 300.

647 Gibson (1988), p. 166.

648 Allen (1977), p. 323.

649 Ibid., p. 324.

650 Diamond (1992), pp. 187–188.

651 Cf. G. P. Murdock and A. Holmberg, cited in Lenski et al. (1995), pp. 114, 119 (this sentence).

652 Cf. Keeley (1996), p. 144.

653 Irwin (1980), p. 325.

654 Cf. e.g. Acsádi & Nemeskéri (1970), p. 187.

655 Cf. e.g. V. L. Smith (1992), p. 3.

656 Cf. F. H. Smith & Paquette (1989), p. 206.

657 Nitecki & Nitecki (1987), pp. 171–172.

658 Cf. V. L. Smith (1992), p. 6.

659 On this latter point, see e.g. ibid.

660 Lee (1968), p. 42.

661 In this regard, cf. Forde (1975), p. 162.

662 Ibid., pp. 158–160, 181.

663 Sollas (1915), p. 458–460.

664 Ibid., p. 497; following quote, ibid., p. 496.

665 Cohen (1989), p. 96.

666 Cf. Ascádi & Nemeskéri (1970), pp. 146, 147.

667 Hole (1992), p. 373.

668 Washburn & Lancaster (1968), p. 299.

669 For this paragraph and the next, cf. Keeley (1996), pp. 12–110 passim.

670 Lorenz (1963), p. 242; next quote, pp. 242–243.

671 R. B. Lee, in Lee et al. (1968), p. 345.

672 Cf. Caspari & Lee (2004).

673 Cf. V. L. Smith (1992), p. 3.

674 Cf. Childe (1975), pp. 189–190.

675 Oakley (1975), p. 33.

676 Gowlett (1992a), p. 132.

677 Cf. Cohen (1977a), p. 197.

678 Photo courtesy of Karl Kilguss.

679 Lorenz (1963), pp. 230–231; next quote, p. 235.

680 Warner (1969), pp. 6 and 122.

681 Rigaud (1989), p. 149.

682 Brues (1960), p. 209.

683 Washburn & Lancaster (1968), p. 298.

684 Forde (1975), p. 157.

685 Ibid., p. 158 (whole paragraph).

686 Ibid., p. 161.

687 Cf. Ponting (1991), p. 29.

688 Cf. Brues (1960), pp. 208, 210–211.

689 Ibid., pp. 209–211, 214.

690 Ponting (1991), p. 66; and Keeley (1996), p. 37.

691 Brues (1960), p. 208.

692 Piedmost in the Czech Republic, as reported in R. Wright (2004), p. 38, with reference to W. Howells and A. Goudie.

693 V. L. Smith (1992), p. 3.

694 Nitecki & Nitecki (1987), p. 148.

695 Cohen (1977a), p. 116, with reference to K. Butzer and R. G. Klein.

696 Tremearne (1912), p. 152.

697 Cf. L. J. Angel, cited in Lenski et al. (1995), p. 109.

698 Acsádi & Nemeskéri (1970), p. 147.

699 J. D. Clark, as cited in Cohen (1977a), p. 102.

700 Jensfelt (2001), p. 15.

701 Cohen (1977a), p. 126.

702 Russell & Russell (1999), p. 72.

703 Acsádi & Nemeskéri (1970), p. 162.

704 Childe (1954), pp. 50–51.

705 In this regard, cf. Harris (1977), pp. 193–194.

706 Childe (1954), p. 51 (whole paragraph).

707 Turner (1985), p. 50.

708 Cf. the dog burial at Usjki on the Kamchatka Peninsula, cited in ibid.

709 Cf. Divale (1972), p. 232 (last point).

710 Cf. Cohen (1977a), p. 161, with reference to P. A. Colvinaux and D. M. Hopkins.

711 Bering Land Bridge, Wikipedia; on the Internet.

712 Cf. Cohen (1977a), p. 161.

713 J. Hey, as cited in Lindström (2005).

714 Cf. Gowlett (1992a), p. 142.

715 Stein & Rowe (1995), p. 307.

716 J. N. Spuhler, as cited in Zegura (1985), p. 12.

717 C. V. Haynes, as cited in Martin (1967), p. 97.

718 M. I. Budyko, as cited in Martin (1973); cf. also J. E. Fitting, referred to in Cohen (1977a), p. 190.

719 Stein & Rowe (1995), p. 307.

720 Cf. Martin (1967), pp. 97, 101; and V. L. Smith (1992), p. 4.

721 Nitecki & Nitecki (1987), p. 215.

722 Picture courtesy of the Archaeological Research Center, South Dakota State Historical Society.

723 Cf. Cohen (1977a), p. 123.

724 Cohen (1989), pp. 112 and 113, citing numerous sources.

725 Walker (2001), p. 586.

726 Ofnet, a 7720-year-old site in Bavaria: D. W. Frayer, as cited in ibid.

727 Walker (2001), p. 587.

728 Keeley (1996), p. 38.

729 Walker (2001), p. 586.

730 H. T. Waterbolk, as cited in Harris (1977), p. 194.

731 Cf. Reed (1977a), p. 942; Binford (1968), p. 332.

732 Harris (1977), p. 188.

733 Cf. e.g. Binford (1968), p. 334.

734 Reed (1977a), pp. 942–943.

735 J. G. D. Clark, as cited in Cohen (1977a), p. 124.

736 Binford (1968), pp. 332–333 (whole paragraph to here).

737 Reed (1977a), p. 942.

738 Suttles (1968), p. 60.

739 Darwin (1874), p. 49.

740 Cf. West (1988), p. xii; quote following, Hesiod, *Works and Days*, ca. ll. 54–59 (p. 38).

741 Cohen (1977a), p. 88.

742 Turnbull (1961), p. 58.

743 Cohen (1977a), p. 89.

744 Ibid., p. 101.

745 Cf. Hardin (1993), p. 118.

746 Cohen (1977a), pp. 86, 87.

747 Foley (1987), p. 269.

748 L. R. Binford, as referred to in Lee & DeVore (1968), p. 12.

749 Cf. Brues (1960),   p. 204.

750 Darwin (1874), pp. 123 and 128.

751 In this regard see Dilworth (2005).

752 Cf. ibid.

753 Cf. Ardrey (1961), p. 169.

754 D. Freeman, as cited in Washburn & Lancaster (1968), p. 300.

755 Ibid. In this regard cf. K. Lorenz, as cited above and in Chapter 7; following quote, Lorenz (1963), p. 39.

756 S. Margolin, as cited in Lorenz (1963), p. 236.

757 Cf. Lewin (1993), pp. 132, 133.

758 Brues (1960), p. 210.

759 Cohen (1977a), pp. 12, 14–15.

760 Ibid., p. 117.

761 Cf. N. Hammond, as referred to in ibid., p. 118; parenthetical quote following, J. Woodburn, as quoted in Hayden (1981), p. 391.

762 P. F. Wilkinson, as noted in Cohen (1977a), p. 120.

763 Cf. Divale (1972), p. 231.

764 Cf. Deevey (1960), p. 202.

765 Birdsell (1968), p. 236.

766 Hassan (1975), p. 43.

767 Divale (1972), p. 224.

768 In this regard, cf. Cohen (1977a), pp. 44–45; and Wilkinson (1973), pp. 33–41.

769 Cf. Binford (1968), p. 331.

770 Wolpoff (1999), p. 581.

771 Martin (1973), p. 972.

772 Cohen (1977a), p. 59.

773 Washburn & Lancaster (1968), p. 297.

774 Much of this and the next paragraph are due to Cohen (1977a), pp. 79–207, with many references.

775 Ibid., p. 83.

776 Sieferle (2001), p. 8.

777 Keeley (1996), p. 86.

778 Sieferle (2001), p. 8.

779 Cohen (1977a), p. 124.

780 Ibid., p. 37 (latter half of this sentence).

781 Harris (1977), p. 192 (whole paragraph to here).

782 Mumford (1967), p. 123.

783 Washburn (1960), p. 17; it may be mentioned that Konrad Lorenz was an ardent advocate of the idea of the self-domestication of man.

784 Cf. Roberts (1989), p. 80.

785 In this regard, cf. ibid., p. 93.

786 Hole (1992), p. 377.

787 Cf. Harris (1977), p. 230.

788 Cf. ibid., p. 186.

789 Cf. F. J. Simoons, cited in ibid., p. 230.

790 Cf. Roberts (1989), pp. 84–86.

791 In this regard cf. M. J. Herskovits, as quoted in Malmberg (1980), p. 73.

792 Reed (1977a), p. 899.

793 Cohen (1977a), p. 135.

794 Cf. Reed (1977a), p. 942; and Binford (1968), p. 333.

795 Cohen (1977a), p. 194.

796 Binford (1968), p. 333.

797 Clark & Haswell (1970), p. 30; cf. also G. P. Murdock, as referred to in Cohen (1977a), p. 35.

798 F. Rätzel, as cited in Clark & Haswell (1970), p. 30; and Deevey (1960), pp. 196–197.

799 Cf. Hole (1968), p. 249.

800 Cf. Cohen (1977a), p. 141.

801 After Drower (1975), p. 539.

802 Cf. Wilkinson (1973), pp. 92–94 (this point and the previous two paragraphs).

803 Clark & Haswell (1970), pp. 55–56 (this sentence).

804 J. F. Eder, as cited in Moran (1982), pp. 276–277.

805 Cf. Clutton-Brock (1987), pp. 15 and 29.

806 Herre & Röhrs (1977), pp. 253, 269, with reference to M. Degerbøl, C. A. Reed, and others.

807 Sieferle (2001), p. 18.

808 Almost 11,000 years ago, according to D. Perkins and others, as cited in Herre & Röhrs (1977), p. 256.

809 Anthony (2007), pp. 60, 61.

810 Herre & Röhrs (1977), p. 263.

811 Reed (1983), p. 527.

812 Cf. Anthony (2007), pp. 185, 224, 341.

813 Herre & Röhrs (1977), p. 269.

814 Childe (1954), p. 90.

815 Clutton-Brock (1987), p. 68.

816 Cf. Herre & Röhrs (1977), p. 252.

817 Clutton-Brock (1987), p. 68.

818 In this regard, see Mumford (1967), p. 135.

819 Hole (1968), p. 249.

820 Cf. Binford (1968), pp. 331, 334, 335.

821 D. R. Harris, as quoted in Gorman (1977), p. 343.

822 Anthony (2007), p. 129.

823 Cf. Binford (1968), p. 335 (whole paragraph to here).

824 M. S. Flannery and C. M. Taylor, as cited in Reed (1977a), p. 900.

825 Malmberg (1980), pp. 73, 75, 76 and 83, with reference to M. J. Herskovits and K. Kenyatta.

826 Lenski et al. (1995), pp. 135–136; see also Cohen (1977b), p. 170.

827 Nationmaster.com; Encyclopedia: Japanese Paleolithic; on the Internet.

828 Hole (1992), pp. 373–374.

829 Cf. Sollas (1915), p. 396; and Schapera (1930), p. 141.

830 Cf. Brues (1960), p. 213.

831 Cf. Mumford (1967), p. 134; and Drower (1975), p. 539.

832 Concerning the steppe, see Anthony (2007), pp. 63, 311.

833 V. G. Childe, as cited in Lenski et al. (1995), p. 175; for a number of these innovations, see Childe (1954), pp. 80, 88 and 90.

834 Keeley (1996), pp. 26–27 (this paragraph to here).

835 Cf. Kottak (1994), p. 216.

836 Stott (1962), p. 366.

837 Diamond (1992), p. 188.

838 Cf. Cohen (1989), p. 112, n. 32, citing numerous sources; and McMichael (1993), p. 90, with reference to J. W. Powles and others.

839 Sieferle (2001), p. 14.

840 Childe (1954), p. 158.

841 Cf. McMichael (1993), p. 91.

842 Ponting (1991), p. 226.

843 Coleman (1986), p. 28.

844 Cf. Lee & DeVore (1968), p. 10.

845 Cf. McMichael (1993), p. 91.

846 M. E. Clark (1989), p. 86.

847 Cf. W. H. McNeill and others, cited in Lenski et al. (1995), p. 174.

848 M. E. Clark (1989), p. 86.

849 Cf. Anthony (2007), p. 137 (this sentence).

850 Cf. Walker (2001), p. 589.

851 The major portion of this paragraph and the next are from Keeley (1996), pp. 31–158, with many references.

852 Walker (2001), p. 586.

853 Divale (1972), pp. 225–226 (last sentence with reference to L. Pospisil).

854 In this regard cf. the first edition of Darwin's (1874), Vol. 1, p. 160; cited in Mayr (1988), p. 79. What follows of this paragraph, and the whole of the next, are from Keeley (1996), pp. 66–178, with references.

855 Sollas (1915), p. 235–236.

856 Tremearne (1912), p. 113.

857 Stott (1962), p. 366.

858 Cf. Keeley (1996), p. 43.

859 Knutsson (1995), p. 88, with reference to R. M. Keesing.

860 Clutton-Brock (1992), p. 381.

861 A. Smith (1776), as cited in Skinner (1979), p. 33.

862 Cf. Anthony (2007), pp. 133, 135.

863 Johnson & Earle (1987), p. 5.

864 *Genesis* 3: 5–23. In this regard see also Hesiod's reference to a 'golden age' in his *Works and Days*, ca. ll. 110–122 (p. 40).

865 Wilkinson (1973), p. 67.

866 R. J. Braidwood, as cited in Binford (1968), p. 320; see also Harris (1977), p. 182.

867 In this regard, cf. Reed (1977b), pp. 4–5.

868 Reed (1977a), p. 941.

869 Sieferle (2001), p. 9. In this regard, see also Boserup (1965), p. 38.

870 Cohen (1977a), p. 23; quote following, Reed (1977a), p. 882.

871 Cf. R. J. Braidwood, as cited in Binford (1968), pp. 319–320; see also Harris (1977), p. 182.

872 Wilkinson (1973), p. 4; next quote, Tremearne (1912), pp. 198–199.

873 Cohen (1977a), p. 26.

874 Warner (1969), p. 138.

875 Sieferle (2001), p. 9 (this sentence).

876 V. L. Smith (1992), p. 10.

877 Cohen (1977a), p. 35, with reference to G. P. Murdock.

878 In these regards see ibid., pp. 8–9, 20, 24, 99, 134–135; and Bronson (1977), pp. 28, 32.

879 Cohen (1977b), pp. 170–171; next quote, Reed (1977a), p. 942.

880 Binford (1968), pp. 334–335; quote following, R. J. Braidwood, as quoted in ibid., p. 321.

881 Concerning *grasses* in this regard, cf. J. R. Harlan and D. Zohary, cited in Cohen (1977a), p. 148.

882 Clutton-Brock (1987), p. 48.

883 Divale (1972), p. 231, with reference to F. Boas, M. Sahlins and others.

884 Cf. Binford (1968), p. 330.

885 L. White, as cited in Cohen (1977a), p. 13.

886 Ibid., p. 14; next quote, p. 279.

887 Ibid., p. 60.

888 As is also suggested on p. 16 of Sieferle (2001).

889 Hassan (1977), p. 599.

890 In this regard, cf. Cohen (1977a), p. 140.

891 C. G. Darwin (1952), Ch. III, p. 49.

892 Clutton-Brock (1987), p. 47 (last sentence).

893 R. Carneiro, as cited in Binford (1968), p. 327.

894 V. L. Smith (1992), p. 8 (this sentence).

895 Diamond (1992), pp. 186–187.

896 Ibid., p. 187 (this sentence).

897 Mumford (1967), p. 137.

898 Cf. Marshall (1976), pp. 359–360.

899 Barnard & Woodburn (1988), pp. 18–19.

900 M. Sahlins, as cited in Hole (1968), p. 251.

901 Clutton-Brock (1987), p. 169.

902 Anthony (2007), p. 136.

903 Ponting (2007), p. 74.

904 McMichael (1993), p. 209.

905 Johnson & Earle (1987), p. 94.

906 Sieferle (2001), pp. 28 and 35.

907 Kottak (1994), p. 250. Next quote, Fisher (1930), pp. 200, 201; see also e.g. Darwin (1874), p. 121.

908 Cf. Knutsson (1995), p. 15, with reference to T. Ljungberg and others.

909 Gowlett (1992a), p. 180.

910 Tylecote (1992), p. 1.

911 Cf. e.g. R. J. Forbes and V. G. Childe, each cited in Lenski et al. (1995), pp. 155, 156.

912 Forbes (1975), p. 572.

913 Childe (1954), p. 85.

914 Ibid., p. 194.

915 Perlin (1989), p. 74.

916 Cf. e.g. L. Aitchison, cited in Lenski et al. (1995), p. 189.

917 *Encyclopedia Britannica* (2005).

918 Anthony (2007), p. 336.

919 Forbes (1975), p. 595.

920 Cf. Childe (1954), p. 87.

921 *Infoplease* (2005).

922 Forbes (1975), p. 597.

923 Sieferle (2001), p. 111.

924 Forbes (1975), p. 595; Tylecote (1992), pp. 47–49.

925 Coghlan (1975), pp. 617 and 620.

926 Cohen (1989), p. 56, n. 4, with reference to M. Colchester.

927 Coghlan (1975), p. 618.

928 Tremearne (1912), p. 199.

929 Harris (1977), p. 230, with reference to E. Isaac.

930 Drower (1975), p. 539.

931 Cf. Hole (1968), p. 246.

932 Reed (1977a), p. 879.

933 Russell & Russell (1999), p. 15.

934 Drower (1975), p. 546.

935 Ibid., p. 548.

936 Kuenen (1980), p. 16.

937 Carson (1962), p. 27 (latter point).

938 This paragraph and most of the following four are from Childe (1975), pp. 187–221; and (1954), pp. 85–93.

939 Cf. Anthony (2007), p. 61.

940 For most of this paragraph to here, see ibid., pp. 224, 302, 397, 402–403 and 503.

941 Cf. W. H. McNeill, cited in Lenski et al. (1995), p. 227.

942 Cf. Anthony (2007), p. 224; and W. H. McNeill, cited in Lenski et al. (1995), p. 227.

943 Lenski et al. (1995), p. 235.

944 Cf. Childe (1954), pp. 116, 262; and De Camp (1960), pp. 59–334 *passim*.

945 Ponting (2007), p. 75 (this sentence); regarding soil erosion in Attica, see also Plato, *Critias*, 111a–d.

946 Cf. e.g. Palmstierna (1967), p. 103; and Mumford (1967), p. 243.

947 Palmstierna (1967), p. 100.

948 Cf. Ponting (2007), p. 70.

949 Carr-Saunders (1922), pp. 257 and 258.

950 Cf. Plato, *Republic*, 5.461c; and Aristotle, *Politics* VII, 1335^{b}20–26.

951 Carr-Saunders (1922), pp. 256 and 269.

952 Ponting (2007), p. 57.

953 W. Lowdermilk, as cited in Eckholm (1976), p. 114.

954 Cf. Clark & Haswell (1970), pp. 211–213.

955 Cohen (1977b), p. 174.

956 J. Steward and L. Faron, as cited in Lenski et al. (1995), p. 152.

957 Diamond (1992), p. 188.

958 Cf. e.g. G. P. Murdock, cited in Lenski et al. (1995), p. 166.

959 This and what follows to the end of the next paragraph are from Childe (1954), pp. 151–202.

960 Ponting (2007), p. 58.

961 Cf. e.g. F. Nussbaum, cited in Lenski et al. (1995), p. 204.

962 Cf. Zeuner (1975), p. 327.

963 Cf. M. Bloch and others, cited in Lenski et al. (1995), p. 195.

964 J. Blum, as cited in ibid., p. 197.

965 Lewin (1993), p. 132.

966 G. P. Murdock, as cited in Lenski et al. (1995), p. 163.

967 Mumford (1967), pp. 192, 213 and 237–238.

968 In this regard, cf. Polgar (1975), p. 22.

969 W. Watson, as cited in Lenski et al. (1995), p. 159.

970 L. Aitchison, as cited in ibid.

971 Cf. R. Turner, cited in ibid., p. 185.

972 Cf. McMichael (1993), p. 90 (this paragraph).

973 Burnet & White (1972), p. 1; WHO 1996, as cited in J. W. Smith et al. (1997), p. 22.

974 Cf. W. Watson, as cited in Lenski et al. (1995), p. 157 (whole paragraph).

975 Cf. Ponting (1991), p. 67; and Carney (2002); following quote, Diamond (1992), p. 296.

976 Chahin (1996), p. 98; cf. also De Camp (1960), p. 62, with reference to J. B. Prichard and A. Champdor; next quote, Homer, *The Iliad*, ll. 523–532.

977 L. Woolley, as cited in Lenski et al. (1995), p. 181.

978 Cf. Cohen (1977a), p. 146.

979 Tremearne (1912), p. 122.

980 De Camp (1960), p. 299.

981 Tylecote (1992), p. 62.

982 Russell & Russell (1999), p. 20.

983 Ponting (2007), p. 268.

984 Russell & Russell (1999), p. 20.

985 Ponting (1991), pp. 161, 162 and 187.

986 Cf. Cohen (1989), p. 123, n. 20.

987 Acsádi & Nemeskéri (1970), p. 209.

988 Carr-Saunders (1922), p. 251.

989 Cf. Anonymous (1804), p. 86.

990 Russell & Russell (1999), pp. 27–28.

991 Ponting (2007), p. 297.

992 J. Carcopino, as cited in Lenski et al. (1995), p. 201.

993 Daly (1977), p. 153.

994 De Camp (1960), pp. 179–180.

995 Russell & Russell (1999), p. 30.

996 Lenski et al. (1995), p. 205 (almost whole paragraph).

997 This paragraph to here: De Camp (1960), pp. 104–173.

998 The horse harness; and The great harness controversy, The Medieval Technology Pages; on the Internet.

999 Russell & Russell (1999), p. 43.

1000 Platt (1980), p. 304.

1001 Ponting (1991), p. 96 (most of this paragraph).

1002 Mumford (1967), plate 25, between pp. 166 and 167.

1003 Russell & Russell (1999), p. 43; cf. also De Camp (1960), p. 229.

1004 Merchant (1980), p. 45.

1005 Sieferle (2001), p. 66.

1006 Merchant (1980), p. 45.

1007 Cf. also M. E. Clark (1989), p. 62; and Cipolla (1978), pp. 50–51, who refers to A. P. Usher and others.

1008 E. J. Kealey, as reported in Russell & Russell (1999), p. 33.

1009 Ibid., p. 35 ; cf. also De Camp (1960), pp. 296, 297.

1010 De Camp (1960), p. 299.

1011 Ibid.

1012 Russell & Russell (1999), p. 46.

1013 Ponting (1991), p. 272.

1014 De Camp (1960), p. 289.

1015 Ibid., p. 322.

1016 Russell & Russell (1999), p. 35.

1017 Cf. De Camp (1960), pp. 306–348.

1018 Ponting (2007), p. 92.

1019 De Camp (1960), p. 251.

1020 Clark & Haswell (1970), p. 66.

1021 Cf. ibid., p. 212.

1022 Ponting (1991), pp. 162 and 163.

1023 Ponting (2007), p. 142.

1024 Clutton-Brock (1987), p. 81.

1025 Sessions (1995), p. xv.

1026 Ponting (2007), p. 138 (this sentence).

1027 R. K. Douglas, as cited in Lenski et al. (1995), p. 199.

1028 Carr-Saunders (1922), p. 261.

1029 In this regard, see Dilworth (2005), p. 146&n.; following quote quoted in Hardin (1993), p. 221.

1030 Russell & Russell (1980), p. 414.

1031 A. Siegfried, as cited in Ziegler (1969), p. 30.

1032 Merchant (1980), p. 48.

1033 As reported by J. W. M. Whiting, in DeVore et al. (1968), p. 337.

1034 Lenski et al. (1995), p. 208.

1035 H. S. Bennett and G. G. Coulton, as cited in ibid., p. 200 (this sentence to here).

1036 Cf. e.g. J. Blum or W. H. Moreland, cited in ibid., p. 197 (most of this paragraph).

1037 J. Blum, as cited in ibid.

1038 Ibid., p. 200.

1039 Ponting (1991), pp. 95, 104, 109.

1040 McMichael (1993), p. 91.

1041 Many of the widespread instances of famine in agrarian societies are recorded in Ponting (2007), pp. 101–108.

1042 Donella Meadows et al. (1972), p. 37.

1043 R. Darnton, as cited in Lenski et al. (1995), p. 198.

1044 Wilkinson (1973), p. 179.

1045 Mumford (1967), p. 271.

1046 Clark & Haswell (1970), p. 142.

1047 Cf. P. A. Sorokin, Q. Wright, and W. Eberhard, as cited in Lenski et al. (1995), pp. 206–207 (this paragraph).

1048 Russell & Russell (1999), p. 11.

1049 Cf. Anthony (2007), p. 341.

1050 Oakley (1951), p. 11.

1051 Grousset (1939), p. ix; much of the content of this subsection is taken from this work.

1052 Cf. De Camp (1960), p. 248.

1053 Russell & Russell (1999), p. 11; quote following, Malthus (1798), pp. 84–85.

1054 Anthony (2007), p. 223.

1055 Forde (1975), p. 163.

1056 Brues (1960), p. 211, with reference to P. E. Klopsteg.

1057 Cf. ibid., p. 214.

1058 Q. Wright (1964), p. 283.

1059 Childe (1975), p. 210.

1060 In this regard cf. Bonar (1885), p. 105.

1061 Cf. ibid.

1062 L. Weiger, as cited in Grousset (1939), p. 58; in this regard see also Dostoyevsky (1879–80), pp. 219–220.

1063 Bonar (1885), p. 105.

1064 De Camp (1960), p. 293.

1065 Ibid., p. 299.

1066 Russell & Russell (1968), pp. 8 and 215.

1067 De Camp (1960), p. 60; Drower (1975), p. 548. Flooding resultant upon the destruction of the irrigation system in Mesopotamia occurred repeatedly after the system first came into existence. In this regard, cf. the first known version of the myth of the great flood in *The Epic of Gilgamesh* (ca. 2000 BC; Penguin Books, 2003).

1068 Diamond (1992), p. 273.

1069 G. Botero, as cited in Bonar (1931), p. 17 (last sentence).

1070 Ponting (2007), p. 173 (this sentence).

1071 Ponting (1991), p. 130.

1072 Ibid., p. 340.

1073 R. H. Tawney, as cited in Lenski et al. (1995), p. 242 (previous two sentences to here).

1074 Mumford (1967), pp. 280, 281.

1075 Dilworth (2008), p. 213.

1076 Cf. Daly (1985), p. 205.

1077 Daly (1987a), p. 186.

1078 R. W. Fogel et al., as cited in Cohen (1989), p. 124.

1079 Ponting (1991), p. 88.

1080 Lenski et al. (1995), p. 242.

1081 Merchant (1980), p. 51.

1082 Cf. e.g. T. Chandler and G. Fox, cited in Lenski et al. (1995), p. 191.

1083 United Nations (1973).

1084 Roberts (1989), pp. 128–129.

1085 Wilkinson (1973), pp. 94–95.

1086 Cf. ibid., pp. 97, 105.

1087 Ibid., p. 96, referring to E. Boserup.

1088 Russell & Russell (1968), p. 210.

1089 Cf. Boserup (1965), pp. 38–39.

1090 W. H. McNeill, as quoted in McMichael (1993), p. 46.

1091 Russell & Russell (1968), pp. 7 and 180.

1092 Ponting (2007), pp. 174–175.

1093 Wilkinson (1973), pp. 75–76.

1094 Ibid., pp. 110–111.

1095 A. Smith (1776), as quoted in Skinner (1979), p. 33.

1096 Wilkinson (1973), p. 63.

1097 Boserup (1965), p. 73.

1098 Cf. Heilbroner (1974), p. 63.

1099 Harris (1977), p. 198, with reference to J. R. Harlan and others.

1100   Cf. Lenski et al. (1995), p. 184.

1101   Cf. ibid., p. 232.

1102   More (1516), p. 130; following quote, p. 66.

1103   In this regard, cf. Boulding (1973), p. xvi.

1104   S. V. Bath, as cited in Ziegler (1969), p. 22.

1105   Mumford (1967), p. 185.

1106   Tainter (1988), p. 121.

1107   Mumford (1967), p. 170; see also p. 244.

1108   Aristotle, *Politics* II, 1270^{a}38–^{b}7; next quote, Wilkinson (1973), pp. 67–68.

1109   Homer, *The Odyssey*, Ch. 9, ll. 36–43.

1110   Cf. Ponting (1991), p. 81.

1111   Cf. Stott (1962), p. 365.

1112   Merchant (1980), p. 63.

1113   Ponting (2007), p. 277.

1114   Perlin (1989), p. 282.

1115   Grousset (1939), p. 309.

1116   Merchant (1980), p. 63 (much of this paragraph).

1117   Perlin (1989), p. 232.

1118   Sieferle (2001), pp. 104 and 198 (this paragraph).

1119   Cf. Wilkinson (1973), p. 101.

1120   Lenski et al. (1995), p. 190; what follows of this paragraph is from Cipolla (1978), pp. 55 and 66–67.

1121   Wilkinson (1973), p. 122.

1122   Thomas (1985), p. 737, with reference to W. T. Jackman.

1123   Sieferle (2001), p. 131.

1124   After A. Pacey, *The Maze of Ingenuity: Ideas and Idealism in the Development of Technology*, London: Allen Lane, 1974.

1125   Cf. Sieferle (2001), p. 118.

1126   Thomas (1985), p. 730; see also Sieferle (2001), p. 118.

1127   Ibid., p. 25 (this sentence).

1128   A. Pacey, as quoted in Russell & Russell (1999), p. 46.

1129   Thomas (1985), p. 748.

1130   Cf. Georgescu-Roegen (1971b), p. 292.

1131   P. Bairoch, as cited in McNeill (2000), p. 223.

1132   N. Tranter, as cited in Stevenson (1993), p. 232.

1133   Cf. J. D. Chambers and G. E. Mingay, as quoted in Thomas (1985), p. 740.

1134   Ibid., p. 750.

1135   H. Jones (1990), p. 8.

1136   McKeown (1976), pp. 158–159.

1137   W. Stanley Jevons, as quoted in Georgescu-Roegen (1971b), p. 295; Sieferle (2001), p. 104.

1138   Russell & Russell (1999), pp. 79–81.

1139   Cf. Thomas (1985), pp. 730–731.

1140   Laqueur (1993), p. 105, with reference to E. A. Wrigley and R. Schofield.

1141   Data from D. Sharp ed., *Annual Abstract of Statistics*, Office for National Statistics (UK), 2007.

1142   After N. Tranter, as cited in Stevenson (1993), p. 232.

1143   Catton (1980), p. 30.

1144   Tanton (1977), pp. 244–245.

1145   Douthwaite (1999), p. 38 (this sentence to here).

1146   Russell & Russell (1968), p. 23.

1147   Hagner et al. eds. (1995), pp. 22–24.

1148   C. G. Darwin (1952), Ch. II, p. 34.

1149   Cf. Douthwaite (1992), p. 36 (this sentence to here).

1150   Donella Meadows et al. (1992), p. 221.

1151   Pettersson (1993), pp. 11–12 (whole paragraph).

1152   Cf. Douthwaite (1992), p. 35.

1153   D. Glass, as cited in H. Jones (1990), p. 67.

1154   Ibid., p. 40.

1155   Donella Meadows et al. (1992), p. 221 (latter half of this sentence).

1156   Cf. Douthwaite (1999), p. 49.

1157   Georgescu-Roegen (1971b), p. 247n., with reference to P. S. Foner and G. Gunton.

1158 Wilkinson (1973), p. 171.

1159 Ponting (1991), p. 186.

1160 Photo courtesy of Pennsylvania Historical & Museum Commission, Drake Well Museum, Titusville, PA.

1161 Cipolla (1978), p. 57.

1162 Ibid., pp. 62, 69, with reference to S. H. Schurr and B. C. Netschert (this sentence).

1163 Sources: ibid.; the *International Energy Annual 2003*; Sieferle (2001); and Aleklett (2002).

1164 Ponting (2007), p. 289.

1165 Cf. Clark & Haswell (1970), p. 213.

1166 Youngquist (1999), p. 6 (this sentence).

1167 Letter from D. Pimentel, quoted in ibid.

1168 Ophuls & Boyoen (1992), p. 56n.

1169 Palmstierna (1967), p. 45.

1170 Thomas (1985), p. 747.

1171 Levin & Levin (2002) (this sentence).

1172 Martin & Guilday (1967), p. 40.

1173 Sessions (1995), p. xv.

1174 Ponting (1991), p. 163.

1175 Sessions (1995), p. xv.

1176 Youth (2003), p. 17.

1177 Diamond (1992), pp. 351–352.

1178 Ponting (2007), p. 170.

1179 Much of this paragraph and the next is from McKeown (1976), pp. 152–161.

1180 L. J. Henderson, as cited in Hardin (1972), pp. 57–58.

1181 Dubos (1966), p. 31; and Ponting (1991), p. 234.

1182 Cf. Cohen (1989), p. 133.

1183 V. J. Knapp. as cited in Hardin (1993), p. 29; cf. also McMichael (1993), p. 92.

1184 Hardin (1993), p. 28.

1185 Cohen (1989), p. 58, n. 17.

1186 McKeown (1976), p. 161.

1187 J. Burnett, as quoted in Thomas (1985), p. 750.

1188 Douthwaite (1999), pp. 51 and 53.

1189 Nierenberg (2006), p. 74, with further references.

1190 Ibid., with further references.

1191 Tanton (1977), pp. 244–245 (previous two sentences to here).

1192 Polgar (1975), p. 22.

1193 Keeley (1996), p. 28.

1194 Cf. e.g. P. Sorokin, cited in Dobkowski & Wallimann (1998), p. 1.

1195 Q. Wright (1964), p. 53.

1196 Russell & Russell (1968), p. 170, with reference to T. Stonier and Q. Wright.

1197 Redclift (1987), p. 72.

1198 Ponting (1991), p. 232.

1199 Burnet & White (1972), pp. 202, 206; see also McNeill (2000), p. 206.

1200 Q. Wright (1964), p. 61.

1201 Ibid., p. 372.

1202 Sternlycke (2005).

1203 Bulletin of the Atomic Scientists, Carnegie Endowment for International Peace, Newsweek (2006).

1204 See Albons (2006), whose source is the Bulletin of the Atomic Scientists, Carnegie Endowment for International Peace; and Pehrson (2005).

1205 Albright & Barbour (1999).

1206 Mishan (1977), pp. 244–245.

1207 Ponting (2007), p. 288.

1208 Ibid. (this sentence to here).

1209 Lenssen (1991), p. 45 (this sentence).

1210 Ponting (2007), p. 288 (this sentence).

1211 Lenssen (2006), p. 34, with further references.

1212 Ponting (2007), pp. 288–289.

1213 Uranium-stocks.net.

1214 Most of this paragraph and the next three is from Wilkinson (1973), pp. 121–172.

1215 Lenssen (2006), p. 35, with further references.

1216 Perlin (1989), p. 346 (this sentence); following quote, Wilkinson (1973), p. 4.

1217 Ponting (1991), p. 284.

1218 Hansson (1996), p. 265.

1219 Ibid., p. 232.

1220 Weber (1920), p. 60.

1221 Wilkinson (1973), p. 172.

1222 Ponting (1991), p. 282.

1223 Sieferle (2001), p. 41.

1224 J. Williams, as quoted in ibid., p. 186; next quote, W. Stanley Jevons, as quoted in ibid., p. 200.

1225 Georgescu-Roegen (1971b), p. 295; following quote, British Secretary of State for the Environment, as quoted in Schumacher (1973), p. 108.

1226 Abernethy (1993), p. 109.

1227 Schumacher (1972), p. 8; quote following, Sieferle (2001), p. 197.

1228 Aleklett (2006a).

1229 Wilkinson (1973), p. 122.

1230 Ibid., p. 176. Following quote, Mishan (1977), p. 40; see also Wilkinson (1973), p. 86.

1231 Cf. Wilkinson (1973), pp. 177–179; quote following, p. 102.

1232 Ellul (1964), p. 52.

1233 von Wright (1986), p. 142; quote following, von Weizsäcker et al. (1997), p. 293.

1234 Wilkinson (1973), p. 186.

1235 Henderson (1981), p. 309.

1236 Cf. Ponting (1991), pp. 400–401 (last sentence); cf. also Dryzek (1994), p. 178.

1237 McNeill (2000), p. 29.

1238 In this regard see Giarini & Loubergé (1978), p. 85.

1239 Wilkinson (1973), p. 129.

1240 Cf. L. I. Dublin and others, as referred to in Cipolla (1978), p. 90.

1241 Cf. Hardin (1993), p. 141. Following quote, Malthus, *An Essay on the Principle of Population* (7th ed., 1872), II. ix, p. 215 (appeared first in 5th ed., 1817); cited in Bonar (1885), p. 185.

1242 Energy data from the *International Energy Annual*, Hanson (2001), and elsewhere.

1243 T. R. Malthus, as quoted in Anonymous (1803), p. 34.

1244 L. Lallemand, 1885, as cited in Langer (1963), pp. 195–196.

1245 In this regard cf. McKeown (1976), p. 162.

1246 Carr-Saunders (1922), p. 202, quoting J. S. Mill.

1247 Wilkinson (1973), pp. 101–102; quote following, p. 91.

1248 C. G. Darwin (1952), Ch. III, p. 43.

1249 Wilkinson (1973), pp. 84, 102, 135, 189–190.

1250 Much in the next five paragraphs is from ibid., pp. 63–199.

1251 Sieferle (2001), p. 192 (this sentence).

1252 Burns (1974), p. 170n.

1253 Cf. Heilbroner (1974), p. 63; following quote, Weber (1920), p. 65.

1254 B. D. Jouvenal, quoted in Burns (1974), p. 170.

1255 Cf. Weber (1920).

1256 Cf. Dilworth (2007), pp. 104–129.

1257 Darwin (1874), p. 102 (quoted earlier).

1258 Cf. Sieferle (2001), p. 29; quote following, Kapp (1950), p. 49&n.

1259 J. M. Clark, as quoted in Kapp (1950), p. 165.

1260 Ibid.; quote following, ibid.

1261 Terry (1995), p. 55.

1262 Cf. Wilkinson (1973), p. 108.

1263 Trainer (1989), p. 129.

1264 Wilkinson (1973), pp. 181, 182; following quote, A. Smith (1790), pp. 215–216 (Pt. IV, Ch. 1, §10).

1265 In this regard see Malthus (1798), pp. 184, 189 and 192–195; and (1830), p. 244.

1266 Douthwaite (1999), p. 81.

1267 Kaufmann (1991), pp. xxvii–xxviii (this sentence to here); quote following,

R. Owen, Report to the County of Lanark (Glasgow, 1821), as quoted in Kapp (1950), p. 34.

1268 Cf. Wilkinson (1973), p. 187, with reference to B. C. Roberts and J. H. Smith. What follows of this paragraph is from ibid., pp. 135–189.

1269 K. Marx, *The Communist Manifesto* (1848), as quoted in Reiman (1996), p. 151; next quote, ibid.

1270 Cf. Hardin (1972), p. 78.

1271 Carr-Saunders (1922), p. 311.

1272 Keeley (1996), p. 204; following quote, Q. Wright (1964), p. 240.

1273 In this regard cf. Russell & Russell (1968), p. 170, with reference to T. Stonier and Q. Wright.

1274 Energy use data from McNeill (2000), Worldwatch, and elsewhere.

1275 Sieferle (2001), p. 44.

1276 Ponting (1991), p. 287.

1277 In this regard, cf. Daly (1977), p. 104.

1278 Deléage (1994), pp. 43–44.

1279 Cf. Wilkinson (1973), p. 103.

1280 Gever et al. (1991), p. 44.

1281 Ellul (1964), p. 271; see also Russell & Russell (1999), p. 86.

1282 Cf. the International Energy Agency, 2005, as presented in Lindahl (2005).

1283 Douthwaite (1999), p. 226, with reference to M. Slesser and others.

1284 Hardin (1993), p. 137.

1285 L. F. Ivanhoe, as quoted in Douthwaite (1999), p. 222.

1286 Aleklett (2005); cf. also Youngquist (1999).

1287 Douthwaite (1999), p. 221, citing J. Gever.

1288 Cf. Wilkinson (1973), p. 5.

1289 Data from Simons & Co. (on the Internet) and elsewhere.

1290 J. Gever as cited in Douthwaite (1999), p. 221; and others.

1291 Hubbert (1976), p. 116.

1292 Young (1992), pp. 8, 10 (most of this paragraph).

1293 Ponting (2007), pp. 324, 325.

1294 Ibid. p. 324.

1295 Cf. W. W. Leontief and others, as quoted in Daly (1992), p. 371.

1296 Picone & Van Tassel (2002).

1297 Cf. Commoner (1971), p. 149.

1298 Brown (1997), p. 37, with reference to FAO and others.

1299 Hower Jordan (2006), p. 29, with further references.

1300 Gever et al. (1991), p. 28.

1301 Kapp (1950), p. 193.

1302 Bartlett (1978), p. 880.

1303 Cf. Halweil (2006a), pp. 22–23, with further references.

1304 Brown (1997), p. 37, with reference to FAO and others.

1305 McNeill (2000), p. 264; data from the Rijkinstituut voor Volksgezondheid en Milieu.

1306 McNeill (2000), p. 26.

1307 Ehrlich & Ehrlich (1990), p. 92, with reference to L. R. Brown.

1308 www3.omu.edu.tr.

1309 Brown (2005), Ch. 4, with reference to the US Department of Agriculture and others.

1310 J. Larsen, as cited in ibid., Ch. 5.

1311 McNeill (2000), p. 48, with reference to D. Pimentel and others.

1312 P. Freund and G. Martin, as cited in ibid., p. 311.

1313 WHO (2007).

1314 Pimentel (1984), p. 205.

1315 Earth Policy Institute (2006).

1316 Brown (2005), p. 91.

1317 Ibid., Ch. 5, with reference to the World Resources Institute and others. (Presumably urban/industrial land is considered desert.)

1318 C. C. Mann, cited in Hawken (1993), p. 205.

1319 R. Wright (2004), p. 184.

1320 Cf. Ehrlich & Ehrlich (1990), p. 96, with reference to L. R. Brown and associates.

1321 Ibid., p. 28, with reference to L. R. Brown and associates.

1322 Cf. Halweil (2002), p. 102, with further references.

1323 M. G. Wolman, as cited in McLaren (1995), p. 12.

1324 R. Lal, as cited in McNeill (2000), p. 48.

1325 D. Nir, as cited in ibid., p. 32.

1326 McLaren (1995), p. 12; Worldwatch (2005), p. 3.

1327 Gardner (2006), p. 102, with further references.

1328 Leslie (1996), p. 54.

1329 Kuenen (1980), p. 14 (this sentence).

1330 McNeill (2000), pp. 48, 229 and 232.

1331 Gardner (2006), p. 102, with further references.

1332 Brown (2005), Ch. 1, with reference to the US Department of Agriculture and the UN.

1333 Brown (2006a).

1334 Anonymous (2007).

1335 Cf. Halweil (2006a), pp. 22–23, with further references.

1336 Douthwaite (1999), p. 315, with reference to L. R. Brown.

1337 Kysar (2000), p. 60, with reference to L. R. Brown (whole paragraph).

1338 Ibid., with further references.

1339 McMichael (1993), p. 218, with reference to L. R. Brown and the World Resources Institute.

1340 Brown (2005), Ch. 3, with reference to FAO, the Worldwatch Institute's *Signposts 2002*, and the UN's *World Population Prospects: the 2002 Revision*.

1341 McMichael (1993), p. 218, with reference to the World Resources Institute.

1342 McNeill (2000), pp. 248–250.

1343 Halweil (2006b), p. 27, with further references.

1344 Ibid., p. 251.

1345 R. A. Myers and B. Worm, as cited in Ch. 5 of Brown (2006b).

1346 R. A. Myers and B. Worm, as cited in ibid.

1347 Palmstierna (1967), pp. 51–52.

1348 Daly (1987b), p. 229.

1349 Daly (1977), p. 42.

1350 D. W. Schindler and S. E. Bayley, as cited in McLaren (1995), p. 13.

1351 Eckholm (1976), p. 134.

1352 Kysar (2000), p. 60.

1353 Ehrlich & Ehrlich (1990), p. 93, with reference to L. R. Brown.

1354 Cf. P. Gleick, and S. Postel, as cited in McNeill (2000), p. 180.

1355 Pimentel (1984), pp. 203–204, with reference to R. L. Ritschard and K. Tsao.

1356 W. Meyer, as cited in McNeill (2000), p. 181.

1357 Cf. ibid., p. 205.

1358 Kysar (2000), p. 60, with reference to D. Pimentel.

1359 M. L'vovich and G. F. White, and I. A. Shiklomanov, as cited in McNeill (2000), p. 121.

1360 Postel (2000), pp. 122–123, with further references.

1361 Cf. Ponting (2007), pp. 284–285 and 261 (most of this paragraph).

1362 Cf. Ponting (1991), p. 367.

1363 Cf. Douthwaite (1999), p. 147.

1364 Cf. Weart (2005).

1365 Cf. e.g. Palmstierna (1967), p. 93.

1366 Sorkin (2006a), p. 42.

1367 Amos (2006).

1368 Bojs (2005), with reference to the EU project, Epica.

1369 R. P. Brennan, as cited in Leslie (1996), p. 62.

1370 Ponting (2007), p. 389.

1371 Sorkin (2006a), p. 42, with further references.

1372 Mastny (2005), p. 88, with further references.

1373 J. Hummels, as cited in Sorkin (2006a), p. 42.

1374 B. B. Fitzharris and associates, as cited in Mastny (2005), p. 88.

1375 Sorkin (2006a), p. 42, with reference to the National Oceanic and Atmospheric Administration.

1376 Sorkin (2006a), p. 43; T. Graedel and P. Crutzen, and D. Salstein, as cited in McNeill (2000), p. 53.

1377 The Ozone Hole; on the Internet.

1378 Ponting (2007), p. 382.

1379 The Ozone Hole; on the Internet.

1380 Pimentel (1984), p. 206, with reference to L. R. Brown.

1381 Hower Jordan (2006), p. 28, with further references.

1382 Ponting (2007), p. 374.

1383 Ibid., p. 375.

1384 Radnet; and Radiation Information Network, Idaho State University; on the Internet.

1385 Ponting (2007), p. 370.

1386 McNeill (2000), p. 312.

1387 Ponting (2007), p. 375.

1388 Radnet, United Kingdom source points; on the Internet.

1389 Friedman (2006).

1390 Hawken (1993), p. 22 (last phrase).

1391 Species extinctions (2005).

1392 V. H. Heywood and R. T. Watson; and R. M. May et al., as cited in McNeill (2000), pp. 262–263.

1393 Cf. F. Günther, cited in Berg (1990), p. 252; Ponting (1991), p. 193; E. O. Wilson, as cited in Hawken (1993), p. 29.

1394 Cf. Levin & Levin (2002).

1395 Cf. E. O. Wilson, as cited in Gowdy (1998), p. 66.

1396 McMichael (1993), p. 29.

1397 Species extinctions (2005).

1398 Cf. McMichael (1993), p. 112.

1399 United Nations (2004).

1400 Baschetti (2005), p. 243.

1401 Mumford (1967), p. 70.

1402 Cf. Deevey (1960), pp. 196–197.

1403 Cf. H. Jones (1990), pp. 211–212.

1404 Sri Lankan development refugees demand implementation of WCD recommendations on Day of Action; on the Internet.

1405 Daly & Cobb (1989), pp. 49–50.

1406 Moore (2004), p. 54.

1407 Dagens Nyheter (Stockholm), 21 February 2002.

1408 Stair (2006), p. 111, citing the UN Development Programme.

1409 Gowdy (1998), p. 80.

1410 Associated Press and Institute for Policy Studies (US), as cited in Moore (2004), p. 54.

1411 C. Lewis and B. Allison, as cited in ibid., p. 56.

1412 Parsons (1977), p. 149.

1413 Daly (1977), p. 160.

1414 N. Chomsky, as cited in Lewis (1998), pp. 48–49.

1415 Weissman (2003).

1416 Hawken (1993), p. 208, with reference to the UN Development Programme's 1992 Human Development Report.

1417 Sarin (2003), p. 88, with further references.

1418 The Economics Policy Institute in Washington, as cited in Brown (2006b), Ch. 6.

1419 UN Human Development Report of 9 September 1998, as cited in R. Wright (2004), p. 128.

1420 Brown (2006b), Ch. 6.

1421 R. Wright (2004), p. 128.

1422 UN Human Development Report, 9 September 1998, as cited in ibid., p. 185.

1423 Reality of Aid 2000, Earthscan
       Publications, 2000, p. 81.
1424 World Bank (2002).
1425 Redclift (1987), p. 57, citing
       L. Timberlake.
1426 Trainer (1989), pp. 82 and 109, with
       reference to the editors, *Monthly
       Review*, April, 1980, and W. Murdoch
       and others (this paragraph to here).
       Following quote, Ponting (1991),
       p. 345.
1427 Kristof (2002).
1428 Moore (2004), p. 85.
1429 Ponting (2007), p. 417.
1430 L. R. Brown, Earth Policy Institute,
       2004.
1431 Dasgupta (1998), p. 80; next quote,
       Self (2000), p. 147.
1432 Ponting (2007), p. 191 (previous three
       sentences).
1433 Young (1992), p. 13 (previous two
       sentences).
1434 Ponting (2007), p. 191.
1435 Ibid., p. 197.
1436 Ponting (2007), p. 340.
1437 Ponting (1991), p. 343 (this
       sentence).
1438 Ibid.
1439 Cf. Aristide (2000), p. 13.
1440 The United Nations Human
       Development Report, 1997, p. 93.
1441 Welch (2000).
1442 Christian Aid (2005).
1443 Trainer (1989), p. 62.
1444 Shah (2007).
1445 K. Annan at the 2002 World Food
       Summit.
1446 Globalis, UN Development
       Programme, cited in *Dagens Nyheter*
       (Stockholm), 30 March 2005; who,
       cited in Brown (2006b), Ch. 6.
1447 The UN's Food and Agriculture
       Organization.
1448 United Nations, *World Population
       Prospects: The 2006 Revision*, NY: 2007.
1449 McMichael (1993), p. 154.
1450 Brown (2005), Ch. 2.
1451 United Nations (2005); Joint UN
       Programme on hiv/AIDS, 2004
       Report on the Global AIDS Epidemic
       (Geneva: 2004), p. 191, as cited in
       Brown (2006b), Ch. 6.
1452 Joint UN Programme on HIV/AIDS,
       2004, as cited in Brown (2006b).
1453 www.whitehouse.gov/onap/facts.html.
1454 Sorkin (2006b), p. 77, with reference
       to UNAIDS.
1455 The UN, as cited in Brown (2006b),
       Ch. 6.
1456 Gore (1993), p. 345.
1457 Ponting (1991), p. 344.
1458 McMichael (1993), p. 323, with
       reference to W. H. Corson.
1459 McLaren (1996), pp. 254-255.
1460 M. Klare, quoted in Trainer (1989),
       p. 148.
1461 Baran (1957), p. xxxi.
1462 Mulama (2007).
1463 Mulama (2005).
1464 I. Gruhn, cited in Lewis (1998), p. 49.
1465 Amnesty International, as cited in
       Moore (2004), p. 85.
1466 Norman (2005).
1467 Cf. D. Summerfield, as cited in Walker
       (2001), p. 581.
1468 Mishan (1977), p. 239.
1469 Hawken (1993), p. 58.
1470 McLaren (1996), p. 255.
1471 Björk (2006), citing P. Stålenheim of
       the peace-research institute Sipri.
1472 Worldwatch (2005), p. 2.
1473 Renner (2006), p. 84, with further
       references.
1474 Ibid., p. 85, with further references.
1475 Stingy Samaritans; on the Internet.
1476 Q. Wright (1964), pp. 54, 293-294.
1477 Scaruffi (2007).
1478 Diamond (1992), pp. 302-304.
1479 P. Leyhausen, as cited in Russell &
       Russell (1968), pp. 90 and 147.

1480 Eckes (1979), p. 123, quoting H. Ickes.

1481 Cf. Douthwaite (1999), p. 228, with reference to the 1990 World Bank Development Report.

1482 McNeill (2000), pp. 360–361.

1483 *Fortune*, 27 July 1992; and *The Economist Book of Vital World Statistics*, 1990, as cited in Hawken (1993), p. 92.

1484 Ponting (2007), p. 29.

1485 Moore (2004), p. 57; quote following, Kapp (1950), p. 200.

1486 McNeill (2000), p. 208.

1487 Burnet & White (1972), p. 232.

1488 Platt McGinn (2003), p. 65.

1489 Ponting (1991), p. 371.

1490 WHO/UNICEF, etc., as cited in Brown (2006b), Ch. 6.

1491 McNeill (2000), p. 203.

1492 WHO 1996, as cited in J. W. Smith et al. (1997), p. 22.

1493 Cf. e.g. Garrett (1994).

1494 *New Scientist*, ca. 1 April 2007.

1495 IRIN/MM (2006).

1496 Terry (1995), p. 85.

1497 C. G. Darwin (1952), Ch. x, p. 178.

1498 Douthwaite (1999), p. 83.

1499 Trainer (1989), p. 41.

1500 Mishan (1977), pp. 169 and 240 (previous two sentences).

1501 Cf. Russell & Russell (1968), p. 226.

1502 Mishan (1977), p. 25.

1503 Trainer (1989), p. 128.

1504 Stair (2006), p. 111.

1505 Hawken (1993), p. 126.

1506 Trainer (1998), pp. 88, 93. See also, e.g. Douthwaite (1992), p. 3; and von Weizsäcker et al. (1997), p. 280.

1507 Douthwaite (1999), *passim* (this paragraph to here).

1508 Abernethy (1993), pp. 200, 205 and 206, with reference to the article Longer hours, in an issue of the *Wall Street Journal* from 1992 (this paragraph to here).

1509 Cf. Terry (1995), pp. 63 and 78, with reference to the article Job drought, *Fortune Magazine*, August 1992.

1510 Abernethy (1993), p. 206.

1511 R. Reich, as cited in Terry (1995), p. 86.

1512 US Bureau of the Census, *Current Population Survey, March 1998*, Table H-3, as cited in Grant (2000), p. 84.

1513 John Ralston Saul, as cited in R. Wright (2004), p. 185.

1514 Terry (1995), pp. 67 and, quoting W. Harman, 158. The following three sentences: pp. 60 and 64.

1515 Douthwaite (1999), p. 84; drawn from J. Schmitt and others.

1516 Daly & Cobb (1989), p. 410.

1517 Douthwaite (1992), p. 81.

1518 Cf. Georgescu-Roegen (1972), p. 18.

1519 Cf. Palmstierna (1967), pp. 93, 99.

1520 Goldsmith et al. (1972), pp. 12, 29; quote following, Dennis Meadows (2004).

1521 Ehrlich & Ehrlich (1990), p. 45.

1522 For a development of this notion, see Dilworth (2008).

1523 I provide such a comparison of what I term the economic and ecological perspectives on sustainable development in ibid., App. vi.

1524 For more concerning which, see Dilworth (2007).

1525 Schumacher (1972), p. 3; next quote, Commoner (1971), p. 186.

1526 In this regard cf. Commoner (1971), p. 449; see also M. E. Clark (1992), pp. 171, 172; and Milbrath (1989), p. 358.

1527 Daly (1987a), p. 185.

1528 Daly & Cobb (1989), p. 198.

1529 Schumacher (1973), pp. 130 and 221.

1530 Schumacher (1968), p. 50.

1531 Palmstierna (1967), pp. 61, 132–133.

1532 Ponting (2007), pp. 331–332.

1533 J. O'Connor (1994), p. 168; quote following, Duhl (1966), p. 41.

1534 Ehrlich & Ehrlich (1990), p. 45.

1535 Baran (1957), p. xxix.

1536 Kapp (1950), p. vii.

1537 Daly (1977), p. 99.

1538 In this regard cf. ibid., p. 103.

1539 Boulding (1966), p. 9.

1540 An index of economic welfare constructed along these lines is included as an appendix to Daly & Cobb (1989); another is in a measure developed by the NGO Redefining Progress (cf. Assadourian, 2006, p. 52).

1541 Cf. Schumacher (1973), pp. 34–35.

1542 Schumacher (1966), p. 44; and (1972), p. 4.

1543 Daly (1977), p. 108.

1544 Cf. e.g. Georgescu-Roegen (1971b), p. 280.

1545 Schumacher (1970), p. 15.

1546 Ibid., pp. 17–18; cf. also Milbrath (1989), p. 27.

1547 Commoner (1971), p. 146; quote following, ibid., p. 261.

1548 Milbrath (1989), p. 27; quote following, E. Goldsmith, in Polunin et al. (1980), p. 593.

1549 Wilkinson (1973), p. 195.

1550 Mishan (1977), p. 15.

1551 Kapp (1950), pp. xiii, 12 and 233.

1552 Ibid., pp. xii and 18; following quote, p. 231.

1553 Commoner (1971), pp. 144 and 157.

1554 Goldsmith et al. (1972), pp. 14 and 16; following quote, Lorenz (1963), p. 38.

1555 Mishan (1977), p. 224.

1556 Forrester (1975), pp. 273-282.

1557 A. and H. Lovins, as quoted in Daly & Cobb (1989), pp. 334–335.

1558 Most of the present paragraph is from Schumacher (1968), pp. 56, 57; and (1973), p. 230.

1559 Reiman (1996), p. 150.

1560 TT-Reuters (2006), the poverty line being taken to be just less than $10,000/year for a single person, and $20,000/year for a family of four.

1561 Ponting (2007), p. 337.

1562 Schumacher (1968), p. 57.

1563 Douthwaite (1999), p. 99.

1564 Schumacher (1973), p. 236; quote following, p. 248.

1565 K. Marx, as quoted in Reiman (1996), p. 152.

1566 Cf. also Arndt (1978), p. 47.

1567 Daly (1977), p. 8.

1568 On neoclassical economics as a science, see Dilworth (2007), Ch. 6; following quote, Commoner (1971), pp. 274-275.

1569 Cf. Kapp (1950), p. 239; following quote, J. M. Clark, as quoted on p. 253.

1570 Ibid., p. 11.

1571 Cf. Arndt (1978), p. 35.

1572 Douthwaite (1999), p. 56.

1573 Cf. Arndt (1978), pp. 36 and 49.

1574 Cf. Dilworth (2007).

1575 Daly (1987a), p. 188; following quote, Schumacher (1970), p. 17.

1576 Mishan (1977), p. 29.

1577 Ibid., pp. 29, 119, 261 and 262.

1578 Daly (1977), p. 153; next quote, pp. 152-153.

1579 In this regard cf. Mishan (1977), p. 223.

1580 Terry (1995), p. 74; following quote, Douthwaite (1999), p. 87.

1581 Cf. Douthwaite (1999), pp. 31 and 87.

1582 Kapp (1950), p. 153.

1583 Baran (1957), p. 73.

1584 Douthwaite (1992), p. 21; following quote, Commoner (1971), p. 260.

1585 Kapp (1950), pp. 218, 219.

1586 Ibid., p. xxiii.

1587 Commoner (1971), p. 144 (most of this paragraph to here).

1588 Terry (1995), pp. 69, 87.

1589 From the files of the General Electric Company, as quoted in Kapp (1950), pp. 177–178; rest of paragraph, pp. 29 and 178.

1590 Terry (1995), p. 64.

1591 Ibid., p. 69.

1592 In this regard, cf. Kapp (1950), p. 192.

1593 Terry (1995), pp. 70 and 71 (this paragraph).

1594 Mishan (1977), p. 26.

1595 Much of this paragraph comes from Arndt (1978), pp. 55–69.

1596 Cf. Goldsmith et al. (1972), p. 30.

1597 On the incoherence of this line, see Dilworth (2008), App. VI.

1598 Trainer (1989), p. 130; following quote, Rio Declaration on Environment and Development (1993), p. 44.

1599 Schumacher (1973), p. 33; following quote, p. 27.

1600 Arndt (1978), p. 143; following quote, K. E. Boulding, as quoted in Ehrlich & Ehrlich (1990), p. 159.

1601 In this regard cf. Georgescu-Roegen (1971b), p. 294n.; quote following, Mishan (1977), p. 259.

1602 J. K. Galbraith, as quoted in Arndt (1978); cf. pp. 125 and 126.

1603 In this regard, cf. Perrings (1994), pp. 111–112; and Milbrath (1989), p. 27.

1604 Hardin (1972), p. 142.

1605 Heilbroner (1974), p. 54; cf. also Goldsmith et al. (1972), p. 26.

1606 Georgescu-Roegen (1971b), p. 304.

1607 Schumacher (1973), p. 117.

1608 F. Berkhout, as cited in McNeill (2000), p. 312.

1609 Schumacher (1973), p. 110.

1610 Schumacher (1972), p. 7; quote following, Daly (1977), p. 12.

1611 Schumacher (1973), p. 110; following quote, p. 114.

1612 Published by Her Majesty's Stationery Office and entitled *Pollution: Nuisance or Nemesis?* Following quote, ibid., pp. 117–119.

1613 Douthwaite (1999), p. 322.

1614 Nuclear decommissioning insurance financial product and method; on the Internet, 2006.

1615 Hardin (1993), p. 151.

1616 Green (2006), p. 2.

1617 The European Nuclear Society; on the Internet, 2006.

1618 Daly (1977), p. 134, with reference to R. G. Kazman and M. C. Day.

1619 Schumacher (1973), p. 89.

1620 Georgescu-Roegen (1971b), p. 303.

1621 Gever et al. (1991), p. 27.

1622 McMichael (1993), pp. 103 and 213.

1623 McNeill (2000), p. 28.

1624 Falk (2006).

1625 Cf. Norgaard (1984), p. 532.

1626 Ponting (1991), p. 371.

1627 McMichael (1993), p. 214, citing P. Weber.

1628 C. G. Darwin (1952), Ch. X, p. 172.

1629 Cf. Prabhakaran Nair (2004).

1630 Ibid.

1631 P. Bairoch, as cited in McNeill (2000), pp. 225–226 (this sentence).

1632 Ibid., p. 222.

1633 M. K. Tolba and O. A. El-Kholy, as cited in ibid., p. 224 (this paragraph to here).

1634 Cf. ibid.

1635 P. Bairoch, as cited in ibid., pp. 225–226 (this paragraph).

1636 Cf. S. D. Ripley, p. 962 of Commoner et al. (1972).

1637 Quoted in Parsons (1977), p. 220.

1638 See e.g. Malthus (1798), pp. 72 and 186–187; and (1830), p. 248. Following quote, Forrester (1971), p. 124; emphasis added.

1639 Ehrlich & Ehrlich (1990), p. 77.

1640 Catton (1980), p. 112 (this paragraph and the next).

1641 Cf. Trainer (1989), p. 107.

1642 N. Polunin, in Pauling et al. (1980), p. 24.

1643 Prabhakaran Nair (2004).

1644 Cf. R. Wright (2004), pp. 45–46. See also Mumford (1967), p. 135, cited in Chapter 5.

1645 Grant (2000), pp. 14 and 15 (previous two sentences); quote following, Environment News Service, as quoted in ibid., p. 15.

1646 Halweil (1999), pp. 22, 23 (previous two sentences).

1647 Ibid., p. 27.

1648 F. Günther, personal communication.

1649 Sandström (2006).

1650 Douthwaite (1999), p. 194.

1651 Cf. Prabhakaran Nair (2004).

1652 D. Pimentel, as cited in Youngquist (1999).

1653 Ponting (1991), p. 371.

1654 Ophuls & Boyoen (1992), p. 23n.

1655 Ponting (1991), p. 371.

1656 Gever et al. (1991), p. 28.

1657 Ponting (1991), p. 371.

1658 Hardin (1993), p. 254.

1659 In this regard, cf. Berg (1990), p. 283.

1660 The major threats, Global Challenges, Emerging Opportunities, Harvard School of Public Health Annual Report 2003; on the Internet.

1661 McNeill (2000), p. 199-200; data from Rijkinstituut voor Volksgezondheid en Milieu/UN Environment Programme; following quote, Mishan (1977), p. 43.

1662 Cf. McKeown (1976), p. 155.

1663 Ponting (1991), pp. 233, 234.

1664 Davis & Mauch (1977), p. 234.

1665 Mishan (1977), p. 230.

1666 Dubos (1966), p. 29 (this sentence); quote following, Darwin (1874), p. 138.

1667 Hubendick (1987), p. 19; following quote, Darwin (1874), p. 139.

1668 McNeill (2000), p. 202.

1669 For the reasoning involved, see H. J. Muller, as reported in Dobzhansky (1962), pp. 293–294.

1670 Ibid., pp. 293–295.

1671 C. G. Darwin (1952), Ch. v, p. 81.

1672 Palmstierna (1967), p. 114.

1673 Garrett (1994), p. 420.

1674 J. W. Smith et al. (1997), p. 21.

1675 Ophuls & Boyoen (1992), p. 23n.

1676 Garrett (1994), p. 419.

1677 McNeill (2000), p. 203&n.

1678 Mishan (1977), pp. 54-55. In the following quote the reference is to H. Koprowski; Georgescu-Roegen (1971b), p. 357.

1679 Edwards (1967), p. 145.

1680 Catton (1980), p. 188.

1681 Douthwaite (1999), p. 42.

1682 Aleklett (2006b).

1683 Cf. Campbell & Laherrère (1998), p. 78.

1684 Ljungqvist & Böhlmark (2004), p. 6; following quote, Schumacher (1973), p. 107.

1685 Schumacher (1972), p. 4.

1686 Daly (1980), p. 379.

1687 Mishan (1977), p. 32.

1688 Pettersson (1993), p. 2.

1689 Wynne-Edwards (1963), pp. 44–45.

1690 As argued for in Dilworth (2008), App. VI.

1691 In this regard cf. Goldsmith et al. (1972), p. 95.

1692 Ponting (2007), p. 285.

1693 Commoner (1971), pp. 172, 177, 259, 268; following quote, Goldsmith et al. (1972), p. 32.

1694 Kuenen (1980), p. 11.

1695 Sieferle (2001), p. 199.

1696 Hanson (2001).

1697 McLaren (1996), p. 255; next quote, Aristotle, *Politics* II, 1265^{b}6–13.

1698 Donella Meadows et al. (1972), p. 178.

1699 Daly (1977), p. 72.

1700 Kuenen (1980), p. 11.

1701 U Thant, as quoted in Hardin (1968), p. 136.

1702 McLaren (1995), p. 10.

1703 McLaren (1996), p. 253.

1704 Cf. Hardin (1993), p. 4.

1705 McLaren (1995), p. 10.

1706 Cf. Hardin (1993), p. 109; quote following, Daly & Cobb (1989), p. 247.

1707 E. Wolf, as cited in Parsons (1977), p. 57.

1708 Ibid., p. 152, with reference to W. Petersen.

1709 Ibid.

1710 Wilkinson (1973), pp. 180, 181.

1711 Hawken (1993), p. 78.

1712 W. Bonger, as referred to in Marxist criminology, Wikipedia; on the Internet.

1713 Reiman (1996), p. 146.

1714 Ibid., p. 151; quote following, p. 174.

1715 Moore (2004), p. 199.

1716 Gardner (2000), p. 151.

1717 Reiman (1996), pp. 142, 146, 171.

1718 W. Bonger, as referred to in Marxist criminology, Wikipedia; on the Internet.

1719 Reiman (1996), p. 148; next quote, ibid., p. 149.

1720 Daly (1977), p. 152; quote following, Lorenz (1963), pp. 228–229.

1721 von Malmborg (2004).

1722 Schumacher (1970), pp. 15–16; and (1966), p. 45.

1723 Schumacher (1973), p. 121.

1724 Cf. Hawken (1993), p. 135.

1725 Baran (1957), p. 12.

1726 Daly (1977), p. 152.

1727 Georgescu-Roegen (1971b), p. 314.

1728 Catton (1980), p. 175.

1729 Cf. Daly & Cobb (1989), p. 290.

1730 E. Goldsmith, in Polunin et al. (1980), p. 606; next quote, Carr-Saunders (1922), p. 183.

1731 Schumacher (1965), p. 141; next quote, Schumacher (1973), p. 134.

1732 K. E. Boulding, p. 966 of Commoner et al. (1972).

1733 Schumacher (1974), p. 159.

1734 Arndt (1978), p. 63.

1735 Malthus (1798), p. 195.

1736 Daly (2002).

1737 Trainer (1989), p. 115; quote following, Kuenen (1980), p. 13.

1738 UN Human Development Report, 9 September 1998, as cited in R. Wright (2004), p. 185.

1739 Schumacher (1971), p. 180.

1740 Shah (2007).

1741 Trainer (1989), pp. 69, 71 (this paragraph to here).

1742 February 2003; as quoted in Weissman (2003).

1743 Cf. Palmstierna (1967), p. 35.

1744 Shah (2007).

1745 J. Browett, as quoted in Trainer (1989), p. 50.

1746 Cf. ibid., pp. 46–47; in this regard see also Schumacher (1971), p. 180.

1747 Trainer (1989), pp. 73–75 and 86 (this paragraph).

1748 C. G. Darwin (1952), Ch. II, p. 33.

1749 McNeill (2000), p. 170.

1750 J. Farley and others, as cited in ibid., p. 171.

1751 Ibid., p. 172.

1752 Trainer (1989), pp. 105, 106.

1753 Dennis Meadows, as quoted in Hopkins (2006). See also Dennis Meadows (2004).

1754 J. W. Smith & Sauer-Thompson (1998), p. 543.

1755 J. W. Smith et al. (1997), p. 154.

References

Abernethy, V. (1993) *Population Politics: The Choices that Shape Our Future*, NY: Plenum Press/Insight Books.
  (1998) Defining the new American community, pp. 101–116 of Dobkowski & Wallimann eds.

Acsádi, Gy. and Nemeskéri, J. (1970) *History of Human Life Span and Mortality*, Budapest: Akadémiai Kiadó.

Albons, B. (2006) En kärnvapenfri värld möjlig på sikt, *Dagens Nyheter* (Stockholm), 1 June, p. 25.

Albright, D. and Barbour, L. (1999) Plutonium watch, ISIS; on the Internet.

Aleklett, K. (2002) Ny oljekris står för dörren, *Svenska Dagbladet* (Stockholm), 24 April.
  (2005) Review of E. N. Luttwak's The truth about global oil supply; on the Internet.
  (2006a) Report from the 7th International Oil Summit in Paris April 7, 2006, Association for the Study of Peak Oil&Gas; on the Internet.
  (2006b) Oil: a bumpy road ahead, *World Watch Magazine*, January/February.

Allen, R. (1977) Towards a primary life-style, pp. 313–329 of Dennis Meadows ed.

Alroy, J. (1999) Putting North America's end-Pleistocene megafaunal extinction in context, pp. 105–143 of MacPhee ed.

Amos, J. (2006) Deep ice tells long climate story, BBC News; on the Internet, 4 September 2006.

Anonymous (1803) The Monthly Review, pp. 21–46 of Pyle ed.
  (1804) The British Critic, pp. 77–89 of Pyle ed.
  (2007) Grain harvest sets record, but supplies still tight, Worldwatch; on the Internet.

Anthony, D. W. (2007) *The Horse, the Wheel and Language*, Princeton and Oxford: Princeton University Press.

Ardrey, R. (1961) *African Genesis*, NY: Atheneum.

Aristide, J.-B. (2000) *Eyes of the Heart: Seeking a Path for the Poor in the Age of Globalization*, Monroe, ME: Common Courage Press.

Aristotle, *The Complete Works of Aristotle*, J. Barnes ed., Princeton: Princeton University Press, 1984.

Arndt, H. W. (1978) *The Rise and Fall of Economic Growth*, Melbourne: Longman Cheshire.

Arsuaga, J. L. and Martínez, I. (1998) *The Chosen Species: The Long March of Human Evolution*, Oxford: Blackwell.

Assadourian, E. (2006) Global economy grows again, pp. 52–53 of Assadourian et al. et al. *Vital Signs 2006-2007 – The Trends that Are Shaping Our Future*, NY and London: W. W. Norton.

Australia in the News (2007); on the Internet.

Baerreis, D. (1980) North America in the early Postglacial, pp. 356–360 of Sherratt ed.

Baran, P. A. (1957) *The Political Economy of Growth*, NY and London: Monthly Review Press.

Barnard, A. and Woodburn, J. (1988) Property, power and ideology in hunter-gathering societies: an introduction, pp. 4–31 of Ingold et al. eds.

Barnowsky, A. D. (1989) The late Pleistocene event as a paradigm for widespread mammal extinction, pp. 235–254 of S. K. Donovan ed., *Mass Extinctions: Processes and Evidence*, NY: Columbia University Press.

Bartholomew, G. A. and Birdsell, J. B. (1953) Ecology and the protohominids, pp. 20–37 of Montagu ed.

Bartlett, A. A. (1978) Forgotten fundamentals of the energy crisis, *American Journal of Physics* **46**: 876–888.

(1994) Reflections on sustainability, population growth, and the environment, *Population and Environment* **16**: 5–35.

Baschetti, R. (2005) Evolutionary, biological origins of morality: implications for research with human embryonic stem cells, *Stem Cells and Development* **14**: 239–247.

Berg, P. G. (1990) *Omsorg om vår planet*, Stockholm: Natur och Kultur.

Binford, L. R. (1968) Post-Pleistocene adaptation, pp. 313–341 of Binford & Binford eds.

(1992) Subsistence – a key to the past, pp. 365–368 of Jones et al. eds.

Binford, S. R. and Binford, L. R. eds. *New Perspectives in Archeology*, Chicago: Aldine, 1968.

Birdsell, J. B. (1968) Some predictions for the Pleistocene based on equilibrium systems among recent hunter-gatherers, pp. 229–240 of Lee & DeVore eds. (1968).

Björk, L. (2006) Färre krig i världen – men fler militärer, *Miljömagasinet* (Sweden), 16 June 2006, p. 4.

Bojs, K. (2005) Isen avslöjar extrema halter av koldioxid, *Dagens Nyheter* (Stockholm), 25 November, p. 10.

Bonar, J. (1885) *Malthus and his Work*, London: Frank Cass & Co., 1966.

(1931) *Theories of Population from Raleigh to Arthur Young*, Bristol: Thoemmes Press, 1992.

Boserup, E. (1965) *The Conditions of Agricultural Growth*, London: George Allen & Unwin.

Boulding, K. E. (1964) *The Meaning of the Twentieth Century*, NY: Harper & Row.

(1966) The economics of the coming spaceship earth, pp. 3–14 of Jarrett ed.

(1970) *Economics as a Science*, NY: McGraw-Hill.

(1973) Introduction to Wilkinson (1973), pp. xiii–xx.

(1980) Spaceship earth revisited, pp. 311–313 of Daly & Townsend eds.

Bramble, D. M. and Lieberman, D. E. (2004) Endurance running and the evolution of *Homo*, *Nature* **432**: 345–352.

Brandon, R. N. and Burian, R. M. eds. *Genes, Organisms, Populations: Controversies over the Units of Selection*, Cambridge, MA and London: MIT Press, 1984.

Bronson, B. (1977) The earliest farming, pp. 23–48 of Reed ed.

Brown, L. R. (1997) The agricultural link: how environmental deterioration could disrupt economic progress, Worldwatch paper 136, Worldwatch Institute.

(2005) *Outgrowing the Earth: The Food Security Challenge in an Age of Falling Water Tables and Rising Temperatures*; on the Internet.

(2006a) World grain stocks fall to 57 days of consumption, Earth Policy Institute; on the Internet.

(2006b) *Plan B 2.0: Rescuing a Planet under Stress and a Civilization in Trouble*; on the Internet.

et al. *Vital Signs 2000 – The Trends that Are Shaping Our Future*, NY and London: W. W. Norton.

Brues, A. (1960) The spearman and the archer – an essay on selection in body build, pp. 202–215 of Montagu ed.

Bulletin of the Atomic Scientists, Carnegie Endowment for International Peace, Newsweek (2006) under the title, 'Nordkorea säger sig vara ett av nio kärnvapenländer,' *Dagens Nyheter* (Stockholm), 10 October 2006.

Burnet, M. and White, D. O. (1972) *Natural History of Infectious Disease*, Cambridge: Cambridge University Press.

Burns, T. (1974) On the rationale of the corporate system, pp. 121–177 of R. Marris ed., *The Corporate Society*, London and Basingstoke: Macmillan.

Campbell, C. J. and Laherrère, J. H. (1998) The end of cheap oil, *Scientific American* **278**: 78–83.

Carney, R. (2002) The chariot: a weapon that revolutionized Egyptian warfare; on the Internet.

Carpenter, C. R. (1958) Territoriality: a review of concepts and problems, pp. 224–250 of A. Roe and G. G. Simpson eds., *Behavior and Evolution*, New Haven: Yale University Press.

Carrier, D. (1984) The energetic paradox of human running and hominid evolution, *Current Anthropology* **25**: 483–495.

Carr-Saunders, A. M. (1922) *The Population Problem: A Study in Human Evolution*, Oxford: Clarendon Press.

Carson, R. (1962) *Silent Spring*, London: Penguin Books, 1965.

Caspari, R. and Lee, S.-H. (2004) *Proceedings of the National Academy of Sciences*, 5 July 2004; from the Internet.

Catton, W. R., Jr. (1980) *Overshoot: The Ecological Basis of Revolutionary Change*, Urbana: University of Illinois Press, 1982.

Chahin, M. (1996) *Before the Greeks*, Cambridge: Lutterworth Press.

Childe, V. G. (1952) *New Light on the Most Ancient East*, London: Routledge & Kegan Paul.

(1954) *What Happened in History*, Harmondsworth: Penguin Books, 1967.

(1975) Rotary motion, pp. 187–215 of Singer et al. eds.

Christian Aid (2005) The economics of failure: the real cost of 'free' trade, *Christian Aid*, 20 June.

Cipolla, C. M. (1978) *The Economic History of World Population*, Sussex: Harvester Press.

Clark, C. and Haswell, M. (1970) *The Economics of Subsistence Agriculture*, 4th edn., London: Macmillan.

Clark, M. E. (1989) *Ariadne's Thread*, NY: St. Martin's Press.

(1992) Critiquing the 'rational' worldview: the parable of the tree-trunk and the crocodile, pp. 169–176 of L. O. Hansson and B. Jungen eds., *Human Responsibility and Global Change*, Gothenburg: University of Gothenburg.

Clutton-Brock, J. (1987) *A Natural History of Domesticated Animals*, London and Cambridge: Cambridge University Press.

(1992) Domestication of animals, pp. 380–385 of Jones et al. eds.

Coghlan, H. H. (1975) Metal implements and weapons, pp. 600–622 of Singer et al. eds.

Cohen M. N. (1975) Archaeological evidence for population pressure in pre-agricultural societies, *American Antiquity* **40**: 471–475.

(1977a) *The Food Crisis in Prehistory*, New Haven and London: Yale University Press.

(1977b) Population pressure and the origins of agriculture: an archaeological example from the coast of Peru, pp. 135–177 of Reed ed.

(1989) *Health and the Rise of Civilization*, New Haven and London: Yale University Press.

Coleman, D. (1986) Population regulation: a long-range view, pp. 14–41 of D. Coleman and R. S. Schofield eds., *The State of Population Theory – Forward from Malthus*, Oxford and NY: Basil Blackwell.

Colinvaux, P. A. (1973) *Introduction to Ecology*, NY: John Wiley & Sons.

Commoner, B. (1971) *The Closing Circle: Confronting the Environmental Crisis*, London: Jonathan Cape, 1972.

et al. (1972) Discussion, pp. 959–983 of M. T. Farvar and J. P. Mitton eds., *The Careless Technology. Ecology and International Development*, NY: Natural History Press.

Cranston, M. (1968) Introduction to Rousseau (1762), pp. 9–43.

Crowley, B. L. (2001) Make way for the Blue Revolution, *Telegraph Journal* (Canada), Sect. A 5, 3 March.

Daly, H. E. (1971) Introduction to *Essays toward a Steady-State Economy*, pp. 11–47 of Daly & Townsend eds.

(1977) Preface to the First Edition, pp. xv–xvi, plus pp. 2–178, of Daly (1992).

(1980) Postscript: some common misunderstandings and further issues
concerning a steady-state economy, pp. 365–382 of Daly & Townsend eds.

(1985) The circular flow of exchange value and the linear throughput of matter-
energy: a case of misplaced concreteness, pp. 195–210 of Daly (1992).

(1987a) The steady-state economy: alternative to growthmania, pp. 180–194 of
Daly (1992).

(1987b) The economic growth debate: what some economists have learned, but
many have not, pp. 224–240 of Daly (1992).

(1990a) Sustainable growth: an impossibility theorem, pp. 267–273 of Daly &
Townsend eds.

(1990b) Sustainable development: from concept and theory toward operational
principles, pp. 241–260 of Daly (1992).

(1992) *Steady-State Economics*, 2nd edn., London: Earthscan.

(2002) Sustainable development: definitions, principles, policies. Invited address
to the World Bank, Washington, 30 April.

and Cobb, J.B. (1989) *For the Common Good*, London: Green Print, 1990.

and Townsend, K.N. eds. *Valuing the Earth: Economics, Ecology, Ethics*, Cambridge, MA
and London: MIT Press, 1993.

Darwin, C. (1859) *The Origin of Species by means of Natural Selection*,
Harmondsworth: Penguin Books, 1968.

(1874) *The Descent of Man*, 2nd edn. (1st edn. 1871), NY: Prometheus Books, 1998.

(1892) *The Autobiography of Charles Darwin, and Selected Letters*, NY: Dover
Publications, 1958.

Darwin, C.G. (1952) *Nästa miljon år*, Stockholm: Bonniers, 1953 (*The Next Million Years*,
NY: Doubleday & Company).

Dasgupta, B. (1998) *Structural Adjustment, Global Trade and the New Political Economy of
Development*, London: Zed Books.

Davis, J. and Mauch, S. (1977) Strategies for societal development, pp. 217–242 of
Dennis Meadows ed.

Dawkins, R. (1989) *The Selfish Gene*, 2nd edn., Oxford: Oxford University Press.

De Camp, L.S. (1960) *The Ancient Engineers*, NY: Dorset, 1990.

Deevey, E.S., Jr. (1960) The human population, *Scientific American* **203**: 195–204.

Deléage, J.-P. (1994) Eco-Marxist critique of political economy, pp. 37–52 of
M. O'Connor ed.

Dersin, D. and Mitchell, P. eds. (1992) *Incas: Lords of Gold and Glory*, Alexandria,
VA: Time-Life Books.

DeVore, I. et al. (1968) Discussions, Part vii, pp. 335–346 of Lee & DeVore eds. (1968).

Diamond, J. (1992) *The Third Chimpanzee*, NY: Harper Perennial, 2006.

Dilworth, C. (1994) Two perspectives on sustainable development, *Population and
Development* **15**: 441–467.

(1997) *Sustainable Development and Decision Making*, Uppsala: Department of
Philosophy, Uppsala University.

(2005) The selfish karyotype – An analysis of the biological basis of morals,
Biology Forum **98**: 125–154.

(2007) *The Metaphysics of Science*, 2nd ed., Dordrecht: Springer.

(2008) *Scientific Progress*, 4th ed., Dordrecht: Springer.

Divale, W. T. (1972) Systematic population control in the Middle and Upper Paleolithic: inferences based on contemporary hunter-gatherers, *World Archaeology* **4**: 222–243.

Dobkowski, M. N. and Wallimann, I. (1998) The coming age of scarcity: an introduction, pp. 1–20 of Dobkowski & Wallimann eds.

eds. (1998) *The Coming Age of Scarcity*, Syracuse, NY: Syracuse University Press.

Dobzhansky, T. (1962) *Mankind Evolving: The Evolution of the Human Species*, New Haven and London: Yale University Press.

Dostoyevsky, F. (1879–80) *The Brothers Karamazov*, R. E. Matlaw ed., NY: W. W. Norton, 1976.

Douthwaite, R. (1992) *The Growth Illusion*, Tulsa: Council Oak Books, 1993.

(1999) *The Growth Illusion* (revised edn.), Devon: Green Books.

Draper, P. (1976) Social and economic constraints on child life among the !Kung, pp. 199–217 of Lee & DeVore eds. (1976).

Drower, M. S. (1975) Water-supply, irrigation and agriculture, pp. 520–571 of Singer et al. eds.

Dryzek, J. S. (1994) Ecology and discursive democracy: beyond liberal capitalism and the administrative state, pp. 176–197 of M. O'Connor, ed.

Dubos, R. (1966) Promises and hazards of man's adaptability, pp. 23–39 of Jarrett ed.

Duhl, L. J. (1966) Mental health in an urban society, pp. 40–43 of Jarrett ed.

Eckes, A. E. (1979) *The United States and the Global Struggle for Minerals*, Austin and London: University of Texas Press.

Eckholm, E. P. (1976) *Losing Ground: Environmental Stress and World Food Prospects*, NY: W. W. Norton.

Edmeades, B. (2006) *Megafauna – First Victims of the Human-Caused Extinction*, Ch. 12; on the Internet.

Edwards, W. E. (1967) The Late-Pleistocene extinction and diminution in size of many mammalian species, pp. 141–154 of Martin & Wright eds.

Ehrlich, P. R. and Ehrlich, A. H. (1990) *The Population Explosion*, NY: Simon & Schuster.

Eldredge, N. (1999) Cretaceous meteor showers, the human ecological 'niche,' and the sixth extinction, pp. 1–15 of MacPhee ed.

Ellul, J. (1964) *The Technological Society*, 2nd edn., NY: Random House/Vintage Books.

Encyclopedia Britannica (2005) Iron and steel industry. History: First smelting and cementation; on the Internet.

Endicott, K. (1988) Property, power and conflict among the Batek of Malaysia, pp. 110–127 of Ingold et al. eds.

Estioko-Griffin, A. and Bion Griffin, P. (1981) Woman the hunter: the Agta, pp. 121–151 of F. Dahlberg ed., *Woman the Gatherer*, New Haven and London: Yale University Press.

Falk, J. (2006) Gott om gifter i svensk mat, *Dagens Nyheter* (Stockholm), 22 September, p. 11.

Fisher, R. A. (1930) *The Genetical Theory of Natural Selection*, Oxford: Oxford University Press, 1999.

Flannery, T.F. and Roberts, R.G. (1999) Late Quaternary extinctions in Australasia, pp. 239–255 of MacPhee ed.

Flew, A. (1970) Introduction to Malthus (1798), pp. 7–56.

Foley, R.A. (1987) *Another Unique Species: Patterns in Human Evolutionary Ecology*, Harlow: Longman Scientific and Technical.

Forbes, R.J. (1975) Extracting, smelting, and alloying, pp. 572–599 of Singer et al. eds.

Forde, D. (1975) Foraging, hunting, and fishing, pp. 154–186 of Singer et al. eds.

Forrester, J.W. (1971) *World Dynamics*, Cambridge, MA: Wright-Allen Press.

(1975) Control of urban growth, pp. 271–284 of his *Collected Papers of Jay W. Forrester*, Cambridge, MA: Wright-Allen Press.

Friedman, J. (2006) Nuclear terrorism – a self-imposed legacy, CommonDreams.org Newscenter, 30 August 2006; on the Internet.

Gardner, G. (2000) Prison populations exploding, pp. 150–151 of Brown et al.

(2006) Deforestation continues, pp. 102–103 of Assadourian et al.

et al. (2003) *State of the World 2003*, London: Earthscan.

Garrett, L. (1994) *The Coming Plague*, NY: Penguin Books.

Georgescu-Roegen, N. (1971a) The entropy law and the economic problem, pp. 75–88 of Daly & Townsend eds.

(1971b) *The Entropy Law and the Economic Process*, Cambridge, MA: Harvard University Press.

(1972) Economics and entropy, *The Ecologist* **2**: 13–18.

(1975) Selections from 'Energy and economic myths,' pp. 89–112 of Daly & Townsend eds.

Gever, J. et al. (1991) *Beyond Oil. The Threat to Food and Fuel in the Coming Decades*, 3rd edn., Colorado: University Press of Colorado.

Giarini, O. and Loubergé, H. (1978) *The Diminishing Returns of Technology: An Essay on the Crisis in Economic Growth*, Oxford: Oxford University Press.

Gibson, T. (1988) Meat sharing as a political ritual: forms of transaction vs. modes of subsistence, pp. 165–179 of Ingold et al. eds.

Goldsmith, E. et al. (1972) *A Blueprint for Survival*, London: Tom Stacey.

Gore, A. (1993) *Earth in the Balance*, NY: Penguin Books.

Gorman, C. (1977) *A priori* models and Thai prehistory: a reconsideration of the beginnings of agriculture in southeastern Asia, pp. 321–355 of Reed ed.

Gowdy, J.M. (1998) Biophysical limits to industrialization, pp. 65–82 of Dobkowski & Wallimann eds.

Gowlett, J.A.J. (1992a) *Ascent to Civilization: The Archaeology of Early Humans*, 2nd edn., NY: McGraw-Hill.

(1992b) Early human mental abilities, pp. 341–345 of Jones et al. eds.

Graham, S. (2003) Jawbone hints at Europe's earliest modern humans, *Scientific American.com*, 23 September 2003; on the Internet.

Grant, L. (2000) *Too Many People: The Case for Reversing Growth*, Santa Ana, CA: Seven Locks Press.

Green, J. (2006) Nuclear power and climate change, November 2006; on the Internet.

Grousset, R. (1939) *The Empire of the Steppes*, NJ: Rutgers University Press, 1970.

Hägerström, A. (1953) *Inquiries into the Nature of Law and Morals*, Uppsala: Almqvist & Wiksell.

Hagner, C. J. et al. eds. (1995) *Anatolia: Cauldron of Cultures*, Richmond, VA: Time-Life Books.

Halweil, B. (1999) The Emperor's new crops, pp. 21–29 of *World Watch* 1999; on the Internet.

(2002) Farmland quality deteriorating, pp. 102–103 of J. N. Abramovitz et al. (2002) *Vital Signs 2002 – The Trends that Are Shaping Our Future*, NY and London: W. W. Norton.

(2006a) Grain harvest flat, pp. 22–23 of Assadourian et al.

(2006b) Fish harvest stable but threatened, pp. 26–27 of Assadourian et al.

Hamburg, D. A. and McCown, E. eds. (1979) *The Great Apes*, Menlo Park, CA: Benjamin/Cummings.

Hanson, J. (2001) Synopsis, 8 March 2001; on the Internet.

Hansson, S. (1996) *Teknikhistoria*, 2nd ed., Lund: Studentlitteratur.

Harako, R. (1981) The cultural ecology of hunting behavior among Mbuti Pygmies in the Ituri Forest, Zaïre, pp. 499–555 of Harding & Teleki eds.

Hardin, G. (1968) The tragedy of the commons, pp. 127–143 of Daly & Townsend eds.

(1972) *Exploring New Ethics for Survival – The Voyage of the Spaceship* Beagle, NY: The Viking Press.

(1993) *Living within Limits. Ecology, Economics and Population Taboos*, NY and Oxford: Oxford University Press.

Harding, R. O. S. and Teleki, G. (1981) Introduction, pp. 1–9 of Harding & Teleki eds. *Omnivorous Primates. Gathering and Hunting in Human Evolution*, NY: Columbia University Press, 1981.

Harpending, H. C. et al. (1998) Genetic traces of ancient demography, *Proceedings of the National Academy of Sciences USA* **95**: 1961–1967.

Harris, D. R. (1977) Alternative pathways toward agriculture, pp. 179–243 of Reed ed.

Hassan, F. A. (1975) Determination of the size, density and growth rate of hunting-gathering populations, pp. 27–52 of Polgar ed.

(1977) The dynamics of agricultural origins in Palestine: a theoretical model, pp. 589–609 of Reed ed.

Hawken, P. (1993) *The Ecology of Commerce*, NY: HarperBusiness.

Hayden, B. (1981) Subsistence and ecological adaptations of modern hunter/gatherers, pp. 344–421 of Harding & Teleki eds.

Haynes, C. V. (1966) Elephant hunting in North America, *Scientific American* **214**: 104–112, June.

Heilbroner, R. L. (1974) *An Inquiry into the Human Prospect*, NY: W. W. Norton.

Heinrich, B. (2001) *Why We Run: A Natural History*, NY: HarperCollins, 2002.

Henderson, H. (1981) *The Politics of the Solar Age,* NY: Anchor Press/Doubleday.

Herre, W. and Röhrs, M. (1977) Zoological considerations on the origins of farming and domestication, pp. 245–279 of Reed ed.

Hesiod, *Theogony, and Works and Days*, Oxford: Oxford University Press, 1988.

Hewlett, B. S. (undated) Cultural diversity among African Pygmies; on the Internet.

Hobbes, T. (1651) *Leviathan*, Harmondsworth: Penguin Books, 1968.

Holdaway, R. N. (1999) Introduced predators and avifaunal extinction in New Zealand, pp. 189–238 of MacPhee ed.

Holdren, J. P. (1990) Energy in transition, *Scientific American* **263**: 157–163, September.

Hole, F. (1968) Evidence of social organisation from western Iran, 8000–4000 B.C., pp. 245–265 of Binford & Binford eds.

(1992) Origins of agriculture, pp. 373–379 of Jones et al. eds.

Holling, C. S. (1994) Ecologist view of Malthusian conflict, pp. 79–103 of K. Lindahl-Kiessling and H. Landberg eds., *Population, Economic Development, and the Environment*, Oxford: Oxford University Press.

Homer, *The Iliad*, R. Fagles tr., Harmondsworth: Penguin Books, 1991.

The Odyssey, E. V. Rieu tr., Harmondsworth: Penguin Books, 1991.

Hopkins, R. (2006) An interview with Dennis Meadows – co-author of *Limits to Growth*; on the Internet.

Howell, N. (1976) The population of the Dobe area !Kung, pp. 137–151 of Lee & DeVore eds. (1976).

Hower Jordan, L. (2006) Pesticide trade shows new market trends, pp. 28–29 of Assadourian et al.

Hubbert, M. K. (1976) Exponential growth as a transient phenomenon in human history, pp. 113–126 of Daly & Townsend eds.

Hubendick, B. (1987) *Människoekologi*, Hedemora, Sweden: Gidlunds Förlag.

Infoplease (2005) Iron Age; on the Internet.

Ingold, T. et al. eds. *Hunters and Gatherers 2*, Oxford: Berg Publishers, 1988.

IRIN/MM (2006) Poliofall hotar utrotningsplaner, *Miljömagasinet* (Sweden), 30 June, p. 5.

Irwin, G. J. (1980) The prehistory of Oceania: colonization and cultural change, pp. 324–332 of Sherratt ed.

Isaac, B. (1987) Throwing and human evolution, *African Archaeological Review* **5**: 3–17.

Isaac, G. Ll. and Crader, D. C. (1981) To what extent were early hominids carnivorous? An archaeological perspective, pp. 37–103 of Harding & Teleki eds.

Jarrett, H. ed. *Environmental Quality in a Growing Economy*, Baltimore: Johns Hopkins Press, 1966.

Jensfelt, A. (2001) Ny brandfackla tänder debatten om människans ursprung, *Svenska Dagbladet* (Stockholm), 14 January, pp. 14–15.

Johnson, A. W. and Earle, T. (1987) *The Evolution of Human Societies*, Stanford: Stanford University Press.

(2000) *The Evolution of Human Societies*, 2nd edn., Stanford: Stanford University Press.

Jones, H. (1990) *Population Geography*, 2nd edn., London: Paul Chapman.

Jones, S. et al. eds. *The Cambridge Encyclopedia of Human Evolution*, Cambridge: Cambridge University Press, 1992.

Kapp, K. W. (1950) *The Social Costs of Private Enterprise*, NY: Schocken Books, 1971.

Kaufmann, R. (1991) Prologue to Gever et al. (1991), pp. xxvii–xli.

Keeley, L. H. (1996) *War Before Civilization*, NY and Oxford: Oxford University Press.

King, M.-C. and Wilson, A. C. (1975) Evolution at two levels in humans and chimpanzees, *Science* **188**: 107–116.

Kirk, R. and Szathmary, E. eds. *Out of Asia*, Canberra: Journal of Pacific History, 1985.

Knutsson, H. (1995) *Slutvandrat? Aspekter på övergången från rörlig till bofast tillvaro*, Uppsala: Societas Archaeologica Upsaliensis.

Konner, M. J. (1976) Maternal care, infant behavior and development among the !Kung, pp. 218–245 of Lee & DeVore eds. (1976).

Kornfeldt, T. (2006) Människan och schimpans närmare än forskare trott, *Dagens Nyheter* (Stockholm), 18 May, p. 10.

Kottak, C. P. (1994) *Anthropology: The Exploration of Human Diversity*, 6th edn., NY: McGraw-Hill.

Kristof, N. D. (2002) Farm subsidies that kill, *New York Times*, 5 July.

Kuenen, D. J. (1980) A biologist's view and warnings, pp. 7–18 of Polunin ed.

Kysar, D. A. (2000) Sustainability, distribution and the macroeconomic analysis of law; on the Internet.

Lack, D. (1954) *The Natural Regulation of Animal Numbers*, Oxford: Clarendon Press.
  (1966) *Population Studies of Birds*, Oxford: Clarendon Press.

Langer, W. L. (1963) Disguised infanticide, pp. 194–197 of G. Hardin ed., *Population, Evolution and Birth Control*, 2nd edn., San Francisco: W. H. Freeman, 1967.

Laqueur, T. (1993) Sex and desire in the Industrial Revolution, pp. 100–123 of O'Brien & Quinault eds.

Laughlin, W. S. (1968) Hunting: an integrating biobehavior system and its evolutionary importance, pp. 304–320 of Lee & DeVore eds. (1968).

Lee, R. B. (1968) What hunters do for a living, or, how to make out on scarce resources, pp. 30–43 of Lee & DeVore eds. (1968).
  (1976) Introduction, pp. 3–24 of Lee & DeVore eds. (1976).
  (1979) *The !Kung San: Men, Women, and Work in a Foraging Society*, Cambridge: Cambridge University Press.
  and DeVore, I. (1968) Problems in the study of hunters and gatherers, pp. 3–12 of Lee & DeVore eds. (1968).
  et al. (1968) Discussion, pp. 343–361 of Binford & Binford eds.

Lee, R. B. and DeVore, I. eds. (1968) *Man the Hunter*, Chicago: Aldine.
  (1976) *Kalahari Hunter-Gatherers – Studies of the !Kung San and Their Neighbors*, Cambridge, MA and London: Harvard University Press.

Lenski, G. et al. (1995) *Human Societies*, 7th edn., NY: McGraw-Hill.

Lenssen, N. (1991) Nuclear waste: the problem that won't go away, Worldwatch Paper 106, Worldwatch Institute.
  (2006) Nuclear power inches up, pp. 34–35 of Assadourian et al.

Leslie, J. (1996) *The End of the World: the Science and Ethics of Human Extinction*, London and NY: Routledge.

Levin, P. S., and Levin, D. A. (2002) The real biodiversity crisis, *American Scientist*, January–February 2002, p. 6; on the Internet.

Lewin, R. (1993) *The Origin of Modern Humans*, NY: Scientific American Library.

Lewis, C.H. (1998) The paradox of global development and the necessary
    collapse of modern industrial civilization, pp. 43–60 of Dobkowski &
    Wallimann eds.

Lindahl, B. (2005) Rejäl ökning av växthus gaser, *Svenska Dagbladet* (Stockholm), 8
    November, p. 16.

Lindström, P.O. (2005) Indianer härstammar från 70 individer, *Miljömagasinet/TT*
    (Sweden), 3 June, p. 16.

Ljungqvist, I. and Böhlmark, A. (2004) Olje- och konsumtionssamhällets brytpunkt,
    2000-Talets Vetenskap **2**: 3–8.

Lorenz, K. (1963) *On Aggression*, London and NY: Routledge, 1966.

MacPhee, R.D.E. ed. (1999) *Extinctions in Near Time: Causes, Contexts, and Consequences*,
    NY: Kluwer Academic/Plenum Publishers.

Macpherson, C.B. (1968) Introduction to Hobbes (1651), pp. 9–63.

Malmberg, T. (1980) *Human Territoriality*, The Hague: Milton Publishers.

Malthus, T.R. *An Essay on the Principle of Population and A Summary View of the Principle*
    of Population, Harmondsworth: Penguin Books, 1970.

    (1798) *An Essay on the Principle of Population*, pp. 59–217 of Malthus.

    (1830) *A Summary View of the Principle of Population*, pp. 219–272 of Malthus.

Marshall, L. (1976) Sharing, talking and giving: relief of social tension among the
    !Kung, pp. 349–371 of Lee & DeVore eds. (1976).

Marshall Thomas, E. (1989) *The Harmless People*, NY: Vintage Books.

Martin, P.S. (1966) Africa and Pleistocene overkill, *Nature* **212**: 339–342.

    (1967) Prehistoric overkill, pp. 75–120 of Martin & Wright eds.

    (1973) The discovery of America, *Science* **179**: 969–974.

    (1984) Prehistoric overkill: the global model, pp. 354–403 of P.S. Martin and
        R.G. Klein eds. *Quaternary Extinctions*, Tucson: University of Arizona Press.

    and Guilday, J.E. (1967) A bestiary for Pleistocene biologists, pp. 1–62 of Martin &
        Wright eds.

    and Leakey, L.S. B (1967) Overkill at Olduvai Gorge, *Nature* **215**: 212–213.

    and Steadman, D.W. (1999) Prehistoric extinctions on islands and continents,
        pp. 17–55 of MacPhee ed.

    and Wright, H.E. eds. (1967) *Pleistocene Extinctions: the Search for a Cause*,
        New Haven: Yale University Press.

Mason, W.A. (1979) Environmental models and mental modes: replication
    processes in the great apes, pp. 277–293 of Hamburg & McCown eds.

Mastny, L. (2005) Global ice melting accelerating, pp. 88–89 of Mastny et al. *Vital*
    Signs 2005 – The Trends that Are Shaping Our Future, NY and London:
    W.W. Norton, 2005.

Mayr, E. (1984) The unity of the genotype, pp. 69–84 of Brandon & Burian eds.

    (1988) *Toward a New Philosophy of Biology*, Cambridge, MA and London: Belknap
        Press.

McKeown, T. (1976) *The Modern Rise of Population*, London: Edward Arnold.

McLaren, D.J. (1995) Humankind: the agent and victim of global change in
    the geosphere-biosphere system, pp. 3–24 of J.A. Leith et al. eds., *Planet*

Earth: Problems and Prospects, Montreal and Kingston: McGill-Queen's University Press.

(1996) Population growth – should we be worried?, *Population and Environment* **17**: 243–259.

McMichael, A.J. (1993) *Planetary Overload: Global Environmental Change and the Health of the Human Species*, Cambridge: Cambridge University Press.

McNeill, J.R. (2000) *Something New under the Sun – An Environmental History of the Twentieth-Century World*, NY and London: W.W. Norton.

Meadows, Dennis L. (2004) 30-year update of *Limits to Growth* finds global society in "overshoot," foresees social, economic, and environmental decline; on the Internet.

ed. *Alternatives to Growth – I: A Search for Sustainable Futures*, Cambridge, MA: Ballinger, 1977.

Meadows, Donella H. et al. (1972) *Limits to Growth: A Report for the Club of Rome's Project on the Predicament of Mankind*, London: Earth Island.

(1992) *Beyond the Limits: Global Collapse or a Sustainable Future*, London: Earthscan.

Media Release (2001) Humans to blame for extinction of Australia's megafauna, University of Melbourne, 8 June.

Merchant, C. (1980) *The Death of Nature*, NY: Harper & Row.

Milbrath, L.W. (1989) *Envisioning a Sustainable Society*, NY: State University of New York Press.

Mishan, E.J. (1977) *The Economic Growth Debate: An Assessment*, London: George Allen & Unwin.

Montagu, M.F.A. (1962) Introduction to Montagu ed., pp. vii–xiii.

ed. *Culture and the Evolution of Man*, NY: Oxford University Press, 1962.

Moore, M. (2004) *Stupid White Men*, London: Penguin Books.

Morales, W.Q. (1998) Intrastate conflict and sustainable development, pp. 245–268 of Dobkowski & Wallimann eds.

Moran, E.F. (1982) *Human Adaptability. An Introduction to Ecological Anthropology*, Boulder: Westview Press.

More, T. (1516) *Utopia*, London: Penguin Books, 1965.

Mulama, J. (2005) Organisationer kräver internationellt vapenavtal, *Miljömagasinet/ IPS* (Sweden), 29 April, p. 5.

(2007) Africa awash with small arms, misery, 20 June 2007; on the Internet.

Mumford, L. (1967) *The Myth of the Machine*, Vol. 1: *Technics and Human Development*, San Diego: Harcourt Brace Jovanovich.

Nabokov, P. (1981) *Indian Running*, Santa Fe: Ancient City Press, 1987.

Nierenberg, D. (2006) Population continues to grow, pp. 74–75 of Assadourian et al.

Nitecki, M.H. and Nitecki, D.V. (1987) *The Evolution of Human Hunting*, NY and London: Plenum Press.

Norgaard, R.B. (1984) Coevolutionary agricultural development, pp. 525–546 of J. Strauss ed., *Economic Development and Cultural Change*, Vol. 32, Chicago: University of Chicago Press.

Norman, H. (2005) Krav på internationellt vapenhandelsavtal, Amnesty Press; on the Internet.

Oakley, K. P. (1951) A definition of man, pp. 3–12 of Montagu ed.

(1975) Skill as a human possession, pp. 1–37 of Singer et al. eds.

O'Brien, P. and Quinault, R. eds., *The Industrial Revolution and British Society*, Cambridge: Cambridge University Press, 1993.

O'Connor, J. (1994) Is sustainable capitalism possible?, pp. 152–175 of M. O'Connor ed.

O'Connor, M. ed. *Is Capitalism Sustainable? Political Economics and the Politics of Ecology*, NY and London: Guilford Press, 1994.

Ophuls, W. and Boyoen, A. S., Jr. (1992) *Ecology and the Politics of Scarcity Revisited: Unravelling the American Dream*, NY: Freeman.

Orwell, G. (1945) *Animal Farm*, London: Secker and Warburg.

Palmstierna, H. (1967) *Plundring, svält, förgiftning*, Stockholm: Rabén & Sjögren, 1971.

Parsons, J. (1977) *Population Fallacies*, London: Elek/Pemberton.

Pauling, L. et al. (1980) Discussion, pp. 19–27 of Polunin ed.

Pehrson, L. (2005) Kärnvapenhotet en realitet, *Dagens Nyheter* (Stockholm), 2 May, p. 25.

Perlin, J. (1989) *A Forest Journey: The Role of Wood in the Development of Civilization*, NY and London: W. W. Norton.

Perrings, C. (1994) Conservation of mass and the time behaviour of ecological-economic systems, pp. 99–177 of P. Burley and J. Foster eds., *Economics and Thermodynamics*, Dordrecht: Kluwer.

Pettersson, R. (1993) Ekonomi och ekologi i historien. Paper delivered at the 1993 Seminar on Economic Growth at Stockholm University.

Picone, C. and Van Tassel, D. (2002) Agriculture and biodiversity loss: industrial agriculture, pp. 99–105 of N. Eldredge ed., *Life on Earth: An Encyclopedia of Biodiversity, Ecology, and Evolution*; from the Internet.

Pimentel, D. (1984) Ecological feedback from economic growth and technological development, pp. 201–213 of A.-M. Jansson ed., *Integration of Economics and Ecology: An Outlook for the Eighties*, Stockholm: Askö Laboratories.

Plato, *The Collected Dialogues of Plato*, E. Hamilton and H. Cairns eds., NJ: Princeton University Press, 1963.

Platt, C. (1980) The rise of temperate Europe, pp. 304–313 of Sherratt ed.

Platt McGinn, A. (2003) Combating malaria, pp. 62–84 of Gardner et al.

Polgar, S. (1975) Population, evolution, and theoretical paradigms, pp. 1–25 of Polgar ed.

ed. *Population, Ecology and Social Evolution*, The Hague and Paris: Mouton, 1975.

Polunin, N. ed. *Growth without Ecodisasters?*, London and Basingstoke: Macmillan, 1980.

et al. (1980) Discussion, pp. 590–610 of Polunin ed.

Ponting, C. (1991) *A Green History of the World*, London: Sinclair-Stevenson.

(2007) *A New Green History of the World*, London: Vintage Books.

Postel, S. (2000) Groundwater depletion widespread, pp. 122–123 of Brown et al.

Prabhakaran Nair, K. P. (2004) Beware biotechnology farming, *The Hindu Business Line*, 15 March.

Pyle, A. ed. *Population: Contemporary Responses to Thomas Malthus*, Bristol: Thoemmes Press, 1994.

Rasmuson, M. (2000) Neandertalare fortsätter att förbrylla och fascinera, *Svenska Dagbladet* (Stockholm), Under strecket, 26 June.

Redclift, M. (1987) *Sustainable Development: Exploring the Contradictions*, London: Methuen.

Reed, C. A. (1977a) Origins of agriculture: discussion and some conclusions, pp. 879–953 of Reed ed.

(1977b) Introduction to Reed ed., pp. 1–5.

(1983) Archeozoological studies in the Near East – a short history (1960–1980), pp. 511–536 of L. S. Braidwood et al. eds., *Prehistoric Archeology along the Zagros Flanks*, Chicago: The Oriental Institute of the University of Chicago.

ed. *Origins of Agriculture*, The Hague and Paris: Mouton, 1977.

Reiman, J. (1996) ... *and the Poor get Prison: Economic Bias in American Criminal Justice*, Boston: Allyn and Bacon.

(2006) Military expenditures keep growing, pp. 84–85 of Assadourian et al.

Renner, M. (2006) Military expenditures keep growing, pp. 84–85 of Assadourian et al.

Richerson, P. J. and Boyd, R. (2001) Built for speed, not for comfort – Darwinian theory and human culture, *History and Philosophy of the Life Sciences* **23**: 423–463; Special Issue: *Darwinian Evolution Across the Disciplines*, pp. 1–53; on the Internet.

Rigaud, J.-P. (1989) From the Middle to the Upper Paleolithic: transition or convergence?, pp. 142–153 of Trinkaus ed.

Rio Declaration on Environment and Development (1993).

Roberts, N. (1989) *The Holocene: An Environmental History*, Oxford: Basil Blackwell.

Rose, F. G. G. (1968) Australian marriage, land-owning groups, and initiations, pp. 200–208 of Lee & DeVore eds. (1968).

Rousseau, J.-J. (1762) *The Social Contract*, London: Penguin Books, 1968.

Russell, C. and Russell, W. M. S. (1968) *Violence, Monkeys and Man*, London: Macmillan.

(1980) Scarcities and societal objectives, pp. 409–428 of Polunin ed.

(1999) *Population Crises and Population Cycles*, London: The Galton Institute.

Sahlins, M. (1968) Notes on the original affluent society, pp. 85–89 of Lee & DeVore eds. (1968).

(1972) *Stone Age Economics*, Chicago: Aldine.

Sandström, M. (2006) Genmodifierad bomull förstörs av insekter, *Dagens Nyheter* (Stockholm), 28 July, p. 11.

Sarin, R. (2003) Rich-poor divide growing, pp. 88–89 of Renner, M. et al. *Vital Signs 2003 – The Trends that Are Shaping Our Future*, NY and London: W. W. Norton.

Scaruffi, P. (2007) 1900–2000: a century of genocides; on the Internet.

Schapera I. (1930) *The Khoisan Peoples of South Africa: Bushmen and Hottentots*, London: Routledge & Kegan Paul, 1960.

Schumacher, E. F. (1965) Social and economic problems calling for the development of intermediate technology, pp. 141–157 of Schumacher (1973).

(1966) Buddhist economics, pp. 38–46 of Schumacher (1973).

(1968) A question of size, pp. 47–58 of Schumacher (1973).

(1970) Peace and permanence, pp. 11–25 of Schumacher (1973).

(1971) The problem of unemployment in India, pp. 171–184 of Schumacher (1973).

(1972) The problem of production, pp. 2–10 of Schumacher (1973).

(1973) *Small is Beautiful*, NY: Harper & Row.

(1974) The age of plenty: a Christian view, pp. 159–172 of Daly & Townsend eds.

Science Daily (1999) Light my fire: cooking as key to modern human evolution, 10 August 1999; on the Internet.

Self, P. (2000) *Rolling Back the Market*, NY: St. Martins.

Sessions, G. (1995) Preface, pp. ix–xxviii of G. Sessions ed., *Deep Ecology for the 21st Century*, Boston and London: Shambhala.

Shah, A. (2007) US and foreign aid assistance, Sustainable development, April 2007; on the Internet.

Sherratt, A. ed. *The Cambridge Encyclopedia of Archaeology*, NY: Crown Publishers/ Cambridge University Press, 1980.

Shostak, M. (1976) A !Kung woman's memories of childhood, pp. 246–277 of Lee & DeVore eds. (1976).

Sieferle, R. P. (2001) *The Subterranean Forest: Energy Systems and the Industrial Revolution*, M. P. Osman tr., Cambridge: White Horse Press.

Silberbauer, G. B. (1981a) pp. 455–498 of Harding & Teleki eds.

(1981b) *Hunter and Habitat in the Central Kalahari Desert*, Cambridge: Cambridge University Press.

Singer, C. et al. eds. *A History of Technology*, Vol. 1, Oxford: Clarendon Press, 1975.

Skinner, A. (1979) Introduction to A. Smith (1776), pp. 11–97.

Smith, A. (1776) *The Wealth of Nations*, Books I–III, London: Penguin Books, 1986.

(1790) *The Theory of Moral Sentiments*, 6th edn., K. Haakonssen ed., Cambridge: Cambridge University Press, 2002.

Smith, F. H. and Paquette, S. P. (1989) The adaptive basis of Neanderthal facial form, with some thoughts on the nature of modern human origins, pp. 181–210 of Trinkaus ed.

Smith, J. W. and Sauer-Thompson, G. (1998) Civilization's wake: ecology, economics and the roots of environmental destruction and neglect, *Population and Environment* **19**: 541–575.

et al. (1997) *Healing a Wounded World*, Westport, CT: Praeger.

Smith, V. L. (1992) Economic principles in the emergence of humankind, *Economic Inquiry* **xxx**: 1–13.

Sollas, W. J. (1915) *Ancient Hunters and their Modern Representatives*, 2nd edn., London: Macmillan.

Sorkin, L. (2006a) Climate change impacts rise, pp. 42–43 of Assadourian et al.

(2006b) HIV/AIDS threatens development, pp. 76–77 of Assadourian et al.

Species extinctions (2005) Species extinctions and human population, Whole Systems Foundation; on the Internet.

Stair, P. (2006) Regional disparities in quality of life persist, pp. 110–111 of Assadourian et al.

Stanford, C. B. (1999) *The Hunting Apes. Meat Eating and the Origins of Human Behavior*, Princeton: Princeton University Press.

Stein, P. L. and Rowe, B. M. (1995) *Physical Anthropology: The Core*, NY: McGraw-Hill.

Sternlycke, H. (2005) Apokalypsen nära, *Miljömagasinet* (Sweden), 12 August, p. 5.

Stevenson, J. (1993) Social aspects of the Industrial Revolution, pp. 229–253 of O'Brien & Quinault eds.

Stott, D. H. (1962) Cultural and natural checks on population-growth, pp. 355–376 of Montagu ed.

Strimmer, K. and Haeseler, A. (1997) Humans and chimpanzees – when did they split?, Technical Report, Department of Zoology, University of Munich; on the Internet.

Strum, S. C. (1981) Processes and products of change: baboon predatory behavior at Gilgil, Kenya, pp. 255–302 of Harding & Teleki eds.

Stuart, A. J. (1999) Late Pleistocene megafaunal extinctions, pp. 257–269 of MacPhee ed.

Sumner, W. G. (1907) *Folkways*, Salem, NH: Ayer Company, 1992.

Suttles, W. (1968) Coping with abundance: subsistence on the Northwest coast, pp. 56–68 of Lee & DeVore eds. (1968).

Tainter, J. A. (1988) *The Collapse of Complex Societies*, Cambridge: Cambridge University Press.

Tanaka J. (1976) Subsistence ecology of central Kalahari San, pp. 98–119 of Lee & DeVore eds. (1976).

Tanton, J. (1977) International migration and world stability, pp. 234–264 of Dennis Meadows ed.

Teleki, G. (1981), The omnivorous diet and eclectic feeding habits of chimpanzees in Gombe National Park, Tanzania, pp. 303–343 of Harding & Teleki eds.

Terry, R. (1995) *Economic Insanity: How Growth-Driven Capitalism Is Devouring the American Dream*, San Francisco: Berrett-Koehler.

Thomas, B. (1985) Escaping from constraints: the industrial revolution in a Malthusian context, *Journal of Interdisciplinary History* **xv**: 729–753.

Trainer, T. (1989) *Developed to Death. Rethinking Third World Development*, London: Green Print.

(1998) Our unsustainable society, pp. 83–100 of Dobkowski & Wallimann eds.

Tremearne, A. J. N. (1912) *The Tailed Head-Hunters of Nigeria*, 2nd edn., London: Seeley, Service & Co..

Trinkaus, E. ed. *The Emergence of Modern Humans: Biocultural Adaptations in the Later Pleistocene*, Cambridge: Cambridge University Press, 1989.

Truswell, A. S. and Hansen, J. D. L. (1976) Medical research among the !Kung, pp. 166–194 of Lee & DeVore eds. (1976).

TT-Reuters (2006) in *Miljömagasinet* (Sweden), 1 September 2006, p. 11.

Turnbull, C. M. (1961) *The Forest People*, London: Pimlico, 1993.

(1968) The importance of flux in two hunting societies, pp. 132–137 of Lee & DeVore eds. (1968).

Turner, C. G. (1985) The dental search for Native American origins, pp. 31–78 of Kirk & Szathmary eds.

Tylecote, R.F. (1992) *A History of Metallurgy*, 2nd edn., London: Institute of Materials.

United Nations (1973) *The Determinants and Consequences of Population Trends*, Vol. 1, NY: United Nations (as cited on the Internet).

(2004) *World Urbanization Prospects: The 2003 Revision*; on the Internet.

(2005) *World Population Prospects: The 2004 Revision*, NY, 2005.

van der Merwe, N.J. (1992) Reconstructing prehistoric diet, pp. 369–372 of Jones et al. eds.

Vidal-Naquet, P. ed. (1987) *Atlas över mänsklighetens historia*, Stockholm: Bonnier Fakta, 1991.

von Malmborg, R. (2004) Kärnkraft mer subventionskrävande än vind, *Miljömagasinet* (Sweden), 22 October, p. 2.

von Weizsäcker, E.U. et al. (1997) *Factor Four: Doubling Wealth, Halving Resource Use*, London: Earthscan.

von Wright, G.H. (1986) *Vetenskapen och förnuftet*, Stockholm: Månpocket.

Walker, P.L. (2001) A bioarchaeological perspective on the history of violence, *Annual Review of Anthropology* **30**: 573–596.

Warner, W.L. (1969) *A Black Civilization: A Social Study of an Australian Tribe* (revised edn.), Gloucester, MA: Peter Smith.

Washburn, S.L. (1960) Tools and human evolution, pp. 13–19 of Montagu ed.

and Lancaster, C.S. (1968) The evolution of hunting, pp. 293–303 of Lee & DeVore eds. (1968).

Watanabe, H. (1968) Subsistence and ecology of northern food gatherers with special reference to the Ainu, pp. 69–77 of Lee & DeVore eds. (1968).

Weart, S. (2005) The discovery of global warming; on the Internet.

Weber, M. (1920) *The Protestant Ethic and the Spirit of Capitalism*, 2nd edn., NY: Charles Scribner's Sons, 1958.

Weissman, R. (2003) Corporate globalization and the global gap between rich and poor, *Multinational Monitor Magazine*, July/August 2003; on the Internet.

Welch, C. (2000) A world in chains, *The Ecologist Report, Globalising Poverty*, September.

West, M.L. (1988) Introduction to Hesiod, pp. vii–xxi.

WHO (2007) World report on road traffic injury prevention; on the Internet.

Whybrow, P.J. (1992) Land movements and species dispersal, pp. 169–173 of Jones et al. eds.

Wilkinson, R.G. (1973) *Poverty and Progress. An Ecological Perspective on Economic Development*, NY: Praeger.

Williams, R. (2001) 'The Science Show,' radio broadcast, 8 September.

Wolpoff, M.H. (1989) The place of the Neanderthals in human evolution, pp. 97–141 of Trinkaus ed.

(1999) *Paleoanthropology*, Boston: McGraw-Hill.

Woodburn J. (1968) An introduction to Hadza ecology, pp. 49–55 of Lee & DeVore eds. (1968).

World Bank (2002) Monterrey: US will 'seek advice on spending aid,' 21 March.

Worldwatch (2005) Vital facts – Selected facts and story ideas from *Vital Signs 2005*, 12 May 2005; on the Internet.

Wright, Q. (1964) *A Study of War*, abridged ed., Chicago and London: University of Chicago Press.

Wright, R. (2004) *A Short History of Progress*, Edinburgh: Canongate.

Wynne-Edwards, V.C. (1963) Intergroup selection in the evolution of social systems, pp. 42–51 of Brandon & Burian eds.

(1964) Untitled comment on R. Maynard Smith's Group selection and kin selection, *Nature* **201**: 1147.

(1965) Self-regulating systems in populations of animals, *Science* **147**: 1543–1548.

(1983) Self-regulation in populations of red grouse, pp. 379–391 of Dupâquier et al. eds., *Malthus Past and Present*, London: Academic Press.

(1986) *Evolution through Group Selection*, Oxford: Blackwell Scientific.

Young, J.E. (1992), Mining the earth, Worldwatch Paper 109, Worldwatch Institute.

Youngquist, W. (1999) The post-petroleum paradigm – and population, *Population and Environment* **20**; on the Internet.

Youth, H. (2003) Watching birds disappear, pp. 14–37 of Gardner et al.

Zegura, S. (1985) The initial peopling of the Americas: an overview, pp. 1–18 of Kirk & Szathmary eds.

Zeuner, F.E. (1975) Domestication of animals, pp. 327–352 of Singer et al. eds.

Ziegler, P. (1969) *The Black Death*, UK: Alan Sutton, 1991.

Index